HANDBOOK OF

Finite State Based Models and Applications

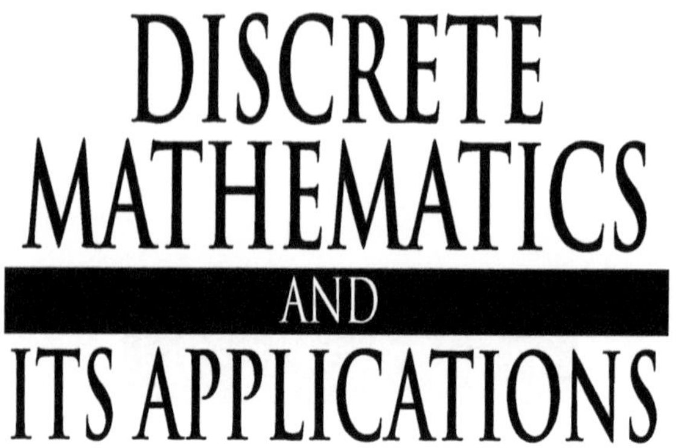

Series Editor
Kenneth H. Rosen, Ph.D.

R. B. J. T. Allenby and Alan Slomson, How to Count: An Introduction to Combinatorics, Third Edition

Juergen Bierbrauer, Introduction to Coding Theory

Katalin Bimbó, Combinatory Logic: Pure, Applied and Typed

Donald Bindner and Martin Erickson, A Student's Guide to the Study, Practice, and Tools of Modern Mathematics

Francine Blanchet-Sadri, Algorithmic Combinatorics on Partial Words

Miklós Bóna, Combinatorics of Permutations, Second Edition

Richard A. Brualdi and Dragoš Cvetković, A Combinatorial Approach to Matrix Theory and Its Applications

Kun-Mao Chao and Bang Ye Wu, Spanning Trees and Optimization Problems

Charalambos A. Charalambides, Enumerative Combinatorics

Gary Chartrand and Ping Zhang, Chromatic Graph Theory

Henri Cohen, Gerhard Frey, et al., Handbook of Elliptic and Hyperelliptic Curve Cryptography

Charles J. Colbourn and Jeffrey H. Dinitz, Handbook of Combinatorial Designs, Second Edition

Martin Erickson, Pearls of Discrete Mathematics

Martin Erickson and Anthony Vazzana, Introduction to Number Theory

Steven Furino, Ying Miao, and Jianxing Yin, Frames and Resolvable Designs: Uses, Constructions, and Existence

Mark S. Gockenbach, Finite-Dimensional Linear Algebra

Randy Goldberg and Lance Riek, A Practical Handbook of Speech Coders

Jacob E. Goodman and Joseph O'Rourke, Handbook of Discrete and Computational Geometry, Second Edition

Jonathan L. Gross, Combinatorial Methods with Computer Applications

Titles (continued)

Jonathan L. Gross and Jay Yellen, Graph Theory and Its Applications, Second Edition

Jonathan L. Gross and Jay Yellen, Handbook of Graph Theory

David S. Gunderson, Handbook of Mathematical Induction: Theory and Applications

Richard Hammack, Wilfried Imrich, and Sandi Klavžar, Handbook of Product Graphs, Second Edition

Darrel R. Hankerson, Greg A. Harris, and Peter D. Johnson, Introduction to Information Theory and Data Compression, Second Edition

Darel W. Hardy, Fred Richman, and Carol L. Walker, Applied Algebra: Codes, Ciphers, and Discrete Algorithms, Second Edition

Daryl D. Harms, Miroslav Kraetzl, Charles J. Colbourn, and John S. Devitt, Network Reliability: Experiments with a Symbolic Algebra Environment

Silvia Heubach and Toufik Mansour, Combinatorics of Compositions and Words

Leslie Hogben, Handbook of Linear Algebra

Derek F. Holt with Bettina Eick and Eamonn A. O'Brien, Handbook of Computational Group Theory

David M. Jackson and Terry I. Visentin, An Atlas of Smaller Maps in Orientable and Nonorientable Surfaces

Richard E. Klima, Neil P. Sigmon, and Ernest L. Stitzinger, Applications of Abstract Algebra with Maple™ and MATLAB®, Second Edition

Richard E. Klima and Neil P. Sigmon, Cryptology: Classical and Modern with Maplets

Patrick Knupp and Kambiz Salari, Verification of Computer Codes in Computational Science and Engineering

William Kocay and Donald L. Kreher, Graphs, Algorithms, and Optimization

Donald L. Kreher and Douglas R. Stinson, Combinatorial Algorithms: Generation Enumeration and Search

Hang T. Lau, A Java Library of Graph Algorithms and Optimization

C. C. Lindner and C. A. Rodger, Design Theory, Second Edition

Nicholas A. Loehr, Bijective Combinatorics

Toufik Mansour, Combinatorics of Set Partitions

Alasdair McAndrew, Introduction to Cryptography with Open-Source Software

Elliott Mendelson, Introduction to Mathematical Logic, Fifth Edition

Alfred J. Menezes, Paul C. van Oorschot, and Scott A. Vanstone, Handbook of Applied Cryptography

Stig F. Mjølsnes, A Multidisciplinary Introduction to Information Security

Jason J. Molitierno, Applications of Combinatorial Matrix Theory to Laplacian Matrices of Graphs

Richard A. Mollin, Advanced Number Theory with Applications

Richard A. Mollin, Algebraic Number Theory, Second Edition

Titles (continued)

Richard A. Mollin, Codes: The Guide to Secrecy from Ancient to Modern Times

Richard A. Mollin, Fundamental Number Theory with Applications, Second Edition

Richard A. Mollin, An Introduction to Cryptography, Second Edition

Richard A. Mollin, Quadratics

Richard A. Mollin, RSA and Public-Key Cryptography

Carlos J. Moreno and Samuel S. Wagstaff, Jr., Sums of Squares of Integers

Goutam Paul and Subhamoy Maitra, RC4 Stream Cipher and Its Variants

Dingyi Pei, Authentication Codes and Combinatorial Designs

Kenneth H. Rosen, Handbook of Discrete and Combinatorial Mathematics

Douglas R. Shier and K.T. Wallenius, Applied Mathematical Modeling: A Multidisciplinary Approach

Alexander Stanoyevitch, Introduction to Cryptography with Mathematical Foundations and Computer Implementations

Jörn Steuding, Diophantine Analysis

Douglas R. Stinson, Cryptography: Theory and Practice, Third Edition

Roberto Togneri and Christopher J. deSilva, Fundamentals of Information Theory and Coding Design

W. D. Wallis, Introduction to Combinatorial Designs, Second Edition

W. D. Wallis and J. C. George, Introduction to Combinatorics

Jiacun Wang, Handbook of Finite State Based Models and Applications

Lawrence C. Washington, Elliptic Curves: Number Theory and Cryptography, Second Edition

DISCRETE MATHEMATICS AND ITS APPLICATIONS
Series Editor KENNETH H. ROSEN

HANDBOOK OF Finite State Based Models and Applications

Edited by

Jiacun Wang

CRC Press is an imprint of the
Taylor & Francis Group, an informa business
A CHAPMAN & HALL BOOK

CRC Press
Taylor & Francis Group
6000 Broken Sound Parkway NW, Suite 300
Boca Raton, FL 33487-2742

First issued in paperback 2016

Version Date: 20120827

ISBN 13: 978-1-138-19935-4 (pbk)
ISBN 13: 978-1-4398-4618-6 (hbk)

Library of Congress Cataloging-in-Publication Data

Handbook of finite state based models and applications / [edited by] Jiacun Wang.
pages cm. -- (Discrete mathematics and its applications)
Summary: "This handbook presents a collection of introductory materials on finite state theories and their applications. It offers an open treatment of the types of algorithms and data structures that are typically retained quietly by companies and provides performance information regarding individual techniques for various domain areas. A complete reference on finite state-based models and their applications, the book can be used by beginners as a quick reference guide and by researchers for an in-depth study of this area"-- Provided by publisher.
Includes index.
ISBN 978-1-4398-4618-6 (hardback)
1. Computer algorithms. 2. Computer progamming. 3. Sequential machine theory. I. Wang, Jiacun, editor of compilation.

QA76.9.A43H364 2013
511.3'5--dc23 2012025781

Visit the Taylor & Francis Web site at
http://www.taylorandfrancis.com

and the CRC Press Web site at
http://www.crcpress.com

Contents

Preface

This handbook presents a collection of introductory materials on finite state theories and their applications. A finite state model is a model of behavior composed of a finite number of states, transitions between those states, and actions. A state describes a behavioral node of the system in which it is waiting for a trigger to execute a transition. Automata are abstract models of machines that perform computations on an input by moving through a series of states or configurations. At each state of the computation, a transition function determines the next state out of a finite number of states that can be reached from the present configuration. Finite state based models, also called finite state machines or finite state automata are used to encode computations, recognize events, or can be used as a data structure for holding strategies for playing games. Basically, they are applicable to all problems that require a finite number of solutions. They have been widely applied to various areas of computer science and engineering, such as transforming text, computation linguistics, search engines, indexing of large bodies of text, speech recognition, neurological system modeling, digital logic circuit design and simulations, and programming languages design and implementation.

The 18 chapters of this handbook were contributed by 32 authors from 12 countries. The first six chapters introduce the fundamentals of automata theory and some widely used automata, including transducers, tree automata, quantum automata, and timed automata. Regular expressions are part of automata theory and provide a concise and flexible means for matching strings of text. Transducers are a sort of state machines that transform an input string to an output string. Tree automata deal with tree structures that are one of the most basic and widely used data structures. Timed automata are suitable for real-time systems modeling and analysis. Quantum automata are a powerful tool to capture various important features of quantum processors. The next two chapters present algorithms for the minimization and incremental construction of finite automata. Chapter 9 introduces Esterel, an automata-based synchronous programming language that has been widely applied to embedded system software development. A few distinguishing applications of automata are showcased in Chapters 10 through 14, including regular path queries on graph-structured data, timed automata in model checking security protocols, pattern-matching, complier design and XML processing. Other finite state based modeling approaches and their applications are presented in Chapters 15 through 18. Petri nets are a high-level model for event-driven systems. The reachability graph of a Petri net is a state transition system. Statecharts, like Petri nets, provide a practical and expressive visual formalism to describe reactive systems. Temporal logic is suitable for system functional property specification, and its application has been promoted by the development of model checking tools. UML state machine diagrams are an object-based variation of Harel's statecharts with the addition of activity diagram notations.

This book is meant to be a comprehensive and complete reference guide for readers interested in finite state based models and their applications. It can be used by beginners as a quick "lookup" reference and can also be used by researchers for in-depth study in this area as a discipline. The intended audiences are academic faculty members, researchers and doctoral students in computer science, software engineering and information systems, first-year graduate students and senior-year undergraduates in computer science, software engineering and information systems, industrial software engineers and industrial information technology practitioners.

I would like to thank professor Ken Rosen, the chief editor of the CRC Press Series on Discrete Mathematics and Its Applications, for encouraging me to work on this project. I would also like to thank all chapter contributors for their dedication to this handbook.

Contributors

Marco Almeida
DCC-FC & LIACC
University of Porto
Porto, Portugal

Omar El Ariss
Department of Computer Science
The Pennsylvania State University
Harrisburg, Pennsylvania

Javier Baliosian
Institute of Computation
University of the Republic in Uruguay
Montevideo, Uruguay

Rana Barua
Stat-Math Unit
Indian Statistical Institute
Kolkata, India

Jan Daciuk
Department of Knowledge Engineering
Gdańsk University of Technology
Gdańsk-Wrzeszcz, Poland

Jin Song Dong
School of Computing
National University of Singapore
Singapore

Zhenhua Duan
Institute of Computing Theory and Technology
Xidian University
Xi'an, China

Olivier Gauwin
Department of Informatics
University of Bordeaux
Bordeaux, France

Jozef Gruska
Faculty of Informatics
Masaryk University
Brno, Czech Republic

Kishan Chand Gupta
Stat-Math Unit
Indian Statistical Institute
Kolkata, India

Mirosław Kurkowski
Institute of Computer and Information Sciences
Czestochowa University of Technology
Czestochowa, Poland

Lvzhou Li
Department of Computer Science
Sun Yat-sen University
Guangzhou, China

Yang Liu
School of Computing
National University of Singapore
Singapore

Hanlin Lu
Department of Computer Science
University of Western Ontario
London, Ontario, Canada

Murali Mani
Department of Computer Science, Engineering and Physics
University of Michigan
Flint, Michigan

Paulo Mateus
Department of Mathematics
Technical University of Lisbon
Lisbon, Portugal

Nelma Moreira
Department of Computer Science
University of Porto
Porto, Portugal

Luiza de Macedo Mourelle
Department of Electronics Engineering and Telecommunications
State University of Rio de Janeiro
Rio de Janeiro, Brazil

Mohammad Reza Mousavi
Department of Mathematics and Computer Science
Eindhoven University of Technology
Eindhoven, the Netherlands

Nadia Nedjah
Department of Electronics Engineering
State University of Rio de Janeiro
Rio de Janeiro, Brazil

Wojciech Penczek
Institute of Computer Science
University of Podlasie
Warsaw, Poland

and

University of Natural Sciences and Humanities
Siedlce, Poland

Viktor K. Prasanna
Ming-Hsieh Department of Electrical Engineering
University of Southern California
Los Angeles, California

Daowen Qiu
Department of Computer Science
Sun Yat-sen University
Guangzhou, China

Rogério Reis
DCC-FC & LIACC
University of Porto
Porto, Portugal

Jun Sun
School of Computing
National University of Singapore
Singapore

Alex Thomo
Department of Computer Science
University of Victoria
Victoria, British Columbia, Canada

Cong Tian
Institute of Computing Theory and Technology
Xidian University
Xi'an, China

S. Venkatesh
Department of Computer Science
University of Victoria
Victoria, British Columbia, Canada

Jiacun Wang
Department of Computer Science and Software Engineering
Monmouth University
West Long Branch, New Jersey

Dina Wonsever
Institute of Computation
University of the Republic in Uruguay
Montevideo, Uruguay

Dianxiang Xu
National Center for the Protection of the Financial Infrastructure
Dakota State University
Madison, South Dakota

Yi-Hua E. Yang
Ming Hsieh Department of Electrical Engineering
FutureWei Technologies
Santa Clara, California

Sheng Yu
Department of Computer Science
University of Western Ontario
London, Ontario, Canada

Yang Zhao
School of Computer Science
Nanjing University of Science and Technology
Nanjing, China

1

Finite Automata

Rana Barua and
Kishan Chand Gupta
Indian Statistical Institute

1.1 Introduction

An *alphabet* Σ is a finite nonempty set of symbols (or letters). A *string* (or *word*) is a finite sequence of symbols from Σ. We generally use u, v, w, x, y, and z to denote strings, and a, b, and c to denote symbols. For example, a and b are symbols and $u = abbaa$ is a string. The length of a string w is denoted by $|w|$. For example, $abbaa$ has length 5. The empty string, denoted by ϵ, is the string consisting of zero symbols and thus $|\epsilon| = 0$. For a string w, w^R denotes the string obtained from w by reversing the order of the symbols. For example, if $w = abcde$, then $w^R = edcba$.

A *language* $\mathcal{L}$ is a set of strings of symbols from some alphabet Σ, that is, $\mathcal{L} \subseteq \Sigma^*$, where Σ^* is the set of all words over an alphabet Σ.

A *prefix* of a string is any number of leading symbols of that string, and a *suffix* is any number of trailing symbols. For example, the string $abbaa$ has prefixes ϵ, a, ab, abb, $abba$, and $abbaa$, its suffixes are ϵ, a, aa, baa, $bbaa$, and $abbaac$. By *concatenation*, one can combine two words to form a new word, whose length is the sum of the lengths of the original words. The result of concatenating a word with the empty word is the original word. The concatenation of *very* and *good* is *verygood*. Juxtaposition is used as the concatenation. For example, if u and v are strings, then uv is the concatenation of these two strings.

Let $\mathcal{L}$, $\mathcal{L}_1$, and $\mathcal{L}_2$ be languages from some alphabet Σ. The concatenation of $\mathcal{L}_1$ and $\mathcal{L}_2$, denoted by $\mathcal{L}_1 \cdot \mathcal{L}_2$, is the set $\{uv : u \in \mathcal{L}_1, v \in \mathcal{L}_2\}$. The *Kleene closure* of $\mathcal{L}$, denoted by $\mathcal{L}^*$, is the set of all strings obtained by concatenating arbitrary number of strings from $\mathcal{L}$. The concatenation of zero string is ϵ and the concatenation of one string is the string itself.

A finite automaton can be thought as a device or a machine with a finite number of internal *states*. Thus, any machine qualifies as a finite automaton. Finite automata have been used to model physical and natural phenomena by biologists, psychologists, mathematicians, engineers, and computer scientists. The first formal use of finite state systems, viz. the neural nets, was made by two physiologists Warren McCulloch and Walter Pitts in 1943 for studying nervous activity [10]. These were later shown to be equivalent to finite automata by S.C. Kleene [8]. Kleene also derived the *regular expressions* for representing *regular* languages—the languages accepted by finite automata. G.H. Mealy [12] and E.F. Moore [13] generalized these to *transducers* by introducing what are now known as the Moore and the Mealy machines. They are finite state machines with output.

Automata theory has been used in a variety of situations. Automata have been applied in areas like lexical analysis in programming language compilation [1], circuit designs [2], text editing, pattern matching [9], and many more. It has also been used to study specific recombinant behaviors of DNA [6].

In this chapter, we study an abstraction of these machines regarded as computing devices. Generally, computability theory deals with devices which have no bounds on the space and time utilized. One such example is the *Turing machine*. The finite automaton that we study is a weak form of a computing machine with limited computing power. We now introduce deterministic finite automata.

1.2 Deterministic Finite Automata

A deterministic finite automaton (DFA) consists of a *finite input tape* with *cells* that hold symbols from some *input alphabet*. There is a tape-head that scans or reads the content of a cell at any given instant. Unlike Turing machines, the tape-head of a finite automaton is allowed to move only from left to right, changing its *state* as it moves right. There is a finite control with internal states that controls the behaviour of the automaton and also, in a sense, functions as memory as it can "store" some information about what has already been read (see Figure 1.1).

The automaton starts by reading the leftmost symbol on the tape in a state called the *initial state*. If at any given instant, the automaton is in state p reading a symbol a on the input tape, the tape-head moves one cell to the right on the tape and enters a state q. The current state and the symbol being read *completely determines* the next state of the automaton.

Thus, a finite automaton is a very limited computing device that, on reading a string of symbols on the input tape, either *accepts* or *rejects* the string depending on the state the automaton is in at the completion of reading the tape. We now give a formal definition of a *finite automaton*.

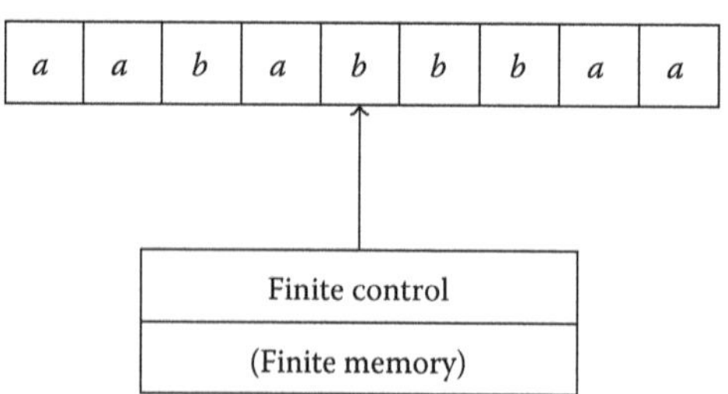

FIGURE 1.1 A finite automaton.

TABLE 1.1 Transition Function as a Table

	a	b
q_0	q_1	q_3
q_1	q_1	q_2
q_2	q_3	q_2
q_3	q_3	q_3

1.2.1 Definition of a Deterministic Finite Automaton

A finite automaton $\mathcal{M}$ consists of

- A nonempty finite set Σ, called the *input alphabet*
- A finite set Q of *states*
- An *initial state* $q_0 \in Q$
- A set of *final* or *accepting* states $F \subseteq Q$
- A *transition function* $\delta : Q \times \Sigma \longrightarrow Q$, which specifies the next state of $\mathcal{M}$ given the current state and the current symbol being scanned

Thus

$$\delta(p, a) = q$$

means that the automaton in state $p \in Q$ scanning or reading the *letter* $a \in \Sigma$ enters the state q and moves the tape-head one cell to the right.

The state transition function can also be represented as a table, known as a *transition table* (see Table 1.1).

1.2.2 Extended Definition of Transition Function

Since we are interested in the behavior of an automaton on an input string, we need to extend the transition function δ to the set of all strings over the input alphabet. Thus, we define $\delta^* : Q \times \Sigma^* \longrightarrow Q$ by induction on the length of strings as follows:

$$\delta^*(p, \epsilon) = p \tag{1.1}$$

$$\delta^*(p, wa) = \delta(\delta^*(p, w), a). \tag{1.2}$$

Here ϵ denotes the null string and wa denotes the string w followed by the letter a. $\delta^*(p, w) = q$ can now be interpreted as follows:

> The automaton in state p, on reading the input string w, will enter the state q when it has finished reading the entire input string w.

1.2.3 Language of a DFA: Regular Language

An input string w is *accepted* by the automaton $\mathcal{M}$ if $\mathcal{M}$, starting from initial state q_0 reading the string w, enters one of the accepting states; otherwise, the string is rejected by the automaton. The set of all strings accepted by $\mathcal{M}$ is called the *language accepted by* $\mathcal{M}$ and is denoted by $\mathcal{L}(\mathcal{M})$. Thus

$$\mathcal{L}(\mathcal{M}) = \{w \in \Sigma^* \ : \ \delta^*(q_0, w) \in F\}. \tag{1.3}$$

A language $\mathcal{L} \subseteq \Sigma^*$ is called *regular* if there is a finite automaton $\mathcal{M}$ over Σ such that

$$\mathcal{L} = \mathcal{L}(\mathcal{M}) \tag{1.4}$$

that is, $\mathcal{L}$ is the language accepted by $\mathcal{M}$.

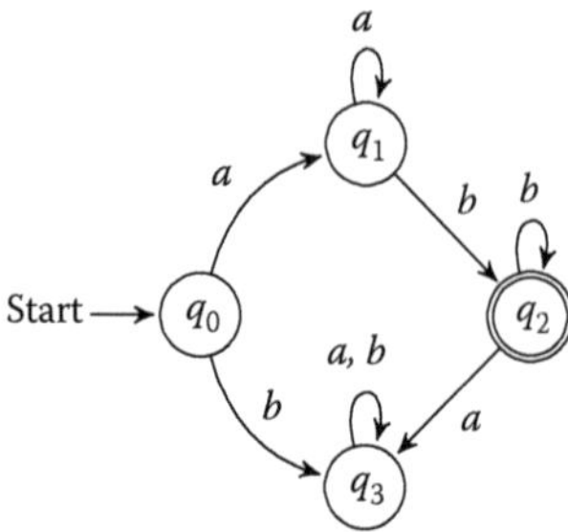

FIGURE 1.2 Transition diagram for Table 1.1.

1.2.4 Transition Diagram

A finite automaton $\mathcal{M}$ can be conveniently represented by a directed labeled graph called its *transition diagram*. The states of $\mathcal{M}$ represent the vertices and there is a directed edge, with label a, from the vertex p to vertex q (where p, q are states of $\mathcal{M}$) iff $\delta(p, a) = q$. In other words, in the state transition diagram of an automaton $\mathcal{M}$, there is an edge labeled a from the vertex with label p to the vertex with label q, if $\mathcal{M}$ in state p reading the letter a enters the state q (and moves one cell to the right). Figure 1.2 gives the state transition diagram of the automaton $\mathcal{M}$ of Table 1.1. Note that accepting states are represented within double circles and the initial state is marked with an incoming $\rightarrow$.

1.2.5 Examples of Finite Automata

Example 1.1:

Let $\mathcal{M} = (Q, \Sigma, \delta, q_0, F)$, where $Q = \{q_0, q_1\}$, $\Sigma = \{a, b\}$, q_0 is the initial state as well as the only accepting state, and δ is the transition function defined by Table 1.2.

The language accepted by $\mathcal{M}$ is the set of all strings which have even number of b's, since $\mathcal{M}$ traversing from the initial state q_0 will come back to q_0, if it traverses horizontally an even number of times (Figure 1.3). Thus, b's contribution in an accepting string is even.

TABLE 1.2 Transition Table for Example 1.1

	a	b
q_0	q_0	q_1
q_1	q_1	q_0

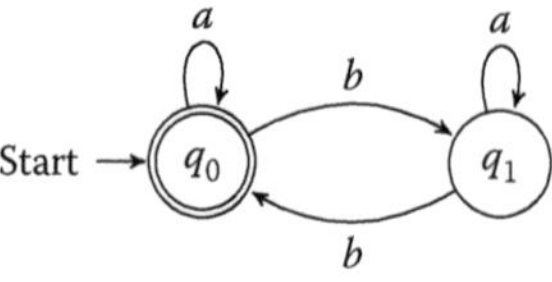

FIGURE 1.3 Transition diagram for Example 1.1.

TABLE 1.3 Transition Table for Example 1.2

	a	b
q_0	q_1	q_2
q_1	q_0	q_3
q_2	q_3	q_0
q_3	q_2	q_1

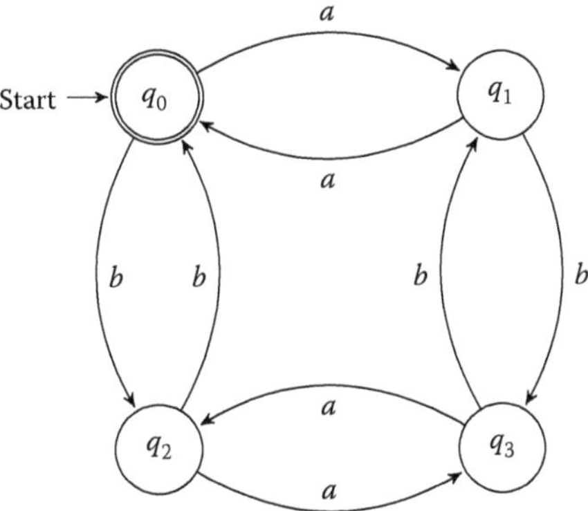

FIGURE 1.4 Transition diagram for Example 1.2.

Example 1.2:

Let $\mathcal{M} = (Q, \Sigma, \delta, q_0, F)$, where $Q = \{q_0, q_1, q_2, q_3\}$, $\Sigma = \{a, b\}$, q_0 is the initial state, $F = \{q_0\}$ is the set of accepting states, and δ is the transition function defined by Table 1.3.

The corresponding transition diagram is given in Figure 1.4.

The language accepted by $\mathcal{M}$ is the set of all strings in which both a and b occur an even number of times. To see this, observe that a string is accepted by $\mathcal{M}$ if, traversing from the state q_0, $\mathcal{M}$ comes back to state q_0. $\mathcal{M}$ can do this if it traverses horizontally an even number of times and also vertically an even number of times. Each horizontal move contributes an a while each vertical move contributes a b.

1.2.6 Equivalence of Automata

When are two automata $\mathcal{M}_1$ and $\mathcal{M}_2$ *equivalent*? A natural answer is, when they have the same "computing power." The computing power is characterized by the strings they accept. Thus, the automata $\mathcal{M}_1$ and $\mathcal{M}_2$ are equivalent if they accept the same language, that is,

$$\mathcal{L}(\mathcal{M}_1) = \mathcal{L}(\mathcal{M}_2). \tag{1.5}$$

1.3 Nondeterministic Finite Automata

The automata that we have already introduced above are *deterministic* in the sense that at a given instant, an automaton $\mathcal{M}$ has exactly one move, that is, the automaton in state p scanning a letter s on the tape will enter exactly one state, say q. One natural extension of such automata would be to allow the automaton

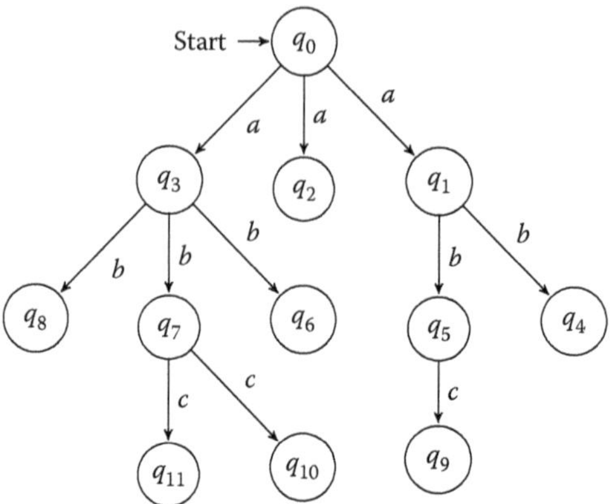

FIGURE 1.5 Transition diagram for NFA.

$\mathcal{M}$ to have several choices (or no choice) for the next move. Such automata are called *nondeterministic.* Thus, the nature of the transition function specifies whether an automaton is deterministic or not. We have the following formal definition of a nondeterministic automaton.

1.3.1 Definition of a Nondeterministic Finite Automaton

A nondeterministic finite automaton (NFA) $\mathcal{M}$ consists of an input alphabet Σ, a finite set of states Q, an initial state $q_0 \in Q$, a set of accepting states $F \subseteq Q$, and a transition function

$$\delta : Q \times \Sigma \longrightarrow \mathcal{P}(Q),$$

where $\mathcal{P}(Q)$ is the power set of Q. Thus, if

$$\delta(p, a) = \{q_1, q_2, \ldots, q_k\},$$

then the automaton $\mathcal{M}$ in state p, reading the letter a on the tape, has the choice of moving into one of the k states $q_1, q_2, \ldots, q_k$. (If $\delta(p, a) = \phi$, then the automaton in state p reading a has no next move, that is, it *halts.*) This means that on input $w = a_0a_1 \ldots a_{n-1}$, the automaton $\mathcal{M}$ (starting from the initial state q_0) will, on reading the first symbol a_0, have several choices of states for the next move. For each state chosen, $\mathcal{M}$, on reading the next letter a_1, again will have several choices for moving into the next state and so on. Thus, we will have a "computation tree" as in Figure 1.5.

It is not very hard to see that the leaf nodes at the lowest level constitute the set of all possible states that the automaton can reach on reading the entire input string. Thus, in our example above, on reading *abc*, the automaton can enter one of the states q_9, q_{10}, q_{11}.

1.3.2 Extended Definition of Transition Function for NFA

As in the case of DFA, the transition function δ of an NFA is extended to all strings by induction as follows:

$$\delta^*(p, \epsilon) = \{p\} \tag{1.6}$$

$$\delta^*(p, wa) = \bigcup_{q \in \delta^*(p,w)} \delta(q, a). \tag{1.7}$$

Thus, $\delta^*(p, w)$ yields the set of all possible states that the automaton $\mathcal{M}$ can reach from the state p on reading w on the tape. A string w is *accepted* by $\mathcal{M}$ if, starting from the initial state q_0, on reading the

entire string w, $\mathcal{M}$ has a choice of states that lead to an accepting state. In other words, there is a *run* of the automaton $\mathcal{M}$ which leads to an accepting state. Let $\mathcal{L}(\mathcal{M})$ denotes the set of all accepting strings. $\mathcal{L}(\mathcal{M})$ is the language accepted by $\mathcal{M}$ and we have

$$\mathcal{L}(\mathcal{M}) = \{w \in \Sigma^* : \delta^*(q_0, w) \cap F \neq \phi\}.$$

One can also refer to the corresponding computation tree to see whether or not a string w is accepted by the NFA $\mathcal{M}$. Starting from the initial state q_0, on input w, consider the computation tree as above. The leaf nodes represent $\delta^*(p, w)$, and w is accepted if, and only if, one of the leaf nodes is an accepting state. If such a leaf node $q_f \in F$ exists, then the path from the root (labeled by q_0) to q_f will constitute an accepting run of the automaton $\mathcal{M}$. It is worthwhile to emphasize the notion of nondeterministic acceptance: a string is accepted if there is at least one accepting computation; it is rejected if all computations reject it.

1.3.3 Transition Diagram and Transition Relation of an NFA

The transition diagram for an NFA is the same as that of a DFA, except that from each node, labeled by a state p, and for each letter $a \in \Sigma$, there may be several edges (or none) leaving p with the same label a. This is not the case with a DFA, where a unique edge with label a will be present.

The transition function δ can be represented by a *transition relation* $\Delta \subseteq Q \times \Sigma \times Q$ as follows:

$$(p, a, q) \in \Delta \quad \text{iff } \delta(p, a) \text{ contains } q.$$

Thus, if $(p, a, q) \in \Delta$, then it means that the automaton $\mathcal{M}$ in state p reading the letter a has the option of entering the state q (moving its tape-head one cell to the right).

1.3.4 Examples of NFA

Example 1.3:

Let $\mathcal{M} = (Q, \Sigma, \delta, q_0, F)$, where $Q = \{q_0, q_1, q_2, q_3\}$, $\Sigma = \{a, b\}$, q_0 is the initial state, q_3 is the only accepting state, and δ is the transition function defined in Table 1.4.

The corresponding transition diagram is given by Figure 1.6.

The language accepted by $\mathcal{M}$ is the set of all strings in which b appears in the third position from the end (e.g., *ababbaa* is accepted by the NFA but *ababaabb* is not).

TABLE 1.4 Transition Table for Example 1.3

	a	b
q_0	$\{q_0\}$	$\{q_0, q_1\}$
q_1	$\{q_2\}$	$\{q_2\}$
q_2	$\{q_3\}$	$\{q_3\}$
q_3	$\emptyset$	$\emptyset$

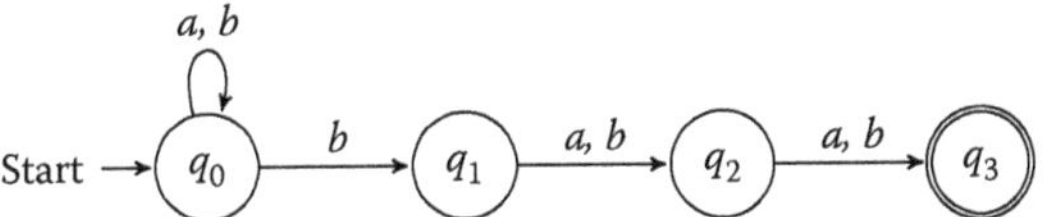

FIGURE 1.6 Transition diagram for Example 1.3.

1.3.5 Equivalence of Deterministic and Nondeterministic Finite Automata

Nondetermistic finite automata were introduced by Rabin and Scott in 1959. Intuitively, it appears that a nondeterministic automaton is more powerful or has more computing power than that of a deterministic one. This is not the case was shown by Rabin and Scott in Ref. [16].

Theorem 1.1:

If $\mathcal{L} \subseteq \Sigma^*$ is accepted by an NFA $\mathcal{M}$, then $\mathcal{L}$ is also accepted by some DFA $\tilde{\mathcal{M}}$. In fact, there is an efficient algorithm for constructing $\tilde{\mathcal{M}}$ from $\mathcal{M}$.

Proof. (Sketch:) Let $\mathcal{M} = (Q, \Sigma, \delta, q_0, F)$ be an NFA accepting $\mathcal{L}$. Construct a DFA $\tilde{\mathcal{M}} = (\tilde{Q}, \Sigma, \tilde{q}_0, \tilde{F}, \tilde{\delta})$ as follows:

- Set of states $\tilde{Q} = \mathcal{P}(Q)$
- Initial state $\tilde{q}_0 = \{q_0\}$
- Set of accepting states $\tilde{F} = \{S \subseteq Q : S \cap F \neq \phi\}$
- Transition function $\tilde{\delta} : \tilde{Q} \times \Sigma \longrightarrow \tilde{Q}$ is defined by

$$\tilde{\delta}(S, a) = \bigcup_{q \in S} \delta(q, a)$$

Note that the states of $\tilde{\mathcal{M}}$ are the subsets of the set of states of the NFA. Thus, if S denotes the set of states that the NFA $\mathcal{M}$ can possibly enter on reading a letter a from the states in T, then the DFA $\tilde{\mathcal{M}}$ enters the state S from its state T on reading a. Consequently, it is not hard to see that a string w is accepted by the automaton $\mathcal{M}$ iff w is accepted by the DFA $\tilde{\mathcal{M}}$. To see this, first observe that on input w (starting from q_0), the set of possible states that $\mathcal{M}$ can enter on reading the first letter is precisely the state $\tilde{\mathcal{M}}$ enters. Similarly, the set of states $\mathcal{M}$ can enter on reading the first two letters of w is just the state $\tilde{\mathcal{M}}$ enters reading the first two letters. Thus, the set of states $\mathcal{M}$ can possibly enter on reading the entire input string w is the state $\tilde{\mathcal{M}}$ enters on reading the corresponding string w. It contains an accepting state if $\mathcal{M}$ accepts w and hence $\tilde{\mathcal{M}}$ accepts by construction. Conversely, if $\tilde{\mathcal{M}}$ accepts w, then one can obtain a computation tree for $\mathcal{M}$ on input w with an accepting state at a leaf node. □

Corollary 1.1:

A language $\mathcal{L} \subseteq \Sigma^*$ is regular iff there is an NFA that accepts $\mathcal{L}$.

Remark:

Clearly, there is an exponential blow-up in the number of states while converting a given NFA to an equivalent DFA. In most practical situations, all the 2^n states of $\tilde{\mathcal{M}}$ may not be required and many of them may be redundant in the sense that they may not be reachable from the initial state or may be "dead" states. However, one can construct examples to show that we cannot do with lesser number of states.

1.4 Properties of Regular Languages

1.4.1 Closure Properties of Regular Languages

We shall obtain some elementary closure properties that regular languages satisfy. The proofs are constructive and give efficient algorithms for obtaining the corresponding automata.

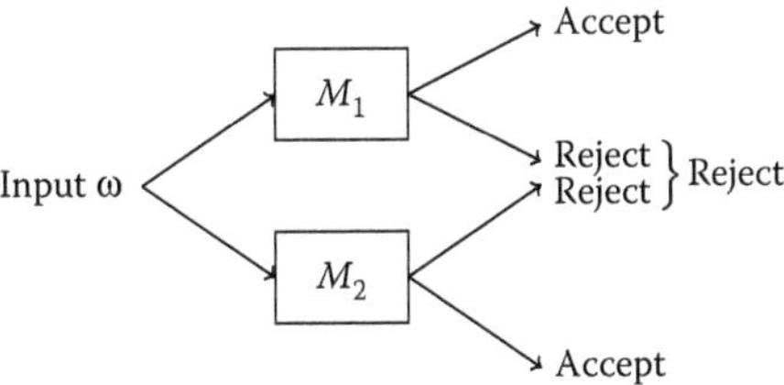

FIGURE 1.7 Construction of Theorem 1.2 for union.

Theorem 1.2:

$\mathcal{REG}$, the class of regular languages, is closed under complementation, finite union, and finite intersection. Consequently, the class $\mathcal{REG}$ forms a *Boolean algebra*.

Proof.

- Let $\mathcal{M}$ be a DFA over Σ accepting a regular language $\mathcal{L} \subseteq \Sigma^*$. Let $\tilde{\mathcal{M}}$ be a DFA which behaves exactly like $\mathcal{M}$ but whose accepting states are precisely the rejecting states of $\mathcal{M}$. Then, clearly, a string $w \in \Sigma^*$ is accepted by $\tilde{\mathcal{M}}$ iff it is rejected by $\mathcal{M}$. Thus, $\tilde{\mathcal{M}}$ accepts $\Sigma^* - \mathcal{L}$.
- The proofs of closure under finite intersection and finite union are similar. We shall consider that of union.
  For closure under finite union, consider two DFAs $\mathcal{M}_1 = (Q_1, \Sigma, \delta_1, q_0^1, F_1)$ and $\mathcal{M}_2 = (Q_2, \Sigma, \delta_2, q_0^2, F_2)$ over Σ accepting $\mathcal{L}_1$ and $\mathcal{L}_2$, respectively. We shall construct a single DFA $\tilde{\mathcal{M}}$ which accepts $\mathcal{L}_1 \bigcup \mathcal{L}_2$. $\tilde{\mathcal{M}}$ combines both $\mathcal{M}_1$ and $\mathcal{M}_2$ by running them in parallel. Thus, given an input string w to $\tilde{\mathcal{M}}$, it runs both $\mathcal{M}_1$ and $\mathcal{M}_2$ on input w and accepts w if either $\mathcal{M}_1$ or $\mathcal{M}_2$ accepts w (see Figure 1.7). Formally, $\tilde{\mathcal{M}} = (\tilde{Q}, \Sigma, \tilde{\delta}, \tilde{q_0}, \tilde{F})$ is constructed as follows. Set
  - $\tilde{Q} = Q_1 \times Q_2$
  - $\tilde{q_0} = (q_0^1, q_0^2)$
  - $\tilde{F} = F_1 \times Q_2 \cup Q_1 \times F_2$
  - $\tilde{\delta}((p,q), a) = (\delta_1(p,a), \delta_2(q,a))$; for all $(p,q) \in \tilde{Q}$ and $a \in \Sigma$

  It is not hard to see that $\mathcal{L}(\tilde{\mathcal{M}}) = \mathcal{L}_1 \bigcup \mathcal{L}_2$. □

Remark:

We have implicitly assumed, without loss of generality, that the regular languages are over the same alphabet Σ. If two regular languages are over two different alphabets, then one can combine them to obtain a common alphabet.

Some more closure properties are obtained below.

Theorem 1.3:

Regular languages are closed under *concatenation*. In other words, if $\mathcal{L}_1$ and $\mathcal{L}_2$ are two regular languages then so is $\mathcal{L}_1 \cdot \mathcal{L}_2$.

Proof. Let $\mathcal{M}_1 = (Q_1, \Sigma, \delta_1, q_0^1, F_1)$ and $\mathcal{M}_2 = (Q_2, \Sigma, \delta_2, q_0^2, F_2)$ be two DFAs accepting $\mathcal{L}_1$ and $\mathcal{L}_2$, respectively. We shall obtain an NFA $\tilde{\mathcal{M}}$ that accepts $\mathcal{L}_1 \cdot \mathcal{L}_2$. $\tilde{\mathcal{M}}$ is obtained by gluing together $\mathcal{M}_1$ and $\mathcal{M}_2$ as shown in Figure 1.8.

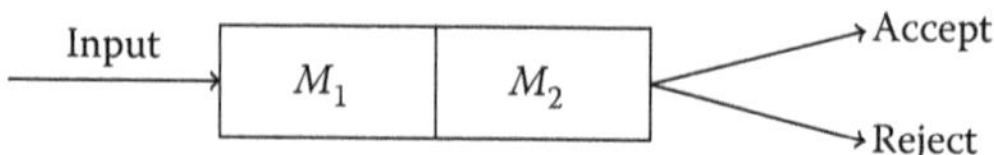

FIGURE 1.8 Construction of Theorem 1.3 for Concatenation.

$\tilde{\mathcal{M}}$ first runs $\mathcal{M}_1$ and as soon as $\mathcal{M}_1$ enters an accepting state on reading a letter of the input string, $\tilde{\mathcal{M}}$ either continues to behave like $\mathcal{M}_1$ or treats the accepting state of $\mathcal{M}_1$ as the initial state of $\mathcal{M}_2$ and runs $\mathcal{M}_2$ on the remaining portion of the string. Formally, $\tilde{\mathcal{M}} = (\tilde{Q}, \Sigma, \tilde{\delta}, \tilde{q_0}, \tilde{F})$ is constructed as follows. Assume without loss of generality that $Q_1 \cap Q_2 = \phi$. Set

- $\tilde{Q} = Q_1 \cup Q_2$
- $\tilde{q_0} = q_0^1$
- $\tilde{F} = F_2$
- $\tilde{\delta}(p, a) = \begin{cases} \{\delta_1(p, a)\} & \text{for } p \in Q_1 - F_1 \\ \{\delta_1(p, a), \delta_2(q_0^2, a)\} & \text{for } p \in F_1 \\ \{\delta_2(p, a)\} & \text{for } p \in Q_2 \end{cases}$

$\tilde{\mathcal{M}}$ behaves as described above and accepts $\mathcal{L}_1 \cdot \mathcal{L}_2$. □

Theorem 1.4:

For every regular language $\mathcal{L}$, $\mathcal{L}^*$ is also regular.
Hence, the class $\mathcal{REG}$ is closed under Kleene closure.

Proof. Let $\mathcal{M} = (Q, \Sigma, \delta, q_0, F)$ be a DFA accepting $\mathcal{L}$. Without loss of generality, we assume that $\mathcal{M}$ never enters the initial state once it leaves. This can easily be ensured by having a copy of the initial state, so that whenever $\mathcal{M}$ tries to enter the initial state it is sent to its copy.

We shall construct an NFA $\tilde{\mathcal{M}} = (\tilde{Q}, \Sigma, \tilde{\delta}, \tilde{q_0}, \tilde{F})$ that accepts $\tilde{\mathcal{L}}$. $\tilde{\mathcal{M}}$ behaves exactly like $\mathcal{M}$ except that when it tries to enter an accepting state, $\tilde{\mathcal{M}}$ has a choice of either entering the accepting state or the initial state of $\mathcal{M}$. The latter choice enables $\tilde{\mathcal{M}}$ to restart $\mathcal{M}$ again. Formally, $\tilde{\mathcal{M}} = (\tilde{Q}, \Sigma, \tilde{\delta}, \tilde{q_0}, \tilde{F})$ is constructed as follows:

- $\tilde{Q} = Q$
- $\tilde{q_0} = q_0$
- $\tilde{F} = \{q_0\}$
- $\tilde{\delta}(q, a) = \begin{cases} \{\delta(q, a)\} & \text{for } \delta(q, a) \notin F \\ \{\delta(q, a), q_0\} & \text{for } \delta(q, a) \in F \end{cases}$

It is now straightforward to show that $\mathcal{L}(\tilde{\mathcal{M}}) = \mathcal{L}^*$. □

Since finite sets are easily seen to be regular, starting from finite sets and applying the operations $\bigcup, \cdot,$ and the Kleene closure $(*)$ finitely many times, one obtains only regular languages. In fact, this characterizes the regular languages as was shown by Kleene in Ref. [8]. One can show the following theorem.

Theorem 1.5: (Kleene)

Every regular language can be derived from finite languages by applying the three operations $\bigcup$ (union), $\cdot$ (concatenation), and $*$ (Kleene closure) a finite number of times.

Remark:

- Kleene's theorem enables us to represent regular languages by means of *regular expressions* defined over an alphabet. This will be studied in detail in Chapter 2.
- The above characterization of regular languages resembles those of *computable functions*, where one begins with simple computable funtions like constant functions, projections, and successor function (called initial functions) and applies certain operations (like composition, primitive recursion) a finite number of times [4].

1.4.2 Pumping Lemma

An interesting property satisfied by regular languages is provided by what is known as the *pumping lemma* for regular languages. It provides a powerful tool for showing nonregularity of certain languages. It can also be used to develop algorithms for testing whether a language accepted by a given automaton is finite or infinite and also whether the language is empty or nonempty.

Theorem 1.6: (The Pumping Lemma)

Let $\mathcal{M}$ be a DFA with exactly n states and let $\mathcal{L} \subseteq \Sigma^*$ be the language accepted by $\mathcal{M}$. Then, for any $x \in \mathcal{L}$ with $|x| \geq n$, we have $x = uvw$ such that

- $v \neq \epsilon$
- $|uv| \leq n$
- for all $l \geq 0$, $uv^l w \in \mathcal{L}$

Proof. Let $\mathcal{M} = (Q, \Sigma, q_0, F, \delta)$ and $x = a_1\ a_2\ \ldots\ a_k$. Thus, $k \geq n$. Suppose

$$\delta(q_i, a_{i+1}) = q_{i+1},$$

for $i = 0, 1, \ldots, k-1$. Thus, $\mathcal{M}$ in state q_0 reading a_1 enters state q_1, and in state q_1 it reads the second letter a_2 and enters state q_2 and so on until it reaches state q_k on reading a_k (Figure 1.9). Since $k \geq n$, the sequence of states $q_0, q_1, \ldots, q_k$ cannot all be distinct. Hence, there exist states q_i, q_j with $i < j$ such that $q_i = q_j$. Choose i, j as small as possible. Set $u = a_1 \ldots a_i$, $v = a_{i+1} \ldots a_j$, and $w = a_{j+1} \ldots a_k$. Clearly, $|v| \geq 1$ as $i < j$.

Thus, starting from the initial state q_0, the DFA $\mathcal{M}$ on reading the prefix u enters the state q_i, then on further reading v enters the state q_j which is the same as q_i. Finally, from $q_j = q_i$, on reading w it enters $q_k \in F$. Since from state q_i the automaton $\mathcal{M}$ on reading v again enters $q_i = q_j$, the substring v may be dropped from or may be introduced any number of times in the input string uvw. Thus, $\mathcal{M}$, starting from q_0, on reading $uv^l w$ will also enter the state q_k which is an accepting state. So, for all $l \geq 0$, $uv^l w \in \mathcal{L}$. Also, by our choice of i, j, the states $q_0, \ldots, q_{j-1}$ are all distinct and so we have $j \leq n$. Thus, $|uv| \leq n$. □

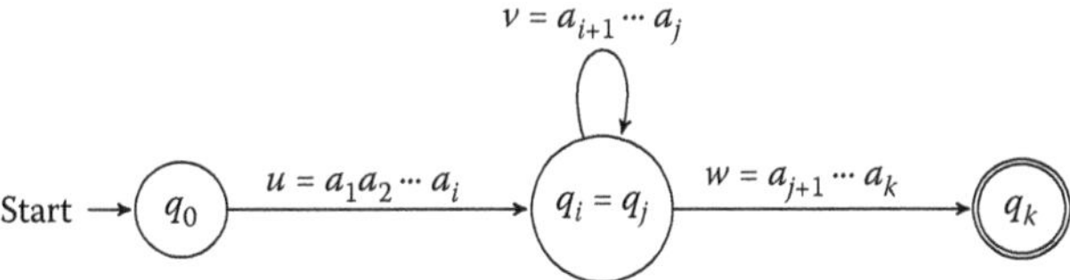

FIGURE 1.9 Construction for pumping lemma.

Arguing as above, we can prove the following.

Corollary 1.2:

Let $\mathcal{M}$ be a DFA with exactly n states and let $\mathcal{L} \subseteq \Sigma^*$ be the language accepted by $\mathcal{M}$. Then for any $zx \in \mathcal{L}$ with $|x| \geq n$, we have $x = uvw$ such that

- $v \neq \epsilon$
- for all $l \geq 0, zuv^l w \in \mathcal{L}$

Using the above corollary, it is not hard to show that the language $\{0^n 1^n : n \geq 1\}$ is *not* regular.

1.4.3 Minimum State Automata

We shall now present a result known as the Myhill–Nerode theorem [14] that not only characterizes regular languages but also implies the existence of a (unique) minimum state DFA for every regular language.

Given a language $\mathcal{L}$, we first associate an *equivalence relation* $R_{\mathcal{L}}$ as follows:

$$xR_{\mathcal{L}}y \text{ iff for every } z, xz \in \mathcal{L} \text{ if, and only if, } yz \in \mathcal{L}.$$

$R_{\mathcal{L}}$ is said to be of *finite index* if the number of equivalence classes is finite. The Myhill–Nerode theorem implies that if $\mathcal{L}$ is regular, then $R_{\mathcal{L}}$ is of finite index. In fact, we have the following theorem.

Theorem 1.7: (Myhill–Nerode)

$\mathcal{L}$ is regular iff the equivalence relation $R_{\mathcal{L}}$ is of finite index.

Proof. Suppose $\mathcal{L}$ is regular. Let $\mathcal{M} = (Q, \Sigma, \delta, q_0, F)$ be a DFA accepting $\mathcal{L}$. Define an equivalence relation $R_{\mathcal{M}}$ as follows:

$$xR_{\mathcal{M}}y \text{ iff } \delta^*(q_0, x) = \delta^*(q_0, y).$$

Clearly, each equivalence class corresponds to a unique state, and since Q is finite, the number of equivalence classes of $R_{\mathcal{M}}$ is finite. We now show that $R_{\mathcal{M}}$ is a refinement of $R_{\mathcal{L}}$, that is, every equivalence class of $R_{\mathcal{M}}$ is contained in some equivalence class of $R_{\mathcal{L}}$. To see this, assume $xR_{\mathcal{M}}y$. Then $\delta^*(q_0, x) = \delta^*(q_0, y)$. So, the state reached by $\mathcal{M}$ from q_0 on reading x is the same on reading y. This easily implies that $xz \in \mathcal{L}$ if, and only if, $yz \in \mathcal{L}$, for every z. Hence we have $xR_{\mathcal{L}}y$. So, the number of equivalence classes of $R_{\mathcal{L}}$ cannot exceed those of $R_{\mathcal{M}}$. Thus, $R_{\mathcal{L}}$ is of finite index.

Conversely, assume $R_{\mathcal{L}}$ is of finite index. First note that if $x \in \mathcal{L}$ and $yR_{\mathcal{L}}x$, then $y \in \mathcal{L}$. This shows that $\mathcal{L}$ is a union of some equivalence classes.

We shall now construct a DFA $\mathcal{M} = (Q, \Sigma, \delta, q_0, F)$ that accepts $\mathcal{L}$. Let $[x]$ denotes the equivalence class of $R_{\mathcal{L}}$ containing x and set

- $Q = \{[x]; x \in \Sigma^*\}$
- $q_0 = [\epsilon]$, where ϵ is the null string
- $F = \{[x] : x \in \mathcal{L}\}$
- $\delta([x], a) = [xa]$

It is not hard to see that the transition function δ is well defined, that is, independent of the choice of the representative x. Also, one can show by induction on the length $|y|$ that

$$\delta^*([x], y) = [xy].$$

In view of these, it follows that

$$\delta^*(q_0, x) = \delta^*([\epsilon], x) = [x].$$

Hence

$$\delta^*(q_0, x) \in F \text{ if, and only if, } [x] \in F \text{ if, and only if, } x \in \mathcal{L},$$

by definition of F. Thus, $\mathcal{L}$ is the language accepted by $\mathcal{M}$. □

Minimum state automaton: A minimum state DFA accepting a regular language $\mathcal{L}$ is given by the automaton $\mathcal{M}$ defined in the proof of the Myhill–Nerode theorem. In fact, it is unique up to a renaming of the states. There are a number of efficient algorithms for minimizing finite automata. Hopcroft [7] gives an $n \log n$ algorithm for minimizing the states of a given DFA (see also Chapter 2).

1.4.4 Applications of Finite Automata

Deterministic finite automata have many practical applications:

- DFA-like code is used by compilers and other language-processing systems to divide an input program into tokens like identifiers, constants, and keywords and to remove comments and white spaces. Lexical analyzers for compilers can be implemented using an NFA or a DFA. During compilation, a transition table can be generated using either an NFA or a DFA. The transition table for an NFA is considerably smaller than that for a DFA, but the DFA recognizes patterns faster than the NFA.
- Command line interpreters often use DFA-like code to process the input command.
- Almost all text processors like awk, egrep, and procmail in Unix use DFA-like code to search a substring in a text file or a given string. Some of the languages, including Perl, Ruby, and Tcl, make use of finite automata.
- Image-processing, speech-processing, and other signal-processing systems sometimes use a DFA-like technique to transform an incoming signal and for data compression.
- A wide variety of finite state systems use DFA-like techniques to track the current state. It includes from industrial processes to video games and many vending machines. They are implemented both in hardware and in software.
- Autonomous automata (variant of automata) are sometime well suited in describing symmetric key cryptosystems (see Ref. [17] for details).

1.5 Variants of Finite Automata

1.5.1 Pushdown Automata

Pushdown automata (PDA) are like the NFAs with control of both an input tape and a stack. The stack provides additional memory beyond the finite amount available in the control. The stack allows a PDA to recognize some nonregular languages. Given an input symbol, current state, and stack symbol, the automaton can follow a transition to another state, and optionally manipulate (push or pop) the stack. A PDA can recognize a nonregular language like $\{0^n 1^n : n \geq 0\}$ because it can use the stack to remember the number of 0s it has seen.

A pushdown automaton (in brief a PDA) $\mathcal{M}$ is a 6-tuple $(Q, \Sigma, \Gamma, \delta, q_0, F)$, where Σ is an input alphabet, Γ is a stack alphabet, Q is a finite set of states, $q_0 \in Q$ is the initial state, $F \subseteq Q$ is the set of accepting states, and a transition function

$$\delta : Q \times (\Sigma \cup \{\epsilon\}) \times \Gamma \longrightarrow \mathcal{P}_{fin}(Q \times \Gamma^*),$$

which specifies the next state as well as the stack symbols to be pushed in place of the top stack symbol. An input string w is accepted if the PDA $\mathcal{M}$ on processing the entire string w empties its stack. The language accepted by $\mathcal{M}$ is the set of all strings that it accepts. Such languages are called *context-free languages.* Context-free languages are also generated by *context-free grammars*, a finite set of rules that specifies how strings should be generated beginning from a start symbol. A typical context-free language is given by

$$\mathcal{L} = \{ww^R : w \in \{a, b\}^*\},$$

where w^R is the string obtained from w by reading the letters of w in the reverse order. A pushdown automaton accepting $\mathcal{L}$ behaves as follows. Given an input of the form ww^R, the PDA pushes symbols onto the stack that correspond to the letters of the prefix w. Having guessed that it has come to the middle of the input, the symbols on the stack will now correspond to w^R when read from the top. So, the PDA starts popping the corresponding symbols until the stack is empty. Notice that the stack becomes empty only when the second half of the input is of the form w^R. It may be mentioned that in the context of PDA nondeterminism indeed makes a difference. The above language cannot be accepted by a deterministic PDA. In fact, a language $\mathcal{L}$ accepted by a deterministic PDA has the prefix property, that is, no proper prefix of a string in $\mathcal{L}$ is in $\mathcal{L}$.

1.5.2 Transducers

A tranducer is another variant of a finite state machine with output from another alphabet and will be studied in detail in Chapter 4. A transducer is a 6-tuple $(Q, \Sigma, \Gamma, \delta, q_0, \lambda)$, where Σ is an input alphabet, Γ is an output alphabet, Q is a finite set of states, $q_0 \in Q$ is the initial state, a transition function $\delta : Q \times \Sigma \longrightarrow Q$, and λ is the output function. If the output function is a function of a state and input alphabet $(\lambda : Q \times \Sigma \rightarrow \Gamma)$, it is known as a *Mealy machine.* If the output function depends only on a state $(\lambda : Q \rightarrow \Gamma)$, it is known as a *Moore machine.* The DFA may be viewed as a special case of transducer where the output alphabet is $\{0, 1\}$ and the state $q = \delta(p, a)$ is accepting iff $\lambda(p, a) = 1$.

Example for Mealy machine: The transition diagram corresponding to sequence detector for 01 or 10 is given below. We use the label a/b on an arc from state p to state q to indicate that $\delta(p, a) = p$ and $\lambda(p, a) = b$ (see Figure 1.10).

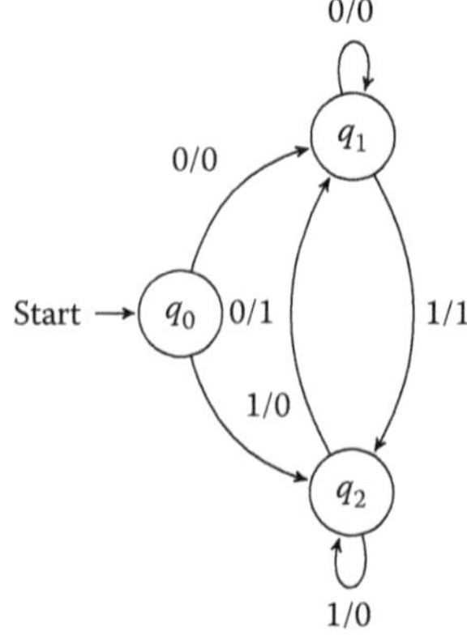

FIGURE 1.10 Mealy machine.

1.5.3 Autonomous Automata

We now introduce another variant of finite automata. Let $\mathcal{M} = (Q, \Sigma, \Gamma, \delta, q_0, \lambda)$ be a transducer. If for any q in Q, $\delta(q, a)$ and $\lambda(q, a)$ do not depend on a, $\mathcal{M}$ is said to be autonomous [17]. We abbreviate the autonomous finite automaton to a 5-tuple $(Q, \Gamma, \delta, q_0, \lambda)$, where δ is a single-valued mapping from Q to Q, and λ is a single-valued mapping from Q to Γ.

An Example of an Autonomous Automata: Let $Q = \Gamma = \mathbb{Z}_5$, $q_0 = 0$ is the initial state. The transition function δ is defined as $\delta(i) = i + 1 \bmod 5$ and the output function λ is defined as $\lambda(i) = i^2 \bmod 5$.

Stream Cipher Modeled as an Autonomous Finite State Machine (FSM): In general, any pseudorandom sequence generator (PSG) can be considered as an (FSM) or some variant of it which may be defined as follows. Suppose $\mathcal{Y}$ and $\mathcal{Z}$ are finite fields (or finite rings) and the elements of $\mathcal{Z}$ are represented by m bits. An FSM is a 5-tuple (S_0, F, G, n, m), where $S_0 \in \mathcal{Y}^n$ is the initial state, $F : \mathcal{Y}^n \times \mathcal{Z} \to \mathcal{Y}^n$ is the state update function, and $G : \mathcal{Y}^n \times \mathcal{Z} \to \mathcal{Z}$ is the output function. At time instant $t \geq 0$, state S_t is represented by an n-tuple $(s_t, s_{t+1}, \ldots, s_{t+n-1})$, where $s_i \in \mathcal{Y}$ and the output $z_t \in \mathcal{Z}$. The state update function and output function are given respectively by

$$S_{t+1} = F(S_t, z_t) \quad \text{and} \quad z_{t+1} = G(S_t, z_t).$$

In modeling practical stream ciphers, in many cases it is seen that, for any S_t, $F(S_t, z_t)$ and $G(S_t, z_t)$ do not depend on z_t. In this situation, we write

$$S_{t+1} = F(S_t) \quad \text{and} \quad z_{t+1} = G(S_t).$$

1.5.4 Turing Machines

The Turing machine (TM), introduced by Alan Turing, is the most powerful of all models of computation. Every other model of computation over the integers has been shown to be equivalent to the TM model. It consists of a two-way infinite tape with cells, a tape-head and a finite control which determines the state of the TM (Figure 1.11). Depending on the letter being scanned by the tape-head and the state of the finite control, the TM

- Prints a possibly new symbol onto the cell
- Moves the tape-head one cell to the right or to the left
- Enters a possibly new state

All these actions take place simultaneously. As in the case of a finite automaton, the input string w occupies $|w|$ consecutive cells of the tape with the tape-head scanning the first letter of w. All other cells are blanks, a blank being denoted by a special symbol B. If after finitely many moves, the TM starting from the initial state enters an *accepting state*, then the input string w is accepted by the TM. Note that a TM may run *forever*. The set of all strings that it accepts is the language accepted by the Turing machine. Such languages are called *recursively enumerable* or r.e. An r.e. language $\mathcal{L} \subseteq \Sigma^*$ is said to be *recursive* if

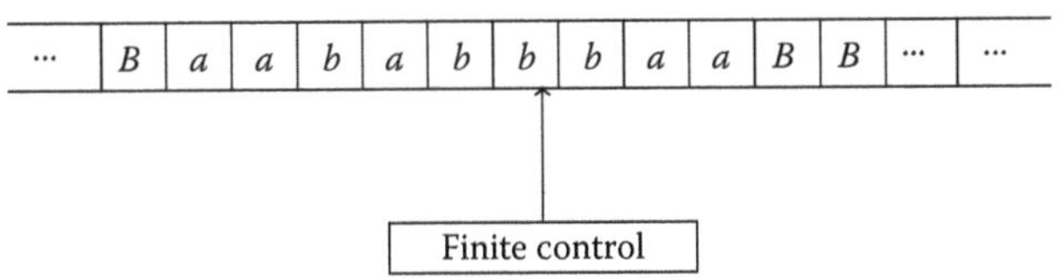

FIGURE 1.11 Turing machine.

both $\mathcal{L}$ and $\Sigma^* - \mathcal{L}$ are r.e. It can easily be shown that $\mathcal{L}$ is recursive if, and only if, there is a TM accepting $\mathcal{L}$ that halts on every input. Thus, one can decide whether a string w belongs to a given recursive language or not.

TMs (as well as other finite state machines) can be encoded by (binary) strings. This encoding enables one to construct a universal TM that can simulate every other TM. The language accepted by a universal TM is called universal r.e. and it "effectively" enumerates all r.e. languages. A universal r.e. language is an example of an r.e. language that is not recursive.

We may restrict the TM by having a one-way infinite tape or generalize by having several tapes (each with a tape-head moving independently) and allow nondeterminism. One may also have an output tape. These restrictions or extensions do not change the computing power. These TMs have the same power as single-tape TMs.

A TM can also be used for computing functions. For example, to compute $(n, m) \longrightarrow n.m$, the input (n, m) will be represented by a string of n zeros and m zeros separated by 1, that is, by the string 0^n10^m. There is a second tape which is a work tape and also used as an *output* tape. For each 0 in the first part of the input, the TM $\mathcal{M}$ will copy the m 0's following the 1 onto the second tape. The 0 read by $\mathcal{M}$ is "marked" or erased, that is, replaced by B. When all the n 0's are marked, we would be having nm 0's onto the output tape.

A function $f(n_1, \ldots, n_k)$ is said to be *computable* or *recursive* if there is a TM that computes f in the sense of the above example.

1.6 Finite Automata and Logic

There is a natural connection between finite state automata and logic. This was first studied by Büchi [3] and Elgot [5]. Automata on infinite objects were studied in the 1960s by Büchi [3], McNaughton [11], and Rabin [15] mainly to investigate decision problems in mathematical logic. An infinite string $\alpha \in \Sigma^\omega$ is accepted by an automaton if it enters an accepting state infinitely often while reading α. For logical definibility, an infinite string α is represented by a model theoretic structure $\underline{\alpha} = (\omega, 0, +1, <, \{Q_a\})$, where the first four is the structure of the natural numbers with zero, the successor function, and the natural ordering, and Q_a gives the positions where a occurs in α. We also allow variables for sets of natural numbers and quantification over them. Such a framework is denoted by S1S and is called *monadic second-order theory with one successor.* Büchi [3] has shown that an ω-language is regular iff it is definable in S1S. For finite strings, we have the following result of Büchi [3] and Elgot [5].

Theorem 1.8:

A language $\mathcal{L} \subseteq \Sigma^*$ is accepted by a finite automaton iff it is definable in S1S (interpreted over finite word structures).

For other connections between automata and logic, see the article by Thomas in Ref. [18].

1.7 Conclusion

We have introduced finite state automata and regular languages and studied some of their important properties. We have also indicated a few applications of these automata. Related finite state machines like pushdown automata, transducers, and TM have been briefly covered. We have also pointed out the relation of finite automata with logic.

Acknowledgment

We thank our colleague Palash Sarkar for some useful comments.

References

1. Aho, A. V., Sethi, R., and Ullman, J. D. 1986. *Compilers—Principles, Techniques and Tools*. Addison-Wesley, Reading, MA.
2. Brzozowski, J. A. and Yoeli, M. 1976. *Digital Networks*. Prentice Hall, Englewood Cliffs, NJ.
3. Büchi, J. R. 1960. Weak second order arithmetic and finite automata. *Z. Math. Logik Grundlagen Math.* 6: 66–92.
4. Davis, M. 1958. *Computability and Unsolvability*. McGraw-Hill, New York.
5. Elgot, C. C. 1961. Decision problems of finite automata design and related arithmetics. *Trans. Am. Math. Soc.* 98: 21–52.
6. Head, T. 1987. Formal language theory and DNA: An analysis of the generative capacity of specific recombinant behaviours. *Bull. Math. Bio.* 49: 737–759.
7. Hopcroft, J. E. 1971. An $n \log n$ algorithm for minimizing the states in a finite automaton. *The Theory of Machines and Computations*. Z. Kohavi (ed.), Academic Press, New York, pp. 181–196.
8. Kleene, S. C. 1956. Representation of events in nerve nets and finite automata. *Automata Studies*. Princeton University Press, Princeton, NJ, pp. 3–42.
9. Knuth, D. E., Morris, J. R., and Pratt, V. R. 1977. Fast pattern matching in strings. *SIAM J. Computing*. 6(2): 323–350.
10. McCulloch, W. and Pitts, W. 1943. A logical calculus of the ideas immanent in nervous activity. *Bull. Math. Biophys.* 5: 115–133.
11. McNaughton, R. 1966. Testing and generating infinite sequences by a finite automaton. *Inform. Contr.* 9: 521–530.
12. Mealy, G. H. 1955. A method for synthesizing sequential circuits. *Bell System Tech. J.* 64: 1045–1079.
13. Moore, E. F. 1956. Gedanken experiments on sequential machines. *Automata Studies*. C. E. Shannon and J. McCarthy (eds), Princeton University Press, Princeton, NJ, pp. 129–153.
14. Nerode, A. 1958. Linear automaton transformations. *Proc. Am. Math. Soc.* 9: 541–544.
15. Rabin, M. O. 1969. Decidibility of second order theories and automata on infinite trees. *Trans. Am. Math. Soc.* 141: 1–35.
16. Rabin, M. O. and Scott, D. 1959. Finite automata and their decision problems. *IBM J. Res. Dev.* 3(2): 115–125.
17. Tao, R. 2009. *Finite Automata and Application to Cryptography*. Tsinghua University Press, Beijing, China.
18. Thomas, W. 1990. Automata on infinite objects. *Handbook of Theoretical Computer Science, Volume B: Formal Models and Sematics*. J. van Leeuwen (ed.), Elsevier and MIT Press, Amsterdam, The Netherlands, pp. 133–192.

2

Large-Scale Regular Expression Matching on FPGA

Yi-Hua E. Yang
FutureWei Technologies

Viktor K. Prasanna
University of Southern California

2.1 Introduction

In formal language theory, a *regular expression* (regex) represents a (potentially infinite-size) set of strings which constitute a *regular language* (Aho and Ullman, 1972). In practice, *regular expression matching* (REM) is the process where a stream of characters is searched for occurrences of all possible strings represented by some regex; we say the regex is *matched against* the input stream. REM has traditionally played a key role in text processing and database filtering. More recently, it has become an essential component in network intrusion detection systems such as Bro (Paxson, 1999) and Snort (Sourcefire, 2011) for deep packet inspection (DPI), as well as in bioinformatics for matching biological sequences (Betel and Hogue, 2002; Fourment and Gillings, 2008).

TABLE 2.1 Magnitude of Scale in Large-Scale REM for DPI

Pattern Length	No. Unique Patterns	Req. Throughput
10–2000 bytes	Up to 2000	∼10 Gbps

Rapid growth of Internet traffic and proliferation of data gathered and exchanged through the Web have made REM a primary bottleneck in applications, where large number of patterns are matched against high bandwidth data. Such workload are usually identified as *large-scale REM*. Table 2.1 shows the magnitude of scale in large-scale REM using DPI as example. In addition to high throughput and large pattern sets, state-of-the-art REM also needs to adapt to pattern updates with consistent throughput against wide range of input streams. The pattern adaptability and throughput consistency require a solution that is efficient in both run time and compile time for matching arbitrary regex patterns. Conventionally, however, the level of performance described in Table 2.1 is either unattainable (Smith et al., 2006; Yu et al., 2006) or achieved only for patterns with limited features (Pasetto et al., 2010). Accelerating large-scale REM in both compile-time and run-time performance with *general* regex patterns remains a challenging problem.

Continual development in semiconductor technologies has produced programmable architectures with massively parallel computation logic, large on-chip memory and high memory bandwidth. One prominent example is the Field-Programmable Gate Arrays (FPGA) technology. In this chapter, we propose the algorithm and architecture for compiling any large-scale REM task into a compact and modular register-transfer level (RTL) circuit targeted for implementation on FPGA. We also discuss several techniques for optimizing performance and resource efficiency of the circuit. Finally, we evaluate the proposed REM solution on FPGA with real-life regexes taken from the official Snort (Sourcefire, 2011) rulesets.

2.2 Background

2.2.1 Regular Expression and Finite Automaton

A regex r over an alphabet Σ is composed of the three basic operators—*concatenation*, *union*, and *Kleene closure*—and the characters (symbols) in Σ.

Definition 2.1:

A *regex* is defined as follows:

1. For each $a \in \Sigma$, $\{a\}$ is the expression consisting of the string of only one symbol a.
2. $\varnothing$ is the expression for *nothing.*
3. $\{\epsilon\}$ is the expression for the *empty string* ϵ.
4. If r_1 and r_2 are expressions for sets of strings S_1 and S_2, respectively, then
    a. r_1r_2 (*concatenation* of r_1 and r_2) is an expression for S_1S_2, the set of any string in S_1 followed by any string in S_2.
    b. $r_1|r_2$ (*union* of r_1 and r_2) is an expression for $S_1 \cup S_2$, the union of S_1 and S_2.
    c. r_1* (*Kleene closure* of r_1) is an expression for S_1^*, the closure of repeating strings from S_1.

Precedence of the three operators are from high to low, Kleene closure, concatenation, and union.

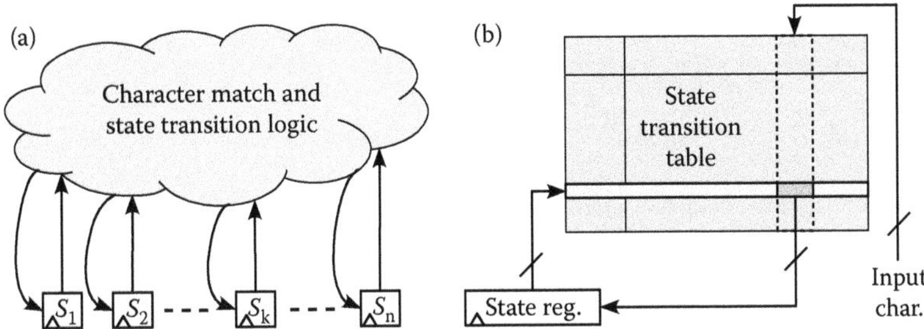

FIGURE 2.1 High-level architectures of (a) NFA and (b) DFA.

All strings that can be generated from a regex r define a *regular set* $R(r)$, which can be *recognized* by a finite automaton, either *nondeterministic* or *deterministic*.* Figure 2.1 shows a high-level comparison of the architectures used by the two types of automata. In the NFA approach, individual regex characters are represented with one-hot encoded states; every input character may incur parallel state transitions, and more than one state can be *active* at any time. The NFA can be further converted into a DFA using the standard *subset construction* procedure (Meyer and Fischer, 1971). Every possible set of concurrently active NFA states are assigned a DFA state number. The transitions between various DFA states, given any input character, are then recorded in a state transition table (STT).

While many simple regexes can be efficiently recognized by a DFA, for some regexes, converting its $O(n)$-state NFA recognizer to a DFA can produce an excessively large STT with $O(2^n)$ states. Such phenomenon is known as the exponential state explosion (Meyer and Fischer, 1971; Yu et al., 2006). In these cases, the DFA can become infeasible to construct and accelerate. Thus, we will focus on the NFA-based approach for implementing regular expression matching (to be defined later in Section 2.2.2) on FPGA.

Definition 2.2:

A *nondeterministic finite automaton* (NFA) recognizing regex r can be formalized as $\mathcal{M}(r) = (Q, \Sigma, \delta, q_0, F)$ where

- Σ is a finite *alphabet* to compose r;
- Q is a finite set of *NFA states*;
- $\delta : Q \times \Sigma \rightarrow 2^Q$ is the *nondeterministic state transition function*, where 2^Q denotes the power set of Q;
- $q_0 \in Q$ is the *initial state* of the finite state control; and
- $F \subseteq Q$ is the set of *final states*.

For each input character, an NFA makes one or more state transitions which are either ϵ-transitions (state transitions that do not consume input characters) or have a label matching the input character value. To avoid backtracking, an NFA implementation usually maintains the activeness of all states. For each input characters, all possible state transitions from the active states are made in parallel. Doing so requires $O(n)$ computation per input character to check for the state activeness of an n-state NFA. In the

* In fact, any finite automaton can also be used to define a regular set, and any regular set can be denoted by at least one regex. Thus, there is an equivalence between regular expression, regular set, and finite automaton (Aho and Ullman, 1972).

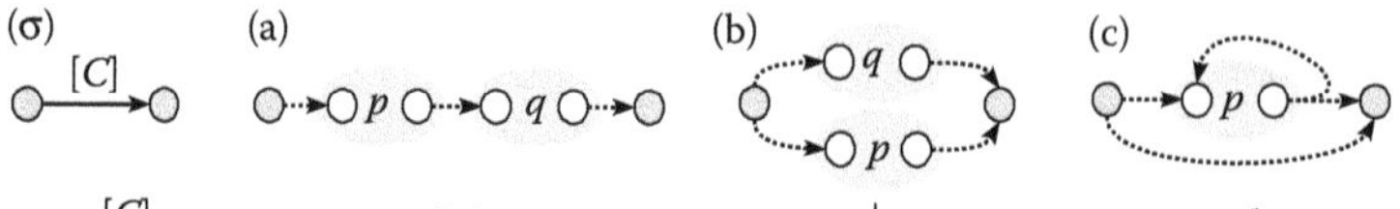

FIGURE 2.2 The McNaughton–Yamada construction. (σ) M-Y character, (a) M-Y concatenation, (b) M-Y union, and (c) M-Y closure.

worst case, the NFA can make $O(n)$ state transitions from each of the $O(n)$ active states and require $O(n^2)$ computation per input character.

Conventionally, an NFA recognizing an arbitrary regex r has been constructed using the McNaughton–Yamada (M-Y) construction (McNaughton and Yamada, 1960).* Figure 2.2 illustrates the M-Y construction rules: each shaded oval represents a subexpression, each dotted arrow an ϵ-transition, and each dark-gray circle an NFA state. Glushkov (1961) also formalized an ϵ-free NFA construction which relies on finding the *first*(x), *last*(x), and *follow*(s, x) functions for every subpattern s and symbol position x in r. In both approaches, the resulting NFA can have an arbitrary structure, making it challenging to map the automaton to a VLSI circuit (Floyd and Ullman, 1982).

Proposition 2.1:

An NFA recognizing multiple regexes $r_1, \ldots, r_n$ concurrently can be constructed by combining the NFAs $\mathcal{M}(r_i) = (Q_i, \Sigma, \delta_i, q_{0,i}, F_i)$, $i = 1 \ldots n$, as follows:

1. Combine the initial states $q_{0,1}, \ldots, q_{0,n}$ of all NFAs into a single initial state q_0.
2. Let $(Q, \delta, F) = (\bigcup_{i=1}^{n} Q_i, \bigcup_{i=1}^{n} \delta_i, \bigcup_{i=1}^{n} F_i)$.

2.2.2 Regular Expression Matching

We refer to the process of recognizing a set of regexes in a continuous stream of input characters as REM. A REM implementation in software or hardware is called a *regular expression matching engine* (REME).

Definition 2.3:

Let regex r defines regular set $R(r)$. Given r and a character stream $X = [x_1 \cdots x_w]$, $x_i \in \Sigma$, $1 \leq i \leq w$, the REM of r *against* X is defined as the process to find and report all substrings in X which are members of $R(r)$. Each substring found is called a *match* of r in X. Formally, REM of r against X finds the following set of matches of r in X:

$$\left\{X_{i,j} = [x_i \cdots x_j] \;\middle|\; 1 \leq i \leq j \leq w,\ X_{i,j} \in R(r)\right\} \tag{2.1}$$

REM of any regex r can be realized as follows: First, we prefix r by "`.*`" to produce r', where "`.`" represents the any-value (wildcard) character and "`*`" is the Kleene closure. Then, we convert r' to the NFA that recognizes r' in the input stream. The NFA continues to process the entire input stream while

* While Figure 2.2 has been credited to McNaughton and Yamada, it is worth noting that McNaughton and Yamada (1960) did not actually show Figure 2.2, nor was it focused on the construction of NFA. Instead, it referenced earlier work (Copi et al., 1958) for a regular expression to "logical nets" construction, which is essentially equivalent to Figure 2.2 when the logical net is abstracted as NFA.

TABLE 2.2 Commonly Used Operators for REM

	Op.	Name	Example	Description
Basic	-	Concatenation	q_1q_2	q_1 followed by q_2
	|	Union	$q_1\|q_2$	Either q_1 or q_2
	*	Kleene closure	$q*$	Repeat q zero or more times
Extended	+	Repetition	$q+$	Repeat q one or more times
	?	Optionality	$q?$	Repeat q zero or one time
	$\{m, n\}$	Constrained rep.	$q\{m, n\}$	Repeat q in m to n times
	^	Start of string	^q	Has q at start of input
	$	End of string	$q$$	Has q at end of input
Character	[...]	Character class	[a − c]	(a|b|c)
Classes	[^...]	Inv. char. class	[^\r\n]	Neither \r nor \n

collecting all characters positions where a final state is reached. The collected character positions are all the j's in Equation 2.1. A similar approach can also be used to match multiple regexes $\{r_1, \ldots, r_m\}$ against an input stream. In this case, each regex r_k is first prefixed by ". *" to produce r'_k, $1 \leq k \leq m$, and converted individually into an NFA. The m NFAs, one for each regex in $\{r'_1, \ldots, r'_m\}$, are then combined as described in Proposition 2.1. The combined NFA can then be used to match all r_k concurrently against the input stream.

In theory, the alphabet Σ of a regex can contain any finite number of symbols. In practice, REM is usually performed over a single-byte (8-bit) alphabet, such as the extended ASCII characters. Recently, it is also common for REM software (Java Regular Expression Package; Regular Expression in ASP.NET; Hazel, 2011) to use multi-byte Unicode characters. For simplicity and without loss of generality, we consider only REM based on single-byte or 8-bit characters. In addition to the three basic operators (concatenation, union, and Kleene closure), most REM softwares also support an extended set of operators offering convenient representations for commonly used subpatterns. Table 2.2 lists the commonly supported regex operators. Each of these operators can be substituted by an equivalent (but longer) subpattern using the three basic operators.*

2.2.3 Using Field-Programmable Gate Array for REM

FPGA technology offers the ability to reconfigure massively parallel on-chip logic and high-speed I/O. Most modern FPGA devices are composed of lookup tables (LUTs) and on-chip block memory (BRAM) based on the static random access memory (SRAM) technology. The LUTs are used to implement any combinational logic; each LUT is paired with one or more flip-flops as state registers. The BRAM provides high-bandwidth memory storage with configurable word width to the LUTs. An SRAM-controlled routing fabric directs signals to appropriate paths to produce the desired architecture. State-of-the-art FPGA devices offer dense logic units, large on-chip memory, and high-frequency operation. They also support high-speed and high-bandwidth interfaces to various external memory and I/O technologies.

FPGAs started out as prototyping devices, allowing convenient and cost-effective development of glue logic connecting discrete ASIC components. As the gate density of FPGA increased, applications of FPGA shifted from the glue logic to a wide variety of high-performance and data-intensive problems. Furthermore, because the functionality of an FPGA device is configured by the on-chip SRAM, it can be

* Some software REM operators such as *backreference* and *recursion* do not produce regular sets. They are not the focus of this dissertation.

altered simply by changing the state of the memory bits. Both features make FPGA a suitable technology for implementing NFA-based REM.

In general, there are three challenges to perform REM on FPGA:

- Match large numbers of regex patterns in parallel.
- Handle arbitrarily structured regexes efficiently.
- Obtain high concurrent matching throughput.

Large number of patterns require more hardware resources, which in the case of FPGA are the amount of LUT, flip-flops (FF), and on-chip block memory (BRAM). In addition, many real-life regexes (e.g., Snort rules (Sourcefire, 2011)) contain nested unions and Kleene closures, long single-character repetitions, arbitrary character classes and multiple character classes in closures or subpattern unions. These regexes are often translated to NFAs with large number (thousands) of states and/or arranged in nontrivial topology. Efficient use of on-chip resources is thus critical to obtaining high regex capacity and achieving maximum clock rate.

To help understand the relation between REM throughput and regex capacity on FPGA, we make the following definition.

Definition 2.4:

The *concurrent throughput* of a REM circuit on FPGA for n regexes is defined as the throughput of the input stream(s) matched by all n regexes concurrently.

It can be seen that the concurrent throughput is determined by three factors: (1) the size/complexity of the regexes; (2) the amount of resource used by the REM circuit; and (3) the achieved clock frequency. In general, to obtain the same level of concurrent throughput, a *larger* set of regex patterns with *more complex* features require proportionally *more* resources (i.e., number of slices). The *concurrent* throughput is different from the *aggregated* throughput, which sums up the throughput of matching individual regexes. For example, a REM circuit matching 5 regexes against a 1 Gbps input may have 5 Gbps aggregated throughput but only 1 Gbps concurrent throughput. When evaluating the performance of a large-scale REM circuit architecture, the concurrent throughput should be used, and the resource usage must be normalized.

2.3 Modular RE-NFA Construction

The basis of our large-scale REM design on FPGA is the modular regex-NFA (RE-NFA) construction process:

- The regex is first parsed from Perl-Compatible Regular Expression (PCRE) to a nested *token list* data structure which identifies all character classes and operators (Section 2.3.1).
- The token list is then converted to a modular RE-NFA using the *modified* McNaughton–Yamada (MMY) construction (Section 2.3.2).

These two steps are discussed in Sections 2.3.1 and 2.3.2, respectively. For each step, we describe both the conventional regex-to-NFA compilation and our RE-NFA construction for clear side-by-side comparison.

For illustration purpose, we will use "`a{4}b*(c[ab]?|d)+f`" as the example regex to describe the RE-NFA construction process.

2.3.1 Parsing PCRE to Token List

As discussed earlier, theoretically each regex defines a regular language. Conventionally, the compilation of a regex starts from parsing the regex into a right-leaned (or left-leaned) parse tree representing the

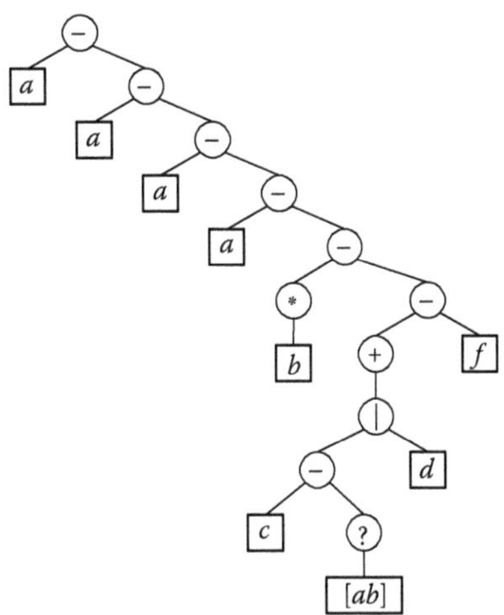

FIGURE 2.3 A right-leaned parse tree for regex "`a{4}b*(c[ab]?|d)+f`".

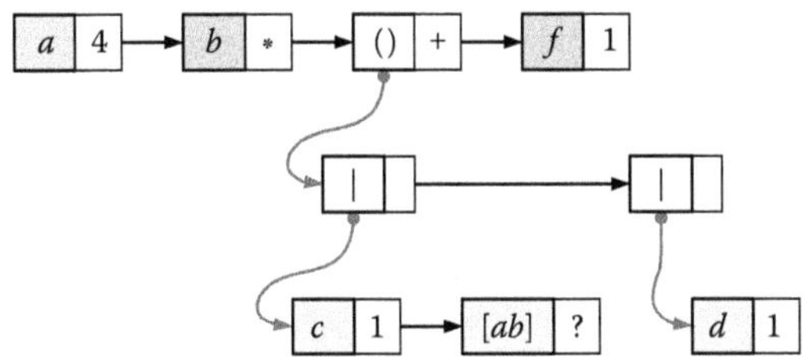

FIGURE 2.4 Token list for regex "`a{4}b*(c[ab]?|d)+f`".

right-linear (or left-linear) production for the regular language (Aho and Ullman, 1972; Floyd and Ullman, 1982). Figure 2.3 shows a parse tree for the example regex. When used for real-world regexes in the PCRE format, however, this approach has several limitations:

1. It is not concise for constrained repetitions. A repetition of $\{n\}$ will be parsed as n identical concatenation nodes. A repetition of $\{m, n\}$ will be converted into an $(n - m)$-level nested union nodes in the parse tree.
2. It creates an unnecessarily deep tree, which in turn results in deep levels of recursion during the following McNaughton–Yamada construction. For example, a union of 10 subregexes will be parsed into 9 consecutive levels of union nodes.
3. It does not accept the PCRE extensions, such as assertions, lookarounds, conditionals, and back-references.

To overcome these limitations, we parse the regex into a *token list*, such as the one shown in Figure 2.4. Compared to a parse tree, the token list is not only more compact and easier to traverse, but also capable of capturing the full PCRE semantics. Structurally, a token list is a two-dimensional linked list of tokens. A *token* in the token list has the following major fields:

Field	Usage and Meaning of the Field
val	Character class; parenthesis of subregex; union-op; a PCRE-extended subregex feature
rep	Constrained repetition; Kleene closure
next	Next token to concatenate or to union
child	Nested subregex or PCRE-extended feature

The *next* and *child* fields are used to chain together all the tokens in the token list. Normally, a concatenate operation is implied between a token and its *next* token; except when the token's *val* field is the "union-op" (|), then a union operation is implied instead. Depending on the *val* field, a token can be one of the four types:

1. Single character class, such as "a" or "`[ab]`".
2. Parentheses over subregex, that is, a placeholder for a "`(...)`" pattern.
3. Union-op to subregex, that is, the local root to a branch in a union.
4. PCRE-extended features including assertion, backreference, and conditional.

A type-1 or type-2 token may have a *rep* field representing the constrained ($\{m, n\}$) or closure (+ or *) repetition; a type-2 or type-3 token always points to a nested token list via the *child* field. The *val* field of a type-4 token stores a PCRE-extended feature, in which case the *child* field points to the corresponding raw subregex. Some of the features (e.g., ^ and $ in Table 2.2) can be implemented by tweaking the RE-NFA (Section 2.3.2); others are either partially ignored (e.g., the conditional operators) or used to cut-short the regex (e.g., the backreference operators) during the following RE-NFA construction.

In our example, the regex "`a{4}b*(c[ab]?|d)+f`" can be parsed as shown in Figure 2.4. The *val* and *rep* fields are the left and right blocks of each token, respectively. The *next* field is represented by the straight right-arrow, and the *child* field by the curved down-arrow. Conceptually, the token list data structure is a combination of the parse tree used in Sidhu and Prasanna (2001) and Floyd and Ullman (1982) and the *opcodes* used in Mitra et al. (2007) and Hazel (2011). Note that although some of these PCRE extensions are nonregular and thus cannot be handled by finite automata, the token list can still capture their semantics.

2.3.2 Modified McNaughton–Yamada Construction

It is well-known that the McNaughton–Yamada (M-Y) construction (Floyd and Ullman, 1982; McNaughton and Yamada, 1960) (Figure 2.2) can be used to convert any regex parse tree into an NFA matching the regex. The resulting NFA, however, may have intermediate nodes and long sequences of ϵ-transitions between labeled transitions. Figure 2.5 shows an NFA for "`a{4}b*(c[ab]?|d)+f`" constructed with the rules in Figure 2.2. A number of states (gray circles) are introduced to "glue" together various character-matching states and labeled transitions. The introduced states only connect to ϵ-transitions. Their presence gives the NFA an irregular structure where there can be variable numbers of ϵ-transitions between two labeled transitions.

In contrast, we use the *modified* McNaughton–Yamada (MMY) construction, outlined in Algorithm 2.1, to convert an arbitrarily structured token list to a modular RE-NFA suitable for implementation on FPGA.

Algorithm 2.1 consists of two mutually recursive subroutines:

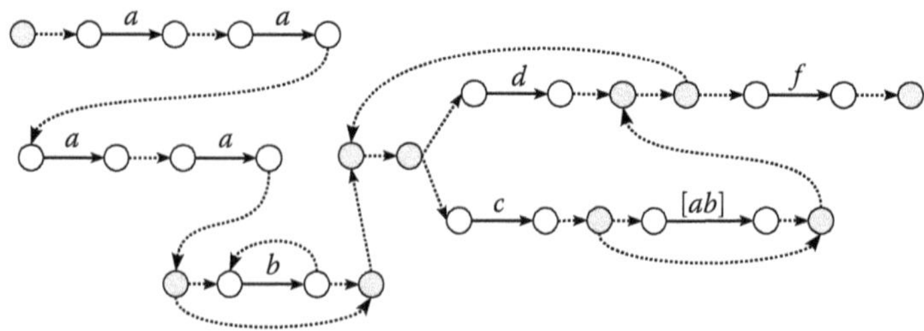

FIGURE 2.5 An NFA for regex "`a{4}b*(c[ab]?|d)+f`" constructed with the rules in Figure 2.2. Dashed lines represent ϵ-transitions. The gray circles are nodes introduced for merging and relaying the ϵ-transitions.

ConvTok handles the conversion of a full token, which may contain unions, closures, and constrained repetitions.

ConvVal handles the conversion of a token's value field, which may be either a character class or a parenthesis.

Algorithm 2.1:

The *modified* McNaughton–Yamada (MMY) construction

t_{cur}	Token currently being converted
$t.\{val,rep,next,child\}$	Token t's *value*, *repetition*, *next* and *child* fields
S_{pre}	Set of fan-in states ϵ-transitioning into t_{cur}
S_{out}	Set of fan-out states ϵ-transitioning out-of t_{cur}
$s.foset$	Set of fan-out states of a resulting state s
create_state (c)	Create a new state matching character class c
delete_state (p)	Delete state p and its fan-in & fan-out relations

```
Subroutine S_out := ConvTok (t_cur, S_pre)
if t_cur.val = "|" then
    while t_cur ≠ null do
        S_out ← S_out ∪ ConvTok (t_cur.child, S_pre)
        t_cur ← t_cur.next
    end while
end if
while t_cur ≠ null do
    if t_cur.rep = "*" or t_cur.rep = "+" then
        p ← create_state (null) {a pseudo-state}
        S_out ← ConvVal (t_cur, S_pre ∪ {p})
        ∀s ∈ S_out, s.foset ← s.foset ∪ p.foset
        delete_state (p)
        if t_cur.rep = "*" then
            S_out ← S_out ∪ S_pre
        end if
    else if t_cur.rep = "{m, n}" then
        for (cnt ← 0 ; cnt < m ; cnt++) do
            S_pre ← ConvVal (t_cur, S_pre)
        end for
        S_out ← S_pre
        for (cnt ← m ; cnt < n ; cnt++) do
            if use fan-in oriented approach then
                S_pre ← ConvVal (t_cur, S_pre)
                S_out ← S_out ∪ S_pre
            else {use fan-out oriented approach}
                S_out ← ConvVal (t_cur, S_out ∪ S_pre)
            end if
        end for
    end if
    t_cur ← t_cur.next
end while
```

SUBROUTINE S_{out} := CONVVAL (t_{cur}, S_{pre})

if $t_{cur}.val =$ "( )" **then**
    $S_{out} \leftarrow$ CONVTOK $(t_{cur}.child, S_{pre})$
else {$t_{cur}.val$ represents a character class}
    $s_{new} \leftarrow$ create_state $(t_{cur}.val)$
    $\forall s \in S_{pre}, s.foset \leftarrow s.foset \cup \{s_{new}\}$
    $S_{out} \leftarrow \{s_{new}\}$
end if

The algorithm starts from a call to CONVTOK on the first regex token and terminates after the last token is converted. The subroutine proceeds in one of the following four main parts:*

1. A union "path" token ($t_{cur}.val =$ "|") is converted by a recursive call to CONVTOK on the token's child ($t_{cur}.child$). The child token receives the same input ϵ-transitions as the parent token; it also contributes its output ϵ-transitions to the parent token. (CONVTOK step 3).
2. For tokens with a "*" or "+" operator, a pseudo state p is created to temporarily represent the source state(s) of the feedback loop. The actual backward ϵ-transitions are made after the token has been converted and its fan-out states have been identified. (CONVTOK steps 9–12).
3. Input ϵ-transitions to a token with the "*" operator are passed (together with the output ϵ-transitions from the token) directly to the subsequent token(s). (CONVTOK step 14).
4. Constrained repetitions are converted directly into NFA states without first represented as multiple unions. (CONVTOK steps 18, 23, or 26).

Two special entities are essential in Algorithm 2.1 for the modular RE-NFA construction. The first is the set of immediate previous states S_{pre}, which contains the source states of all fan-in ϵ-transitions to the token currently being converted (t_{cur}). Identifying S_{pre} helps to collapse a long sequence of ϵ-transitions produced by the original M-Y construction into a single ϵ-transition in the MMY construction. The second special entity is the pseudo state p created at CONVTOK step 9 for the closure tokens. The pseudo state works as a temporary placeholder for the source states of the feedback loop of a closure operator. This is needed to resolve the circular dependency when converting a closure token, whose S_{pre} depends on the fan-out states (S_{out}) of the very closure token being converted. After the closure token has been converted recursively (CONVTOK step 10), the pseudo state is replaced by the actual set of source states (CONVTOK step 12) (see Figure 2.6).

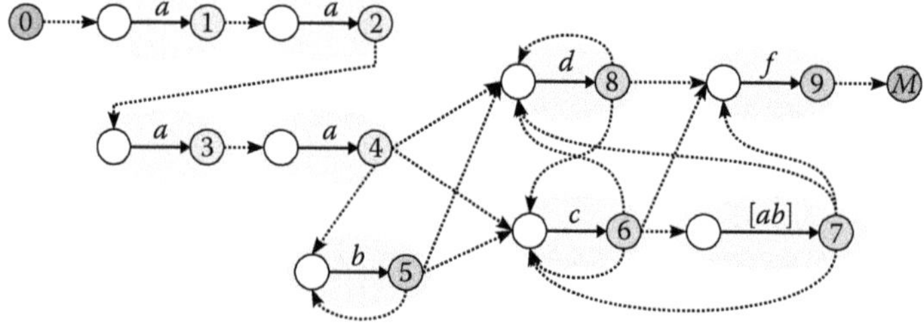

FIGURE 2.6 Modular RE-NFA for regex "a{4}b*(c[ab]?|d)+f". The straight dashed lines represent ϵ-transitions that make forward progress in the regex; the curved dashed lines represent backward (due to closure operators) and crosstalk (due to union operators) ϵ-transitions.

* For clear presentation error checking and corner cases are omitted.

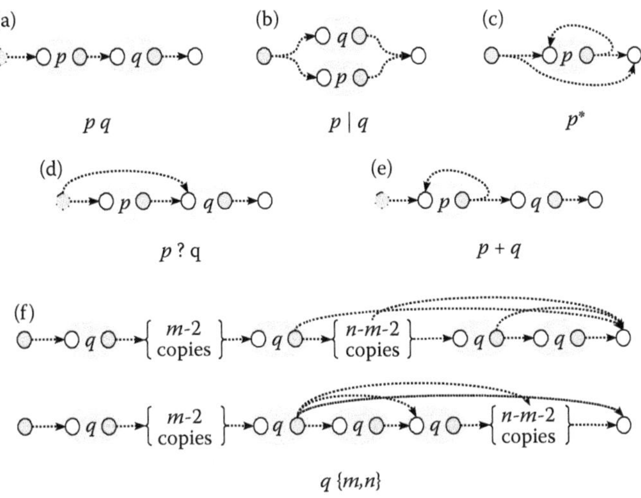

FIGURE 2.7 Graphical representation of the *modified* McNaughton–Yamada (MMY) construction rules. A letter (p or q) represents a token. Each shaded oval represents a (recursively constructed) RE-NFA for the token. Dashed lines represent ϵ-transitions. Dark and white circles represent (sets of) states making and receiving ϵ-transitions, respectively. (a) MMY-concatenation, (b) MMY-union, (c) MMY-closure, (d) MMY-qm, (e) MMY-plus, and (f) MMY-repeat.

Graphically, the MMY construction can be described by the six construction rules (a)–(f) in Figure 2.7. For each construction rule, the left-most (dark) dashed circle corresponds to the S_{pre} in Algorithm 2.1, whereas the right-most (white) dashed circuit corresponds to the set of states that receive ϵ-transitions from S_{out}. To ensure applicability of the MMY rules to the edge tokens, extra START and MATCH states are added to the RE-NFA as the initial "source" and final "sink" of ϵ-transitions, respectively. In the context of Algorithm 2.1, the START state constitutes the first S_{pre} and the MATCH state collects the last S_{out} during the recursive token list conversion. In practice, these two states can also be used to handle the special "^" and "$" operators (see Table 2.2), respectively. A regex *with* a "^" in the beginning means its START state is *only* active at the first input character; otherwise, the START state is active at every input character. Similarly, a regex *with* a "$" at the end means its MATCH state is checked *only* at the last input character; otherwise, the MATCH is checked at every input character.

2.3.3 Properties of the MMY Construction

Lemma 2.1:

Given any regex composed of the *character matching* (σ), *concatenation* ($\cdot$), *union* ($|$), and *Kleene closure* ($*$) operators, an NFA constructed by the *modified* McNaughton–Yamada (MMY) construction in Figure 2.7 is equivalent to one constructed by the *original* McNaughton–Yamada (M-Y) construction in Figure 2.2.

Proof. Both MMY and M-Y construct the same finite automaton for matching any character class (rule σ in Figure 2.2). For the three basic operators, we show that, given the same p and q as input, the MMY rules in Figure 2.7a–c produce effectively equivalent finite automata to the M-Y rules in Figure 2.2a–c, respectively.

For simple notation, we use p and q to denote two regexes as well as the respective finite automata matching the regexes. Let $R(p)$ and $R(q)$ be the regular sets of strings generated from p and q, respectively.

Given p and q as input to the MMY (Figure 2.7) and M-Y (Figure 2.2) constructions:

1. Figures 2.7a and 2.2a both construct NFA generating the concatenation of $R(\text{p})$ and $R(\text{q})$.
2. Figures 2.7b and 2.2b both construct NFA generating the union of $R(\text{p})$ and $R(\text{q})$.
3. Figures 2.7c and 2.2c both construct NFA generating the Kleene closure of $R(\text{p})$.

Thus, both MMY and M-Y constructions are effectively equivalent with respect to character matching and the three basic operators. □

Note that in Algorithm 2.1, no extra state, other than the pseudo state, is produced to propagate ϵ-transitions in Algorithm 2.1. The pseudo state is created only temporarily and will be deleted after the corresponding closure token is converted. The only place where a nonpseudo state is created is in ConvVal step 4, where the state is always assigned a valid character-matching label $t_{cur}.val$. It is thus guaranteed that (1) every state is associated with (reached via) a labeled transition, and (2) any state transition goes through exactly one ϵ-transition followed by one labeled transition to reach the next state. Figure 2.6 shows the example modular RE-NFA.

Also note there are two approaches to the constrained repetition in Figure 2.7f: a *fan-in oriented* (upper) approach and a *fan-out oriented* (lower) approach. The two approaches are described in Algorithm 2.1 from ConvTok steps 23–26. They differ in how the feedforward ϵ-transitions are connected in the last $(n-m)$ repeating part of $\text{p}\{m,n\}$. While functionally equivalent, the two approaches can result in different circuit-level tradeoff on FPGA. The fan-in oriented approach induces at least $(n-m)$ fan-in ϵ-transitions at the state following the constrained repetition. The fan-out oriented approach increases the fan-out of the mth repeating state by $(n-m)$ times.

When the value of $(n-m)$ is high, it may be advantageous to break $\text{p}\{m,n\}$ into $k>0$ segments: $\text{p}\{m_1,n_1\}\text{p}\{m_2,n_2\}\cdots\text{p}\{m_k,n_k\}$ where $m=\sum_{i=1}^{k} m_i$, $n=\sum_{i=1}^{k} n_i$ and $m_i, n_i > 0$, and apply both fan-in and fan-out oriented approaches alternately to various segments. When $(n-m)=1$, both fan-in and fan-out oriented approaches become identical. This includes the "?" operator, which is equivalent to $m=0$ and $n=1$.

2.4 REM Circuit Construction

2.4.1 State Update Module

As shown in Figure 2.6, the modular RE-NFA consists primarily of pairs of nodes that match a character label. The number of node pairs equals the number of matching characters in the regex. When mapped to REM circuit on FPGA, each of these node pairs constitutes a *state update module* matching a single character class. Within each node pair, the right node corresponds to a 1-bit state register; the left node corresponds to an input ϵ-transition aggregator. The ϵ-transitions between various node pairs specify the interconnection between the corresponding state update modules. At run time, the state register is set to '1' if and only if *some* input signal is '1' *and* the input character matches the character label of the state update module. Input character can be matched elsewhere (see Section 2.4.2); only the 1-bit match result is sent to the state update module. Thus, character matching is inherently uncoupled from state transition in our REM circuit architecture.

Fundamentally, the state update module performs a logic OR operation on the input state transition signals, followed by a logic AND with the character matching result. Each state update module is configured with the number of input ports as a parameter, determined by the number of "previous" states that immediately ϵ-transition to the state update module. We design two types of state update modules: (i) one accepting normal character matching (Figure 2.8a); (ii) the other accepting negated character matching (Figure 2.8b). This allows us to instantiate only one character matching circuit for both a character class and its negation.

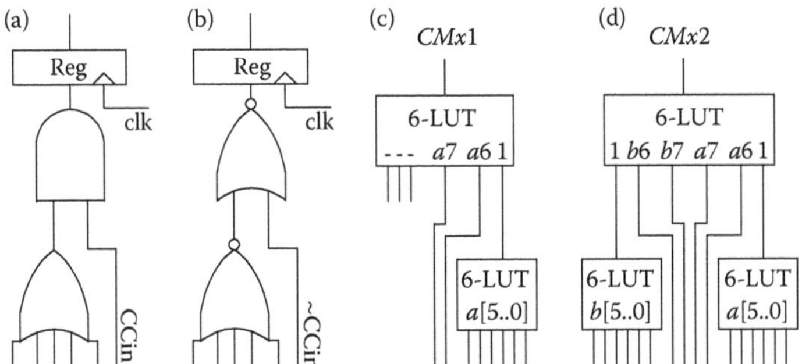

FIGURE 2.8 6-LUT optimized circuit elements: state update modules with (a) normal and (b) inverted character matching inputs; compact logic for matching (c) one-value and (d) two-value simple character classes.

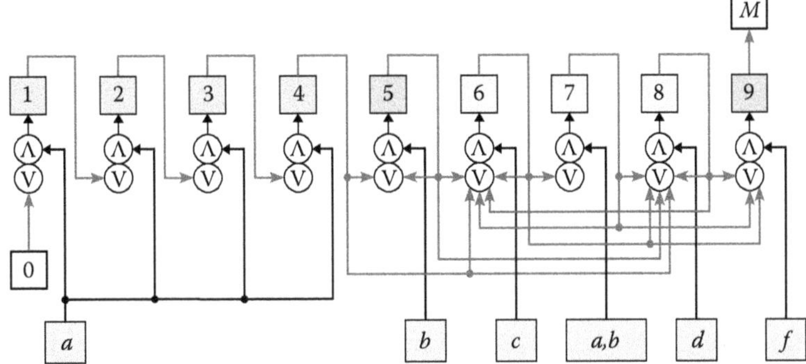

FIGURE 2.9 A basic REM circuit for regex "`a{4}b*(c[ab]?|d)+f`". The $\wedge$ and $\vee$ symbols represent logic AND and OR gates, respectively.

Figure 2.9 shows a REM circuit for the example regex. On modern FPGAs with 6-input LUT, a k-input OR followed by a 2-input AND can be realized by a single LUT if $k \leq 5$, or two LUTs if $6 \leq k \leq 10$. Since most RE-NFA states (>95%) from real-world regexes have less than 10 fan-in transitions, the REM circuit architecture spends only one or two LUTs per state in the majority of cases.

2.4.2 Character Matching and Classification

An important feature available in most REM softwares (e.g., Perl-Compatible Regular Expression, PCRE) is the use of character classes. A character class is a convenient and efficient way to encode a union of multiple character symbols from the alphabet. Efficient implementation of character class matching can significantly reduce the complexity of the REM circuit.* For example, the regex "`[ac][ab]{n}`" can be matched by an $(n+1)$-state NFA with character classes `[ac]` and `[ab]`. The equivalent regex "`(a|c)(a|b){n}`" would have to be matched by $2n+2$ states with characters `a`, `b`, and `c`; the number of inter-state ϵ-transitions is also doubled. In general, expanding a character class of size v to individual characters increases the number of NFA states and state transitions by v times.

We observe that, for 8-bit characters, any character classification can be fully specified by 256 bits, one for the inclusion of each 8-bit value. Thus, w arbitrary character classes can be fully implemented on a block memory (BRAM) of $256 \times w$ bits, all matched in parallel by one BRAM access. Furthermore, if two RE-NFA states accept the same character class, they can share the same character classification output.

* The modular RE-NFA produced in Section 2.3 fully captures the character class semantics, where a state can match an arbitrary character class.

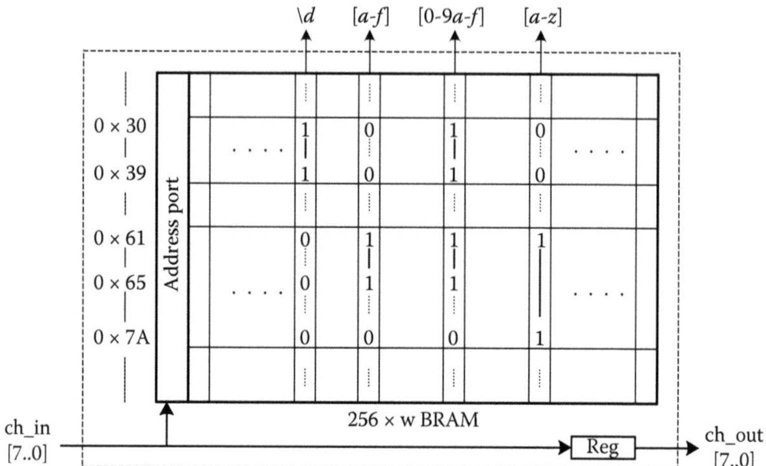

FIGURE 2.10 BRAM-based character classifier for *w* complex character classes.

Figure 2.10 illustrates an example character classifier for `\d`, `[a-f]`, `[0-9a-f]`, and `[a-z]` (among a few unspecified others).

By default, BRAM-based character classifier is only used to match *complex* character classes which would otherwise require much logic resource to implement. We define a character class to be complex if it consists of more than 2 character values. For simple character classes that consist of only one or two character values, efficient matching circuit can be realized in the logic resource on FPGA, as shown in Figure 2.8c or Figure 2.8d, respectively. Matching simple characters in logic reduces the utilization of on-chip BRAM, which can be used for buffering or other purposes.

2.4.3 Differences from Previous Approach

While our REM circuit architecture is functionally similar to the prior work (Bispo et al., 2006; Hutchings et al., 2002; Mitra et al., 2007; Sidhu and Prasanna, 2001; Yamagaki et al., 2008), there are three subtle but important differences:

1. State register occurs *after*, rather than *before*, the character matching. This allows the character matching latency to overlap with the state transition latency, resulting in higher clock rates.
2. The modular RE-NFA can be converted automatically into a modular circuit on FPGA. As is discussed later, individual circuits can be *spatially stacked* to match multiple input characters per clock cycle.
3. We do *not* minimize logic at the HDL level. In our REM circuit, the same signal can be produced by different state update modules. We leave such minimization to the synthesis software, which will apply its own optimization on the circuit to meet the implementation constraints.

In the next section, we discuss various optimizations which further improve the throughput performance and resource efficiency of the basic REM circuit. These optimizations are made possible due to the modular RE-NFA and compact state update module designs.

2.5 REM Circuit Optimizations

2.5.1 Spatial Stacking for Multicharacter Matching

Previous designs of multicharacter matching (Yamagaki et al., 2008) adopted *temporal* transformations at the NFA-level, where state transitions are extended forward in time (clock cycle) to construct a new

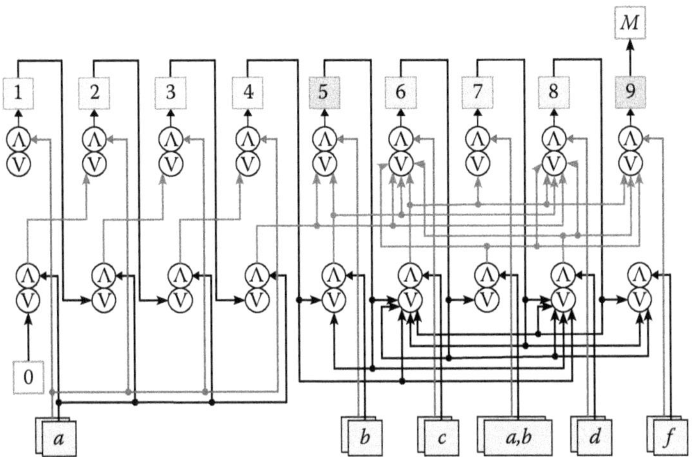

FIGURE 2.11 A 2-character matching circuit for regex "`a{4}b*(c[ab]?|d)+f`".

NFA with multicharacter state transitions. In contrast, we adopt a circuit-level *spatial stacking* technique to construct multicharacter matching circuits. An example 2-character matching circuit for the example regex is shown in Figure 2.11. A formal procedure is described in Algorithm 2.2.

Algorithm 2.2:

Multicharacter matching circuit construction

Input: An n-state m_1-character matching circuit C_{m_1} and an n-state m_2-character matching circuit C_{m_2} for some regex.

Output: An n-state m-character matching circuit C_m, $m = m_1 + m_2$, for the same regex.

for each $i \in \{1, \ldots, n-1\}$ **do**
    Suppose state output i connects to state inputs $\{i_1, i_2, \ldots, i_t\}$
    Remove C_{m_1} state register i;
    Forward the AND gate output directly to state output i.
    Disconnect C_{m_1} state output i from all state inputs of C_{m_1};
    Re-connect C_{m_1} state output i to C_{m_2} state inputs $\{i_1, i_2, \ldots, i_t\}$.
    Disconnect C_{m_2} state output i from all state inputs of C_{m_2};
    Re-connect C_{m_2} state output i to C_{m_1} state inputs $\{i_1, i_2, \ldots, i_t\}$.
    The combined state i receives $(m_1 + m_2)$ consecutive character matching signals: first m_1 by the C_{m_1} part; last m_2 by the C_{m_2} part.
end for
Merge the START states of C_{m_1} and C_{m_2} into a single START state.
Merge the MATCH states of C_{m_1} and C_{m_2} into a single MATCH state by OR-ing the original MATCH states.

For an n-state RE-NFA with max state fan-out d, each application of Algorithm 2.2 requires $O(n \times d)$ time and produces a circuit taking $O(n \times m \times d^2)$ area. Applying the algorithm on two copies of the same m-character matching circuit generates a $2m$-character matching circuit. Thus recursively, an n-state

m-character matching circuit can be constructed in $O(n \times d \times \log_2 m)$ time, starting from a one-character matching circuit for the n-state RE-NFA.

Lemma 2.2:

The circuit constructed by Algorithm 2.2 over two copies of a single-character matching circuit is a valid two-character matching circuit.

Proof. First note that a state machine at any moment is completely described by its current state values and the next state transitions. Suppose we perform Algorithm 2.2 on two identical one-character input circuits, C_A and C_B, for the same regular expression, but *without* removing the state registers of C_A (i.e., do not perform step 3 of Algorithm 2.2):

1. The algorithm does not add, remove, or reorder the states in either C_A or C_B, nor does it modify the logic used to produce any state value.
2. From C_A's outputs to C_B's inputs, state transitions are preserved by step 4.
3. From C_B's outputs to C_A's inputs, state transitions are also preserved by step 5.
4. The state transition labels of both C_A and C_B are kept intact by step 6.

The combined circuit, with the same state values and state transitions, would function exactly the same as an original one-character input state machine at every clock cycle. Removing the state registers of C_A simply merges the two-cycle pipeline of the combined circuit into one cycle, resulting in a circuit of 2-character input per clock cycle. □

Theorem 2.1:

Algorithm 2.2 can be used to construct a valid m-character matching circuit for any $m > 1$.

Proof. By the same arguments as for Lemma 2.2, a circuit constructed by Algorithm 2.2 over an m-character matching circuit and a single-character matching circuit is a valid $(m+1)$-character matching circuit. By induction, any circuit constructed by Algorithm 2.2 with an integer $m > 1$ is a valid m-character matching circuit. □

Compared to the previous approaches, our multicharacter construction is simpler, more flexible, and more efficient with the larger number of inputs in the LUTs in modern FPGAs. The resulting circuit can be optimized automatically and effectively by the FPGA synthesis tools at the circuit level. While a higher value of m generally lowers the achievable clock rate, overall throughput still improves dramatically by the ability to match multiple characters per clock cycle (see Figures 2.15 and 2.16).

2.5.2 Parallel SRL for Single-Character Repetition

A common PCRE feature used by many real-world regexes is single-character repetition, where a character class is repeated for a fixed number of times. In a straightforward implementation, each repeating instance requires a state register and its associated state update logic, which can occupy much resource on FPGA for a repetition of over a few hundred times. Alternatively, the repetition can be implemented as shift-register lookup tables (SRLs), which are much more resource-efficient than individual registers on FPGA (Bispo et al., 2006). However, the SRL approach proposed in Bispo et al. (2006) produces only single-character matching circuits. For example, while states 3–6 in Figure 2.9 can be easily implemented as a shift register of length 4, the same states in Figure 2.11 involve interleaving state transitions and cannot be implemented by the simple shift register architecture.

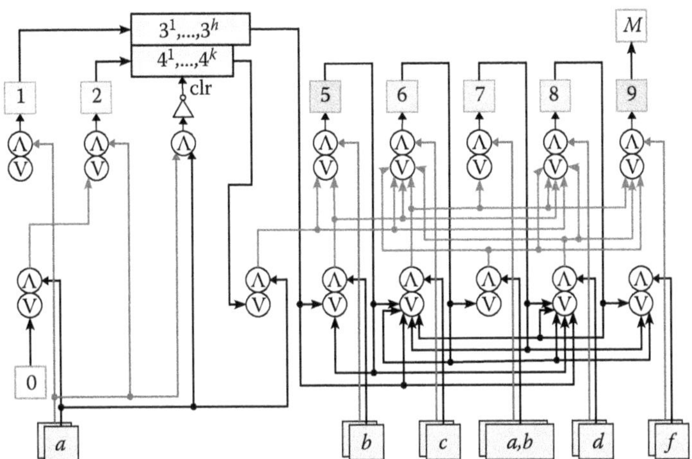

FIGURE 2.12 A 2-character matching circuit for regex "`a{4}b*(c[ab]?|d)+f`" with a 2-way parallel SRL. The character 'a' is matched a total $(h+k+2)$ times.

We design a *parallel SRL* (pSRL) architecture for single-character repetitions with multicharacter matching. A pSRL of width $m > 1$ consists of m *tracks*. Each track is implemented as a shift register, has its own `shift_in` port, state storage and `data_out` port. All tracks share the same `clear` signal, but may not have the same length. In every clock cycle, each track accepts one bit from its `shift_in` port and shifts its internal states by one bit-position toward the `data_out` port. To construct an m-character matching circuit for a single-character repetition of $r \geq 2m$ times, we connect the output of the first m repeating states to the `shift_in` ports of an m-track pSRL. The ith pSRL track, $0 \leq i \leq m-1$, has a state storage of $\lceil \frac{r-m-i}{m} \rceil$ bits, with its `data_out` equivalent to the original kth repeating state, $k = i + m \times \lceil \frac{r-m-i}{m} \rceil$. The entire pSRL is cleared when any of the m input characters does not match the repeated character class.

Figure 2.12 shows a 2-character matching circuit with pSRL. To better demonstrate the resource saving, we increase the repetition amount (r) from 5–15 in the example. Note that the pSRL architecture is only effective in reducing resource usage on FPGA when $r > 2m$ for an m-character matching circuit. In practice, we have hundreds of instances of large (>100) single-character repetitions in Snort-rules regexes, some of them repeating over 1000 times.

2.5.3 Multipipeline Priority Grouping

In many real-world REM scenarios for deep packet inspection, a character input stream is usually matched by a small set of regexes. Thus, a large-scale REM solution consisting of thousands of matched regex patterns can be naturally grouped into several tens of regex groups, each matching only a few tens of regexes.

In our large-scale REM design we use a 2-dimensional *pipelining* structure as illustrated in Figure 2.13. Our design supports up to 8 pipelines, each pipeline having up to 16 stages and each stage having 32 regexes. In total up to 4096 RE-NFAs can be matched in parallel. Each pipeline matches the input characters with its own BRAM-based complex character classifier. Within a pipeline, all the RE-NFAs are grouped in stages with lower-numbered stage having higher priority. Input characters as well as the complex character classifications are buffered and forwarded through the stages. A match output from a low-number (high-priority) stage overrides those from all higher-number stages at the same clock cycle. Within each stage, match outputs of all RE-NFAs can either be joined (logic-AND'd) or prioritized, depending on how the group of regexes are matched.

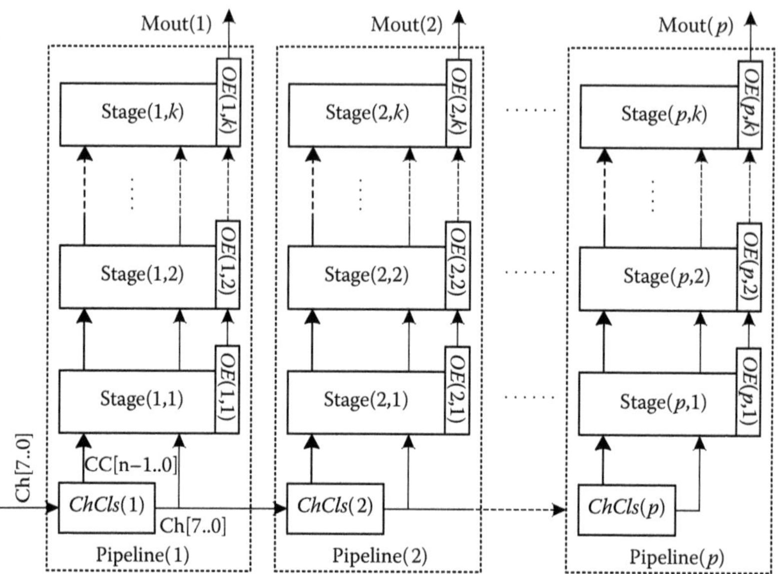

FIGURE 2.13 Multi-pipeline architecture (p pipelines, each k stages).

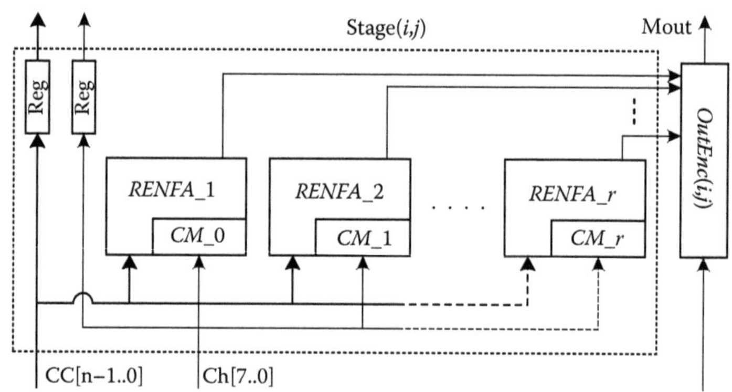

FIGURE 2.14 Stage architecture with a priority output encoder.

Figure 2.14 shows the architecture of a stage with separate character matching circuits (`CM`). Conceptually, all RE-NFAs are separate from one another; practically, the circuit construction backend can exploit the common prefix and shared character matching among various RE-NFAs in the same group to improve resource efficiency. All RE-NFAs in the same stage are prioritized by the output encoder (`OutEnc`) to produce at most one matching output per clock cycle. To maximize flexibility, a stage receives two types of character inputs: (1) a set of character classification results (`CC[n-1..0]`) propagated from a previous stage; (2) the 8-bit input character (`Ch[7..0]`) generating these classification results.

All RE-NFAs in a pipeline share the same complex character classifier, which greatly reduces the amount of BRAM required for character classification. When constructing the BRAM-based character classifier, a function is called to compare each state's character class to the character class entries (columns of 256 bits) collected in BRAM so far. If the current character class is the same as or a simple negate of an existing character class, then the output of the existing entry is simply wired to the current state. This procedure has time complexity $O(n \times w)$ with n total number of states and w *distinct* character classes in *all* RE-NFAs. The space complexity is just $256 \times w$. In the worst case, w could be linear in n; in practice,

w tends to grow much more slowly than $O(\log n)$. Prudent grouping of RE-NFAs into pipelines can further reduce the growth of w with respect to n.

Using multiple pipelines increases the number of independent match outputs per clock cycle. On the other hand, doing so also reduces the potential resource sharing in the (complex) character classifier. Subject to the I/O constraint, a "flat" multi-pipeline favors more concurrent match outputs, whereas a "tall" multi-pipeline favors more common characters.

2.6 Implementation Procedures on FPGA

2.6.1 Regex from Snort Rules

We used regexes from Snort rules (published in February 2010) for our RE-NFA implementation example. A Snort rule is described in two parts: (i) a *header* part that specifies the 5-field classification for the packet (type, source IP, destination IP, source part, destination port) and (ii) a payload part that specifies the pattern of the packet payload.

For fair evaluation, we counted identical regexes as one (i.e., all regexes are syntactically unique). We also removed regexes that are too short (<10 states), as well as a handful of very long regexes (>2500 states). The short regexes are better compiled as DFAs running on multi-core processors, whereas the long regexes are very rare and could be handled as exceptions.

Some Snort rules use "non-regular" regex features such as *backreference*—the ability to refer to a previously matched string as the string pattern to match against the following input. In our implementation, for any regex with backreference, the part of the regex after the first backreference is ignored. Out of over 2600 *distinct* regexes we took from Snort rules, less than 100 (<4%) use backreferences, mainly in the *netbios* and *web-client* rule sets. Our construction also ignores the PCRE lookahead/lookbehind predicates.

The entire set of 2630 regexes is too large to fit onto a single state-of-the-art FPGA. Instead, we partition them into six subsets as follows:

wax1 Part 1 of the *web-activex* rule set.
wax2 Part 2 of the *web-activex* rule set.
wax3 Part 3 of the *web-activex* rule set.
osp From *oracle*, *spyware-put* and a few small sets.
wsv From *web-**, *smtp* and *voip* rule sets, etc.
bem From *backdoor*, *exploit*, and *misc* rule sets, etc.

The six regex sets can be categorized into two groups. The "*wax*" sets contain more Kleene closure and union operators, but no constrained repetition. The other three sets (*osp*, *wsv* and *bem*) have much fewer closures and unions but a fair amount of constrained repetitions. Within each group, various regex sets have very similar statistics, as shown in Table 2.3. A "size-k" state ($k = 2, 3, \ldots$) accepts k incoming transitions from the same or other states. In general, a size-k state is more complex than a size-h state if $k > h$.

TABLE 2.3 Statistics of the Implemented Snort-Rules Regexes

Name	# Pat.	# States	% Size-2	% Size-3	% Size-4+
wax1	399	34,043	14.0	1.46	3.50
wax2	399	35,485	13.8	1.45	3.37
wax3	382	32,944	14.1	1.38	3.71
osp	565	44,341	5.76	0.15	0.10
wsv	332	37,032	7.33	0.86	0.21
bem	553	67,964	3.57	0.21	0.14
Total	2630	251,809	8.74	0.79	1.52

2.6.2 Regex Complexity Metrics and Estimates

The one point that we will emphasize here is that *all regular expressions are **not** created equal.* Thus, when measuring the performance of a REM architecture, it is critical to take into account the intrinsic complexity of the regex being matched. We say a regex is more complex when it is converted into a more complex RE-NFA quantified by the following metrics:*

State count Total number of states in the RE-NFAs.
State size The number of states that immediately transition to any state in the RE-NFAs.
Fan size The number of states that any state in the RE-NFAs immediately transitions to.

The *state count* was used in Floyd and Ullman (1982) to describe the area requirement of the NFA-based REM circuit; in our RE-NFA architecture, a higher state count translates to more state update modules. The *state size* and *fan size* are affected by the use of union and closure operators. A higher *state size* translates to a larger `OR` gate to aggregate the input transitions, directly affecting the size of the state update module. A higher *fan size* increases the signal routing distance and complexity.

Theoretically, the *maximum* state size and fan size would determine the largest state update module and the slowest signal route, respectively; in practice, circuit-level optimizations performed by the synthesis tools are often able to compensate the effects of a few states with large numbers of fan-ins or fan-outs. On the other hand, high *average* state size and fan size can make the REM circuit more complex on the whole and thus effectively impact its overall performance.

Owing to the modular structure of our RE-NFA, we can fairly accurately estimate these three metrics for every RE-NFA state with a specified multicharacter matching value (m). We design two types of the *size estimation function* for the ith state of a RE-NFA, $S(i)$, as follows:

- For ordinary state update module,

$$S(i) = \left\lceil n(i) \times \frac{m}{L} \right\rceil \tag{2.2}$$

- For single-character repetition in pSRL,

$$S(i) = (m+1) \times \left\lceil \frac{m}{L} \right\rceil + \left\lceil \frac{r(i)}{R} \right\rceil \tag{2.3}$$

where $n(i)$ is the number of fan-in ϵ-transitions to the ith state, m is the multi-character matching number, L and R are the device-dependent constants specifying the size of each LUT and SRL, respectively, and $r(i)$ is the repeating count of the single-character repetition covering the ith state. Note that only one size estimate is produced for multiple states in the same single-character repetition.

Note that Equations 2.2 and 2.3 are derived directly from the circuit architectures in Sections 2.5.1 and 2.5.2, respectively; they also do not account for the character matching logic. In practice, for large REM circuits the total number of register-LUT pairs used tend to increase faster than $n(i) \times \lceil \frac{m}{L} \rceil$ due to resources used for signal routing.

Note also that both equations offer only approximate resource estimations; the actual resource usage of each REM circuit is highly dependent on the optimizations performed by the synthesis and place-and-route tools. Nevertheless, the size estimate functions can still be useful when we need to quickly compare the relative complexities of a REM solution on FPGA matching different sets of regexes. The number of LUTs required for character matching, on the other hand, can be calculated from the circuit architecture in Figure 2.8c and d.

---

* The character classes used in a regex can also affect the regex complexity. This factor, however, is eliminated in our RE-NFA by the use of BRAM-based character classification.

2.6.3 Implementation Results

As evidenced from Section 2.6.1 above, all regexes are *not* created equally. Thus, we defined the following metric to fairly evaluate the performance of the proposed REM solution across various different regex sets:

Definition 2.5:

The *throughput efficiency* of a REM circuit on FPGA, in units of Gbps**state*/LUT, is defined as the concurrent throughput of the circuit divided by the average number of LUTs used per state.*

Multiple circuits were constructed for each of the 6 regex sets in Table 2.3, matching $m = 2, 4, 6$, and 8 characters per cycle. Each circuit was organized as two pipelines having 5–9 stages per pipeline. The targeted device was Virtex 5 LX-220, synthesized (*xst*) and place-and-routed (*par*) using Xilinx ISE 11.1 with either the "maximize speed" or "minimize area" option.

Figure 2.15 shows the throughput performance and resource usage, respectively, of the circuits matching the *wax1*, *wax2*, and *wax3* regex sets. {`Fbase`, `Tbase`} plot the {Frequency, Throughput} of the "base" circuits where regexes are arbitrarily grouped into stages; {`Fsort`, `Tsort`} plot those of the "sorted" circuits where regexes are organized into stages of roughly equal size (in number of states). Similarly, {`Nbase`, `Ebase`} plot the {No. FF-LUT pairs, throughput efficiency} of the base circuits, while {`Nsort`, `Esort`} the sorted circuits. Because there is no constrained repetitions in these three regex sets, pSRL was not used. An "`_s`" suffix indicates the circuit is optimized for speed when synthesized; an "`_a`" suffix indicates optimized for area. Table 2.4 gives an overview of the notations used in Figures 2.15 and 2.16.

Higher multicharacter matching reduced clock rate but increased matching throughput; it also required proportionally more resource on FPGA. As a result, throughput efficiency did not change significantly between $m = 2$ and $m = 8$. Sorting the regexes to balance the number of states across stages had a small but noticeable impact on throughput, since size-balanced stages would theoretically reduce the length of the longest signal path. More interestingly, while optimizing for area (`*_a`) resulted in lower LUT usage, for $m = 8$ optimizing for area actually resulted in higher throughput than optimizing for speed (`*_s`) for the *wax1* and *wax3* regex sets, as shown in Figure 2.15. One explanation is that with $m = 8$, the state update modules became very big, making signal routing a much more dominant factor in determining the clock period. Thus, reducing the circuit size, which helps to shorten the signal route, also contributes more significantly to increase the circuit speed.

Figure 2.16 shows the throughput performance and resource usage, respectively, of the circuits matching the *osp*, *wsv*, and *bem* regex sets. Similar conventions were used for the plots in these figures, except the "`Xpsrl`" labels which indicate the circuits were optimized with pSRL for matching constrained repetitions (Section 2.5.2). On the other hand, the regexes were *not* sorted in any particular way.

Like before, we observed similar throughput increase with higher multicharacter matching, as well as higher throughput optimizing for area than optimizing for speed when $m = 8$. In addition, we find throughput efficiency to slightly increase with higher multicharacter values and peak around $m = 6$. This is expected since the targeted FPGA had 6-input LUTs. As shown in Figure 2.16, applying pSRL to the circuits significantly reduced their resource usage by up to 40%. Interestingly, only when the circuits were optimized for area did pSRL make a big difference. The smaller circuits with pSRL also had visible speed advantage than the base circuits without pSRL.

---

* Throughput alone does not account for the length and complexity of the regexes. LUT efficiency alone, on the other hand, does not reflect clock frequency or multicharacter matching.

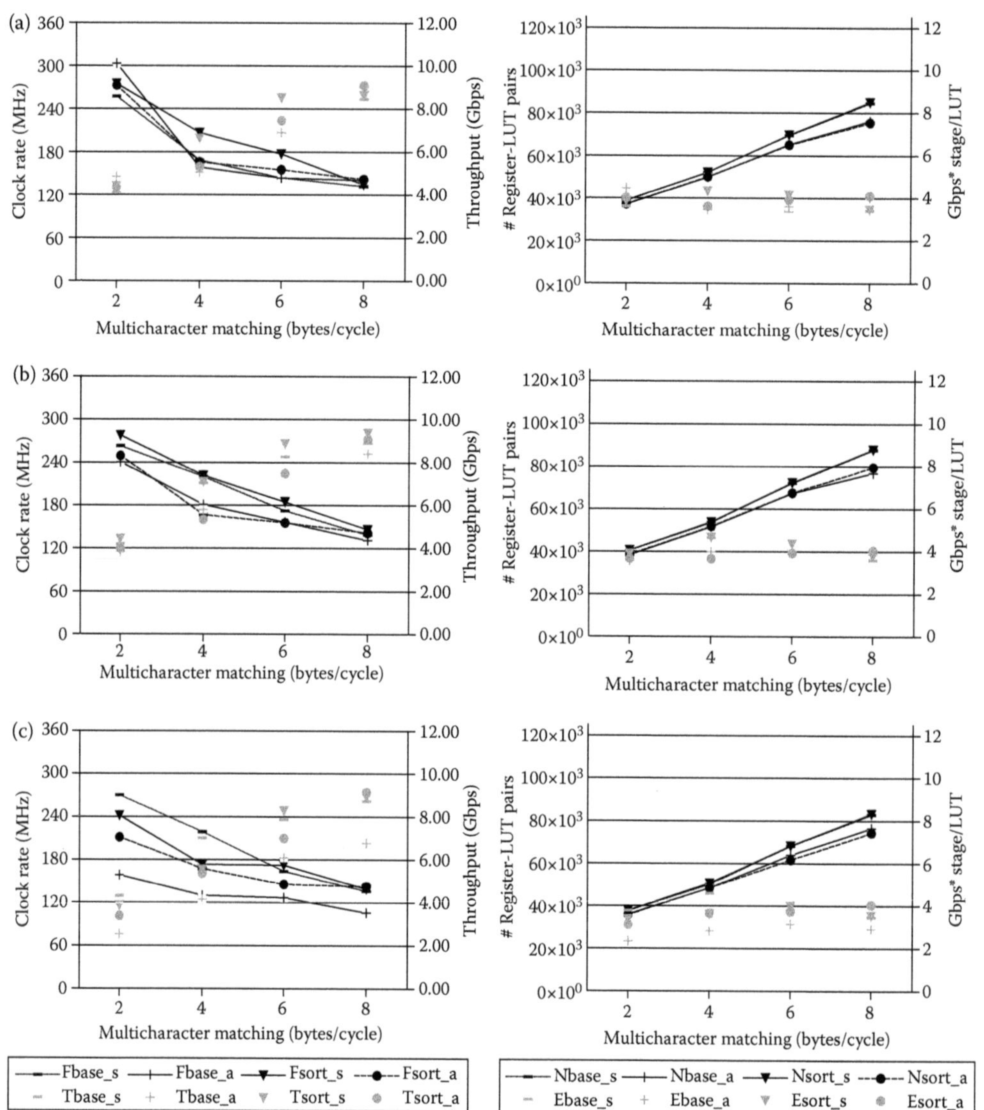

FIGURE 2.15 Performance of our REM circuits for regexes with no constrained repetition. (a) *wax1*, (b) *wax2*, and (c) *wax3* regex sets (see Table 2.4 for the meaning of the label notations).

TABLE 2.4 Notations Used in Figures 2.15 and 2.16

	{A}{B}_{C}				
	{A}		{B}		{C}
F	Frequency	base	Arbitrary grouping	_a	Area optimized
T	Throughput	sort	Equal # regex states	_s	Speed optimized
N	# FF-LUTs	psrl	Optimized with pSRL		
E	Efficiency				

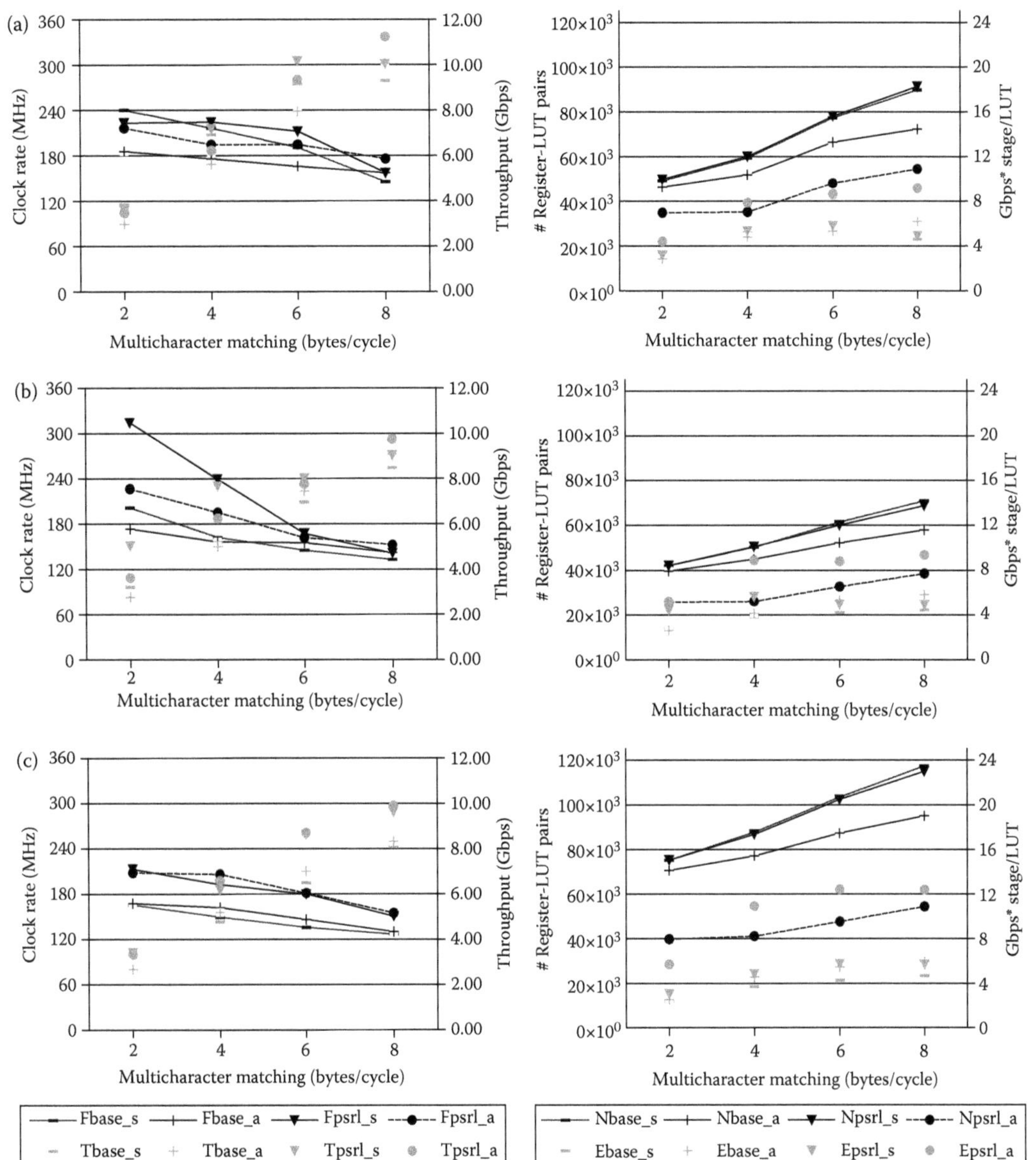

FIGURE 2.16 Performance of our REM circuits for regexes with some constrained repetitions. (a) *osp*, (b) *wsv*, and (c) *bem* regex sets (see Table 2.4 for the meaning of the label notations).

2.7 Related Work

Hardware implementation of NFA-based REM was first studied by Floyd and Ullman (1982). They showed that, when translating directly to integrated circuits, an n-state NFA requires no more than $O(n)$ circuit area. Sidhu and Prasanna (2001) later proposed a self-configuration algorithm to translate an arbitrary regex directly into its matching circuit on FPGA. Both studies are based on the same NFA architecture with the McNaughton–Yamada construction.

Large-scale REM was first considered in Hutchings et al. (2002), where character inputs are pipelined and broadcast in a tree structure. It also proposed an automatic REM circuit construction using JHDL, in contrast to the self-configuration approach used in Sidhu and Prasanna (2001). Automatic REM circuit construction in VHDL was proposed in Bispo et al. (2006) and Mitra et al. (2007). In Bispo et al. (2006), each regex was first tokenized and parsed into a hierarchy of basic NFA blocks, then transformed into VHDL using a bottom-up scheme. In Mitra et al. (2007), a set of scripts were used to compile regexes into op-codes, convert op-codes into NFAs, and construct the NFA circuits in VHDL.

An algorithm for construction multi-stride NFA was proposed in Yamagaki et al. (2008) to transform an n-state NFA accepting 2^i input characters per cycle into an equivalent NFA accepting 2^{i+1} input characters per cycle, $i \geq 0$, in $O(n^3)$ time. A technique using shift-register lookup tables (SRL) for implementing single-character repetitions was proposed in Bispo et al. (2006). Other techniques such as common-prefix extraction and centralized character matching were used to further reduce resource usage. The resulting circuit, however, accepts only one input character per clock cycle. In Lin et al. (2006), the authors explored the idea of pattern *infix* sharing to reduce the number of lookup tables (LUTs) required to match similar patterns.

Although the implementation of backreference was mentioned in Bispo et al. (2006), its approach, which uses the regex pattern rather than the matched input for backreference matching, actually defeats the original purpose of making backreferences. The first real backreference implementation on FPGA is done by Mitra et al. (2007), where block memory (BRAM) is utilized to store the referenced strings. The number of characters in all backreferences is limited by the amount of BRAM available, and backtracking is not supported due to the excessive memory size and logic complexity required by the feature.

2.8 Conclusion

We presented a compact architecture for high-performance large-scale REM on FPGA. We described the conversion from regex to token list, the construction of modular RE-NFA, and the mapping to register-transfer level circuit on FPGA. The resulting circuit can be further optimized by techniques such as multicharacter matching, parallel SRL, shared BRAM-based character classifier and multi-pipeline priority grouping. The designs achieve significantly higher matching throughput and throughput efficiency than previous state-of-the-art over comprehensive sets of regexes from Snort rules.

References

A. V. Aho and J. D. Ullman, *The Theory of Parsing, Translation, and Compiling*. Prentice-Hall, Inc., Englewood Cliffs, NJ, 1972. ISBN 0-13-914556-7.

D. Betel and C. W. Hogue, Kangaroo—A pattern-matching program for biological sequences. *BMC Bioinformatics*, 3(20): 31 July 2002.

J. Bispo, I. Sourdis, J. M. P. Cardoso and S. Vassiliadis, Regular expression matching for reconfigurable packet inspection. In *Proc. IEEE Intl. Conf. on Field Programmable Technology (FPT'06)*, pp. 119–126, Bankok, Thailand, Dec. 2006.

I. M. Copi, C. C. Elgot and J. B. Wright, Realization of events by logical nets. *Journal of ACM*, 5:500181–196, April 1958. ISSN 0004-5411. Available at: http://doi.acm.org/10.1145/320924.320931.

R. W. Floyd and J. D. Ullman, The compilation of regular expressions into integrated circuits. *Journal of ACM*, 29(3): 603–622, 1982. ISSN 0004-5411.

M. Fourment and M. R. Gillings, A comparison of common programming languages used in bioinformatics. *BMC Bioinformatics*, 9(82): 5 Feb. 2008.

V. Glushkov, The abstract theory of automata. *Russian Math. Surveys*, 16: 1–53, 1961.

P. Hazel, *PCRE–Perl Compatible Regular Expression*. Available at: http://www.pcre.org/, Jan. 2011.

B. L. Hutchings, R. Franklin and D. Carver, Assisting network intrusion detection with reconfigurable hardware. In *Proc. IEEE Sym. on Field-Programmable Custom Computing Machines (FCCM'02)*, p. 111, Napa, California, 2002. ISBN 0-7695-1801-X.

Java regular expression package: java.util.regex. Available at: http://java.sun.com/j2se/1.4.2/docs/api/java/util/regex/package-summary.html.

C.-H. Lin, C.-T. Huang, C.-P. Jiang and S.-C. Chang, Optimization of regular expression pattern matching circuits on FPGA. In *Proc. Conf. on Design, Automation and Test in Europe (DATE'06)*, pp. 12–17, Leuven, Belgium, 2006. European Design and Automation Association. ISBN 3-9810801-0-6.

R. McNaughton and H. Yamada Regular expressions and state graphs for automata. *IEEE Transactions on Computation*, 9(1): 39–47, March 1960.

A. R. Meyer and M. J. Fischer, Economy of description by automata, grammars, and formal systems. *In Proc. 12th Sym. on Switching and Automata Theory (SWAT'71)*, pp. 188–191, Washington, DC, USA, 1971. IEEE Computer Society.

A. Mitra, W. Najjar and L. Bhuyan, Compiling PCRE to FPGA for accelerating SNORT IDS. In *Proc. ACM/IEEE Sym. on Architecture for Networking and Communications Systems (ANCS'07)*, pp. 127–136, New York, NY, USA, 2007. ISBN 978-1-59593-945-6.

D. Pasetto, F. Petrini and V. Agarwal, Tools for very fast regular expression matching. *Computer*, 43:50–58, 2010. ISSN 0018-9162.

V. Paxson, Bro: A system for detecting network intruders in real-time. *Computer Networks*, 31(23–24):2435–2463, 1999. Available at: http://www.icir.org/vern/papers/bro-CN99.pdf.

Regular Expressions in ASP.NET. Available at: http://msdn.microsoft.com/en-us/library/ms972966.aspx, b.

R. Sidhu and V. Prasanna, Fast regular expression matching using FPGAs. In *Proc. IEEE Sym. on Field-Programmable Custom Computing Machines (FCCM'01)*, pp. 227–238, Rohnert Park, California, 2001.

R. Smith, C. Estan and S. Jha, Backtracking algorithmic complexity attacks against a NIDS. In *Proc. 22nd Annual Computer Security Applications Conference (ACSAC'06)*, pp. 89–98, Miami Beach, Florida, Dec. 2006.

Sourcefire : Snort. http://www.snort.org/, Apr. 2011.

N. Yamagaki, R. Sidhu and S. Kamiya, High-speed regular expression matching engine using multi-character NFA. In *Proc. Intl. Conf. on Field Programmable Logic and Applications (FPL'08)*, pp. 697–701, Heidelberg, Germany, Aug. 2008.

F. Yu, Z. Chen, Y. Diao, T. V. Lakshman and R. H. Katz, Fast and memory-efficient regular expression matching for deep packet inspection. In *Proc. ACM/IEEE Sym. on Architecture for Networking and Communications Systems (ANCS'06)*, pp. 93–102, San Jose, California, 2006. ISBN 1-59593-580-0.

3

Finite State Transducers

Javier Baliosian
and Dina Wonsever
University of the Republic in Uruguay

3.1 Introduction

Finite state transducers (FSTs) are a sort of finite state machines that, in addition to the notion of accepting or rejecting a string, generate an output string, being Moore and Mealey machines the first antecedent of this model. Indeed, FSAs can be seen as FSTs with the same input and output symbol in each transition. But while deterministic FSAs are equivalent to nondeterministic ones, this property does not hold for transducers. In addition, closure properties are more restrictive than those that hold for FSAs.

A widely used extension to FSTs is the addition of a weight to each transition. The class of machines thus defined is called Weighted Finite State Transducers (WFST). They are used in many applications such as natural language processing, network management, and bioinformatics.

In this chapter, we will present an FSTs overview. First, in Section 3.2 we will introduce the classic letter transducers with their most relevant characteristics and operations. Later, in Section 3.3 we will get into more detail in the particularly useful class of sequential transducers and in Section 3.4 we will present transducer with weights. Finally, in Section 3.5 we will mention some of the most common FST applications and some Fuzzy Logic-like extensions used in the field of networks management.

3.2 An Introduction to Classic FSTs

We can ideally picture FSTs as a machine that operates on discrete steps and that for each step it basically:

- Reads and consumes the next symbol from an input tape,
- Chooses a new state that depends on the current state and the symbol just read and,
- Writes a string on an output tape.

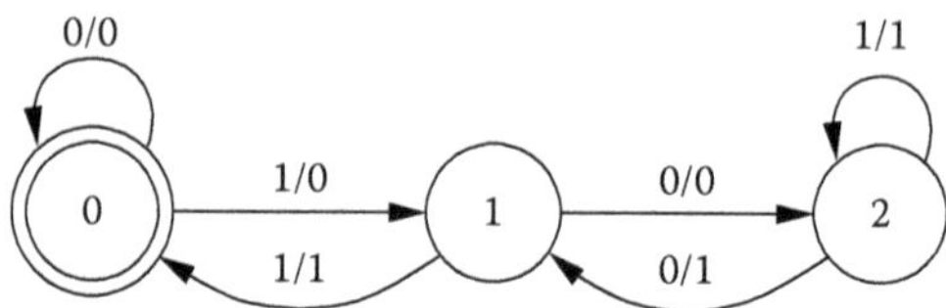

FIGURE 3.1 Transducer T_{d3} representing division by 3 of binary numbers. (From E. Roche and Y. Schabes. *Finite-State Language Processing*. Technical report, MIT Press, Cambridge, MA, 1997. With permission.)

Some additional features have to be taken into account:

- The machine starts its operation in an initial state (or in anyone of a set of initial states, according to some definitions).
- There is a set of final states, the machine accepts the input string if it is in a final state when the input string is exhausted.
- ϵ transitions are allowed, in the input or/and in the output tape. This means that it is not possible to consume any input symbol or not to write any output symbol in a computation step.
- The behavior of an FST is usually abstracted by a transition function. For each pair *(state, input symbol)* the function indicates a set of pairs *(new state, output symbol)*. The transition function is a partial function; if there is no transition defined for the current *(state, input symbol)* the machine blocks and the input string is rejected.

There is a variant to this model without final states. In this case, the machine's function is just to emit an output string corresponding to an input. This is equivalent to considering that all states are final.

FSTs determine a class of relations on strings. Under a relational view, we focus on the set of *(input string, output string)* pairs such the FST accepts the input string generating the output string in the output tape.

For instance, the relation given by the transducer T_{d3} of Figure 3.1 contains exactly all pairs such that the first element is a multiple of 3, and such that the second element is the quotient of the division by three of the first element (e.g., 11,01).

As for FSAs, FSTs get their strength from various closure properties. FSTs are closed under *union*, *inversion*, *composition* and with some restrictions under *intersection* and *complement*. The meaning of these operations is the same as of binary relations' operations.

To illustrate the flexibility induced by the closure properties, suppose one wants to build a transducer that computes the division by 9. There are two possible ways of tackling the problem. The first consists of building the transducer from the beginning. The second consists of building the transducer T_{d3} encoding the division by three and then composing it with itself since $T_{d9} = T_{d3} \circ T_{d3}$. The transducer T_{d9}, computed by composing T_{d3} with itself is shown in Figure 3.2.

The two examples depicted before are extracted from [RS97].

The FST definition vary slightly from author to author. In this book, we are presenting what is called a *letter transducer* [RS97] in which an FST consumes only one symbol per edge (including ϵ), and produces one symbol, or ϵ, each time an edge is crossed.

Finite State Transducers

Definition 3.1:

A finite state transducer (FST) M is a tuple $(\Sigma_1, \Sigma_2, Q, i, F, E)$ where:

- Σ_1 is a finite alphabet (the input alphabet),
- Σ_2 is a finite alphabet (the output alphabet),

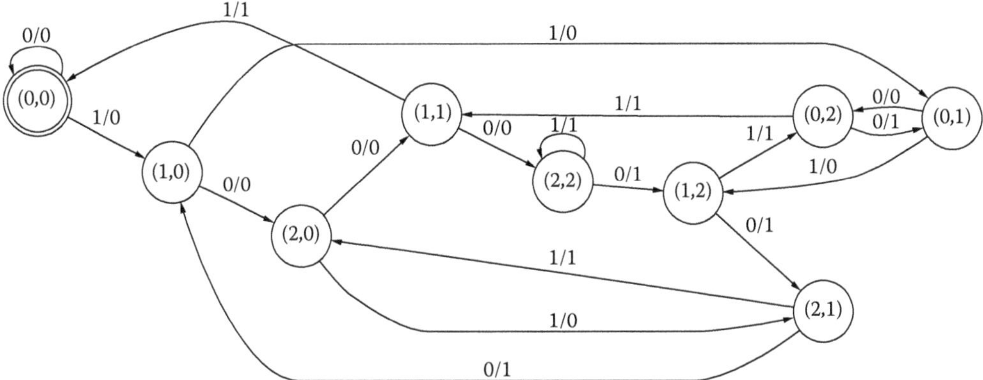

FIGURE 3.2 Transducer T_{d9} representing division by 9 of binary numbers. (From E. Roche and Y. Schabes. *Finite-State Language Processing*. Technical report, MIT Press, Cambridge, MA, 1997. With permission.)

- Q is a finite set of states,
- $i \in Q$ is an initial state,
- $F \subseteq Q$ is a set of final states.
- E is a set of edges such that $E \subseteq Q \times \Sigma_1 \cup \{\epsilon\} \times \Sigma_2 \cup \{\epsilon\} \times Q$.

The FST just defined is a nondeterministic device, considering its input: for each state and input symbol pair there is a set of possible new states and output symbols. A FST can be seen as a labeled graph, where vertex are states and each edge is labeled with a pair *(input symbol, output symbol)*. A path on this graph is a sequence of consecutive edges, an accepting path is a path that starts in the initial state and ends in a final state.

Definition 3.2: (Path on a FST)

A path in a FST $(\Sigma_1, \Sigma_2, Q, i, F, E)$ is a sequence $e_1, e_2, \ldots, e_n$ of edges, such that for each $e_i = (p_i, a_i, b_i, q_i)$ $q_i = p_{i+1}$, for $i = 1, \ldots, n-1$.

Definition 3.3: (Accepting path)

An accepting path in a FST $(\Sigma_1, \Sigma_2, Q, i, F, E)$ is a path $e_1, e_2, \ldots, e_n$ such that $p_1 = i$ and $q_n \in F$. The string $a_1, a_2, \ldots, a_n$ if the input accepted string and the string $b_1, b_2, \ldots, b_n$ is the output accepted string.

Relation Induced by an FST

Finite automata are associated with languages: the language associated with an automata is the set of all strings that are accepted by the automata. In a similar way, FSTs are associated with pairs of input–output strings such that there is an accepting path labeled with this pair.

Rational Relation

Let τ be a mapping $\tau : \Sigma_1^* \rightarrow 2^{\Sigma_2^*}$ between strings in Σ_1^* and (sets of) strings in Σ_2^*. τ is a rational transduction if there exists an FST T such that if T accepts $w \in \Sigma_1^*$ then the associated output string

x belongs to $\tau(w)$. In this case, we say that the relation with set of pairs $R = (w, \tau(w))$ is a rational (or regular) relation.

3.2.1 Operations

FSTs get their strength from the closure properties under some interesting operations. These operations are, the classic regular operations, concatenation, union and Kleene star and some specific to transducers such as projection (mapping regular relations to regular languages), inversion and composition. Composition is central for the use of transducers in different applications. It bears some resemblance to automata intersection in the way of construction of the transducer for the composition of two rational relations, but rational relations are not closed under intersection.

In the following, we introduce some of the most important FST operations.

Union

The union of two transducers is a transducer whose associated relation is the union of the relations defined by each of the two original transducers.

The simplest way to construct a transducer T for the union of transducers $T1$ and $T2$ is to define a new initial state i' and $\epsilon : \epsilon$ transitions to the initial states of $T1$ and $T2$.

Given

$$T1 = (\Sigma_{1,1}, \Sigma_{1,2}, Q_1, i_1, F_1, E_1)$$
$$T2 = (\Sigma_{2,1}, \Sigma_{2,2}, Q_2, i_2, F_2, E_2)$$

the following transducer

$$T = (\Sigma_1, \Sigma_2, Q, i, F, E)$$

with:

- $\Sigma_1 = \Sigma_{1,1} \cup \Sigma_{1,2}$
- $\Sigma_2 = \Sigma_{2,1} \cup \Sigma_{2,2}$
- $Q = Q_1 \cup Q_2 \cup \{i\}$
- i as initial state
- $F = F_1 \cup F_2$
- $E = E_1 \cup E_2 \cup \{(i, \epsilon, \epsilon, i_1), (i, \epsilon, \epsilon, i_2)\}$

is the union transducer.

This means that the new transducer will recognize the union of sets of strings recognized for both original transducers and produce the same strings that the original ones would produce.

Inversion

The inversion of a relation R is a new relation R^{-1} where the elements in each pair are swapped.

If a relation R is rational, so is the inverse relation R^{-1}. The following construction based on a transducer T for the relation R proves this:

$$T = (\Sigma_1, \Sigma_2, Q, i, F, E)$$
$$T^{-1} = (\Sigma_2, \Sigma_1, Q, i, F, E')$$

where there is an edge (p, b, a, q) in E' if the edge $(p, a, b, q) \in E$.

Projections

The first and second projections of a rational relation R are the languages L_1 and L_2 whose strings are the first and second component, respectively, of the relation R. These languages are regular, this can be observed by constructing the two following FSA $A1$ and $A2$ from the FST T associated with R.

$$T = (\Sigma_1, \Sigma_2, Q, i, F, E)$$
$$A1 = (Q, \Sigma_1, i, F, \delta_1)$$

where $\delta_1(q, x)$ contains p for each p, y such that $(q, x, y, p) \in E$.

$$A2 = (Q, \Sigma_1, i, F, \delta_2)$$

where $\delta_2(q, y)$ contains p for each p, x such that $(q, x, y, p) \in E$.

Intersection

The intersection of two rational relations is not necessarily a rational relation. For instance, the intersection of the rational relations

$$R1 = \{(c^n, a^n b^m)\}, m \geq 0$$
$$R2 = \{(c^n, a^m b^n)\}, n \geq 0$$

is the relation

$$R3 = \{(c^n, a^n b^n)\}, n \geq 0$$

$R3$ is not a rational relation, as its second projection, $\{a^n b^n\}$, is not a regular language (example from [RS97]).

But in some cases, basically when the transducers are ϵ-free letter transducers, the intersection of two transducers is well defined and results in a transducer that defines the relation resulting from the intersection of the relations of the two original transducers. It is one of the most important and powerful operations on automata.

In the classical case, the intersection of two given automata M_1 and M_2 is constructed by considering the cross product of states of M_1 and M_2.

A transition $((p_1, p_2), \sigma, (q_1, q_2))$ exists in the intersection iff the corresponding transition (p_1, σ, q_1) exists in M_1 and (p_2, σ, q_2) exists in M_2.

Composition

The meaning of composition here is the same as for any other binary relations:

$$R_1 \circ R_2 = \{(x, z) \mid (x, y) \in R_1, \quad (y, z) \in R_2\}$$

This can be seen as a chain of events processing: the symbols outgoing from the first transducer are used as the input of the second one. In the basic case, the composition of two transducers M_1 and M_2 is built considering the cross product between the states of M_1 and M_2.

A transition $((p_1, p_2), \sigma_i, \sigma_o, (q_1, q_2))$ exists iff there is some σ such that $(p_1, \sigma_i, \sigma, q_1)$ exists in M_1 and $(p_2, \sigma, \sigma_o, q_2)$ exists in M_2. A good example was depicted before in Section 3.2, with the division by 9 (see Figure 3.2) as the composition of two divisions by 3 (see Figure 3.1). However, the presence of ϵ input labels make the things a little bit more complicated. For example, if we use, as is, the algorithm described above to compute $B' \circ A'$ (see Figure 3.3a and b) the resultant FST will consume an ϵ and produce a b, which is not what we expected.

The solution given by Mohri [MP96] was to transform the transducers to be composed adding innocuous self-loops with $\epsilon : \epsilon$ labels on each node of them (see Figure 3.3c and d), and then the composition

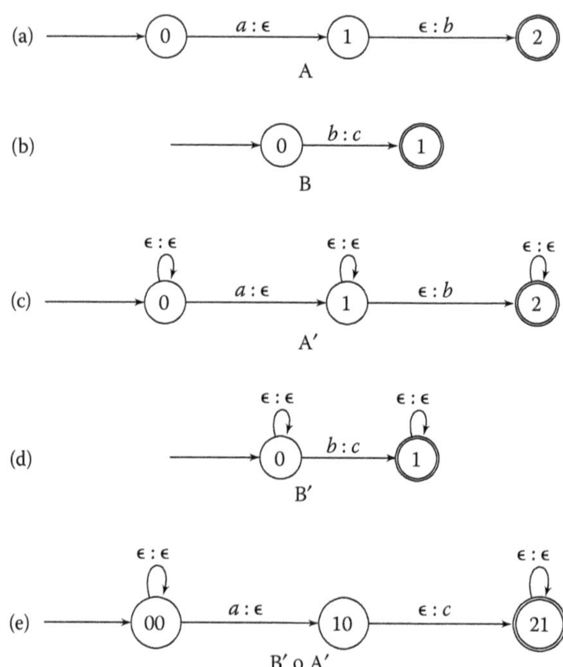

FIGURE 3.3 A composition algorithm for FSTs with ϵ labels.

is performed on the new transducers considering the ϵ symbol as any other symbol in the process. This leads to the intermediate transducer in Figure 3.3e which, after eliminating unreachable nodes and $\epsilon:\epsilon$ labels, is the composition transducer that we were expecting. Actually, Mohri's solution is a bit more ingenious and efficient (yet less intuitive). It uses different types of ϵ labels in such a way that, after the composition algorithm, there is no unreachable nodes and $\epsilon:\epsilon$ labels. See [MP96] for the details. Here, we described the most naive manner of composing FSTs, however, there is more sophisticated work on this. For example, Allauzen et al. [AM09b] have developed a technique to perform a chain of more than two transducers compositions without producing the intermediate transducers, saving processing time and memory.

3.3 Deterministic Transducers

One of the main reasons to use the word "transducer" instead of "function" when we are speaking of finite state machines with an input and an output is that FSTs may have a finite, or possibly infinite, number of outputs for a single input.

For example, when the input is an *a*, the transducer in Figure 3.4 may produce indistinctly a *c* or a *b*.

However, *deterministic* or *functional* transducers are a category of transducers that present no ambiguities in the output producing a single output for each input. Among these particular FSTs there is

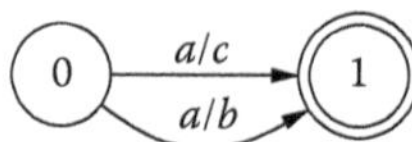

FIGURE 3.4 Ambiguous transducer. (From E. Roche and Y. Schabes. *Finite-State Language Processing.* Technical report, MIT Press, Cambridge, MA, 1997. With permission.)

a particular class named *sequential transducers*. In a sequential transducer, at any state of it, at most one outgoing edge has a label with a given element of the alphabet. Output labels might be strings, including the empty string ϵ. However, the empty string is not allowed in input labels. It is easy to see that a deterministic transducer can be modified to be sequential, this has caused that, in the literature and in this book, the term deterministic has been often used as a synonym of sequential. The determinization of a transducer actually means, in this book, the transformation of it into a sequential transducer.

Sequential transducers are computationally very interesting, because their computation load does not depend on the size of the transducer but rather only on the length of the input since that computation consists of following the only possible path corresponding to the input string and in writing consecutive output labels along this path.

A formal definition of a sequential transducer is:

Definition 3.4: (Sequential Transducer [RS97])

A sequential finite state transducer (FST) T [Eil74] is a 6-tuple $(\Sigma_1, \Sigma_2, Q, i, \delta, \sigma)$ where:

- Σ_1 is a finite alphabet (the input alphabet),
- Σ_2 is a finite alphabet (the output alphabet),
- Q is a finite set of states,
- $i \in Q$ is an initial state,
- δ is the state transition function, $\delta : Q \times \Sigma_1 \rightarrow Q$
- σ is the emission function, $\sigma : Q \times \Sigma_1 \rightarrow \Sigma_2^*$

An interesting particularity is that, because there is only one possible path to follow when the transducer is evaluated, it can be considered that all states are final and the output can be produced at the same time that the input is consumed.

This property is computationally very interesting, sadly, most of the transductions cannot be represented by a sequential transducer. However, there is a less restrictive class of transducers named subsequential transducers that help with this problem.

Definition 3.5: (Subsequential Transducer [RS97])

A subsequential finite state transducer (FST) T is a 8-tuple $(\Sigma_1, \Sigma_2, Q, i, F, \delta, \sigma, \rho)$ where:

- Σ_1 is a finite alphabet (the input alphabet),
- Σ_2 is a finite alphabet (the output alphabet),
- Q is a finite set of states,
- $i \in Q$ is an initial state,
- $F \subseteq Q$ is a set of final states,
- δ is the state transition function, $\delta : Q \times \Sigma_1 \rightarrow Q$
- σ is the emission function, $\sigma : Q \times \Sigma_1 \rightarrow \Sigma_2^*$
- ρ is the final state emission function, $\rho : F \rightarrow \Sigma_2^*$

A subsequential transducer is a kind of deterministic-on-input transducer. As Roche and Schabes say, we could think that input strings are followed by a special symbol $ not belonging to the input alphabet, we add a new final state and transitions labeled with the final emission strings from the former final states to this state. It follows that the relation defined by the transducer is a functional one: for each input string there is only one output string. Notice that the ability of emitting strings rather than symbols allows to avoid ϵ on input for some cases. Obviously, the inversion operation is not allowed: the inverse relation does not have to be functional.

Moreover, sequential transducers are a kind of subsequential transducers where all states are final and the emission function value for final states is ϵ.

Finally, subsequential transducers lead us to a much more useful class of transductions: those that can be represented by a *p-subsequential* transducer.

Definition 3.6: (*p*-subsequential Transducer [RS97])

A p-subsequential finite state transducer (FST) T is a 7-tuple $(\Sigma_1, \Sigma_2, Q, i, F, \delta, \sigma, \rho)$ where:

- Σ_1 is a finite alphabet (the input alphabet),
- Σ_2 is a finite alphabet (the output alphabet),
- Q is a finite set of states,
- $i \in Q$ is an initial state,
- $F \subseteq Q$ is a set of final states,
- δ is the state transition function, $\delta : Q \times \Sigma_1 \rightarrow Q$
- σ is the emission function, $\sigma : Q \times \Sigma_1 \rightarrow \Sigma_2^*$
- ρ is the final state emission function, $\rho : F \rightarrow 2^{\Sigma_2^*}$

A p-subsequential transducer is a kind of quasi-deterministic on input transducer. It is similar to the subsequential model, with the difference that there is a finite set of strings (at most p strings for each final state) on the output alphabet that can be concatenated to the generated output string. It is important to clarify the behavior of the final state emission function: it should be applied only when the input string is exhausted. Otherwise, on the presence of arcs outgoing from final states to other states in the transducer, the quasi-determinism of the machine does not hold any more. So, the transducer behavior exhibits two important features: (i) the transducer's operation is deterministic on input, that is, there is only one possible next state for each pair (state, input symbol) and (ii) the transducer can cope with a limited amount of ambiguity, necessary for natural language applications.

It can be shown that p-subsequential transducers are closed under composition: the composition of a p-subsequential transducer with a q-subsequential transducer is a pq-subsequential transducer. They are also closed under union, the union of a p-subsequential transducer and a q-subsequential transducer is a $(p+q)$-subsequential transducer. In the algorithm for the construction of a union transducer, the set of states is the Cartesian product of the sets $Q1 \cup \{-\}$ and $Q2 \cup \{-\}$.

Not all transducers can be determinized or transformed in an equivalent p-subsequential transducer. Algorithms for determinization (if possible) and minimization can be seen in detail in [RS97] and [MS96].

3.4 Weighted Finite State Transducers

3.4.1 Introduction to WFSTs

Weighted finite-state transducers (WFST) extend FSTs by adding a weight (usually a real number) to transitions. This weight can be interpreted and operated upon in different ways, depending on the application on hand. For instance, the weights are costs in graph distance problems, or probabilities for transitions in diverse natural language applications.

A primary operation on a WFST is its application to an input string. It proceeds as in the un-weighted case, starting in an initial state and consuming symbols from the input string that agree with the input label of a transition. The output label of this transition is emitted and the machine enters a new state,

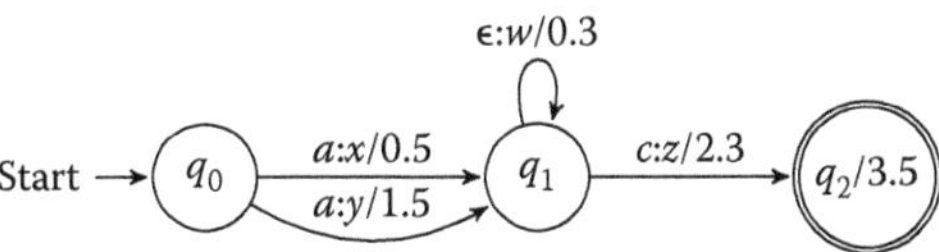

FIGURE 3.5 A weighted finite state transducer.

as specified by the transition. If the input string is exhausted and the transducer is in a final state, it is accepted and an output string, consisting of the concatenation of all output symbols of the undertaken transitions, is emitted. This is the output that corresponds to an accepting path for the input string. In a weighted transducer, there is also a weight associated with this accepting path. It is obtained by operating upon the weights found in the transitions that compose the accepting path. If weights are costs, they will be added; if weights are probabilities, the adequate operation (under a Markovian assumption) is weight product. So, an operation for treating weights along a path has to be defined in the set where weights come from.

Figure 3.5 illustrates the basic elements of a WFST. As for the unweighted case, the edges are labeled with a pair of *input symbol, output symbol*, with ϵ allowed in input or in output. But in this case there is also a weight (in the example, a real number) on each edge. A final weight (3.5) appears also in the final state.

For the input string *ac* strings of the form xw^*c and yw^*c are produced. The weight associated with each *(input, output)* pair depends on the particular set and operations defined for weights.

Also, there can be in a transducer multiple accepting paths for an unique *(input, output)* pair of strings. Each path generates a weight, as stipulated by the operation on paths. But another operation is needed for composing weights on multiple paths. This operation will be referred to as the accumulation operation. If weights are costs and we are interested in minimum cost path, the operation is MIN. In the case of probabilities, weights should be added upon.

The set of weights has to support then two operations: product and accumulation. Some intuitive properties such as associativity and distributive are desirable. This leads to a kind of algebraic structure for weights: the semiring. Abstracting the particular set of weights over the structure of a semiring provides an unified treatment for different sets of weights and operations for WFSTs.

WFSTs have diverse applications in the field of Natural Language Processing, such as automatic speech recognition (ASR), spelling correction, part of speech tagging. They provide also a formal frame for some useful algorithms in bioinformatics, such as finding homologous via minimum edit distances.

In what follows we define formally semirings and WFSTs and present some useful operations and properties.

3.4.2 Semirings

A semiring is a set K with two binary operations $+$ and $\cdot$, addition or accumulation and product, such that:

- $(K, +)$ is a commutative monoid ($+$ is associative and commutative and has an identity element) with identity 0
  1. $(a+b)+c = a+(b+c)$
  2. $0+a = a+0 = a$
  3. $a+b = b+a$
- $(K, \cdot)$ is a monoid with identity element 1:
  1. $(a \cdot b) \cdot c = a \cdot (b \cdot c)$
  2. $1 \cdot a = a \cdot 1 = a$
- Product distributes over addition:
  1. $a \cdot (b+c) = (a \cdot b) + (a \cdot c)$

2. $(a+b) \cdot c = (a \cdot c) + (b \cdot c)$

- 0 annihilates K, with respect to product:
  1. $0 \cdot a = a \cdot 0 = 0$

Examples:

- Boolean semiring: The set of boolean values $B = \{0, 1\}$ constitutes a semiring $(B, \wedge, \vee, 0, 1)$. Taking weights over this semiring, with an assignment of weight 1 to every transition, is equivalent to having an unweighted transducer. The weight on a path is always 1 and the weight on a set of paths is also 1. That means that weights are not significant in this case.
- Natural and real semirings: Perhaps, the simplest example of a semiring is the set of natural numbers with usual addition and product, $N = (N, +, ., 0, 1)$. Also are semirings the sets of positive real numbers $(R_{+}, +, ., 0, 1)$ and the set of positive rational numbers with the usual sum and product operations $(Q_{+}, +, ., 0, 1)$. The real semiring is suitable for representing probabilities. These are examples of semirings that are not rings, but every ring (real numbers, rational numbers) is also a semiring
- String semiring: It is also possible to define a restricted form of a semiring over the set of strings extended with ∞ as the maximum length string, taking as weights the strings themselves, with string concatenation as product and longest common prefix (suffix) as the sum operation obtaining a left (right) semiring, as the distributive only holds at the left (right).
- Tropical semiring: The set R_{+} of positive reals, extended with $+\infty$ as the maximum element (and unit element for addition (min)) and with min as the addition operation and the usual sum on reals as the product operation form a semiring known as the tropical semiring . The tropical semiring offers an algebraic generalization to shortest paths problems on graphs. In WFSTs, a usual situation is that weights represent probabilities and a desired result is the most probable path in a graph. It is used when operating with negative log probabilities, these are added on a path and the best result is the minimum length path (highest probability).

3.4.3 WFST Definitions

Definition 3.7: (Weighted Finite State Transducers)

A weighted finite state transducer (WFST) T on a semiring K is a 8-tuple $(Q, \Sigma, \Omega, E, i, F, \lambda, \rho)$ where:

- Q is a finite set of states,
- Σ is an input alphabet,
- Ω is an output alphabet,
- $E \subseteq Q \times \Sigma \cup \{\epsilon\} \times \Omega \cup \{\epsilon\} \times K \times Q$ is a finite set of transitions
- $i \in Q$ is the initial state,
- $F \subseteq Q$ is a set of final states,
- λ is an initial weight,
- $\rho \subseteq F \times K$ is a final weight function.

A path in a WSFT T is a consecutive set of transitions. Let t be a transition in $T, t = (q, a, b, w, p)$, we denote by $p(t)$ its origin (q), by $n(t)$ its final node (p), by $i(t)$ the input symbol (a), by $o(t)$ the output symbol (b) and by $w(t)$ its weight.

A path P from state u to state v is a sequence of transitions $t_1, t_2, \ldots, t_n$ such that $t_1 = u, t_n = v, n(t_i) = p(t_{i+1})$ for $i = 1, \ldots, n-1$. The input string s_i corresponding to path P is the concatenation of the input symbols in the transitions of P, $s_i = i(t_1), \ldots, i(t_n)$.

Similarly, the output string s_o corresponding to path P is the concatenation of the output symbols in the transitions of P, $s_o = o(t_1), \ldots, o(t_n)$.

The transducer T establishes a relation between input and output strings. Two strings s_i and s_o are related if there is a path P going from the initial state i to a final state $f \in F$ such that its input and output strings are s_i and s_o, respectively. For a specific pair (*input string*, *output string*) there can be more than one path in these conditions. The relation on strings induced by a WFST is a weighted relation, computed following the operations of the semiring where the WFST is defined.

3.4.4 Operations

The most useful operation for WFST is composition. It proceeds as in the unweighted case, but special care has to be taken with ϵ moves. In some cases, ϵ transitions can be removed, but this is not the case in different length relations, with ϵ appearing in the input or the output side of some transitions. In this case, the composition algorithm for FST may give rise to several alternative paths with the same weight, only one of these paths has to be considered. This is managed by a special filter transducer.

As shown in [MP96] regulated WFST are closed under union, concatenation and Kleene-closure. A WFST is regulated is the output weight associated with a pair of strings is well defined (unique). WFSTs without ϵ cycles are always regulated. As it occurs with FST, WFST are not closed under intersection (and, consequently, under difference or complement), but Weighted Finite Automata are closed under intersection, and, provided that the semiring has a negation operation, they are also closed under complement and difference. Mohri [MP96] shows that the closure property holds for a restricted class of difference: the difference between a weighted finite automata and unweighted (traditional FSA) one.

3.4.5 Algorithmic Issues

Traditional FSAs are efficient devices that can be determinized and minimized (minimum number of states). This situation does not hold for transducers. Determinization (including ϵ removal) is not always possible [RS97] and minimization requires a deterministic machine in input. However, some interesting applications involve acyclic transducers (morphology and dictionaries, automatic speech recognition) and in this case determinization is possible.

An important feature is the possibility of computing transducers operations in a lazy *on demand* way. It happens that the entire transducer resulting from composition operations is sometimes not necessary. Avoiding the complete construction of the transducer resulting from a cascade of composition operations results in a significative reduction in time and memory requirements. This is the police of the *Opensfst* toolkit (www.openfst.org), with lazy implementation of main operations whenever it is possible.

3.5 Applications

3.5.1 Natural Language Processing

Natural Language Processing has been a main field of application for finite state techniques. Very large dictionaries representation and access, with morphology representation and morphological recognition, have encountered in transducers' technology the way of encoding the relation between an inflected form and a canonical form plus morphosyntactic features. These resources integrate smoothly with others finite state components. Also, a finite state calculus, that can be seen as an extension of the language of regular expressions to regular relations, has been developed.

Probabilistic language processing finds a convenient implementation methodology based on weighted FSAs and transducers. Successful applications include automatic speech recognition systems (ASR), statistical machine translation (SMT), handwriting and OCR systems, information extraction, part of speech (POS) tagging and chunking.

In what follows, we present in more detail some of these applications. We also present some common basic modules used in different applications, such as language models and Hidden Markov Models, which find a transducers' representation and integrates as such in a transducers' composition cascade.

Dictionaries and Morphology

Very large dictionaries can be represented by FSTs, usually unweighted, mapping an input word to its lemma and morphological properties. An efficient lookup is ensured, as these are acyclic transducers and algorithms for determinization and minimization are, therefore, applicable [RS07].

A dictionary may contain all attested forms in a particular language, or it may be composed by base forms and rules for flexible morphology (i.e., number and gender variation, verbal conjugation). It may be also convenient to represent the derivative morphological paradigm of the language (affixes, suffixes) as this is a productive mechanism and unknown words may arise from a known root by a derivative process. These rules are represented by independent transducers, and rule application is resolved by transducers composition [Sil97].

Language Models

Probabilistic language processing has made an extensive use of weighted automata and transducers. In first place, a very basic component of different language processing applications, a language model, is conveniently represented as a weighted automata. This is usually an n-gram model constructed from a big corpus, where the probability of occurrence of one word given the previous one (bigrams) or given the two previous words (trigrams) are represented. This representation is useful when the language model is composed with another components. Language models are usually refined by the application of smoothing techniques, in order to eschew 0 probabilities for unseen n-grams. The overall is still represented by a weighted automata, and weights are estimated form n-grams occurrences in very large corpora.

The Noisy Channel Model

The noisy channel model [KMD$^+$09] has been used in many applications in natural language processing. Its origin is in Information Theory, and it refers to the noise or errors appearing in a channel during the transmission of an original message X that is received after transmission as a distorted message Y. The noise is modeled as a probability distribution $p(Y/X)$.

Given the received message Y, we are interested in the original message X. One way is to find the X that maximizes the probability of Y. That is:

$$X = \underset{X}{argmax}\, p(Y/X)$$

This situation arises in various applications: machine translation, part of speech tagging, chunking. As the conditional probability may be hard enough to evaluate, a different solution has been proposed. In what follows, we rely mainly on Knight et al. [KMD$^+$09]. Suppose that we are translating from Spanish (S) to English (E), by a process described by the probability distribution $P(E/S)$. But we consider that our Spanish sentence is in fact a distorted English sentence that we want to recover. By Bayes rule:

- $p(E/S) = \dfrac{p(S/E)p(E)}{p(S)}$ and we want to obtain
- $\hat{E} = \underset{E}{argmax}\, p(S/E)p(E)$, dropping the given sentence S in maximization

In this way, the problem is split in two subproblems, $p(S/E)$ and $p(E)$. The second distribution, $p(E)$ is what is called a language model, that is, the probability of well-formed English sentences. The probabilities are computed from very large corpora, and a language model is a reusable basic component for different natural language systems. The first distribution, $p(S/E)$, is the translation model. For both of them weighted FSA and transducers based methods have been proposed.

Regular Relations Calculus

The construction of FSTs, able to handle a huge amount of language information is too hard to be held by hand. Kaplan and Kay, and Kartunnen at Xerox (xfst)[KGK97], have proposed a language for the definition of regular relations and compilers for the automatic derivation of finite transducers from the expressions in this language. This language includes the regular expressions for regular languages and there are also some specific operations for regular relations. Some extensions [MS96] have also been proposed for weighted regular relations. In what follows, we present the basic constructions and some rewrite rules for regular relations.

Operations for Regular Languages and Relations

We review the basic operations in *xfst* [KGK97]. In what follows, r and s are two regular expressions over the alphabet Σ, f and g are two regular relations over the same alphabet and x and y are regular languages or relations:

1. $x|y$-UNION - string or pair of strings in x or in y
2. xy-CONCATENATION - string in x followed by a string in y or pair concatenation (component by component)
3. x^*-KLEENE CLOSURE - concatenation of any finite number (including 0) of strings or pairs in x
4. $\neg r$-COMPLEMENT - all strings in Σ^* and not in r
5. $r\&s$-INTERSECTION - all strings in r and in s
6. $r - s$-DIFFERENCE - all strings in r and not in s
7. $r \times s$-CROSSPRODUCT - pairs of strings with first element in the regular language r and second element in the regular language s
8. $f \circ g$-COMPOSITION - composition of the transduction f with the transduction g

Operations 1–3 are applicable to regular expressions and regular relations, while operations 4–6 can only be applied safely to regular expressions, as regular relations are not closed under complement and intersection. In addition to the basic operations, other high level operators have been defined. The most paradigmatic is the REPLACE operator family ([Kar00]):

> A $\rightarrow$ B-Unconditional replacement of A (upper language, i.e., input string) by B (lower language, i.e., output string).
> A $\rightarrow$ B...C-Enclosing A (upper language) between an element of B and an element of C (in the lower language). Usually B and C are atomic. This is a markup rule.
> A $\rightarrow$ B/C...D-Replacement of A by B whenever it is preceded by C and followed by D.

Different variants arise if we consider that the replacement or markup can be optional, or that it can proceed left to right or right to left. The interesting point here is that these rewrite rules are more expressive than the basic operations but they are defined on top of these operations, sometimes in a rather complex way [KK94]. The basic expressions of these rules are compiled into transducers, which are in turn composed according to the designed application.

3.5.1.1 Hidden Markov Models and FSTs

A hidden Markov model is a statistical multivariate model for sequential data with two classes of variables: (i) *observed* variables and (ii) *latent* or *hidden* variables [Bis06].

In Figure 3.6, the variables Y_i are hidden, while the gray nodes for variables X_i indicate that these variables are observed. The model reflects a Markovian assumption: each hidden variable directly depends on the previous one and is independent of the others given the immediate neighbors. The transition probabilities $p(Y_i/Y_{i-1})$ are usually constrained to be equal. Figure 3.6 is a schematic representation. Usually, in NLP applications, the hidden variables are discrete, taking values on a finite set $C = \{c_1, c_2, \ldots, c_n\}$.

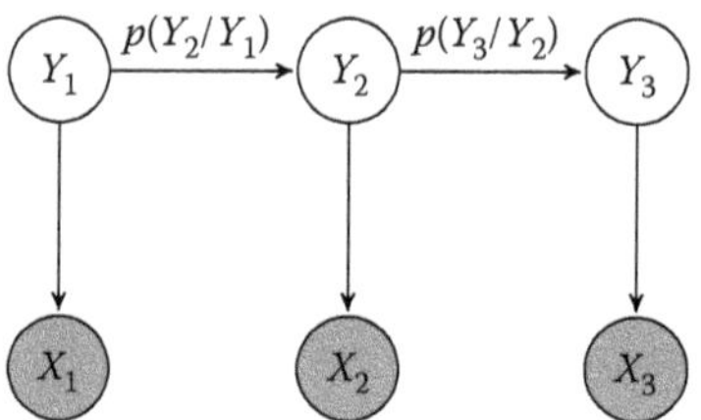

FIGURE 3.6 Schematic representation of a Hidden Markov Model.

So, there is a state transition automata between two consecutive stages, that can be deployed as in Figure 3.6, with a graph with n nodes for each variable Y_i.

To complete the model, the state variables emit the observed symbols with a distribution that depends on their value: for each value of the (identical) states variables there is a probability distribution relating this value with the set of possible values for the observed variables.

The decoding task for hidden Markov models consists in the calculus of the most probable sequence of values for the hidden variables given a sequence of values for the observed variables. This task can be modeled as an WFSA, and thus included in an automata cascade via the composition operation [JM08].

3.5.2 Bioinformatics

The weighted FSTs calculus offers a general framework for some interesting problems in computational biology, such as the search of similarities in biological sequences and the construction of evolutionary models. The similarity of two or more biological sequences is usually measured through sequence alignment. A sequence alignment is a superposition between two sequences obtained by an appropriate sequence of character deletion, insertion, or substitution operations. Each operation is assigned a cost and the objective is to find the minimal cost sequence of operations leading to superposition. Given two sequences, the number of possible sequences of operations is exponential, so a naive algorithm that examines all possible alignments will take exponential time.

An alignment scoring function can be defined by a WFST and general algorithms can be used to efficiently compute the best alignments between sequences, or sets of sequences represented by finite automata.

The full system of transducers operations and optimizations allows a systematic combination of elementary machines and to obtain probabilistic models for multiple, evolutionary related sequences.

3.5.3 Network Management Applications

It is commonly accepted that the main obstacle currently facing the Information and Communications Technologies (ICT) industry is a looming software complexity crisis. The need to integrate several heterogeneous environments to one or more computing systems, and beyond a single company's boundaries, introduces levels of complexity never seen before. In particular, the management of communications systems is in a pretty difficult situation; a highly diverse set of hardware and software must interact in even more heterogeneous environments.

As vision of the solution, IBM's manifesto on *Autonomic Computing* [IBM01] offers the holistic approach of Autonomic Computing or Autonomic Communications in the case of networks. An autonomic system embeds the complexity into the infrastructure and manages it with little or no need for recognition or effort on the part of the user. Users are thus allowed to focus on their tasks, not on the means of accomplishing them, which results in greater productivity and enhanced work experience.

3.5.3.1 Enabling Autonomy: Policy-Based Management

One of the many obstacles in the path to Autonomic Communications is related to the policy-based foundations of autonomicity: high-level policies must be modeled in a way that may be efficient and easily implemented, and the conflicts between policy-rules should be resolved in an autonomous and instantaneous manner.

Those policy rules are used by a Policy-based Network Manager which, basically, listen for some events of the network, such as links coming down or the quality of a transmission degrading, feed with them and with the relevant network status some sort of reasoning engine and, produces some action or "order" to be enforced on the configuration of the managed network devices with the purpose of adapt, optimize or heal the network following the lines imposed by the policies.

Then, policy-rules may be seen as event-condition-action rules such as the one below in which a video-class connection is rerouted when the the network conditions of the current route are not good for video streaming.

```
if |variation of latency| > 20 ms then
re-route Video-class connection;
```

An important problem with policy-based managers is that rules may be contradictory or, in other words, may have conflicts. A policy conflict will occur when the objectives of two or more policies cannot be met simultaneously. Resolving this situation is a difficult task; it relies on certain form of knowledge of the semantics of policies, a battery of presumptions to prioritize some policies over others, and in certain cases, on human decision.

In the following, we describe a FST-based policy model that uses the determinization operation mentioned in Section 3.3 to resolve policy conflicts.

Policy-Rules Modeling

If we look at the structure of a basic transducer with only one edge, we can see the same structure than in a *if–then* clause. For example, if we think in the *if–then* format of the rule mentioned above, we can see this sentence as a transducer (not necessarily a finite-state one) that receives information about one connection's jitter (i.e., variation of latency) and, depending on this value, produces a certain re-routing action. The next step is to consider such a transducer as a finite-state one.

This FST has only one edge, its input label represents the type of event the rule is waiting for and the condition that must fulfill the input event (i.e., the jitter value) and the output label the condition with which the output action must conform to. This can be seen graphically in Figure 3.7.

The idea of FST labels defining classes of symbols (events in our case) instead only one symbol, was formalized by van Noord and Gerdemann [vNG01]. They defined a new kind of transducers called *finite state transducers with predicates and identities* (PFST) in which labels defining only one symbol were replaced by predicates.

Now, we can relate the concept of ambiguity on transducers with the ambiguity between policy rules: in front of the same input, two or more potentially contradictory outputs are produced; here we are in front of a potential policy conflict.

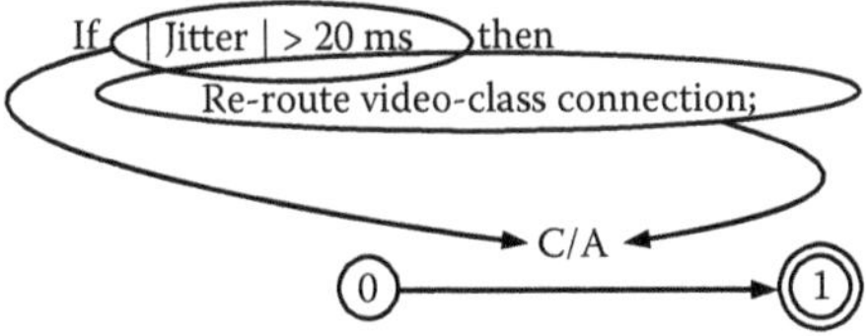

FIGURE 3.7 Correspondence between if–then constituents and FST elements.

3.5.3.2 Network Management Extension for FSTs

In the literature, there are several heuristics to resolve policy conflicts. The most common strategy is to give an explicit priority to each policy. This is the most straightforward solution but has the implicit problem of assigning and maintaining a coherent set of priorities to each rule of a potentially massive system. Another option is to use the idea of the FST determinization as a manner to eliminate ambiguity together with a metric of the specificity of a condition for a given event and state.

Tautness Functions

A policy has a condition delimiting a region where a given event can or cannot lay in. When such an event is inside two or more overlapping regions, what is called a "modality" conflict may arise. We are concerned about how tautly a condition fits on an event instead of how far from the border it is. Then, our preferred condition will be that which is the most "taut" around the event under consideration.

In order to quantitatively represent the aforementioned "tautness," we will define a metric called *tautness function* (TF), so that the more taut a condition is, the closer its TF is to 0.

To provide an intuitive example of TF, let us assume that one policy specifies *users* and another policy specifies *administrators*, who are special kind of users. For an action attempted by an administrator, the second policy should define a TF that is closer to 0 than the first policy.

Definition 3.8:

A *tautness function* associated with a condition c, denoted τ_c, establishes a mapping from $E \times C$ to the real interval $[-1, 1]$ where:

- E is the set of possible network events or attempted actions,
- C is the set of policy conditions,
- $\tau_c(e) \in [-1, 0) \Leftrightarrow$ *e not satisfies c*,
- $\tau_c(e) \in (0, 1] \Leftrightarrow e$ satisfies c,
- $\tau_c(e) = 1 \Leftrightarrow \forall e \in E$, *e satisfies c*,

considering, inside the condition c, the subject, target or any other condition such as temporal constraints.

An Algebra for Tautness Functions

The logic we use for tautness functions is inspired by the algebra used with **fuzzy sets**.

Definition 3.9: **(Disjunction)**

The TF of the *disjunction* of two policy conditions A and B with TFs $\tau_A$and τ_B, respectively, is defined as the maximum of the two individual TFs:

$$\tau_{A \vee B} = max(\tau_A, \tau_B)$$

This operation is the equivalent of the OR operation in Boolean algebra.

Definition 3.10: **(Conjunction)**

The TF of the *conjunction* of two policy conditions A and B with TFs $\tau_A$and τ_B, respectively, is defined as the minimum of the two individual TFs:

$$\tau_{A \wedge B} = min(\tau_A, \tau_B)$$

Definition 3.11: (Negation)

The TF of the *negation* of a policy condition A with TFs τ_A, is defined changing the sign of the specified TF:

$$\tau_{\neg A} = -\tau_A$$

This operation is the equivalent of the NOT operation in Boolean algebra.

Definition 3.12: (Tauter-than)

The *tauter-than* operation ($\rightarrow_\tau$) is an ad hoc operation created to simplify the notation involved in the determinization process which will be explained in Section 3.5.4. The TF of the *tauter-than* of two policy conditions A and B with TFs τ_A and τ_B, respectively, is defined as

$$\tau_{A \rightarrow_\tau B} = \begin{cases} \tau_A, & \text{if } \tau_A < \tau_B \\ -1, & \text{else} \end{cases}$$

This operation is intended especially for the distance concept behind our TFs. It tell us when A fits more tautly around an event than B.

Definition 3.13: (AsTautAs)

The *astautas* operation ($\rightleftarrows_\tau$), like *tauter-than*, was created to simplify the notation involved in the determinization process. The TF of the *as-taut-as* of two policy conditions A and B with TFs $\tau_A$and τ_B, respectively, is defined as

$$\tau_{A \rightleftarrows_\tau B} = \begin{cases} \tau_A, & \text{if } \tau_A = \tau_B \\ -1, & \text{else} \end{cases}$$

Transducers with Tautness Functions and Identities

Let us start defining *finite state recognizers with TFs* (TFFSR) because they are the starting point for presenting the corresponding transducers and because, in some cases, transducers should be "reduced" to recognizers in order to compute some operations.

The mechanics of processing an incoming stream of events with a TFFSR is as follows: starting at an initial state, an event is consumed if there is a transition with a positive tautness function value evaluated for that event. The new current state will be the one at the end of the chosen transition. Then the process is repeated. We can see a TFFSR as a set of rules that accept a certain pattern of events.

Definition 3.14:

A finite state recognizer with tautness functions (TFFSR) M is a tuple (Q, E, T, Π, S, F) where:

- Q is a finite set of states,
- E is a set of events,
- T is a set of tautness functions over E,
- Π is a finite set of transitions $Q \times T \cup \{\epsilon\} \times Q$,

- $S \subseteq Q$ is a set of start states,
- $F \subseteq Q$ is a set of final states.

The relation $\widehat{\Pi} \subseteq Q \times E^* \times Q$ is defined inductively:

- for all $q \in Q, (q, \epsilon, q) \in \widehat{\Pi}$.
- for all $(p, \epsilon, q) \in \Pi, (p, \epsilon, q) \in \widehat{\Pi}$.
- for all $(q_0, \tau, q) \in \Pi$ and for all $e \in E$, if $\tau(e)$ then $(q_0, e, q) \in \widehat{\Pi}$.
- if $(q_o, x_1, q_1) \in \widehat{\Pi}$ and $(q_1, x_2, q) \in \widehat{\Pi}$ then $(q_0, x_1x_2, q) \in \widehat{\Pi}$.

The "language" of events $L(M)$ accepted by M is defined to be $\{w \in E^* \mid q_s \in S, q_f \in F, (q_s, w, q_f) \in \widehat{\Pi}\}$

An extension could be defined to allow the recognizer to deal with strings of events in each transition, in order to make some operations easier and produce more compact transducers.

A TFFSR is called ϵ-free if there are no $(p, \epsilon, q) \in \Pi$. For any given TFFSR, there is an equivalent ϵ-free TFFSR. It is straightforward to extend the corresponding algorithm for classical automata to TFFSRs. We will assume this equivalence afterward for the sake of simplicity.

Definition 3.15:

A finite state transducer with tautness functions and identities (TFFST) M is a tuple (Q, E, T, Π, S, F) where:

- Q is a finite set of states,
- E is a set of symbols,
- T is a set of tautness functions over E.
- Π is a finite set of transitions $Q \times (T \cup \{\epsilon\}) \times (T \cup \{\epsilon\}) \times Q \times \{-1, 0, 1\}$. The final component of a transition is a sort of "identity flag" used to indicate when an incoming event must be replicated in the output.*
- $S \subseteq Q$ is a set of start states,
- $F \subseteq Q$ is a set of final states.
- For all transitions $(p, d, r, q, 1)$ it must be the case that $d = r \neq \epsilon$.

Note that we are assuming that the input and output sets of tautness functions are the same, and the same for input and output set of events. This could be refined but does not diminish the generality of this work.

We define the function *str* from $T \cup \{\epsilon\}$ to 2^{E^*}.

$$str(x) = \begin{cases} \{\epsilon\} & if\ x = \epsilon \\ \{s \mid s \in E, x(s)\} & if\ x \in T \end{cases}$$

If $\tau \in T$ and $str(x)$ is a singleton set, then the transitions (p, τ, τ, q, i) where $i \in \{-1, 0, 1\}$ are equivalent. The relation $\widehat{\Pi} \subseteq Q \times E^* \times E^* \times Q$ is defined inductively:

- for all $p, (p, \epsilon, \epsilon, p) \in \widehat{\Pi}$.
- for all $(p, d, r, q, 0), x \in str(d), y \in str(r), (p, x, y, q) \in \widehat{\Pi}$.
- for all $(p, d, r, q, -1), x \in str(d), y \in str(r), x \neq y, (p, x, y, q) \in \widehat{\Pi}$.
- for all $(p, \tau, \tau, q, 1)$ and $x \in str(\tau), (p, x, x, q) \in \widehat{\Pi}$.
- if $(q_o, x_1, y_1, q_1) \in \widehat{\Pi}$ and $(q_1, x_2, y_2, q) \in \widehat{\Pi}$ then $(q_0, x_1x_2, y_1y_2, q) \in \widehat{\Pi}$.

* The negative identity value is to express the obligatory difference between input and output, this is needed to compute the complement of a TFFST.

The relation $R(M)$ accepted by a TFFST M is defined to be $\{(w_d, w_r) \mid q_s \in S, q_f \in F, (q_s, w_d, w_r, q_f) \in \widehat{\Pi}\}$.

An extension could be defined to let the transducer deal with strings of events in each transition.

3.5.4 Operations on TFFST

Most of the operations on TFFSTs such as *union, intersection, composition*, and *Kleene closure* are straightforward extensions of the FST operations seen above in Section 3.2.1 or PFST operations and can be seen in detail in [BS04], however, the TFFST *determinization* is a somehow complicated procedure and has a particularly important meaning in PBNM. The *identity* operation it is also particular of a system that what consumes and produces event notifications and remote device re-configuration orders. Below, we will introduce both operations.

Identity

As with letter transducer and recognizers, there is an identity operation for TFFSTs that transforms a tautness functions-based recognizer (TFFSR) into a TFFST. However, if we apply an straightforward extension of the letter transducer to TFFSTs we can have some inconveniences and, therefore the idea of an "identity flag" has to be introduced.

The identity relation for a given language of events L is $id(L) = \{(w, w) \mid w \in L\}$. Further on, we will see that a *right* will be an identity relation applied to the incoming authorized event. For a given TFFSR $M = (Q, E, T, \Pi, S, F)$, the identity relation is given by the TFFST $M' = (Q, E, T, \Pi', S, F)$ where $\Pi' = \{(p, \tau, \tau, q, 1) \mid (p, \tau, q) \in \Pi\}$. Note that several events could be positive under τ so the "identity flag" is set to 1 in order to force the event produced to be the same that the one that entered. A more in-depth explanation of this can be found in [vNG01] but one difference from the identity defined for PFSTs, is that we are defining $a -1$ value for the identity flag.

Symbols "<" and ">" around TFs in the labels express identity between the input and the output, and symbols "[" and "]" express difference.

Determinization

This is the most important operation for our objectives, but also the one that exhibits the highest computational cost. To solve the conflicts between a set of rules, we must determinize the transducer representing that set in order to eliminate ambiguities.

We will say that a TFFST M is *deterministic* if M has a single start state, if there are no states $p, q \in Q$ such that $(p, \epsilon, x, q, i) \in \Pi$, and if for all states p and events e there is at most one transition (p, τ_d, x, q, i) such that $\tau_d(e)$ is positive. If a TFFST M is deterministic then the process of computing the output events for a given stream of events ω as defined by M, can be implemented efficiently. This process is linear in ω, and independent of the size of M.

In order to extend the determinization algorithm of PFST [vNG01] to TFFST, we must take into account the case where, although an event satisfies two conditions, one of these conditions fits more tautly than the other. We must also take into account the case where both conditions have the same specificity. In this way, all possible combinations between TFs are computed as with PFSTs, but instead of using conjunctions only, we use *conjunction, tauter-than*, and *as-taut-as* operations defined in Section 3.5.3.2.

For now, we will assume that the output part of a transition contains a sequence of predicates; this implies an extension to the above-defined TFFSTs, which will not be described in this thesis. We will also assume that there are no ϵ input functions, since an equivalent transducer could be computed for one that does contain ϵ input functions [RS97].

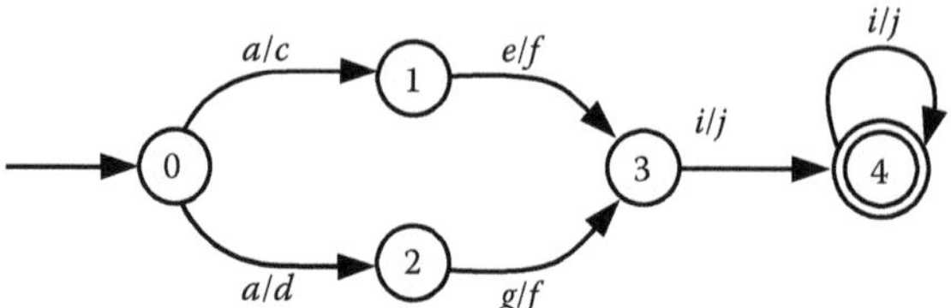

FIGURE 3.8 Before determinization.

In this algorithm, outputs are delayed as much as possible. This is because a local ambiguity may not be such if we look at the whole transducer and realize that only one path would be possible until the final state. This is the case of the ambiguity shown in state 0 in Figure 3.8.

The algorithm maintains sets of pairs $Q \times T*$. Each of these sets corresponds to a state in the determinized transducer. In the example of Figure 3.8, once an event that satisfies condition a is read, we can be at states 1 and 2, with pending outputs c and d, then $P = \{(1, c), (2, d)\}$. In order to compute the transitions leaving such a set of pairs P, we compute (as in PFSTs)

$$Trans^P(\tau_d) = \{(q, xy) \mid (p, x) \in P, (p, \tau_d : y, q) \in \Pi\}$$

In the example, $Trans^P(e) = \{(3, cf)\}$ and $Trans^P(g) = \{(3, df)\}$. Let T' be the TFs in the domain of $Trans^P$. For each split of T' into $\tau_1 \ldots \tau_i$ and $\neg\tau_{i+1} \ldots \neg\tau_n$, and each possible order of $\tau_1 \ldots \tau_i$ we have a proto-transition with the operator $\rightarrow_\tau$ and another with the operator $\rightleftarrows_\tau$ between two consecutive TFs τ_j, τ_{j+1}:

$$(P, \tau_1 \ldots \tau_j \rightarrow_\tau \tau_{j+1} \ldots \tau_i \wedge \neg\tau_{i+1} \wedge \ldots \neg\tau_n, A)$$
$$(P, \tau_1 \ldots \tau_j \rightleftarrows_\tau \tau_{j+1} \ldots \tau_i \wedge \neg\tau_{i+1} \wedge \ldots \neg\tau_n, A)$$

where A is $[Trans^P(\tau_1), \ldots, Trans^P(\tau_i)]$. In the example, we have proto-transitions:

$$(P, e \wedge \neg g, [(3, cf)])$$
$$(P, g \wedge \neg e, [(3, df)])$$
$$(P, e \rightarrow_\tau g, [(3, cf), (3, df)])$$
$$(P, g \rightarrow_\tau e, [(3, df), (3, cf)])$$
$$(P, e \leftrightarrows_\tau g, [(3, cf), (3, df)])$$
$$(P, g \leftrightarrows_\tau e, [(3, cf), (3, df)])$$

The last two proto-transitions are equivalent and one should be erased. A transition is created from proto-transitions putting in its output the longest common prefix of TFs in the target pairs (ϵ in the example). Before removing the longest common prefix, the sequences of output TFs should be packed, putting together the sequences associated with the same target state. This is done using a combination of conjunction and disjunction in the case of a *tauter-than* relation or including an `xor` with an `error` event in the case of a *as-taut-as*. This means that, in the case that no decision can be made, an `error` is marked.* Thus, two pairs of target states and TF sequences (p, s_1) and (p, s_2) can be combined into a single pair (p, s) iff (p, s_1) is in $A[j]$ and (p, s_1) is in $A[j+1]$ and $s_1 = \tau_1 \ldots \tau_i \ldots \tau_n$, $s_2 = \tau_1 \ldots \tau_i' \ldots \tau_n$ and $s = \tau_1 \ldots (\tau_i \vee (\tau_i \wedge \tau_i')) \ldots \tau_n$ if the operator between τ_j and τ_{j+1} is $(\rightarrow_\tau)$, or $s = \tau_1 \ldots (Error \oplus (\tau_i \wedge \tau_i')) \ldots \tau_n$ if the operator between τ_j and τ_{j+1} is $(\leftrightarrows_\tau)$. In the example, the third proto-transition is

* This would be useful to detect nonresoluble conflicts and report them to an operator.

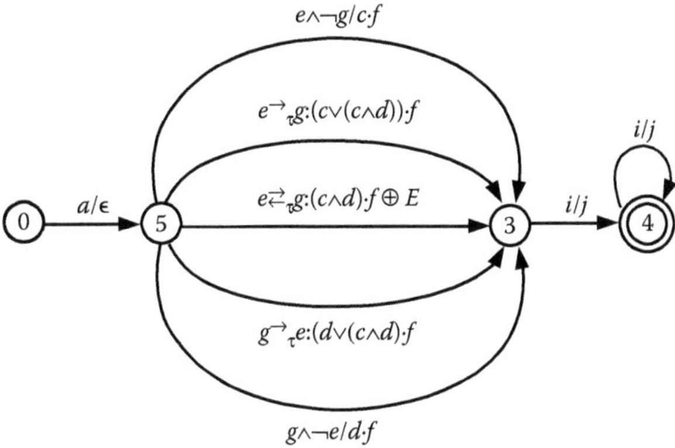

FIGURE 3.9 After determinization procedure.

packed in

$$(P, e \rightarrow_{\tau} g, \{(3, (c \vee (c \wedge d))f)\})$$

and the fourth

$$(P, e \leftrightarrows_{\tau} g, \{(3, (Error \oplus (c \wedge d))f)\})$$

as can be seen in Figure 3.9.

Despite the fact that $a \vee (a \wedge b) \equiv a$, we have chosen to use the expression on the left because we are working with events that must sometimes be replicated in the output. This expression is closer to the behavior of the actual implementation when computing identities.

3.5.5 TFFST Semantics in Policy-Based Management

The entities defined so far will be used to represent policies. Then, policies will be represented as TFFSTs and as such, they will be combined using the above-defined operations to represent a free-of-conflicts system behavior.

Using the terminology of Rei project [Jos03], policy objects are *rights, prohibitions, obligations* and *dispensations*, in addition, it is possible to express *constraints* on those policies. Below, we will show how to model obligations and constraints. More information can be find in [BS04].

Obligations

A rule expressing that when an event fulfills a certain condition, a given action shall be executed is called an "obligation." Obligations are represented as a transducer with a main link with the event in the input and the action in the output.

Typically, the incoming event will report the occurrence of a fact and the outgoing event will order the execution of a given action. However, other combinations are possible as well, for instance, in front of certain pattern of incoming events some can be replicated in the output and others not, in this way the transducer works also as a filter of relevant events or patterns of events working in a similar manner to the event-cancellation monitors mentioned in [CLN03].

To express an action as a consequence of a set of events, you can build a transducer like the one in Figure 3.10. This transducer models the rule:

Policy 3.policy: If the user dial-up (d in the figure) and the system send the order of charge (c in the figure) then, the action connect (k in the figure) should be executed.

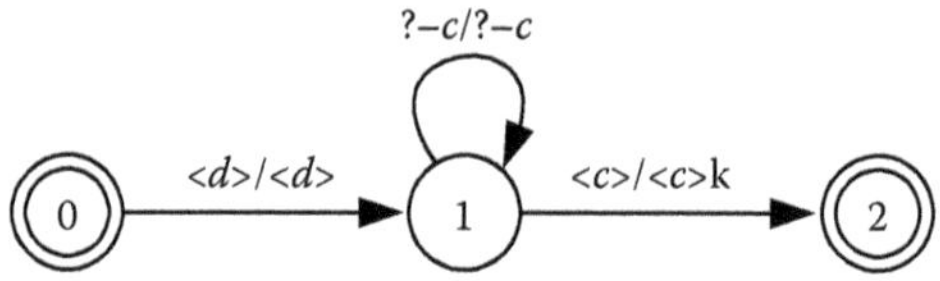

FIGURE 3.10 TFFST model for the obligation in Policy 3.5.5.

For the sake of simplicity, we disregard the fact that events can arrive without order, and we do not include other possible events before and after the sequence of interest in the model. The symbol "?" represents the TF associated to the "all events" condition.

Constraints

Sometimes, in a policy-based system, we want to express the idea that no matter what happens some actions cannot be performed under some conditions or that a couple of actions cannot be enforced at the same time, for example, a call cannot be charged and, at the same time, canceled due to an error. These ideas are implemented through "constraints" or "meta-policies". In the FST-based model, constraints are expressed using the composition operation introduced in Section 3.2.1. Once all rights, prohibitions, obligations, and dispensations are represented in a single transducer, the transducer representing constraints should be composed after it.

To see how constraints work, let us re-assume the *dial-up-charge-connect* example in Section 3.5.5. If we rely only on Policy 3.5.5, in the event of an error in the connection, the user will be charged anyway and this must not be allowed. One possible solution, among others, is to create the following constraint:

Policy 3.policy: If an error (e in the figure) occurs charge action should not be triggered.

The transducer shown in Figure 3.11 represents this constraint.

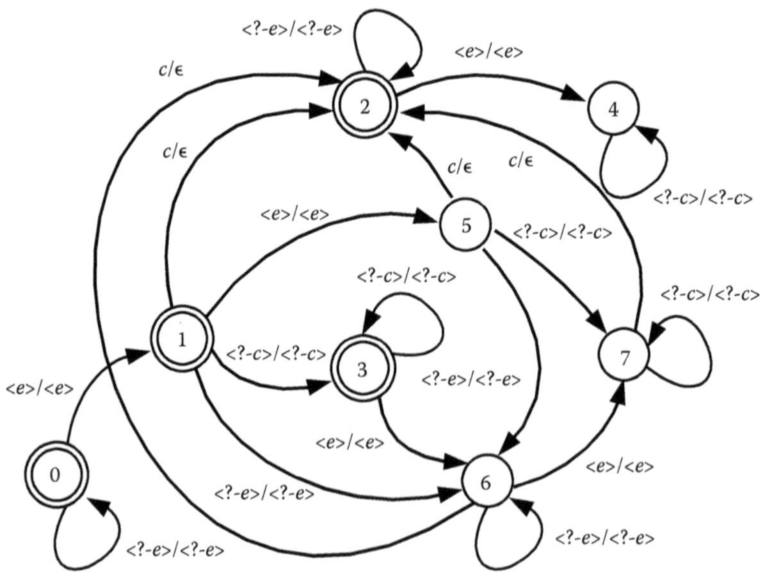

FIGURE 3.11 TFFST model for the constraint in Policy 3.5.5.

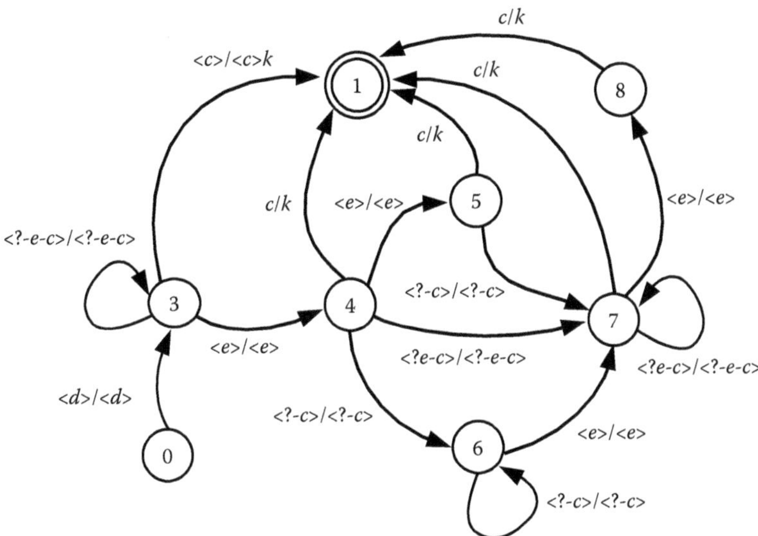

FIGURE 3.12 TFFST model for composition of Policies 3.5.5 and 3.5.5.

Computing a *composition* of both transducers results in the transducer shown in Figure 3.12, where what is underlying is that all possible system responses are computed in advance when a dynamic conflict like this occurs.

References

[AM09b] C. Allauzen and M. Mohri. N-way composition of weighted finite-state transducers. *Int. J. Found. Comput. Sci.*, 613–627, 2009.

[Bis06] C. M. Bishop. *Pattern Recognition and Machine Learning (Information Science and Statistics).* Springer-Verlag, New York, Inc., Secaucus, NJ, USA, 2006.

[BS04] J. Baliosian and J. Serrat. Finite state transducers for policy evaluation and conflict resolution. In *Proceedings of the Fifth IEEE International Workshop on Policies for Distributed Systems and Networks (POLICY 2004)*, pp. 250–259, Yorktown Heights, New York, June 2004.

[CLN03] J. Chomicki, J. Lobo, and S. Naqvi. Conflict resolution using logic programming. *IEEE Transactions on Knowledge and Data Engineering*, 15(1):245–250, Jan/Feb 2003.

[Eil74] S. Eilenberg. *Automata, Languages, and Machines*, Volume A, Academic Press, New York, 1974.

[IBM01] IBM. Autonomic Computing Manifesto. On line, October 2001. Available at: www.research.ibm.com/autonomic/.

[JM08] D. Jurafsky and J. H. Martin. *Speech and Language Processing (2nd Edition).* Pearson Prentice Hall, Upper Saddle River, New Jersey, 2008.

[Jos03] A. Joshi. A policy language for a pervasive computing environment. In *Policies for Distributed Systems and Networks, 2003. Proceedings. POLICY 2003. IEEE 4th International Workshop on*, pp. 63–74, Lake Como, Italy, 2003.

[Kar00] L. Karttunen. Applications of finite-state transducers in natural-language processing. In *Implementation and Application of Automata. Lecture Notes in Computer Science* Vol. 2088, pp. 34–46. Springer-Verlag, 2000.

[KGK97] L. Karttunen, T. GaÃ¡l, and A. Kempe. Xerox Finite-State Tool. Technical report, Xerox Research Centre Europe, Grenoble, June 1997.

[KK94] R. M. Kaplan and M. Kay. Regular models of phonological rule systems. *Computational Linguistics*, 20(3):331–378, September 1994.

[KMD+09] K. Knight, J. May, M. Droste, W. Kuich, and H. Vogler. *Applications of Weighted Automata in Natural Language Processing*, pp. 571–596. Springer, Berlin, 2009.

[MP96] M. Mohri, F. C. N. Pereira, and M. Riley. Weighted automata in text and speech processing. In *Proceedings of the 12th Biennial European Conference on Artificial Intelligence (ECAI-96)*, Workshop on Extended finite state models of language. Budapest, Hungary, 1996. ECAI.

[MS96] M. Mohri and R. Sproat. An efficient compiler for weighted rewrite rules. In *ACL '96: Proceedings of the 34th annual meeting on Association for Computational Linguistics*, pp. 231–238, Morristown, NJ, USA, 1996. Association for Computational Linguistics.

[RS97] E. Roche and Y. Schabes. *Finite-State Language Processing*. Technical report, MIT Press, Cambridge, MA, 1997.

[RS07] B. Roark and R. Sproat. *Computational Approaches to Morphology and Syntax (Oxford Surveys in Syntax & Morphology)*. Oxford University Press, USA, 2007.

[Sil97] M. Silberztein. *The Lexical Analysis of Natural Languages*, pp. 175–203. Cambridge, Mass./London, MIT Press, 1997.

[vNG01] G. van Noord and D. Gerdemann. Finite state transducers with predicates and identities. *Grammars*, 4(3):263–286, December 2001.

4
Tree Automata

Olivier Gauwin
University of Bordeaux

4.1 Introduction

The tree structure is one of the most basic and widespread data structure. Recent developments motivated by XML languages (Bray et al., 2008) have raised new interest for trees. Trees are also a genuine structure for modeling systems runs, as for instance program executions. Early works on trees (Thatcher and Wright, 1968; Doner, 1970) were motivated by algebraic properties of structures and decidability of related logics, as first investigated by Büchi on strings. Hopefully, many of the properties of regular (string) languages can be lifted to trees. A notion of tree automaton can be defined, which expressiveness is exactly captured by monadic second-order formulas: such regular tree languages are closed under Boolean operations, tree automata can be minimized, and a pumping lemma can be applied.

In this chapter, we present tree automata and related results, with a focus on links with logic. We only consider *finite* trees, for which every node is *labeled* by a symbol from a finite alphabet, and where children of each node are *ordered*. We first investigate *ranked* trees, where every symbol a uniquely determines the number of children of nodes labeled by a. We present tree automata for ranked trees and study their closure properties. A tree automaton recognizes a set of trees, called its language. We extend this definition, so that a tree automaton can recognize a tree relation. Recognizability of tree relations is preserved under first-order operations using tree variables. We present a monadic second-order logic using node variables, capturing the expressiveness of tree languages recognizable by tree automata. Finally, we present tree automata for unranked trees, that is, trees for which the number of children is not determined by the label of a node. Here, three automata notions are presented: automata using encodings into ranked trees, hedge automata, and visibly pushdown automata. These three automata definitions are in fact all equivalent in expressiveness to a monadic second-order logic for unranked trees. Some references to related surveys and results conclude this chapter.

4.2 Tree Automata for Ranked Trees

We begin this chapter with the case of ranked trees, the first kind of tree structures has been investigated formally.

4.2.1 Ranked Trees

Ranked trees rely on a *ranked alphabet* $\sigma = (\Sigma, ar)$, where $\Sigma = \{a, b, c, \dots\}$ is a finite set of *symbols*, and ar maps each symbol $a \in \Sigma$ to a positive integer called its *arity*. This arity is the number of children allowed for each symbol. We write Σ_n for the set of symbols of arity n: $\Sigma_n = \{a \in \Sigma \mid ar(a) = n\}$. The set $\mathcal{T}_\sigma$ of *ranked trees* over the ranked alphabet $\sigma = (\Sigma, ar)$ is the least set containing all the finite terms $a(t_1, \dots, t_n)$ for every $a \in \Sigma$ with $ar(a) = n$ and $t_1, \dots, t_n \in \mathcal{T}_\sigma$. This definition presents $\mathcal{T}_\sigma$ as a set of terms, while our goal was to define a set of trees. Indeed, in this chapter, these objects are the same: a term $a(t_1, \dots, t_n)$ represents a tree, with a root labeled by a, and having n children: $t_1, \dots, t_n$. Trees are ordered: The first child of the root is the root of t_1, while the nth child is the root of t_n. For convenience, we do not write empty parentheses for symbols with arity 0. Note also that we exclude the empty tree from the set of ranked trees, as it is usually not handled by tree automata.

Example 4.1:

Consider the ranked alphabet $\sigma_{abcd} = (\Sigma, ar)$ where $\Sigma = \{a, b, c, d\}$ and $\Sigma_0 = \{d\}$, $\Sigma_1 = \{c\}$ and $\Sigma_2 = \{a, b\}$. Some trees are presented in Figure 4.1. Tree $t_1 = a(d, b(c(d), d))$ belongs to $\mathcal{T}_\sigma$ but not $t_2 = a(d, a(b(d), d))$, because the arity of b should be 2. Tree $t_3 = a(d, b(b(d), d))$ is not ranked: b has not a fixed arity in this tree.

The set of *nodes* N_t of a ranked tree t is a prefix-closed subset of $\mathbb{N}^*$, where $\mathbb{N} = \{1, 2, \dots\}$ is the set of nonnegative integers. It is inductively defined:

$$N_{a(t_1,\dots,t_n)} = \{\epsilon\} \cup_{1 \le i \le n} i \cdot N_{t_i}$$

ϵ is the root node, and is also the empty symbol for concatenation. With each node π of t, we associate a *label* $\lambda_t(\pi) \in \Sigma$ using the following definition: $\lambda_{a(t_1,\dots,t_n)}(\epsilon) = a$, and $\lambda_{a(t_1,\dots,t_n)}(i \cdot \pi') = \lambda_{t_i}(\pi')$. A node labeled by a is sometimes called a-node in the sequel. The *descendants* of a node $\pi \in N_t$ are the nodes $\pi \cdot \pi' \in N_t$ for $\pi' \in \mathbb{N}^*$.

Example 4.2:

The set of nodes of the tree t_1 in Figure 4.1a is $N_{t_1} = \{\epsilon, 1, 2, 2\cdot1, 2\cdot1\cdot1, 2\cdot2\}$. The node $2\cdot1$ is labeled by c, that is, $\lambda_{t_1}(2\cdot1) = c$.

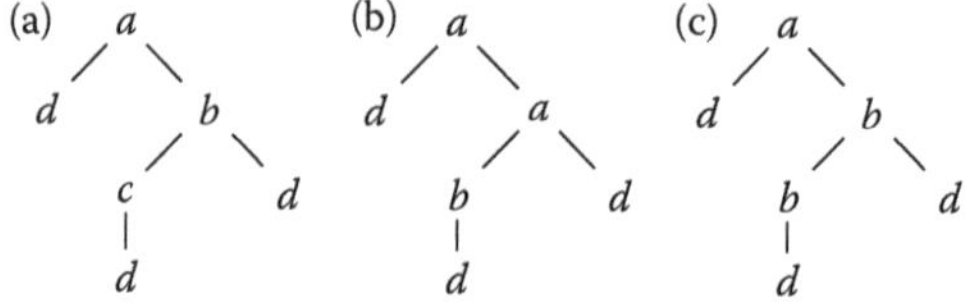

FIGURE 4.1 Examples of trees. (a) $t_1 \in \mathcal{T}_\sigma$, (b) $t_2 \notin \mathcal{T}_\sigma$, and (c) $t_3 \notin \mathcal{T}_\sigma$.

4.2.2 Tree Automata

A *tree automaton* (TA for short) is a tuple (σ, Q, F, δ) where $\sigma = (\Sigma, ar)$ is a ranked alphabet, Q is a finite set of *states*, $F \subseteq Q$ is the set of *final* states, and δ is a finite set of rules. A rule is a tuple of the form $(q_1, \ldots, q_n, a, q)$ with $a \in \Sigma_n$ and $q, q_1, \ldots, q_n \in Q$, that we write $(q_1, \ldots, q_n) \xrightarrow{a} q$.

A *run* of a TA $A = (\sigma, Q, F, \delta)$ on a ranked tree $t \in \mathcal{T}_\sigma$ is a function $\rho : N_t \to Q$ assigning a state to every node of t according to the rules of A: For every $\pi \in N_t$, $(\rho(\pi \cdot 1), \ldots, \rho(\pi \cdot n)) \xrightarrow{a} \rho(\pi) \in \delta$ where $a = \lambda_t(\pi) \in \Sigma_n$. This definition is bottom-up, as the state of a node is determined by the states of its children. A run ρ of A on t is *accepting* if $\rho(\epsilon) \in F$. The tree t is *accepted* by the TA A if there is an accepting run of A on t. The set of trees accepted by A is called the *language* of A and is written as $\mathcal{L}(A)$. A tree language $L \subseteq \mathcal{T}_\sigma$ is *recognizable* if there exists an TA A such that $\mathcal{L}(A) = L$. Two TAs are *equivalent* if their languages are the same.

A tree automaton $A = ((\Sigma, ar), Q, F, \delta)$ is *complete* if, for every symbol $a \in \Sigma_n$ and every n-tuple $(q_1, \ldots, q_n) \in Q^n$, there exists $q \in Q$ such that $(q_1, \ldots, q_n) \xrightarrow{a} q \in \delta$. Every TA can be transformed into an equivalent complete automaton by adding a (non-final) sink state, and missing rules to this state.

Example 4.3:

We illustrate the definitions with a TA A_{odd} recognizing the set L_{odd} of trees over σ_{abcd} (introduced in Example 4.1) having an odd number of a-nodes. This automaton will use two states $Q = \{q_{ev}, q_{od}\}$ where q_{ev} (resp. q_{od}) will be assigned to nodes having an even (resp. odd) number of a-labeled descendants. Hence, the only final state is q_{od}. The set of rules is the following:

$$
\begin{array}{llll}
() \xrightarrow{d} q_{ev} & (q_{ev}) \xrightarrow{c} q_{ev} & (q_{od}) \xrightarrow{c} q_{od} & \\
(q_{ev}, q_{ev}) \xrightarrow{b} q_{ev} & (q_{ev}, q_{od}) \xrightarrow{b} q_{od} & (q_{od}, q_{ev}) \xrightarrow{b} q_{od} & (q_{od}, q_{od}) \xrightarrow{b} q_{ev} \\
(q_{ev}, q_{ev}) \xrightarrow{a} q_{od} & (q_{ev}, q_{od}) \xrightarrow{a} q_{ev} & (q_{od}, q_{ev}) \xrightarrow{a} q_{ev} & (q_{od}, q_{od}) \xrightarrow{a} q_{od}
\end{array}
$$

A run of A_{odd} on a tree t is illustrated in Figure 4.2 by underlined states. This run maps the root to the final state q_{od}, so the run is accepting, and $t \in \mathcal{L}(A_{odd})$.

Example 4.4:

Now we want to build a TA $A_{\neg ac}$ recognizing all trees over σ_{abcd} such that there is no a-node having a c-child. An intuitive way to do this, is to store the label in the state. Here, we do not have to know the label in all cases, we just have to know whether it is a c (via state q_c) or not (via state $q_{\neg c}$). For a-nodes,

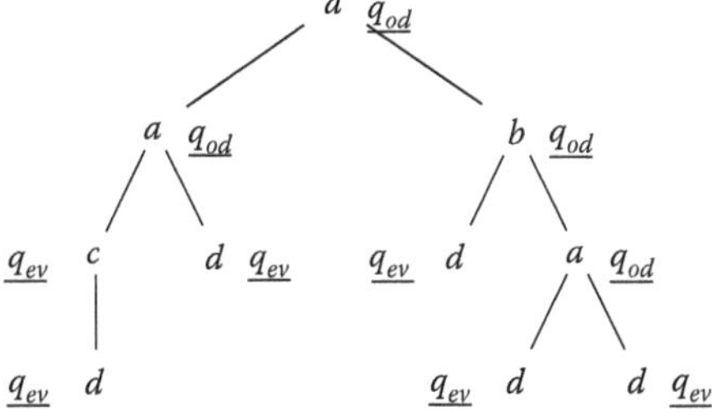

FIGURE 4.2 A run of the TA A_{odd} on a tree.

rules check that both incoming states are $q_{\neg c}$:

$$() \xrightarrow{d} q_{\neg c} \qquad (q_c) \xrightarrow{c} q_c \qquad (q_{\neg c}) \xrightarrow{c} q_c \qquad (q_{\neg c}, q_{\neg c}) \xrightarrow{a} q_{\neg c}$$
$$(q_{\neg c}, q_{\neg c}) \xrightarrow{b} q_{\neg c} \quad (q_{\neg c}, q_c) \xrightarrow{b} q_{\neg c} \quad (q_c, q_{\neg c}) \xrightarrow{b} q_{\neg c} \quad (q_c, q_c) \xrightarrow{b} q_{\neg c}$$

The final states of $A_{\neg ac}$ are $\{q_c, q_{\neg c}\}$. A tree having an a-node with a c-child will have no corresponding run of $A_{\neg ac}$, and thus no accepting run: When reading such an a-node, no rule can be applied.

4.2.3 Determinism

A TA (σ, Q, F, δ) is *deterministic* if for every pair of rules $(q_1, \ldots, q_n) \xrightarrow{a} q$ and $(q_1, \ldots, q_n) \xrightarrow{a} q'$ in δ, we have $q = q'$. Thus, a deterministic TA (called dTA in the sequel) has at most one run per tree: At every step of the run, there is no choice between rules to apply.

Theorem 4.1:

For every TA, there exists an equivalent dTA.

Proof. The determinization procedure is very similar to the subset construction for words. Consider the TA $A = (\sigma, Q, F, \delta)$, and the TA $D = (\sigma, Q_D, F_D, \delta_D)$ which states are subsets of Q and final states are those containing a state of F: $Q_D = 2^Q$ and $F_D = \{S \subseteq Q \mid S \cap F \neq \emptyset\}$. Rules of D are defined so that the state S assigned to a node π during a run is exactly the set of states q of A for which a run of A on the subtree of t rooted at π assigning q to the root exists (this can be easily checked by an induction on the structure of terms):

$$(S_1, \ldots, S_n) \xrightarrow{a} S \in \delta_D \text{ iff } S = \{q \mid \exists q_1 {\in} S_1 \ldots \exists q_n {\in} S_n \ (q_1, \ldots, q_n) \xrightarrow{a} q \in \delta\}$$

Hence, the language is preserved, and D is deterministic. □

4.2.4 Top-Down Determinism

The definitions of rules and runs of TAs are bottom-up. A run starts at the leaves, where rules of the form $() \xrightarrow{a} q$ are applied, then propagates until the root, where the membership to final states is tested.

In the literature, a top-down tree automaton (σ, Q, I, δ) is defined as the exact symmetrical object with respect to TAs. A run ρ of a top-down tree automaton starts at the root with one of the initial states $I \subseteq Q$ (corresponding to final states of TAs) and follows rules of the form $q \xrightarrow{a} (q_1, \ldots, q_n)$: For every node $\pi \in N_t$ with $\lambda_t(\pi) = a \in \Sigma_n$, there is a rule $\rho(\pi) \xrightarrow{a} (\rho(\pi{\cdot}1), \ldots, \rho(\pi{\cdot}n))$ in δ.

In fact, top-down tree automata are just another way of writing TAs. This is strictly equivalent: For every TA, one can build a top-down tree automaton recognizing the same language, and conversely. We just have to consider initial states as final states, and revert each rule. For this reason we only consider (bottom-up) TAs in the sequel.

The only difference between bottom-up and top-down tree automata is determinism. A top-down tree automaton is deterministic if for every pair of rules $q \xrightarrow{a} (q_1, \ldots, q_n)$ and $q \xrightarrow{a} (q'_1, \ldots, q'_n)$, we have $(q_1, \ldots, q_n) = (q'_1, \ldots, q'_n)$. While every TA has an equivalent dTA, there exist top-down tree automata that cannot be determinized.

Proposition 4.1:

Deterministic top-down tree automata are strictly less expressive than TAs.

Proof. Let L be the set of trees over alphabet σ_{abcd} where all a-nodes have a c-child and a d-child (but in any order). This language is recognizable, but there is no deterministic top-down tree automaton accepting it. Indeed, a deterministic top-down tree automaton accepting the trees $a(c(d),d)$ and $a(d,c(d))$ (belonging to L) also accepts $a(d,d)$ (not belonging to L): Let $q \xrightarrow{a} (q_1, q_2)$ be the rule applied at the root (this is the same rule for the three mentioned trees). As $a(d,c(d))$ (resp. $a(c(d),d)$) is accepted, there is a run on d starting from q_1 (resp. from q_2). Hence, $a(d,d)$ is also accepted by this automaton. While L cannot be recognized by a deterministic top-down automaton, it can be recognized by a nondeterministic one, proving its recognizability. A nondeterministic top-down automaton can *guess*, for each a-node, whether c is the first or second child, and *check* it when reading children. Let A be such an automaton, with states $\{q_*, q_c, q_d\}$, the only initial state being q_*, and the rules being:

$$\begin{array}{llll} q_* \xrightarrow{a} (q_c, q_d) & q_* \xrightarrow{a} (q_d, q_c) & q_* \xrightarrow{b} (q_*, q_*) & \\ q_c \xrightarrow{c} (q_*) & q_d \xrightarrow{d} () & q_* \xrightarrow{c} (q_*) & q_* \xrightarrow{d} () \end{array}$$

□

4.2.5 Closure Properties

4.2.5.1 Boolean Operations

Recognizable languages enjoy closure properties against all Boolean operations.

Proposition 4.2:

If $L \subseteq \mathcal{T}_\sigma$ is recognizable, then $\mathcal{T}_\sigma \setminus L$ is recognizable.

Proof. Let $A = (\sigma, Q, F, \delta)$ be a TA recognizing L. We assume w.l.o.g. (see Theorem 4.1) that A is deterministic and complete, so that it admits exactly one run per tree in $\mathcal{T}_\sigma$. Then, the dTA $(\sigma, Q, Q \setminus F, \delta)$ recognizes the language $\mathcal{T}_\sigma \setminus L$. Indeed, $t \notin \mathcal{L}(A)$ iff the unique run ρ of A on t satisfies $\rho(\epsilon) \notin F$. □

Note that the above construction may involve an exponential blow-up, as it requires determinism.

Proposition 4.3:

If $L_1, L_2 \subseteq \mathcal{T}_\sigma$ are recognizable languages, then $L_1 \cup L_2$ is recognizable.

Proof. Let $A_1 = (\sigma, Q_1, F_1, \delta_1)$ and $A_2 = (\sigma, Q_2, F_2, \delta_2)$ be TAs recognizing, respectively, L_1 and L_2. We assume w.l.o.g. that $Q_1 \cap Q_2 = \emptyset$. Then, the TA $(\sigma, Q_1 \cup Q_2, F_1 \cup F_2, \delta_1 \cup \delta_2)$ recognizes $L_1 \cup L_2$, as this TA checks whether a run on a tree exists for either A_1 or A_2. □

As $L_1 \cap L_2 = \mathcal{T}_\sigma \setminus (\mathcal{T}_\sigma \setminus L_1 \cup \mathcal{T}_\sigma \setminus L_2)$, we immediately get the closure by intersection using Propositions 4.2 and 4.3.

Corollary 4.1:

If L_1 and L_2 are recognizable languages over σ, so is $L_1 \cap L_2$.

The proofs for closure under union and intersection rely on time-consuming procedures: Intersection requires complementation, and thus determinization and completion. Moreover, determinism is lost when performing union.

The *synchronized product* is a way to avoid such drawbacks. This construction relies on the parallel computation of two runs. Consider two TAs $A_1 = (\sigma, Q_1, F_1, \delta_1)$ and $A_2 = (\sigma, Q_2, F_2, \delta_2)$. Let $A_\cup = (\sigma, Q, F_\cup, \delta)$ and $A_\cap = (\sigma, Q, F_\cap, \delta)$ be the TAs where $Q = Q_1 \times Q_2$, $F_\cup = \{(q_1, q_2) \in Q \mid q_1 \in F_1 \text{ or } q_2 \in F_2\}$, $F_\cap = F_1 \times F_2$, and $((q_1, q'_1), \ldots, (q_n, q'_n)) \xrightarrow{a} (q, q') \in \delta$ iff $(q_1, \ldots, q_n) \xrightarrow{a} q \in \delta_1$ and $(q'_1, \ldots, q'_n) \xrightarrow{a} q' \in \delta_2$. These definitions allow to compute simultaneously a run of A_1 and a run of A_2. We get: $\mathcal{L}(A_\cap) = \mathcal{L}(A_1) \cap \mathcal{L}(A_2)$ and $\mathcal{L}(A_\cup) = \mathcal{L}(A_1) \cup \mathcal{L}(A_2)$. Determinization is not required here, and determinism is preserved: $A_\cap$ and $A_\cup$ are deterministic, whenever A_1 and A_2 are.

4.3 Logics over Ranked Trees

Recognizable languages of ranked trees form a Boolean algebra. In this section, we extend these basic closure properties to n-ary relations over trees instead of sets of trees (which can be considered as unary relations over trees). This will lead to a first-order logic allowing to quantify over trees. We introduce the monadic second-order logic, and prove the classical result that this logic exactly captures recognizable languages.

4.3.1 Recognizable Relations over Trees

Let Ω be a countable set of ranked alphabets. An n-ary *tree relation* R over $(\sigma_1, \ldots, \sigma_n) \in \Omega^n$ is a set of tuples of trees: $R \subseteq \mathcal{T}_{\sigma_1} \times \cdots \times \mathcal{T}_{\sigma_n}$.

4.3.1.1 Overlays and Recognizable Relations

The notion of recognizability for tree relations also relies on tree automata. Thus, a first step is to associate a tree language with a relation.

We define the overlay of two trees (over different alphabets) as a new tree obtained by overlapping them. If their structures differ, we align them in a leftmost manner and use a fresh symbol $\diamond$, as illustrated in Figure 4.3. More precisely, for two alphabets $\sigma_i = (\Sigma_i, ar_i)$ with $i \in \{1, 2\}$, let $\sigma_1 \times_\diamond \sigma_2 = (\Sigma_\diamond, ar_\diamond)$ where $\Sigma_\diamond = ((\Sigma_1 \uplus \{\diamond\}) \times (\Sigma_2 \uplus \{\diamond\})) \setminus \{(\diamond, \diamond)\}$, and, for $(a, b) \in \Sigma_\diamond$, $ar_\diamond((a, b)) = \max\{ar_1(a), ar_2(b)\}$ with $ar_1(\diamond) = ar_2(\diamond) = 0$. Let $t_1 \in \mathcal{T}_{\sigma_1}$ and $t_2 \in \mathcal{T}_{\sigma_2}$. The *overlay* of $t_1 = a(t_1, \ldots, t_k)$ and $t_2 = b(t'_1, \ldots, t'_l)$ is the tree $t_1 \circledast t_2 \in \mathcal{T}_{\sigma_1 \times_\diamond \sigma_2}$ inductively defined by

$$a(t_1, \ldots, t_k) \circledast b(t'_1, \ldots, t'_l) = \begin{cases} (a, b)(t_1 \circledast t'_1, \ \ldots, \ t_l \circledast t'_l, \ t_{l+1} \circledast \diamond, \ \ldots, \ t_k \circledast \diamond) & \text{if } l \leq k \\ (a, b)(t_1 \circledast t'_1, \ \ldots, \ t_k \circledast t'_k, \ \diamond \circledast t'_{k+1}, \ \ldots, \ \diamond \circledast t'_l) & \text{otherwise} \end{cases}$$

the tree $\diamond \circledast t$ being obtained from t by relabeling every node π by $(\diamond, \lambda_t(\pi))$, and symmetrically for $t \circledast \diamond$. This definition naturally extends to overlays of n-tuples of trees: for $(t_1, \ldots, t_n) \in \mathcal{T}_{\sigma_1} \times \cdots \times \mathcal{T}_{\sigma_n}$, $t_1 \circledast \ldots \circledast t_n = ((t_1 \circledast t_2) \circledast \ldots \circledast t_n) \in \mathcal{T}_{\sigma_1 \times_\diamond \cdots \times_\diamond \sigma_n}$. For a tree relation R over $\sigma_1, \ldots, \sigma_n$, the overlay of R is the set of overlays of tuples in R: $ovl(R) = \{t_1 \circledast \ldots \circledast t_n \mid (t_1, \ldots, t_n) \in R\}$. If a tree automaton A is such that $\mathcal{L}(A) = ovl(R)$, we say that A *recognizes* R. A tree relation R is *recognizable* if it is recognized by a tree automaton. For conciseness, we write *recognizable relation* instead of *recognizable tree relation*. If every tuple in R relates trees with identical structures, that is, if $ovl(R)$ does not contain symbol $\diamond$, we say that R is a *relabeling*.

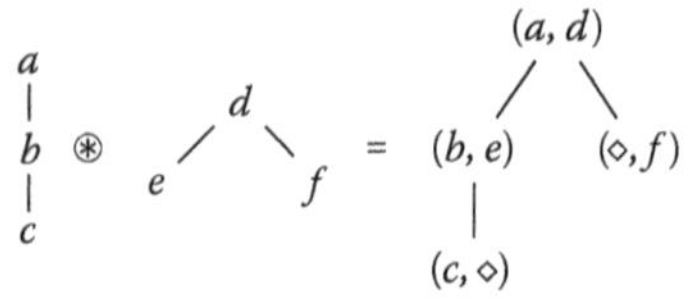

$$a(b(c)) \circledast d(e, f) = (a,d)\big((b,e)((c,\diamond)),\ (\diamond,f)\big)$$

FIGURE 4.3 Example of overlay.

4.3.1.2 Closure Properties

We study closure properties of recognizable relations. As we will relate relations over different alphabets, we first introduce operators that manipulate relations component-wise: projection and cylindrification.

4.3.1.2.1 Projection

The *ith projection* (for $1 \leq i \leq n$) of the n-ary relation $R \subseteq \mathcal{T}_{\sigma_1} \times \cdots \times \mathcal{T}_{\sigma_n}$ is the $(n-1)$-ary relation $\Pi_i(R) \subseteq \mathcal{T}_{\sigma_1} \times \cdots \times \mathcal{T}_{\sigma_{i-1}} \times \mathcal{T}_{\sigma_{i+1}} \times \cdots \times \mathcal{T}_{\sigma_n}$ defined by

$$(t_1, \ldots, t_{i-1}, t_{i+1}, \ldots, t_n) \in \Pi_i(R) \quad \Leftrightarrow \quad \exists t_i \in \mathcal{T}_{\sigma_i}\ (t_1, \ldots, t_n) \in R$$

Proposition 4.4:

If R is an n-ary recognizable relation, then, for every $1 \leq i \leq n$, $\Pi_i(R)$ is a recognizable relation.

Proof. Let A be a TA recognizing R. We assume w.l.o.g. that all states in A are accessible: for each state q there exists a tree with a run assigning q to the root. A TA A_i recognizing $\Pi_i(R)$ is obtained by keeping the same states as A, and projecting each rule of A:

$$(q_1, \ldots, q_\ell) \xrightarrow{\Pi_i(a)} q \in \delta_{A_i} \quad \Leftrightarrow \quad (q_1, \ldots, q_k) \xrightarrow{a} q \in \delta_A$$

where $\Pi_i((a_1, \ldots, a_n)) = (a_1, \ldots, a_{i-1}, a_{i+1}, \ldots, a_n)$ and ℓ is the maximal arity for symbols $a_1, \ldots, a_{i-1}, a_{i+1}, \ldots, a_n$. □

4.3.1.2.2 Cylindrification

While projection reduces the arity of a relation, cylindrification allows to add, copy, and swap (but not delete) components of a relation. Let $[m]$ denotes the set of integers i such that $1 \leq i \leq m$, and let θ be a function $\theta : [m] \to [m]$ such that $m > 0$ and $[n] \subseteq \theta([m])$. The θ-*cylindrification* of the n-ary relation R is the m-ary relation $\zeta_\theta \subseteq \mathcal{T}_{\sigma_{\theta(1)}} \times \cdots \times \mathcal{T}_{\sigma_{\theta(m)}}$ defined by

$$(t_{\theta(1)}, \ldots, t_{\theta(m)}) \in \zeta_\theta(R) \quad \Leftrightarrow \quad (t_1, \ldots, t_n) \in R$$

Proposition 4.5:

For every function $\theta : [m] \to [m]$ such that $m > 0$ and $[n] \subseteq \theta([m])$, and every n-ary recognizable relation R, the relation $\zeta_\theta(R)$ is recognizable.

Proof. Cylindrification $\zeta_\theta(R)$ can be obtained progressively using three features: (1) adding one component, that is $(t_1, \ldots, t_{n+1}) \in R' \Leftrightarrow (t_1, \ldots, t_n) \in R$, (2) checking whether two components i, j are equal, that is $(t_1, \ldots, t_n) \in R' \Leftrightarrow ((t_1, \ldots, t_n) \in R$ and $t_i = t_j)$, and (3) permuting components. Corresponding features on overlay languages are easy to implement by tree automata. □

4.3.1.2.3 Boolean Operations

Intersection, union, and complement can also be defined at the relations level. Given two n-ary relations $R_1, R_2 \subseteq \mathcal{T}_{\sigma_1} \times \cdots \times \mathcal{T}_{\sigma_n}$, $R_1 \cap R_2$ (resp. $R_1 \cup R_2$) is the n-ary relation over the same alphabets, containing tuples in both R_1 and R_2 (resp. tuples in R_1 or R_2). The relation $\overline{R_1}$ is also a n-ary relation over these alphabets, defined by: $\overline{R_1} = \mathcal{T}_{\sigma_1} \times \cdots \times \mathcal{T}_{\sigma_n} \setminus R_1$.

Proposition 4.6:

If $R_1, R_2 \subseteq \mathcal{T}_{\sigma_1} \times \cdots \times \mathcal{T}_{\sigma_n}$ are recognizable relations, then $R_1 \cap R_2$, $R_1 \cup R_2$ and $\overline{R_1}$ are also recognizable.

Proof. Indeed, $ovl(R_1 \cap R_2) = ovl(R_1) \cap ovl(R_2)$ and $ovl(R_1 \cup R_2) = ovl(R_1) \cup ovl(R_2)$ are recognizable by closure properties of recognizable languages (Corollary 4.1 and Proposition 4.3). Furthermore, $ovl(\overline{R_1}) = \{t_1 \circledast \ldots \circledast t_n \mid (t_1, \ldots, t_n) \in \mathcal{T}_{\sigma_1} \times \cdots \times \mathcal{T}_{\sigma_n} \setminus R_1\}$ can be recognized by a tree automaton checking that the input is an overlay, but not in R_1 (for conciseness we omit its construction). □

4.3.1.3 First-Order Logic

4.3.1.3.1 *Syntax*

Given the aforementioned closure properties, we define a logic over tree relations that preserves recognizability of relations. For instance, if R_1 and R_2 are recognizable, then the relation $R'(t_1, t_2) = \exists t\ R_1(t, t_1, t_2) \wedge \neg R_2(t_2, t)$ is also recognizable: $R_1(t, t_1, t_2) \wedge \neg R_2(t_2, t)$ is recognizable by closure under Boolean operations, while R' is recognizable thanks to closure under projection.

Let $\mathcal{R}$ be a countable set of relation symbols, where each symbol $r \in \mathcal{R}$ has an arity n and a signature $(\sigma_1, \ldots, \sigma_n)$ in Ω^n. We define FO[$\mathcal{R}$] as the set of formulas of the following form:

$$\phi ::= r(t_1, \ldots, t_n) \mid \phi \wedge \phi' \mid \neg\phi \mid \exists t \in \mathcal{T}_\sigma\ \phi$$

where $r \in \mathcal{R}$, $t, t_1, \ldots, t_n$ belong to an infinite countable set of variables, $\sigma \in \Omega$, and each formula ϕ is well-typed, in the following sense: with each variable t in ϕ, we can assign a unique *type* $\sigma \in \Omega$ such that for all existential quantifications $\exists t \in \mathcal{T}_{\sigma'}\ \phi$ over t, $\sigma' = \sigma$, and for all relation symbols $r(t_1, \ldots, t_n)$ in which t appears at the ith component, the signature $(\sigma_1, \ldots, \sigma_n)$ of r is such that $\sigma_i = \sigma$.

4.3.1.3.2 *Semantic*

A formula in FO[$\mathcal{R}$] is interpreted relatively to an assignment μ of every variable according to its type ($t \in \mathcal{T}_\sigma$ if t has type σ), and an interpretation ν of the relation symbols: for every $r \in \mathcal{R}$, $\nu(r) \subseteq \mathcal{T}_{\sigma_1} \times \cdots \times \mathcal{T}_{\sigma_n}$ if r has signature $(\sigma_1, \ldots, \sigma_n)$. Satisfiability $\nu, \mu \models \phi$ of ϕ under ν, μ is inductively defined by

$$\begin{array}{ll}
\nu, \mu \models r(t_1, \ldots, t_n) & \text{iff } (\mu(t_1), \ldots, \mu(t_n)) \in \nu(r) \\
\nu, \mu \models \phi \wedge \phi' & \text{iff } \nu, \mu \models \phi \text{ and } \nu, \mu \models \phi' \\
\nu, \mu \models \neg\phi & \text{iff } \nu, \mu \not\models \phi \\
\nu, \mu \models \exists t \in \mathcal{T}_\sigma\ \phi & \text{iff there exists } t' \in \mathcal{T}_\sigma \text{ such that } \nu, \mu[t \leftarrow t'] \models \phi
\end{array}$$

Moreover, it is natural to consider the semantic of a formula ϕ as a new relation. Given $\phi \in$ FO[$\mathcal{R}$] with an ordering of its free variables $t_1, \ldots, t_n$, and an interpretation ν of relation symbols,

$$R_{\phi(t_1, \ldots, t_n), \nu} = \{(\mu(t_1), \ldots, \mu(t_n)) \mid \nu, \mu \models \phi\}$$

is a subset of $\mathcal{T}_{\sigma_1} \times \cdots \times \mathcal{T}_{\sigma_n}$ (where σ_i is the type of t_i), hence a tree relation.

Theorem 4.2:

Let $\phi \in$ FO[$\mathcal{R}$] and an ordering $t_1, \ldots, t_n$ of its free variables. If $\nu(r)$ is recognizable for all $r \in \mathcal{R}$, then $R_{\phi(t_1, \ldots, t_n), \nu}$ is recognizable.

Proof. The proof is by induction on the structure of formulas. If $\phi = r(t_1, \ldots, t_n)$, then $R_{\phi(t_1,\ldots,t_n),v} = v(r)$ is recognizable. If $\phi = \phi' \wedge \phi''$, we have:

$$\begin{aligned} R_{\phi(t_1,\ldots,t_n),v} &= \{(\mu(t_1), \ldots, \mu(t_n)) \mid v, \mu \models \phi' \wedge \phi''\} \\ &= \{(\mu(t_1), \ldots, \mu(t_n)) \mid v, \mu \models \phi'\} \cap \{(\mu(t_1), \ldots, \mu(t_n)) \mid v, \mu \models \phi''\} \\ &= R_{\phi'(t_1,\ldots,t_n),v} \cap R_{\phi''(t_1,\ldots,t_n),v} \end{aligned}$$

so by induction hypothesis and Proposition 4.6, $R_{\phi(t_1,\ldots,t_n),v}$ is recognizable. We assumed here that ϕ, ϕ', and ϕ'' have the same free variables. In the general case, we use cylindrification to reorder the components (this does not affect recognizability thanks to Proposition 4.5). Similarly, $R_{\neg\phi(t_1,\ldots,t_n),v}$ is recognizable because recognizable relations are closed under complement (Proposition 4.6), and $R_{\exists t \in \mathcal{T}_\sigma\ \phi(t_1,\ldots,t_n),v}$ is recognizable because projection preserves recognizability (Proposition 4.4). □

Of course, Theorem 4.2 still holds if $v(r)$ is recognizable only for relation symbols r used in ϕ. Using some fixed relations and a unique alphabet, one can extend Theorem 4.2 to get an equivalence, that is a tree relation is first-order definable (over this precise signature) iff it is recognizable (Benedikt et al., 2007).

4.3.2 Monadic Second-Order Logic

The expressiveness of recognizable tree languages is exactly captured by the Monadic Second-Order logic (MSO for short), as proved by Thatcher and Wright (1968) and Doner (1970) in the late 1960s. MSO logic allows first-order variables over nodes, second-order variables over sets of nodes (and quantification over them), label tests, moves along child axes, membership testing, and Boolean operations. This logic is sometimes called WSkS, the weak second-order logic with k successors: *weak* denotes a quantification over finite sets, the k successors being binary predicates ch_k accessing to the kth child of a node.

Let $\sigma = (\Sigma, ar)$ be a ranked alphabet with maximal arity $m = \max_{a \in \Sigma} ar(a)$. Predicates of the MSO logic, we consider, rely on testing labels and moving along child axes. Hence, we consider the relational signature Λ composed by unary relation symbols lab_a with $a \in \Sigma$, and binary relation symbols ch_i with $1 \leq i \leq m$. A MSO[Λ] formula is a formula obtained by using the following grammar:

$$\phi ::= lab_a(x) \mid ch_i(x, y) \mid \phi \wedge \phi' \mid \neg\phi \mid \exists x\ \phi \mid \exists X\ \phi \mid x \in X$$

where $a \in \Sigma$, $1 \leq i \leq m$, and x, y (resp. X) are members of an infinite countable set of first-order (resp. second-order) variables. For convenience, we write $\phi(\vec{x}, \vec{X})$ where $\vec{x}$ (resp. $\vec{X}$) is the set of first-order (resp. second-order) free variables in ϕ. The term *monadic* relates to the arity of second-order predicates $X : x \in X$ is syntactic sugar for $X(x)$.

A formula $\phi(\vec{x}, \vec{X}) \in$ MSO[Λ] is interpreted over a tree t under an assignment μ mapping each free first-order variable $x \in \vec{x}$ to a node of t, and each second-order variable $X \in \vec{X}$ to a subset of N_t. Satisfiability $t, \mu \models \phi$ is inductively defined by

$$\begin{array}{ll} t, \mu \models lab_a(x) & \text{iff } \lambda_t(\mu(x)) = a \\ t, \mu \models ch_i(x, y) & \text{iff } \mu(y) = \mu(x) \cdot i \\ t, \mu \models \phi \wedge \phi' & \text{iff } t, \mu \models \phi \text{ and } t, \mu \models \phi' \\ t, \mu \models \neg\phi & \text{iff } t, \mu \not\models \phi \\ t, \mu \models \exists x\ \phi & \text{iff there exists } \pi \in N_t \text{ s.t. } t, \mu[x \leftarrow \pi] \models \phi \\ t, \mu \models \exists X\ \phi & \text{iff there exists } S \subseteq N_t \text{ s.t. } t, \mu[X \leftarrow S] \models \phi \\ t, \mu \models x \in X & \text{iff } \mu(x) \in \mu(X) \end{array}$$

For convenience, we may use parentheses and the usual Boolean operators in the sequel, for disjunction $\vee$, xor $\oplus$, implication $\Rightarrow$, and so on. Note that for closed formulas (i.e., formulas without free variables), no assignment is needed.

Example 4.5:

Let root$(x) = \neg\exists y\, ch_1(y,x)$. This MSO formula checks whether a node is the root of the tree: $t,\mu \models \text{root}(x)$ iff $\mu(x) = \epsilon$. Similarly, leaf$(x) = \neg\exists y\, ch_1(x,y)$ checks that a node is a leaf: $t,\mu \models \text{leaf}(x)$ iff $\mu(x)$ has no children.

Example 4.6:

We can define a closed MSO formula ϕ_{odd}, which will be satisfied by all trees $t \in L_{\text{odd}} \subseteq \mathcal{T}_{\sigma_{abcd}}$ having an odd number of a-nodes (see Example 4.3): $t \models \phi_{\text{odd}}$ iff $t \in L_{\text{odd}}$. The formula is based on a variable X that exactly characterizes all nodes having an odd number of a-descendants. The tree is in L_{odd} iff the root belongs to this set.

$$\begin{aligned}\phi_{\text{odd}} = \exists X\, \forall x\, (&\text{root}(x) \Rightarrow (x \in X)) \wedge \\ &(lab_a(x) \Rightarrow \exists y\, \exists z\, ch_1(x,y) \wedge ch_2(x,z) \wedge (x \in X \Leftrightarrow \neg(y \in X \oplus z \in X)))\wedge \\ &(lab_b(x) \Rightarrow \exists y\, \exists z\, ch_1(x,y) \wedge ch_2(x,z) \wedge (x \in X \Leftrightarrow (y \in X \oplus z \in X))) \wedge \\ &(lab_c(x) \Rightarrow \exists y\, ch_1(x,y) \wedge (x \in X \Leftrightarrow y \in X)) \wedge \\ &(lab_d(x) \Rightarrow (x \notin X))\end{aligned}$$

To compare the expressive power of MSO and tree automata, models of MSO formulas have to be converted into trees, instead of pairs t,μ of tree and assignment. This can be achieved by relabeling t, so that a Boolean is added for each free variable. We denote by $\mathbb{B} = \{0,1\}$ the set of Booleans. Let $\phi(\vec{x},\vec{X}) \in$ MSO[Λ] over alphabet $\sigma = (\Sigma, ar)$, and let $n = |\vec{x}| + |\vec{X}|$ be the total number of free variables in ϕ. We use the notation $(x_1,\ldots,x_i,X_{i+1},\ldots,X_n)$ for $(\vec{x},\vec{X})$. An MSO[Λ] formula ϕ defines the following *tree relation* over $\mathcal{T}_\Sigma \times \mathcal{T}_{\mathbb{B}}^n$: $\mathrm{R}_{\phi(x_1,\ldots,x_i,X_{i+1},\ldots,X_n)} = \{(t, t_{\mu(x_1)},\ldots,t_{\mu(x_i)},t_{\mu(X_{i+1})},\ldots,t_{\mu(X_n)}) \mid t,\mu \models \phi\}$ where $t_{\mu(x_j)} \in \mathcal{T}_{\mathbb{B}}$ is obtained from t by relabeling it in the following way: for $\pi \in N_t$, $\lambda_{t_{\mu(x_j)}}(\pi) = 1$ if $\pi = \mu(x_j)$, and 0 otherwise. For second-order variables, $\lambda_{t_{\mu(X_j)}}(\pi) = 1$ if $\pi \in \mu(X_j)$, and 0 otherwise.* The *language of* $\phi(\vec{x},\vec{X})$ is $\mathcal{L}(\phi(\vec{x},\vec{X})) = ovl(\mathrm{R}_{\phi(\vec{x},\vec{X})})$. A tree language $L \subseteq \mathcal{T}_\sigma$ is said *regular* if $L = \mathcal{L}(\phi)$ for some closed formula $\phi \in$ MSO[Λ].

Example 4.7:

Let $\phi(x_1, X_2)$ be an MSO[Λ] formula over alphabet σ_{abcd}. Figure 4.4 illustrates the definitions with a tree $t \in \mathcal{T}_\sigma$ and the tree $t \circledast t_{\mu(x_1)} \circledast t_{\mu(X_2)} \in \mathcal{T}_\sigma$ corresponding to the assignment $\mu(x_1) = 1$ and $\mu(X_2) = \{2, 2\cdot1\cdot1\}$. Then $t \circledast t_{\mu(x_1)} \circledast t_{\mu(X_2)} \in \mathcal{L}(\phi(x_1,X_2))$ iff $t,\mu \models \phi$.

Theorem 4.3: (Thatcher and Wright, 1968; Doner, 1970)

A tree language is regular iff it is recognizable.

Proof. Suppose that L is recognizable, and let $A = (\sigma, Q, F, \delta)$ be a TA recognizing L. We can build an MSO formula ϕ_A encoding accepting runs of A. If $Q = \{q_1,\ldots,q_n\}$, we use the second-order

* As $t_{\mu(x_j)}$ is a ranked tree, each label should have a fixed arity. This is easily fixed using labels (β, n) with arity n (with $n \le \max_{a\in\Sigma} ar(a)$), but we omit this detail for clarity.

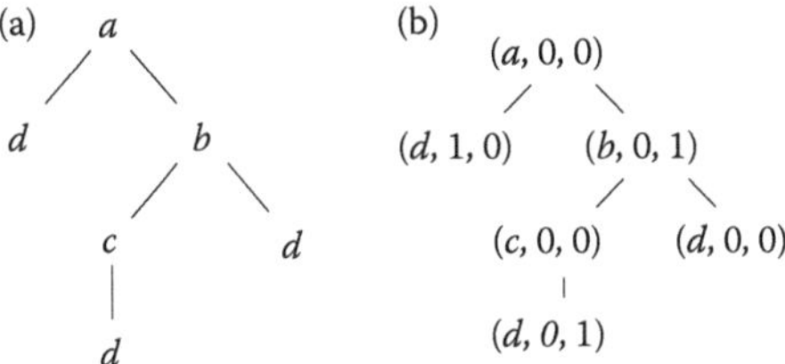

FIGURE 4.4 Example of a tree t annotated with an assignment μ. (a) $t \in \mathcal{T}_\sigma$ and (b) $t \circledast t_{\mu(x_1)} \circledast t_{\mu(x_2)}$.

variables $X_{q_1}, \ldots, X_{q_n}$.

$$\phi_A = \exists X_{q_1} \ldots \exists X_{q_n} \text{ partition}(X_{q_1}, \ldots, X_{q_n}) \wedge \exists x \left(\text{root}(x) \wedge \bigvee_{q \in F} x \in X_q \right) \wedge$$
$$\forall x \bigwedge_{a \in \Sigma} \bigwedge_{q_i \in Q} \left(\begin{array}{c} lab_a(x) \wedge \\ x \in X_{q_i} \end{array} \right) \Rightarrow \bigvee_{(q_{i_1}, \ldots, q_{i_k}) \xrightarrow{a} q_i \in \delta} \bigwedge_{1 \le j \le k} \left(\begin{array}{c} \exists y_j \\ ch_j(x, y_j) \wedge \\ y_j \in X_{q_{i_j}} \end{array} \right)$$

where *partition* checks that exactly one state is assigned to every node:

$$\text{partition}(X_{q_1}, \ldots, X_{q_n}) = \forall x \left(\bigvee_{1 \le i \le n} x \in X_{q_i} \wedge \bigwedge_{1 \le i < j \le n} \neg \left(x \in X_{q_i} \wedge x \in X_{q_j} \right) \right)$$

and ϕ_A checks also that a final state is assigned to the root, and the last line asserts that at each node, a rule of A is applied. Hence, we have $\mathcal{L}(A) = \mathcal{L}(\phi_A)$.

For the converse, we prove by induction on ϕ that $\mathrm{R}_{\phi(\vec{x}, \vec{X})}$ is recognizable, using closure properties of recognizable relations.

- $\phi(x) = lab_a(x)$. The TA $A_{lab_a(x)} = (\sigma, \{\top, \bot\}, \{\top\}, \delta)$ with the following rules in δ recognizes $\mathrm{R}_{lab_a(x)}$:

$$\underbrace{(\bot, \ldots, \bot)}_{ar(b)} \xrightarrow{(b,0)} \bot \text{ for all } b \in \Sigma$$

$$\underbrace{(\bot, \ldots, \bot)}_{ar(a)} \xrightarrow{(a,1)} \top$$

$$\underbrace{(q_1, \ldots, q_n)}_{ar(b)} \xrightarrow{(b,0)} \top \text{ for all } b \in \Sigma, \text{ whenever } q_i = \top \text{ for exactly one } i$$

- $\phi(x,y) = ch_i(x,y)$. The TA $A_{ch_i(x,y)} = (\sigma, \{\top, \bot, q_{ch}\}, \{\top\}, \delta)$ with the following rules in δ recognizes $\mathrm{R}_{ch_i(x,y)}$:

$$\underbrace{(\bot,\ldots,\bot)}_{ar(a)} \xrightarrow{(a,0,0)} \bot \text{ for all } a \in \Sigma$$

$$\underbrace{(\bot,\ldots,\bot)}_{ar(a)} \xrightarrow{(a,0,1)} q_{ch} \text{ for all } a \in \Sigma$$

$$\underbrace{(q_1,\ldots,q_n)}_{ar(a)} \xrightarrow{(a,1,0)} \top \text{ for all } a \in \Sigma, \text{ with } q_i = q_{ch} \text{ and } q_j = \bot \text{ for all } j \neq i$$

$$\underbrace{(q_1,\ldots,q_n)}_{ar(a)} \xrightarrow{(a,0,0)} \top \text{ for all } a \in \Sigma, \text{ whenever } q_j = \top \text{ for exactly one } j$$

- $\phi(\vec{x},\vec{X}) = \phi'(\vec{x},\vec{X}) \wedge \phi''(\vec{x},\vec{X})$. By definition, $\mathrm{R}_{\phi(\vec{x},\vec{X})} = \mathrm{R}_{\phi'(\vec{x},\vec{X})} \cap \mathrm{R}_{\phi''(\vec{x},\vec{X})}$, so by induction hypothesis and Proposition 4.6, R_{ϕ} is recognizable.*
- $\phi(\vec{x},\vec{X}) = \neg\phi'(\vec{x},\vec{X})$. Here, $ovl(\mathrm{R}_{\neg\phi'(\vec{x},\vec{X})}) = L_{\text{valid}(\vec{x},\vec{X})} \setminus \mathcal{L}(\phi(\vec{x},\vec{X}))$, where $L_{\text{valid}(\vec{x},\vec{X})}$ is the set of trees $t * \mu$ where $t \in \mathcal{T}_\sigma$ and μ is an assignment of variables $\vec{x},\vec{X}$. In other terms, $L_{\text{valid}(\vec{x},\vec{X})}$ is the set of trees over $\Sigma \times \mathbb{B}^{|\vec{x}|+|\vec{X}|}$ for which there is exactly one 1-node on each of the $\vec{x}$ components. The recognizability of $L_{\text{valid}(\vec{x},\vec{X})}$ is trivial. By induction hypothesis, Proposition 4.2 and Corollary 4.1, $\mathrm{R}_{\neg\phi'(\vec{x},\vec{X})}$ is recognizable.
- $\phi(\vec{x},\vec{X}) = \exists x\ \phi'(\vec{x}',\vec{X}')$ or $\phi(\vec{x},\vec{X}) = \exists X\ \phi'(\vec{x}',\vec{X}')$. We assume that $(\vec{x}',\vec{X}')$ is obtained from $(\vec{x},\vec{X})$ by inserting x (resp. X) at the ith position: If this is not the case, we add a cylindrification step. Then $\mathrm{R}_{\phi(\vec{x},\vec{X})} = \Pi_i(\mathrm{R}_{\phi'(\vec{x}',\vec{X}')})$, and this relation is recognizable by induction hypothesis and Proposition 4.4.
- $\phi(x,X) = x \in X$. Then $ovl(\mathrm{R}_{\phi(x,X)})$ is the language of trees with symbols in $\Sigma \times \mathbb{B} \times \mathbb{B}$, such that the first Boolean component has exactly one node labeled by 1, and for this node, the second component must also be labeled by 1 (but without restrictions for other nodes). This language is recognized by the tree automaton $(\sigma, \{q\}, \{q\}, \delta)$ where δ is the set of rules $\underbrace{(q,\ldots,q)}_{ar(a)} \xrightarrow{(a,\beta_1,\beta_2)} q$ for $a \in \Sigma$ and $\beta_1, \beta_2 \in \mathbb{B}$ such that $\beta_1 \Rightarrow \beta_2$.

This concludes the proof of Theorem 4.3. □

4.4 Tree Automata for Unranked Trees

We now investigate unranked trees, a family of tree structures more related to applications like XML and program verification. Most results on ranked trees can be lifted to unranked trees.

4.4.1 Unranked Trees and Hedges

Unranked trees are similar to ranked ones, but their symbols have no fixed arity. Hence, an *unranked alphabet* (called alphabet in the sequel) is a finite set Σ of symbols, and the set of *unranked trees over* Σ is the least set $\mathcal{T}^u_\Sigma$ such that if $t_1,\ldots,t_n \in \mathcal{T}^u_\Sigma$ (with $n \geq 0$) and $a \in \Sigma$, then $a(t_1,\ldots,t_n) \in \mathcal{T}^u_\Sigma$. Note that n is arbitrary, but children are ordered.

* We assume here that ϕ' and ϕ'' have the same free variables, but in the general case we can rearrange components using cylindrification (Proposition 4.5).

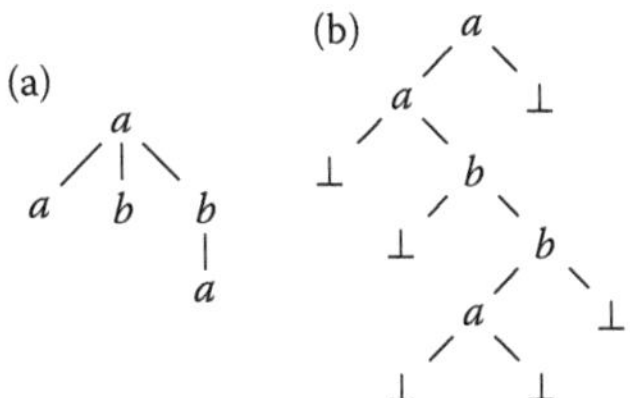

FIGURE 4.5 The *fcns* (first-child next-sibling) encoding. (a) A tree $t \in \mathcal{T}^u_\Sigma$ and (b) *fcns*(t).

Like for ranked trees, an unranked tree $t \in \mathcal{T}^u_\Sigma$ can be considered as a prefix-closed set of nodes N_t together with a labeling function λ_t, inductively defined by: $N_{a(t_1,\ldots,t_n)} = \{\epsilon\} \cup \bigcup_{1\leq i \leq n} i{\cdot}N_{t_i}$, and $\lambda_{a(t_1,\ldots,t_n)}(\epsilon) = a$ and $\lambda_{a(t_1,\ldots,t_n)}(i{\cdot}\pi) = \lambda_{t_i}(\pi)$.

A *hedge* over an unranked alphabet Σ is a sequence of trees $(t_1, \ldots, t_k)$ with $k \geq 0$, and $t_i \in \mathcal{T}^u_\Sigma$ for $1 \leq i \leq k$. The set of hedges over Σ is written $\mathcal{H}_\Sigma$. Note that the hedge (t) is different from the tree t. We will sometimes consider the empty hedge (), but never the empty tree. Hedges are also known as *forests* in the literature.

4.4.2 Tree Automata over Encodings

While unranked trees are more related to hedges than ranked trees, it is useful to consider transforming unranked trees to ranked ones, in order to transfer theoretical results.

Rabin's first-child next-sibling encoding *fcns* (Rabin, 1969) translates trees over an unranked alphabet Σ to ranked trees over the ranked alphabet $\sigma_\perp = (\Sigma \uplus \{\perp\}, ar_\perp)$ where $ar_\perp(\perp) = 0$ and $ar_\perp(a) = 2$, for $a \in \Sigma$. This encoding is defined by the following rules, and illustrated in Figure 4.5b. For convenience, we first encode hedges into binary trees using $fcns_\mathcal{H}$:

$$fcns_\mathcal{H}(()) = \perp$$
$$fcns_\mathcal{H}((a(t'_1, \ldots, t'_m), t_2, \ldots, t_k)) = a(fcns_\mathcal{H}(t'_1, \ldots, t'_m)\,,\, fcns_\mathcal{H}((t_2, \ldots, t_k)))$$

Then, we simply use $fcns_\mathcal{H}$ on unary hedges: for $t \in \mathcal{T}^u_\Sigma$, $fcns(t) = fcns_\mathcal{H}((t))$. We extend the notation to languages $L \in \mathcal{T}^u_\Sigma$ of unranked trees: $fcns(L) = \{fcns(t) \mid t \in L\}$.

A first way to define automata recognizing languages of unranked trees is to use tree automata (for ranked trees) over encodings: A language $L \in \mathcal{T}^u_\Sigma$ of unranked trees is *recognizable* if *fcns*(L) is recognizable.

Example 4.8:

Consider the language L_0 of unranked trees over $\Sigma = \{a, b\}$ such that the root and all first-children are a-nodes, all right siblings of a-nodes are b-nodes, and b-nodes are leaves. In other terms, L_0 is the least set containing a, and for which whenever $t \in L_0$, then $a(t, \underbrace{b, \ldots, b}_{n}) \in L_0$ for every $n \geq 0$. L_0 is recognizable since $fcns(L_0) = \mathcal{L}(A_0)$ for the TA $A_0 = (\sigma_\perp, \{q_\perp, q_a, q_b\}, \{q_a\}, \delta)$ where the seven rules in δ are: $() \xrightarrow{\perp} q_\perp$, $(q, q') \xrightarrow{a} q_a$ and $(q_\perp, q') \xrightarrow{b} q_b$ with $(q, q') \in \{q_\perp, q_a\} \times \{q_\perp, q_b\}$.

Proposition 4.7:

Recognizable languages of unranked trees are closed under union, intersection, and complement.

Proof. Immediate from the definition and Propositions 4.2, 4.3, and Corollary 4.1. □

Note that a second encoding, namely the curryfication of terms, can be used instead of the *fcns* encoding. Curryfication is based on a binary operator on unranked trees, adding a tree as last child of the root. *Stepwise tree automata* (Carme et al. 2004) are exactly tree automata over this encoding.

4.4.3 Hedge Automata

An unranked tree $a(t_1, \ldots, t_k)$ is defined as a label $a \in \Sigma$ and a hedge $(t_1, \ldots, t_k) \in \mathcal{H}_\Sigma$. This leads to a second kind of automata, namely hedge automata (Brüggemann-Klein et al. 2001). These automata are inspired by the syntax of Document Type Definitions (DTDs), used in XML in order to specify schemas, that is, sets of trees that are considered as well-formed for an application. Roughly speaking, a DTD is a set of rules: $a \to e$ assigning for each label, $a \in \Sigma$ a regular expression, e specifying which strings can be formed by labels of children of an a-node. For instance, $a \to b.c^*$ indicates that the first child of an a-node is always labeled by b, and the other (potential) children are c-nodes.

A *hedge automaton* is a tuple (Σ, Q, F, δ) where Σ is an unranked alphabet, Q a finite set of states, $F \subseteq Q$ the set of final states, and δ a set of rules of the form $L \xrightarrow{a} q$ with $a \in \Sigma$, $q \in Q$, and $L \subseteq \Sigma^*$ is a regular (word) language over Q. A run of a hedge automaton $A = (\Sigma, Q, F, \delta)$ on an unranked tree $t \in \mathcal{T}^u_\Sigma$ is a function $\rho : N_t \to Q$ such that, for every node $\pi \in N_t$, there is a rule $L \xrightarrow{\lambda_t(\pi)} \rho(\pi) \in \delta$ such that $\rho(\pi{\cdot}1) \cdots \rho(\pi{\cdot}k) \in L$, where k is the number of children of π. A run ρ is accepting if $\rho(\epsilon) \in F$. A tree t is accepted by a hedge automaton A if there exists an accepting run of A on t. The language $\mathcal{L}(A)$ of a hedge automaton A is its set of accepted trees.

Example 4.9:

The language L_0 defined in Example 4.8 is recognized by the hedge automaton $(\Sigma, \{q_a, q_b\}, \{q_a\}, \delta)$ where δ is composed of two rules: (i) $L_a \xrightarrow{a} q_a$ and $L_\emptyset \xrightarrow{b} q_b$ where $L_\emptyset$ is the empty language and (ii) L_a is the set of words satisfying the regular expression $\epsilon + q_a{\cdot}q_b^*$. Both $L_\emptyset$ and L_a are regular.

The notion of recognizability of hedge automata is the same as recognizability using tree automata over *fcns* encoding.

Proposition 4.8:

A language of unranked trees is recognizable iff it is accepted by a hedge automaton.

The proof, left as exercise, relies on the effective translation of hedge automata into tree automata over *fcns* encoding, and conversely. Hence, hedge automata benefit from the same closure properties. We have seen that, for DTDs, recognizable languages L used in rules were specified by regular expressions. They could also be defined by other means, as for instance finite state (word) automata.

4.4.4 Pushdown Automata

The third kind of models of automata for unranked trees comes from the effective input of such automata. For most applications like XML documents (resp. program execution flows), the input is not a tree but a sequence of opening and closing tags <a> and </a> (resp. calls and returns). Such a sequence corresponds to the pre-order traversal of the corresponding tree: it can be seen as a well-nested sequence of Σ-typed parentheses. This shows the strong links with Dyck languages and balanced grammars (Berstel and Boasson, 2002). In terms of automata, these languages are context-free, and thus can be recognized using pushdown automata. However, such automata are much more expressive than needed, and do not benefit from nice closure properties. Instead, restricted models appeared in order to recover closure properties. The dominant model among them is the visibly pushdown automaton (VPA) introduced in Alur and Madhusudan (2004).* The term "visibly" denotes a partition of the input symbols Σ into two separate sets: calls and returns.† In the XML terminology, these correspond to opening tags <a> and closing tags </a>. A prior and similar model is the pushdown forest automaton (Neumann and Seidl, 1998). Streaming tree automata are the specialization of VPAs to trees (Gauwin et al., 2008).

A VPA is a tuple $(\Sigma, \Gamma, Q, I, F, \delta)$ where Σ is a finite set of symbols, partitioned into call symbols Σ_c and return symbols Σ_r, Γ is a finite set of stack symbols, Q is a finite set of states, including initial states $I \subseteq Q$ and final states $F \subseteq Q$, and δ is a set of rules $q \xrightarrow{a:\gamma} q'$ with $q, q' \in Q$, $a \in \Sigma$ and $\gamma \in \Gamma$.

The set of *well-nested words* over the partitioned alphabet $\Sigma = \Sigma_c \uplus \Sigma_r$ is the least set containing the empty word ϵ and words uu' and cur whenever u and u' are well-nested, $c \in \Sigma_c$ and $r \in \Sigma_r$. A *run* of a VPA over a well-nested word $w \in \Sigma^*$ is a pair $(q, q') \in Q^2$ inductively defined as follows: (1) if $w = uu'$ with u, u' well-nested words, then there exists q'' such that (q, q'') is a run on u and (q'', q') is a run on u'; and (2) if $w = cur$ with $c \in \Sigma_c$, $r \in \Sigma_r$ and u well-nested, then there exists a run (q_u, q'_u) on u and $\gamma \in \Gamma$ such that $q \xrightarrow{c:\gamma} q_u \in \delta$ and $q'_u \xrightarrow{r:\gamma} q' \in \delta$. A run is accepting if it belongs to $I \times F$. A well-nested word w is accepted by a VPA if there is a run of this VPA on this word. The set of accepted well-nested words of a VPA A is denoted $\mathcal{L}(A)$.

In order to use VPAs as (unranked) tree acceptors, we have to convert unranked trees into well-nested words. From an unranked alphabet Σ, we define the partitioned alphabet $\Sigma_c \uplus \Sigma_r$ where $\Sigma_c = \{a_c \mid a \in \Sigma\}$ and $\Sigma_r = \{a_r \mid a \in \Sigma\}$. The *linearization* of $t \in \mathcal{T}^u_\Sigma$ is the well-nested word $lin(t)$ inductively defined by: $lin(a(t_1, \ldots, t_n)) = a_c \cdot lin(t_1) \cdots lin(t_n) \cdot a_r$. We extend this notation to tree languages: for $L \subseteq \mathcal{T}^u_\Sigma$, $lin(L) = \{lin(t) \mid t \in L\}$.

Example 4.10:

The language $lin(L_0)$, with L_0 defined in Example 4.8, is recognized by the VPA $(\Sigma, \{\gamma\}, \{q_0, q_1\}, \{q_0\}, \{q_0\}, \delta)$ where δ is the set of rules: $q_0 \xrightarrow{a_c:\gamma} q_0$, $q \xrightarrow{b_c:\gamma} q_0$, $q \xrightarrow{a_r:\gamma} q_0$, and $q_0 \xrightarrow{b_r:\gamma} q_1$ with $q \in \{q_0, q_1\}$. State q_0 is used for calls, and for a-returns, while state q_1 is only used for b-returns.

Proposition 4.9:

A language $L \subseteq \mathcal{T}^u_\Sigma$ of unranked trees is recognizable iff $lin(L) = \mathcal{L}(A)$ for some VPA A over the partitioned alphabet $\Sigma_c \uplus \Sigma_r$.

The proof is by back-and-forth translation between both automata models, as detailed for instance in Gauwin (2009). Hence, VPAs (as tree acceptors) are closed under Boolean operations. Additionally, all VPAs can be determinized.

* See also the similar model of nested word automaton (Alur and Madhusudan, 2009).

† The original definition relies on a three-sets partition by adding internal symbols. They are, however, useless when using VPAs on trees. For the same reason, we only define runs over balanced well-nested words.

4.4.5 Logics over Unranked Trees

4.4.5.1 Recognizable Relations

Recognizable (unranked tree) relations can be defined in a way very similar to the ranked case. The overlay operator $\circledast$ is extended to unranked trees in the same way, that is, by a leftmost overlap of nodes. Then, relations built over a first-order formula where relation symbols are interpreted by recognizable relations are still recognizable. We refer the reader to Gauwin (2009), Benedikt et al. (2007) for further details.

4.4.5.2 MSO

The MSO logic based on predicates lab_a (for $a \in \Sigma$), fc (first child) and ns (next sibling) exactly captures recognizable (unranked) tree languages. More precisely, a MSO[Λ_u] formula is obtained using grammar:

$$\phi ::= lab_a(x) \mid fc(x,y) \mid ns(x,y) \mid \phi \wedge \phi' \mid \neg\phi \mid \exists x\ \phi \mid \exists X\ \phi \mid x \in X$$

where $a \in \Sigma$, and x, y (resp. X) belong to an infinite countable set of first-order (resp. second-order) variables. The semantic of such a formula is defined like formulas in MSO[Λ], except for binary predicates:

$$t, \mu \models fc(x,y) \text{ iff } \mu(y) = \mu(x)\cdot 1$$
$$t, \mu \models ns(x,y) \text{ iff } \mu(x) = \pi\cdot i \text{ and } \mu(y) = \pi\cdot(i+1) \text{ for some } \pi \in N_t \text{ and } i > 0$$

An unranked tree language $L \subseteq \mathcal{T}^u_\Sigma$ is said *regular* iff there exists a closed formula $\phi \in$ MSO[Λ_u] such that $L = \{t \mid t \models \phi\}$.

Example 4.11:

The language L_0, as defined in Example 4.8, is regular. It is the set of models of the formula $\phi_0 = \forall x\ (lab_a(x) \Rightarrow fc_a(x) \wedge ns_b(x)) \wedge (lab_b(x) \Rightarrow leaf(x) \wedge ns_b(x))$ where $fc_a(x) = \forall y\ fc(x,y) \Rightarrow lab_a(y)$ and $ns_b(x) = \forall y\ ns(x,y) \Rightarrow lab_b(y)$, and $leaf(x) = \neg\exists y\ fc(x,y)$.

Theorem 4.4:

An unranked tree language is regular iff it is recognizable.

Proof. By definition, an unranked tree language L is recognizable iff $fcns(L)$ is. By Theorem 4.3, this is equivalent to the existence of an MSO[Λ] formula ϕ such that $t \in fcns(L)$ iff $t \models \phi$. Trees in $fcns(L)$ are binary, so ϕ only uses binary predicates ch_1 and ch_2. It is possible (but technical) to translate ϕ into an MSO[Λ_u] formula ϕ_u such that $t \models \phi_u$ iff $fcns(t) \models \phi$. Conversely, an MSO[Λ_u] formula ϕ_u can be translated to an MSO[Λ] formula ϕ such that $t \models \phi$ iff $t = fcns(t')$ with $t' \models \phi_u$. □

4.5 Further Readings

This chapter mainly focuses on the links between tree automata and logic. The reference book (Comon et al. 2007) provides a large overview of results related to tree automata. In particular, tree automata benefit from a pumping lemma, a minimization algorithm based on a Myhill–Nerode theorem, an efficient algorithm for testing emptiness, equivalent forms of regular expressions and grammars. These properties hold for ranked trees, but not all of them hold for unranked trees. In particular, VPAs do not have a unique minimal equivalent dVPA, and pumping arguments can be applied vertically and horizontally.

Some surveys worth to be mentioned here: about logics over trees (Libkin, 2006), and automata and XML (Schwentick, 2007).

Tree automata described in this chapter are all expressively equivalent to regular sets of trees. Other automata over finite trees with different expressiveness also exist, as for instance tree-walking automata (Aho and Ullmann, 1971; Bojańczyk, 2008) and automata over data trees (Segoufin, 2006). These specific automata aim at capturing features of the XPath query language (and related modal logics), like data values comparison.

References

A. V. Aho and J. D. Ullmann. Translations on a context-free grammar. *Information and Control*, 19: 439–475, 1971.

R. Alur and P. Madhusudan. Visibly pushdown languages. In *36th ACM Symposium on Theory of Computing*, pp. 202–211. ACM Press, New York, NY, 2004.

R. Alur and P. Madhusudan. Adding nesting structure to words. *Journal of the ACM*, 56(3):1–43, 2009.

M. Benedikt, L. Libkin, and F. Neven. Logical definability and query languages over ranked and unranked trees. *ACM Transactions on Computational Logics*, 8(2), April 2007.

J. Berstel and L. Boasson. Balanced grammars and their languages. In *Formal and Natural Computing*, Vol. 2300 of *Lecture Notes in Computer Science*, pp. 3–25. Springer-Verlag, New York, NY, 2002.

M. Bojańczyk. Tree-walking automata. Tutorial at LATA'08, 2008.

T. Bray, J. Paoli, C. M. Sperberg-McQueen, E. Maler, and F. Yergeau. Extensible Markup Language (XML) 1.0 (Fifth Edition), November 2008. Available at: http://www.w3.org/TR/2008/REC-xml-20081126/.

A. Brüggemann-Klein, D. Wood, and M. Murata. Regular tree and regular hedge languages over unranked alphabets: Version 1, April 07 2001.

J. Carme, J. Niehren, and M. Tommasi. Querying unranked trees with stepwise tree automata. In *19th International Conference on Rewriting Techniques and Applications*, Vol. 3091 of *Lecture Notes in Computer Science*, pp. 105–118. Springer-Verlag, New York, NY, 2004.

H. Comon, M. Dauchet, R. Gilleron, C. Löding, F. Jacquemard, D. Lugiez, S. Tison, and M. Tommasi. Tree automata techniques and applications, October 2007. Available at http://people.cs.aau.dk/~srba/courses/FS-07/tata.pdf

J. E. Doner. Tree acceptors and some of their applications. *Journal of Computer and System Science*, 4:406–451, 1970.

O. Gauwin, J. Niehren, and Y. Roos. Streaming tree automata. *Information Processing Letters*, 109(1):13–17, December 2008.

O. Gauwin. *Streaming Tree Automata and XPath.* PhD thesis, Université Lille 1, 2009.

L. Libkin. Logics over unranked trees: An overview. *Logical Methods in Computer Science*, 3(2):1–31, 2006.

A. Neumann and H. Seidl. Locating matches of tree patterns in forests. In *Foundations of Software Technology and Theoretical Computer Science*, Vol. 1530 of *Lecture Notes in Computer Science*, pp. 134–145. Springer-Verlag, New York, NY, 1998.

M. O. Rabin. Decidability of second-order theories and automata on infinite trees. *Transactions of the American Mathematical Society*, 141:1–35, 1969.

T. Schwentick. Automata for XML—A survey. *Journal of Computer and System Science*, 73(3):289–315, 2007.

L. Segoufin. Automata and logics for words and trees over an infinite alphabet. In Z. Ésik, editor, *Computer Science Logic*, Vol. 4207 of *Lecture Notes in Computer Science*, pp. 41–57. Springer, Berlin, Germany, 2006.

J. W. Thatcher and J. B. Wright. Generalized finite automata with an application to a decision problem of second-order logic. *Mathematical System Theory*, 2:57–82, 1968.

5

Timed Automata

Jun Sun, Yang Liu, and
Jin Song Dong
National University of Singapore

5.1 Introduction

Real-world systems often depend on quantitative timing. Real-time systems differ from untimed systems in that their behavioral correctness relies not only on the results of their computations but also on when the results are produced. For instance, a pacemaker should not only function in the detection of an abnormal heart condition but it must also react within a small time framework. Modeling and verification of real-time systems are important research topics which have practical implications. System modeling is of great importance as well as highly nontrivial. The choice of modeling language is an important factor in the success of the entire system analysis or development. The language should cover several facets of the requirements and the model should reflect exactly (up to abstraction of irrelevant details) an existing system or a system to be built. The language should have a semantic model suitable to study the behaviors of the system and to establish the validity of desired properties. The latter is extremely important for real-time systems in order to avoid undecidability in system verification.

A number of modeling languages have been proposed for real-time systems, for example, Timed Automata [AD94; LV96], Algebra of Timed Processes [NS94], Timed Calculus of Communicating Systems [Yi91], Timed Communicating Sequential Processes [Sch00], and so on. Though each language has its unique characteristics and thus suits for one particular class of systems, research on model checking real-time systems have been mostly focused on Timed Automata or a variant named Timed Safety Automata [HNSY94]. Several model checkers have been developed with Timed Automata being the core of their input languages [LPW97; BDM$^+$98; TAKB96]. Timed Automata-based model checkers have been successfully applied to industry systems [HSLL97; LMNS05].

In this chapter, we introduce Timed Automata and its extensions. The rest of the chapter is organized as follows. Section 5.2 defines the syntax and operational semantics of Timed Automata. Section 5.3 shows how to verify Timed Automata models. In particular, we show two ways of abstracting Timed Automata which imply decidability of reachability analysis. Section 5.4 presents an approach to extend original Timed Automata so as to model hierarchical systems, that is, timed patterns. Section 5.5 then shows an alternative approach for modeling and verifying real-time systems.

5.2 Timed Automata Syntax and Semantics

In the original theory of Timed Automata [AD94], a Timed Automaton is a finite-state Büchi automaton equipped with a finite set of real-valued clocks. Timed Automata are powerful in modeling real-time systems by explicitly manipulating clock variables. Real-time system behaviors are captured by clock constraints on transitions and setting or resetting clocks. Progress properties in Timed Automata are enforced by Büchi acceptance conditions. That is, a set of the states are marked as accepting and only those system executions which visit an accepting state infinitely are considered valid. Later, an intuitive *local* notion of progress is introduced in Timed Safety Automata [HNSY94]. In Timed Safety Automata, an automaton is allowed to stay in a state only if the local invariants associated with the state is satisfied. Due to its intuitiveness, Timed Safety Automata has been adopted by several tools, including the popular UPPAAL model checkers [LPW97]. In the following, we follow the literature and refer them as *Timed Automata.*

Let $\mathbb{R}_+$ be the set of nonnegative real numbers. For a set of clocks C, the set $\Phi(C)$ of clock constraint δ is defined inductively by $\delta := x \sim n | \neg\delta | \delta_1 \wedge \delta_2$, where $\sim \in \{=, \leq, \geq, <, >\}$; x and y are clocks in C, and $n \in \mathbb{R}_+$ is a constant. Without loss of generality (Lemma 4.1 of [AD94]), we assume that n is an integer constant. The set of downward closed constraints obtained with $\sim \in \{<, \leq\}$ is denoted as $\Phi_{\leq,<}(C)$. A clock valuation v for a set of clocks C is a function which assigns a real value to each clock. A clock valuation v satisfies a clock constraint δ, written as $\delta \models v$, if and only if δ evaluates to true using the clock values given by v. For $n \in \mathbb{R}_+$, let $v + d$ denote the clock valuation v' such that $v'(c) = v(c) + d$ for all $c \in C$. For $X \subseteq C$, let $[X \mapsto 0]v$ denote the valuation v' such that $v'(c) = v(c)$ for all $c \notin X$ and $v'(x) = 0$ for all $x \in X$.

Definition 5.1: (Timed Automata)

A Timed Automaton $\mathcal{A}$ is a tuple $(S, I, \Sigma, C, L, \rightarrow)$, where S is a finite set of states; $I \subseteq S$ is a set of initial states; Σ is an alphabet; C is a finite set of clocks; $L : S \rightarrow \Phi_{\leq,<}(C)$ is a function which associates an invariant with each state; $\rightarrow: S \times \Sigma \times \Phi(C) \times 2^C \times S$ is a labeled transition relation. □

For simplicity, we write $s \overset{e,\delta,X}{\rightarrow} s'$ to denote that $(s, e, \delta, X, s') \in\rightarrow$. Intuitively, the transition is fired only if δ is satisfied and e occurs. Afterwards, clocks in X are reset to zero. Timed Automata are often presented graphically, as shown in the following example.

Example 5.1:

Figure 5.1 presents a Timed Automaton graphically. The Timed Automaton models a railcar door. For the sake of readability, the states are labeled with names, for example, *Open*, *Close*, and *ToOpen*. We

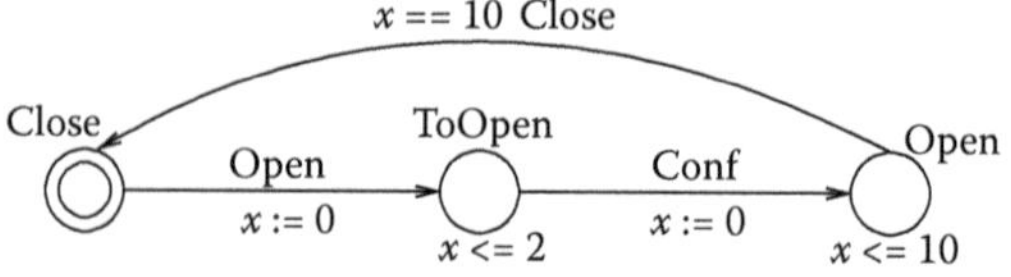

FIGURE 5.1 An example Timed Automaton.

assume the door can be closed instantly and hence there is no state named *ToClose*. Initially, the control is at state *Close*, indicated by the double-lined circle. The transition is fired once event *open* occurs (note that the transition is unguarded), the control goes to state *ToOpen* and clock x is reset. Within 2 time units (constrained by the state invariant), the control goes to the state *Open*. Event *conf* takes place along with the transition. The control remains at state *Open* for 10 time units (i.e., passengers have 10 time units for boarding) before control goes to state *Close*. □

A configuration of a Timed Automaton is a pair (s, v), where $s \in S$ is a state and v is a clock valuation. Intuitively, s is the state where the control is currently at and v gives the current values of the clocks. Because clock values are real numbers, the number of configurations is always infinite. The operational semantics of a Timed Automaton is an infinite-state labeled transition system.

Definition 5.2: (Semantics of Timed Automata)

The semantics of a Timed Automaton $\mathcal{A} = (S, I, \Sigma, C, L, \rightarrow)$ is a labeled transition system (LTS) $\mathcal{L} = (S_l, I_l, \Sigma \cup \mathbb{R}_+, \rightarrow_l)$ such that S_l is a set of configurations; $I_l \subseteq S_l$ is a set of initial configurations such that: for all $(s, v) \in I_l$, $s \in I$ and $v(c) = 0$ for all $c \in C$; and $\rightarrow_l$: $S_l \times \Sigma \cup \mathbb{R}_+ \times S_l$ is a labeled transition relation defined by the following rules: for all $(s, v) \in S_l$,

- $(s, v) \xrightarrow{n}_l (s, v + n)$, where $n \in \mathbb{R}_+$ if $L(s) \vDash v + n$, that is, elapsing of n time units is allowed as long as the invariant labeled with the state is satisfied.
- $(s, v) \xrightarrow{e}_l (s', [X \mapsto 0]v)$ if $s \xrightarrow{e,\delta,X} s'$ and $v \vDash \delta$ and $[X \mapsto 0]v \vDash I(s')$, that is, a transition is fired if the guard condition is satisfied and the invariant labeled with the target state is satisfied after resetting the clocks. □

Based on the above definition, we define the untimed and timed language of a Timed Automaton. Let $s \overset{a}{\leadsto} s'$ denote that there exists $n \in \mathbb{R}_+$ such that $s \xrightarrow{n}_l s_0$ and $s_0 \xrightarrow{a}_l s'$. An untimed trace of $\mathcal{A}$ is a finite or infinite event sequence $a_0, a_1, \cdots$ such that there exists a sequence of configurations $s_0, s_1, \cdots$ such that $s_i \overset{a_i}{\leadsto} s_{i+1}$ for all i. The untimed language of $\mathcal{A}$ is the set of untimed traces of $\mathcal{A}$. A timed trace of $\mathcal{A}$ is a finite or infinite sequence of pairs $(n_0, a_0), (n_1, a_1), \cdots$ such that there exists a sequence of configurations $s_0, s_1, \cdots$ such that $s_i \xrightarrow{n_i}_l s_0$ and $s_0 \xrightarrow{a}_l s'$ for all i. The timed language of $\mathcal{A}$ is the set of timed traces of $\mathcal{A}$.

5.3 Verifying Timed Automata

Besides its simplicity in system modeling, Timed Automata is well designed for automatic system analysis, in particular, model checking. Properties of Timed Automata models can be stated in many different forms, reachability condition [ABBL03], temporal logic [LP85], timed language inclusion [OW04], and so on. Different properties often lead to different model checking algorithms. In the following, we show how to verify Timed Automata models against one of the most common as well as most important properties reachability, that is, whether a system (in Timed Automata) can reach certain state. A survey on other kinds of properties is given at the end of the section.

For a Timed Automaton $\mathcal{A} = (S, I, \Sigma, C, L, \rightarrow)$ with operational semantics $\mathcal{L} = (S_l, I_l, \rightarrow_l)$, $\mathcal{A}$ reaches a configuration (s, v) if and only if (s, v) is a reachable state in $\mathcal{L}$. Reachability analysis in Timed Automata is decidable. Given a finite-state automaton, reachability can be solved using standard algorithms like depth-first-search or breadth-first-search. This problem is more complex in the setting Timed Automata, however, because $\mathcal{L}$ has infinitely many states, it is not possible to build $\mathcal{L}$ and decide whether a configuration (s, v) is reachable or not. The remedy is to build an abstraction transition system $\mathcal{L}'$ such that not only $\mathcal{L}'$ has only finitely many states but also it preserves reachability. That is, (s, v) is reachable in $\mathcal{L}$ if and only if certain state is reachable in $\mathcal{L}'$ so that reachability can be solved by building $\mathcal{L}'$. Because $\mathcal{L}_r$ only has finitely many states, this implies that a state of $\mathcal{L}'$ groups many or even infinitely many states of $\mathcal{L}$. Two ways of grouping configurations have been studied and applied in the literature.

5.3.1 Region Graph

The first observation [AD94] is that clock valuations in the same *region* are *equivalent* and, therefore, can be grouped. Formally, two clock valuations v and v' are in the same region if and only if

- For all $c \in C$, either the integral part of $v(c)$ and the integral part of $v'(c)$ are the same, or both $v(c)$ and $v'(c)$ are greater than n_c, which is the largest integer n such that $c \sim n$ is subformula of some clock constraint appearing in $\mathcal{A}$
- For all $x, y \in C$ with $v(x) \leq n_x$ and $v(y) \leq n_y$, the fractional part of $v(x)$ is less than or equal to that of $v(y)$ if and only if the fractional part of $v'(x)$ is less than or equal to that of $v'(y)$
- For all $c \in C$ with $v(c) \leq n_c$, the fractional part of $v(c)$ is 0 if and only if that of $v'(c)$ is 0

Given a clock valuation v, let $[v]$ denotes the unique region containing v. By a simple argument, it can be shown that there are only finitely many regions for a finite number of clocks.

Example 5.2:

Assume there are two clocks x and y, and n_x is 1 and n_y is 2. Figure 5.2 shows all the regions, in the forms of corner points (intersections), line segments, and open areas. □

Intuitively, two configurations in the same state and region, for example, (s, v) and (s, v') such that $[v] = [v']$, are *equivalent* because for any clock constraint δ in $\Phi(C)$, $\delta \models v$ if and only if $\delta \models v'$. Thus, instead of building $\mathcal{L}$, a region graph can be built.

Definition 5.3: (Region Graph)

A region graph $\mathcal{R}$ of a Timed Automaton $\mathcal{A} = (S, I, \Sigma, C, L, \rightarrow)$ is a transition system $(S_r, I_r, \rightarrow_r)$ such that S_r is a finite set of regions; $I_r = \{(i, v) | i \in I \wedge \forall c : C.\ v(c) = 0\}$ is the set of the initial regions; and $\rightarrow_r$: $S_r \times S_r$ is a (unlabeled) transition relation defined by the following rule: $(s, [v]) \rightarrow_r (s', [v'])$ if and only if there exists $x \in \Sigma \cup \mathbb{R}_+$ such that $(s, v) \xrightarrow{x} (s', v')$. □

The following theorem establishes the soundness and completeness of reachability analysis based on region graphs. That is, reachability analysis of infinite-state $\mathcal{L}$ is reduced to reachability analysis of finite-state $\mathcal{R}$. The latter can be solved by standard reachability analysis based depth-first-search or breath-first-search.

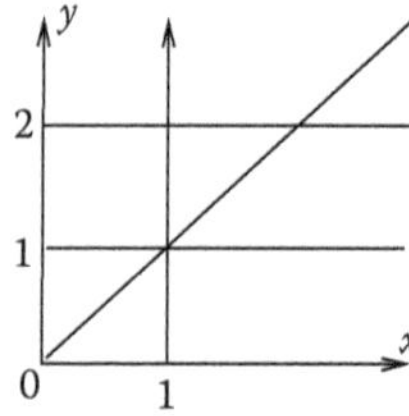

FIGURE 5.2 Regions for a system with two clocks with bound 1 and 2.

Theorem 5.1:

[AD94] For a Timed Automaton $\mathcal{A}$, let $\mathcal{R}$ be its region graph. A configuration (s, v) is reachable in $\mathcal{A}$ if and only if $(s, [v])$ is reachable in $\mathcal{R}$. □

Several other verification problems such as untimed language inclusion (i.e., whether a Timed Automaton allows more untimed traces than the other), language emptiness [AD94] as well as timed bisimulation [Cer92] can be solved based on region graphs.

5.3.2 Zone Graph

However, the problem with region graphs is the potential explosion in the number of regions, which is *exponential* in the number of clocks as well as the maximal constants appearing in the guards of $\mathcal{A}$. A more efficient way of grouping configurations is based on the notion of *zones* and *zone graphs*.

A zone is a clock constraint, or strictly speaking, the maximal set of clock valuations satisfying a clock constraint. In an abuse of notations, zones and clock constraints are used interchangeably in the following. In practice, zones give a coarser and thus more compact abstraction of $\mathcal{L}$. An abstraction configuration with zones is a pair (s, δ) such that $s \in S$ is a state and δ is a zone. Let $\delta^{\uparrow}$ denotes the zone $\{v + d | \delta \models v \wedge d \in \mathbb{R}_{+}\}$, that is, the zone obtained by arbitrary delay. Let $[X \mapsto 0]\delta$ denote the the zone $\{[X \mapsto 0]v | \delta \models v\}$, that is, the zone obtained by resetting clocks in X.

Definition 5.4: **(Zone Graph)**

A zone graph $\mathcal{Z}$ of a Timed Automaton $\mathcal{A} = (S, I, \Sigma, C, L, \rightarrow)$ is a transition system $(S_z, I_z, \rightarrow_z)$ such that S_z is a set of abstract configurations; $I_z = \{(i, \bigwedge_{c \in C} c = 0) | i \in I\}$ is the set of the initial abstract configurations; and $\rightarrow_z$: $S_z \times S_z$ is a (unlabeled) transition relation defined by the following rules:

- $(s, z) \rightarrow_z (s, z^{\uparrow} \wedge L(s))$ if $z^{\uparrow} \wedge L(s)$ is not false. That is, the automaton can delay at state s as long as its invariant is satisfied.
- $(s, z) \rightarrow_z (s', [X \mapsto 0](z \wedge \delta) \wedge L(s'))$ if $s \xrightarrow{e,\delta,X} s'$ and $[X \mapsto 0](z \wedge \delta) \wedge L(s')$ is not false. That is, the automaton can fire a transition as long as the guard is satisfied and after resetting the clocks, the invariant of the target state is satisfied. □

In order to build zone graphs, it is necessary to *efficiently* determine whether two zones are equivalent or a zone is false or not. The former allows us to decide whether an abstract configuration has been visited before during reachability analysis based on zone graph. The latter allows us to decide whether an abstract configuration is feasible or not. This is achieved by Difference Bound Matrices (DBM). It is well known that a zone can be equivalently represented as a DBM [Dil89; BLP+99]. Let $c_1, c_2, \cdots, c_n$ denote n clocks. Let c_0 denotes a dummy clock whose value is always 0. A DBM representing a clock constraint contains $n+1$ rows, each of which contains $n+1$ elements. Entry (i, j) in the matrix, denoted by D^i_j, represents the upper bound on difference between clock c_i and c_j, i.e., $c_i - c_j \leq D^i_j$. A DBM, thus, represents the following constraint:

$$\forall i : 0 \mathinner{..} n.\ \forall j : 0 \mathinner{..} n.\ c_i - c_j \leq D^i_j$$

The bound on difference between c_i and c_j is captured by $-D^j_i \leq t_i - t_j \leq D^i_j$. Because c_0 is always 0, we have $-D^0_i \leq c_i \leq D^i_0$ which gives the bound on value of clock c_i. In the following, we briefly introduce the

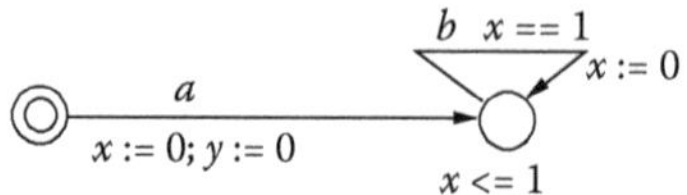

FIGURE 5.3 A Timed Automaton generating infinitely many zones.

relevant zone operations/properties and its corresponding DBM implementation. Interested readers can refer to [Dil89; BLP+99; BY03] for details on DBM.

- *Zone equivalence:* In theory, there are infinitely many different clock constraints representing the same zone. For instance, the clock valuations for $0 \leq t_1 \leq 3 \wedge 0 \leq c_1 - c_2 \leq 3$ and $0 \leq c_1 \leq 3 \wedge 0 \leq c_1 - c_2 \leq 3 \wedge c_2 \leq 1000$ are exactly the same and hence they represent the same zone. Zones represented as DBMs can be systematically compared if they are in their *canonical* forms. A DBM is said to be in its canonical form if and only if every entry D_j^i is the tightest bound on difference between clock c_i and c_j. An important property of DBM is that there is a relatively efficient procedure to compute a unique canonical form. If the clocks are viewed as vertices in a weighted graph and the clock difference as the label on the edge connecting two clocks, the tightest clock difference is the shortest path between the respective vertices. Floyd–Warshall algorithm [Flo62], thus, can be used to compute the tightest bound on clock differences and hence the canonical form. Note that the complexity of Floyd–Warshall algorithm is cubic in the number of clocks and, therefore, it is desirable to reduce the number of clocks.
- *Zone satisfiability:* It is essential to check whether a zone is empty or not. A zone is empty if and only if its DBM representation, in its canonical form, contains an entry D_i^i such that $D_i^i < 0$. Intuitively, it means that clock c_i is constrained to satisfy $c_i - c_i < 0$, which is impossible. Furthermore, it can be shown that a DBM in its canonical form represents an empty zone if and only if D_0^0 is negative.

Further, in order to build the zone graph, it is necessary to show that the number of zones is finite. Unfortunately, this is not true. Figure 5.3 shows a Timed Automaton which has an infinite zone graph. Intuitively, this is because the value of clock y is unbounded. However, observe that zone $x = 1 \wedge y = 1$ is equivalent to zone $x = 1 \wedge y = 2$ (in fact, it is equivalent to zone $x = 1 \wedge y = n$ for any n) for this automaton. Intuitively, this is because once the valuation of a clock (or the differences between two clocks) is larger than n_c, then its exact value becomes irrelevant. This leads to the idea of *zone normalization* [Rok93]. Given a clock constraint represented by a DBM, first the canonical form of the DBM is computed. Next, the canonical DBM is updated as follows so as to get a normalized zone: for all entry D_j^i, if $D_j^i > n_{c_i}$, then set D_j^i to be n_{c_i}. It can be shown that there are only finitely many normalized zones [Rok93]. The zone graph with zones replaced by its normalization is called a normalized zone graph.

Lastly, it is necessary to show that it is sound and complete to perform reachability analysis based on normalized zone graph, which is stated in the following theorem.

Theorem 5.2:

[Pet99] For a Timed Automaton $\mathcal{A}$, let $\mathcal{Z}$ be its normalized zone graph. A configuration (s, v) is reachable in $\mathcal{A}$ if and only if (s, δ), where $\delta \models v$ is reachable in $\mathcal{Z}$.

Example 5.3:

Figure 5.4 shows a simple Timed Automaton with an unreachable state. The following shows the DBMs which represents the clock constraint $x \leq 3 \wedge x = 5$. The one on the right is in its canonical form. Because D_0^0 is negative, the DBM is false and, therefore, the right-most state is not reachable.

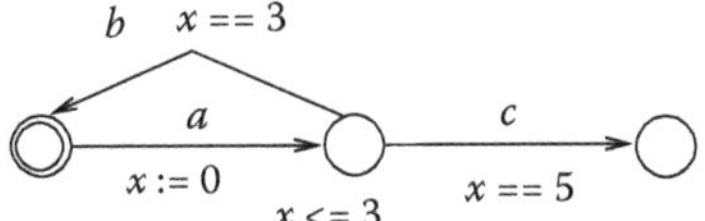

FIGURE 5.4 A simple Timed Automaton with an unreachable state.

	c_0	x
c_0	0	−5
x	3	0

=

	c_0	x
c_0	−1	−5
x	3	−2

□

An alternative way of specifying a property is through a finite-state automaton. The verification problem is then whether the set of untimed event sequences of a Timed Automaton is a subset of those of the finite-state automaton. It has been proved that the untimed traces of a Timed Automaton are preserved in its region graph [AD94] or zone graph [Dil89]. It follows then that untimed language inclusion checking between two Timed Automata are decidable [AD94]. There are, however, some negative results on Timed Automata. Timed language inclusion checking, that is, whether the timed language of an implementation Timed Automaton is a subset of those of a specification Timed Automaton, is undecidable [AD94]. Unlike finite-state automata, Timed Automata can not be determinized or complemented, that is, the complement may not be a Timed Automaton. There have been much research interest in finding the subset of Timed Automata which makes timed language inclusion checking decidable. For instance, it has been shown that the problem becomes decidable if the specification is deterministic [AD94], or contains only one clock [OW04], or is γ-clock bounded [BBBB09], and so on. The problem of identifying the maximum subset, however, remains an open problem.

5.3.3 Verification Tool Support

Along the theory development, efficient verification tools are developed for Timed Automata. Established ones include CoSPAN [TAKB96], Kronos [BDM$^+$98], and UPPAAL [LPW97]. CoSPAN takes in two Timed Automata (one as model and one as property). The correctness of the model against the property is proved by showing a timed simulation relation [TAKB96] between the model automata and the property automata. Kronos supports the verification of Timed Automata against real-time temporal logic (TCTL). Among them, UPPAAL has been proved to be the state-of-the-art model checker for (extended) Time Automata.

UPPAAL is an integrated tool environment for modeling, simulation, and verification of real-time systems, jointly developed by Uppsala University, Sweden and Aalborg University, Denmark. The tool consists of three main parts: (i) a description language, (ii) a simulator, and (iii) a model-checker. The modeling language supported by UPPAAL is an extension of Timed Automata with additional features, including shared integer variables, urgent channels, and committed states. Variables in UPPAAL have bounded discrete integers, and considered as part of the state. The simulator is a validation tool which enables examination of possible dynamic executions of a system during early design (or modeling) stages and thus provides an inexpensive mean of fault detection prior to verification by the model-checker which covers the exhaustive dynamic behavior of the system. The model-checker in UPPAAL is to check invariant and liveness properties by exploring the state-space of a system, that is reachability analysis in terms of symbolic states represented by constraints. The query language of UPPAAL, used to specify properties to be checked, is a subset of CTL (computation tree logic).

5.4 Timed Automata Patterns

Timed automata are essentially *flat* finite-state machine equipped with clocks. Modeling practical systems, however, often require additional modeling features (often as syntactic sugars) to capture a variety of system features. There have been a number of extension of original Timed Automata. For instance, UPPAAL's input language supports parallel composition of Timed Automata which communicates through synchronous channels or shared variables, urgent or atomic states, diagonal clock constraints (i.e., in the form of $c_1 - c_2 \leq n$), and so on. UPPAAL and its language have been used to model and analyze many real-time systems. Later, Timed Automata is further extended with *timed patterns* so as to model fully hierarchical systems [DHQ+08]. In the following, we present a number of common timed patterns. The first, and arguably most useful, timed pattern is parallel composition [LPW97], where multiple Timed Automata run in parallel. In process algebras literature, various parallel composition operators have been proposed to model different aspects of concurrency, among which the most noticeable is CCS [Mil97] and CSP [Hoa85]. In UPPAAL, the CCS parallel composition operator is adopted. Here we give a different definition such that communication among different automata is through CSP-style event synchronization [Hoa85].

Definition 5.5: (Parallel Composition)

Let $\mathcal{A}_i = (S_i, I_i, \Sigma_i, C_i, L_i, \rightarrow_i)$, where $i \in \{1, 2\}$ be two Timed Automata, where $C_1 \cap C_2 = \emptyset$. Parallel composition of $\mathcal{A}_1$ and $\mathcal{A}_2$, written as $\mathcal{A}_1 \parallel \mathcal{A}_2$, is a Timed Automaton $(S, I, \Sigma, C, L, \rightarrow)$ such that $S = S_1 \times S_2$; $I = \{(i, j) | i\, I_1 \wedge j \in I_2\}$; $\Sigma = \Sigma_1 \cup \Sigma_2$; $C = C_1 \cup C_2$; for all $(s_1, s_2) \in S$, $L((s_1, s_2)) = L_1(s_1) \wedge L_2(s_2)$; $\rightarrow$ is the smallest transition relation satisfying the following rules:

- If $s_1 \overset{a,\delta,X}{\rightarrow}_1 s_1' \wedge a \notin \Sigma_2$, then $(s_1, s_2) \overset{a,\delta,X}{\rightarrow} (s_1', s_2)$ for all $s_2 \in S_2$.
- If $s_2 \overset{a,\delta,X}{\rightarrow}_2 s_2' \wedge a \notin \Sigma_1$, then $(s_1, s_2) \overset{a,\delta,X}{\rightarrow} (s_1, s_2')$ for all $s_1 \in S_1$.
- If $s_1 \overset{a,\delta,X}{\rightarrow}_1 s_1'$ and $s_2 \overset{a,\delta',X'}{\rightarrow}_2 s_2'$, then $(s_1, s_2) \overset{a,\delta \wedge \delta',X \cup X'}{\rightarrow} (s_1', s_2')$. □

A transition of the composition is either a local transition of either $\mathcal{A}_1$ or $\mathcal{A}_2$, or a synchronization on a common event of $\mathcal{A}_1$ and $\mathcal{A}_2$. By definition, parallel composition is communicative and distributive. Graphically, if a Timed Automaton is represented as a triangle, then parallel composition is drawn as shown in Figure 5.5. Parallel composition of multiple Timed Automata, or equivalently, a network of Timed Automata, can be used to model many systems, including industrial examples [HSLL97; LMNS05; LPW01]. For many Timed Automata-based tools, including UPPAAL, parallel composition is the only supported timed pattern.

Models based on a network of flat Timed Automata have the advantage of a simple architecture which translates to efficient system analysis. Nonetheless, designing and verifying compositional real-time systems are becoming an increasingly important task due to the widespread applications and increasing complexity of real-time systems. High-level requirements for real-time systems are often stated in terms of *deadline, time out*, and *timed interrupt* [LW97; DMF99; LPW01]. In industrial case studies of real-time

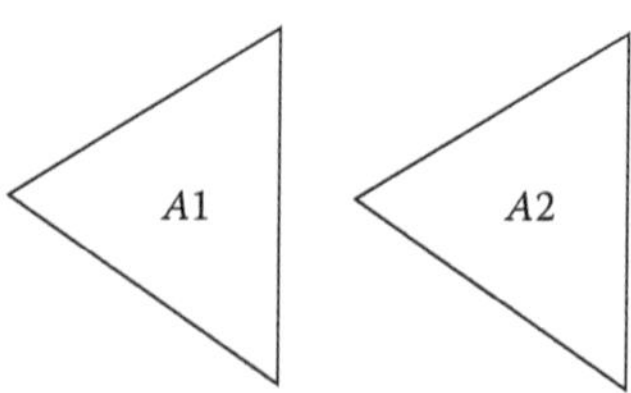

FIGURE 5.5 Parallel composition of Timed Automata.

system verification, system requirements are often structured into phases, which are then composed in many different ways. Unlike Statecharts equipped with clocks [HG97] or timed process algebras [NS94; Yi91; Sch00], timed automata lack high-level compositional patterns for hierarchical design. Users often need to manually cast those terms into a set of clock variables with carefully calculated clock constraints. This process is tedious and error-prone. The remedy is to define additional timed patterns.

Definition 5.6: (Event Prefixing)

Let $\mathcal{A} = (S, I, \Sigma, C, L, \rightarrow)$ be a Timed Automaton. Let a be an event. The event-prefixing pattern, that is, the Timed Automaton in which a precedes any action of $\mathcal{A}$, written as $a \rightarrow \mathcal{A}$, is a Timed Automaton $(S', I', \Sigma', C', L', \rightarrow')$ such that $init' \notin S$ is a fresh state; $S' = S \cup \{init'\}$; $I' = \{init'\}$; $\Sigma' = \Sigma \cup \{a\}$; $C' = C$; and

- For all $s \in S'$, $L'(s) = L(s)$ if $s \in S$; $L'(s) = true$ otherwise.
- $\rightarrow'$ contains all transitions in $\rightarrow$ and additional ones: for all $init \in I$, $init' \overset{a,true,\emptyset}{\rightarrow'} init$). □

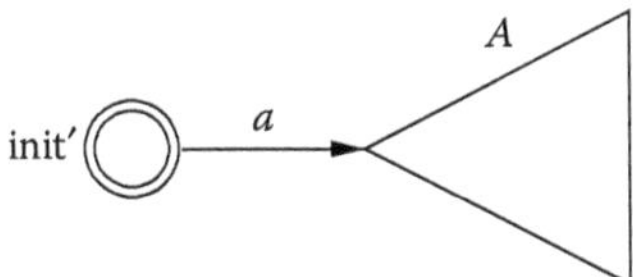

Event prefixing is a commonly used pattern, for example, a task is preceded by some event. In this pattern, event a (which may be a synchronization barrier or variable assignment) must be engaged before certain task (which is modeled as $\mathcal{A}$) must be carried out. The fresh state $init'$ serves as the new initial state. It is connected to the initial states of $\mathcal{A}$ via a transition labeled with a.

Definition 5.7: (Delay)

The delay pattern, written as $delay(n)$, where $n \in \mathbb{R}_+$, is a Timed Automaton $(S, I, \Sigma, C, L, \rightarrow)$ such that $S = \{s_1, s_2, s_3\}$; $I = \{s_1\}$; $\Sigma = \emptyset$; $C = \{x\}$; and

- $L(s_1) = urgent$ and $L(s_2) = x \leq d$ and $L(s_3) = urgent$.
- $\rightarrow$ contains two transitions: $s_1 \overset{\tau,true,\{x\}}{\rightarrow} s_2$ and $s_2 \overset{\tau,x=t,\emptyset}{\rightarrow} s_3$. □

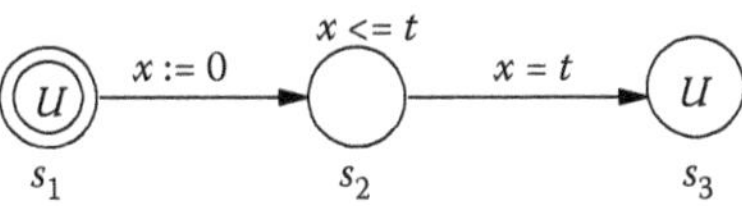

This pattern is used to delay the execution (of the system) by exactly n time units, for example, one of the common requirements for timed systems. The state invariant *urgent* is a syntactic sugar which is equivalent to $c = 0$, where c is a clock which has been reset prior to enter the state. The idea is to prevent time elapsing at a state. Event τ is a special event which is *invisible*.

Definition 5.8: (Choice)

Let $\mathcal{A}_i = (S_i, I_i, \Sigma_i, C_i, L_i, \rightarrow_i)$, where $i \in \{1, 2\}$ be two Timed Automata. An external choice of $\mathcal{A}_1$ and $\mathcal{A}_2$, written as $\mathcal{A}_1 \Box \mathcal{A}_2$, is a Timed Automaton $(S, I, \Sigma, C, L, \rightarrow)$, where $S = S_1 \cup S_2$; $I = I_1 \cup I_2$; $\Sigma = \Sigma_1 \cup \Sigma_2$; $C = C_1 \cup C_2$; and

- $L(s) = L_1(s)$ for all $s \in S_1$ and $L(s) = L_2(s)$ for all $s \in S_2$.
- $\rightarrow$ contains all transitions in $\rightarrow_1$ and $\rightarrow_2$. □

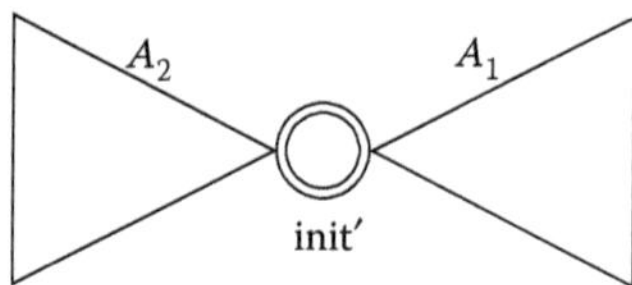

This pattern models a unconditional choice between two Timed Automata.

Definition 5.9: (Sequential Composition)

Let $\mathcal{A}_i = (S_i, I_i, \Sigma_i, C_i, L_i, \rightarrow_i)$, where $i \in \{1,2\}$ be two Timed Automata. Sequential composition of $\mathcal{A}_1$ and $\mathcal{A}_2$, written as $\mathcal{A}_1; \mathcal{A}_2$, is a Timed Automaton $(S, I, \Sigma, C, L, \rightarrow)$, where $S = S_1 \cup S_2$; $I = I_1$; $\Sigma = \Sigma_1 \cup \Sigma_2$; $C = C_1 \cup C_2$; and let $F = \{s \in S_1 \mid \nexists a, \delta, X, s'. \ s_1 \overset{a,\delta,X}{\rightarrow}_1 s'\}$ be the set of deadlocking states in $\mathcal{A}_1$,

- $L(s) = L_1(s)$ for all $s \in S_1 \setminus F$; $L(s) = L_1(s) \wedge urgent$ for all $s \in F$; and $L(s) = L_2(s)$ for all $s \in S_2$.
- $\rightarrow$ contains all transitions in $\rightarrow_1$ and $\rightarrow_2$ and in addition: for all $f \in F$ and $i \in I_2, f \overset{\tau,true,\emptyset}{\rightarrow} i$. □

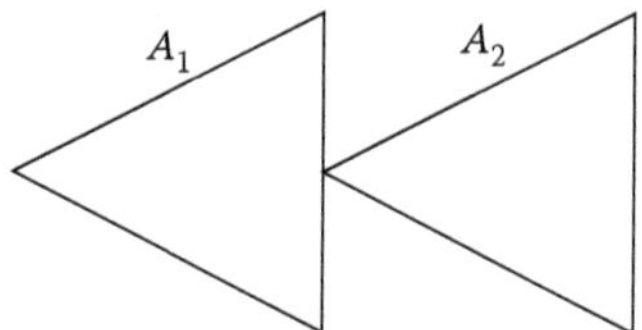

Given two Timed Automata $\mathcal{A}_1$ and $\mathcal{A}_2$, the above shows the sequential composition. The idea is to connect the *final* states of the first automaton to the initial states of the second. Further, the final states of the first automaton are marked as urgent so that the control transfers to the second automaton without any delay after the first automaton terminates. Sequential composition is commonly use to accomplish two tasks/jobs in order. Note that if A_1 is nonterminating, the states in $\mathcal{A}_2$ are not reachable and, therefore, the composition is exactly $\mathcal{A}_1$.

Definition 5.10: (Time Out)

Let $\mathcal{A}_i = (S_i, I_i, \Sigma_i, C_i, L_i, \rightarrow_i)$, where $i \in \{1,2\}$ be two Timed Automata. Let $n \in \mathbb{R}_+$. Let $init' \notin S_1 \cup S_2$ be a fresh state and $x \notin C_1 \cup C_2$ be a fresh clock. The time-out pattern, written as $\mathcal{A}_1 \rhd_n \mathcal{A}_2$, is a Timed Automaton $(S, I, \Sigma, C, L, \rightarrow)$, where $S = S_1 \cup S_2 \cup \{init'\}$; $I = \{init'\}$; $\Sigma = \Sigma_1 \cup \Sigma_2$; $C = C_1 \cup C_2 \cup \{x\}$; and

- $L(init') = urgent$ and $L(s) = L_1(s) \wedge x \leq t$ for all $s \in I_1$ and $L(s) = L_1(s)$ for all $s \in S_1 \setminus I_1$ and $L(s) = L_2(s)$ for all $s \in S_2$.
- $\rightarrow$ contains all transitions in $\rightarrow_1$ and fun_2 and additionally: for all $init_1 \in S_1$; $init_2 \in S_2$, $init' \overset{\tau,true,\{x\}}{\rightarrow} init)$ and $init_1 \overset{\tau,x=t,\emptyset}{\rightarrow} init_2$. □

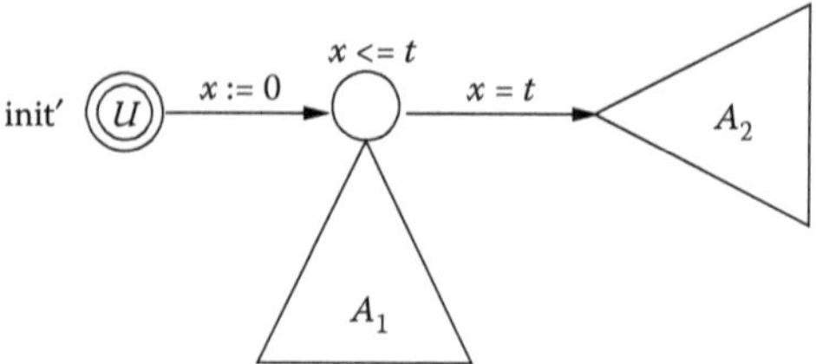

Time out is a common behavior patterns in real-time systems, which is partially evidenced by different ways proposed to model it [Gro; Bow99; Sch00]. The fresh clock x is reset along the transition from the fresh initial state to an initial in $\mathcal{A}_1$. Each initial state in $\mathcal{A}_1$ is constrained to make a move no later than n time units. If the control moves out the initial state before t time units, the system behaves as prescribed by $\mathcal{A}_1$. Otherwise, after exactly t time units, $\mathcal{A}_2$ takes over the control. If $\mathcal{A}_1$ may make a move at exactly time n, the system nondeterministically fires one of the transitions and prevents the other from happening.

Definition 5.11: (Timed Interrupt)

Let $\mathcal{A}_i = (S_i, I_i, \Sigma_i, C_i, L_i, \rightarrow_i)$, where $i \in \{1,2\}$ be two Timed Automata. Let $n \in \mathbb{R}_+$. Let $init' \notin S_1 \cup S_2$ be a fresh state and $x \notin C_1 \cup C_2$ be a fresh clock. The timed-interrupt pattern, written as $\mathcal{A}_1 \triangle_n \mathcal{A}$, is a Timed Automaton $(S, I, \Sigma, C, L, \rightarrow)$, where $S = S_1 \cup S_2 \cup \{init'\}$; $I = \{init'\}$; $\Sigma = \Sigma_1 \cup \Sigma_2$; $C = C_1 \cup C_2 \cup \{x\}$; and

- $L(init') = urgent$ and $L(s) = L_1(s) \wedge x \leq t$ for all $s \in S_1$ and $L(s) = L_2(s)$ for all $s \in S_2$.
- $\rightarrow$ contains all transitions in $\rightarrow_1$ and $\rightarrow_2$ and additionally: for all $init_1 \in I_1$ and $init_2 \in I_2$ and $s_1 \in S_1$, $init' \overset{\tau,true,\{x\}}{\rightarrow} init_1$ and $s_1 \overset{\tau,x=t,\emptyset}{\rightarrow} init_2$. □

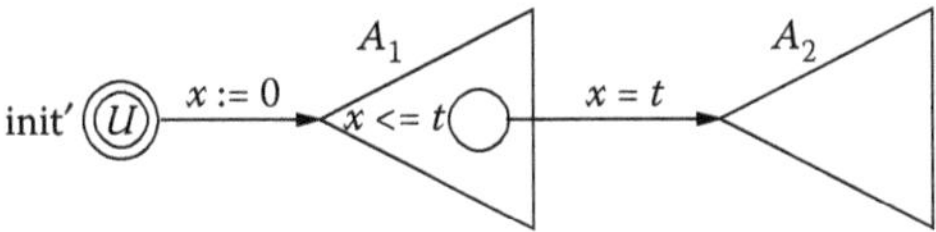

The pattern is composed of two automata, $\mathcal{A}_1$ and $\mathcal{A}_2$. Intuitively, this pattern states that $\mathcal{A}_2$ takes over control after $\mathcal{A}_1$ starts executing for exactly n time units. The fresh state $init'$ and clock x is used similarly as in the previous patterns. Every state in A_1 is constrained to make a move before n time units. After n time units elapse, the control transfers from one of the state in $\mathcal{A}_1$ (not necessary a final state) to the initial state in $\mathcal{A}_2$. Note that the interruption is preemptive, that is, transitions in $\mathcal{A}_1$ are prevented from happening once n time units have elapsed.

Definition 5.12: (Repetition)

Let $\mathcal{A} = (S, I, \Sigma, C, L, \rightarrow)$ a Timed Automaton. Let $F = \{s \in S | \nexists a, \delta, X, s'. s \overset{a,\delta,X}{\rightarrow} s'\}$ be the set of deadlocking states in $\mathcal{A}$. The repetition pattern, written as *repeat* $\mathcal{A}$, is a Timed Automaton $(S, I, \Sigma, C, L, \rightarrow')$, where $\rightarrow'$ contains all transitions in $\rightarrow$ and additionally $f \overset{\tau,true,\emptyset}{i}$ for all $f \in F$ and $i \in I$.

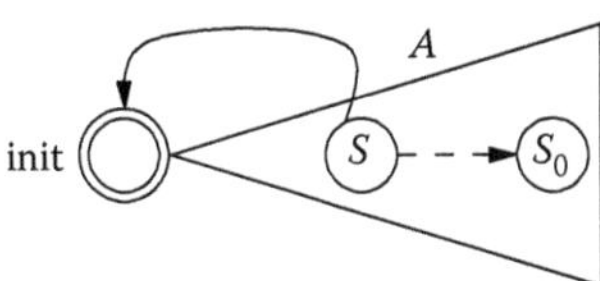

Repetition is used to introduce infinite behaviors, which is commonly used for specifying nonterminating reactive systems. Given a timed automaton $\mathcal{A}$, the pattern is constructed by connecting all final states in $\mathcal{A}$ to an initial state. The dotted arrow represents an imaginary transition out of a final state. In the resultant automaton, such transitions are redirected to the initial state of A.

Definition 5.13: (Deadline)

Let $\mathcal{A} = (S, I, \Sigma, C, L, \rightarrow)$ be a Timed Automaton. Let $n \in \mathbb{R}_+$, $init' \notin S_1 \cup S_2$ be a fresh state and $x \notin C_1 \cup C_2$ be a fresh clock. The deadline pattern, written as *deadline*(A, n), is a Timed Automaton $(S', I', \Sigma', C', L', \rightarrow')$, where $S' = S \cup \{init'\}$; $I' = \{init'\}$; $\Sigma' = \Sigma$; $C' = C \cup \{x\}$; and

- $L'(init') = urgent$ and $L'(s) = L(s) \wedge xleqt$ for all $s \in S$.
- $\rightarrow'$ contains all transitions in $\rightarrow$ and additionally: for all $i \in I$, $init' \xrightarrow{\tau, true, \{x\}}{}' \; i$. □

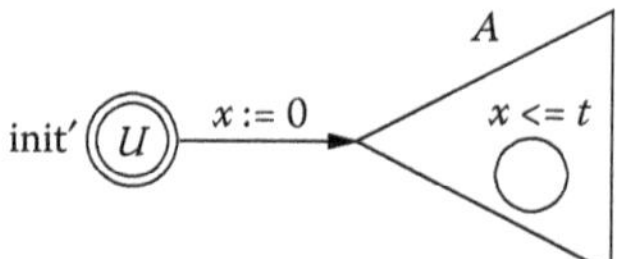

The above pattern is used to capture the requirement that some task must be finished by certain time. It is more toward a requirement than a design. Given a Timed Automaton $\mathcal{A}$, if it is constrained to finish within n time units, all states in $\mathcal{A}$ is labeled with an invariant $x \leq n$ in which x is a fresh clock which is reset upon the control enters the automaton. The left-most state is the fresh state *init*$'$. The local invariant $x \leq n$ covers each state of the timed automaton and thus $\mathcal{A}$ must terminate no later than n time units. This pattern may introduce timelocks [BG06]. In particular, there might be *time-actionlocks* (i.e., situations in which neither time nor event transitions can be performed) or *zeno-timelocks* (i.e., situations in which time is unable to pass beyond a certain point, but events continue to be performed). Detecting and resolving timelocks are nontrivial tasks (refer to [BG06] for sufficient conditions and sufficient-and-necessary conditions for timelock-freeness).

The above cover many of the common timed patterns. Nonetheless, they are by no means complete. New useful patterns can be introduced or composed from the existing ones. For instance, a *periodic* pattern which repeats a task every n time units (the hidden requirement is that the task must terminate before n time units) can be defined by composing the *deadline* and *repetition* patterns. It is important that the Timed Automata patterns combine with the bottom-up composibility because complex real-time systems may be naturally modeled as collections of subcomponents at different abstraction levels. In the bottom-up design process, a subcomponent of reasonable complexity (and is flatten) may be modeled as a Timed Automaton. The system can then be naturally composed from the modeling of the subcomponents. The Timed Automata patterns provide a user-friendly templates to build such hierarchical modeling. In the top-down design process, the modeling of a complex system can be generated by refining the components step-by-step using appropriate patterns until it is simple enough to be modeled as a flattened Timed Automaton. As a reasonable price to pay, system design based on Timed Automata patterns may require extra states or clocks, which could lead to unnecessary state space explosion.

5.5 Stateful Timed CSP

Recently, an alternative approach for modeling and verifying hierarchical complex real-time systems is proposed [SLDZ09]. Instead of Timed Automata, real-time systems are modeled in a language named *Stateful Timed CSP*. Stateful Timed CSP extends Timed CSP [Sch00] with language constructs to manipulate data structures and data operations in order to support real-world applications. It supports a rich set of timed process constructs to capture timed system requirements, for example, *wait*, *deadline*, *timeout*, and *timed interrupt*. Different from Timed Automata, the timed process constructs rely on *implicit* clocks.

Furthermore, Stateful Timed CSP is designed for model checking and comes with complete tool support, namely PAT (Process Analysis Toolkit) [SLDP09].

5.5.1 Stateful Timed CSP Syntax

In this section, we introduce Stateful Timed CSP and its informal semantics. A Stateful Timed CSP model contains a finite set of global variables and a timed process. A variable can be of a pre-defined type like Boolean, integer, array of integers, or any user-defined data type defined in a C# library*. The time process models the control logic of the system using a rich set of process constructs. A process can be defined by the grammar presented in Figure 5.6.

Process *Stop* does nothing but idling. Process *Skip* terminates, possibly after idling for some time. Process $e \rightarrow P$ engages in event *e* first and then behaves as *P*. Note that *e* may serve as a synchronization barrier, if combined with parallel composition. In order to seamlessly integrate data operations, we allow sequential programs to be attached with events. Process $a\{program\} \rightarrow P$ performs data operation *a* (i.e., executing the sequential *program* while generating event *a*) and then behaves as *P*. The *program* may be a simple procedure updating data variables (written in the form of $a\{x := 5; y := 3\}$) or a complicated sequential program. A conditional choice is written as **if** (b) $\{P\}$ **else** $\{Q\}$. If *b* is true, then it behaves as *P*, else it behaves as *Q*. Process $P|Q$ offers an (unconditional) choice between *P* and *Q*. Process $P; Q$ behaves as *P* until *P* terminates and then behaves as *Q* immediately. $P \setminus X$ hides occurrences of events in *X*. Parallel composition of two processes is written as $P \parallel Q$, where *P* and *Q* may communicate via multiparty event synchronization (following CSP rules [Hoa85]) or shared variables.

A number of timed process constructs can be used to capture common real-time system behavior patterns. A timed process construct is often associated with a *parameter d*, where *d* denotes an integer constant. Given a model, the maximum parameter of the timed processes is called the clock ceiling.

- Process $Wait[d]$ idles for exactly *d* time units.
- In process P $timeout[d]$ Q, the first observable event of *P* shall occur before *d* time units elapse (since process P $timeout[d]$ Q is activated). Otherwise, *Q* takes over control after exactly *d* time units.
- Process P $interrupt[d]$ Q behaves exactly as *P* until *d* time units, and then *Q* takes over. In contrast to P $timeout[d]$ Q, *P* may engage in multiple observable events before it is *interrupted*.

$P = Stop$	- inaction
$\mid Skip$	- termination
$\mid e \rightarrow P$	- event prefixing
$\mid a\{program\} \rightarrow P$	- data operation prefixing
$\mid$ **if** (b) $\{P\}$ **else** $\{Q\}$	- conditional choice
$\mid P \mid Q$	- general choice
$\mid P \setminus X$	- hiding
$\mid P; Q$	- sequential composition
$\mid P \parallel Q$	- parallel composition
$\mid Wait[d]$	- delay*
$\mid P\ timeout[d]\ Q$	- timeout*
$\mid P\ interrupt[d]\ Q$	- timed interrupt*
$\mid P\ within[d]$	- timed responsiveness*
$\mid P\ deadline[d]$	- deadline*
$\mid Q$	- process referencing

FIGURE 5.6 Process constructs.

* Refer to PAT user manual on how to define a type.

- Process P *within*$[d]$ must react within d time units, that is, an observable event must be engaged by process P within d time units. Note that P *within*$[d]$ puts a constraint on P. Urgent event prefixing [Dav93], written as $e \twoheadrightarrow P$, is defined as $(e \rightarrow P)$ *within*$[0]$, that is, e must occur as soon as it is enabled.
- Process P *deadline*$[d]$ constrains P to terminate, possibly after engaging in multiple observable events, before d time units.

A process may be given a name, written as $P \widehat{=} Q$, and then referenced through its name. Recursion is allowed by process referencing. Additional process constructs (e.g., while or periodic behaviors) can be defined using the above processes. Different from Timed Automata, clock variables are made implicit. For instance, given a process P *timeout*$[d]$ Q, it is intuitively clear that an implicit clock should start whenever the process is activated. Because clocks are implicit, they cannot be compared with each other directly.

Example 5.4:

Let δ and ϵ be two constants such that $\delta < \epsilon$. Fischer's mutual exclusion algorithm is modeled as a model $(V, v_i, Protocol)$. V contains two variables *turn* and *counter*. The former indicates which process attempted to access the critical section most recently. The latter counts the number of processes accessing the critical section. Initial valuation v_i maps *turn* to -1 (which denotes that no process is attempting initially) and *counter* to 0 (which denotes that no process is in the critical section initially). Process *Protocol* is defined as follows:

$$
\begin{array}{ll}
Protocol & \widehat{=}\ Proc(0) \parallel Proc(1) \parallel \cdots \parallel Proc(n) \\
Proc(i) & \widehat{=}\ \textbf{if}\ (turn = -1)\ \{\ Active(i)\ \}\ \textbf{else}\ \{\ Proc(i)\ \} \\
Active(i) & \widehat{=}\ (update.i\{turn := i\} \rightarrow Wait[\epsilon])\ within[\delta]; \\
& \quad \textbf{if}\ (turn = i)\ \{ \\
& \quad\quad cs.i\{counter := counter + 1\} \rightarrow \\
& \quad\quad exit.i\{counter := counter - 1; turn := -1\} \rightarrow Proc(i) \\
& \quad \}\ \textbf{else}\ \{ \\
& \quad\quad\quad Proc(i) \\
& \quad \}
\end{array}
$$

where n is a constant representing the number of processes. Process *Proc*(i) models a process with a unique integer identify i. If *turn* is -1 (i.e., no other process is attempting), *Proc*(i) behaves as specified by process *Active*(i). In process *Active*(i), first *turn* is set to be i (indicating that the i-process is now attempting) by action *update.i*. Note that *update.i* must occur within δ time units (captured by *within*$[\delta]$). Next, the process idles for ϵ time units (captured by *Wait*$[\epsilon]$). It then checks whether *turn* is still i. If so, it enters the critical section and leaves later. Otherwise, it restarts from the beginning.

Quantitative timing plays an important role in this algorithm to guarantee mutual exclusion, that is, mutual exclusion is not guaranteed if $\delta \geq \epsilon$. In order to verify mutual exclusion, one way is to show that $counter \leq 1$ is always true. Readers are encouraged to model the algorithm in Timed Automata and compare the differences. □

5.5.2 Stateful Timed CSP Semantics

Before presenting the operational semantics of Stateful Timed CSP, we define the notion of a configuration to capture the global system state during the execution, which is referred to as *concrete configurations*. This terminology distinguishes the notion from the state space abstraction and abstract configurations, which will be introduced in Section 5.5.4.

A concrete system configuration is a pair (V, P), where V is a variable valuation function and $P \in \mathcal{P}$ is a process. For simplicity, an empty valuation is written as $\emptyset$. A transition of the system is written in the form $(V, P) \xrightarrow{x} (V', P')$ such that $x \in \Sigma_\tau \cup \mathbb{R}_+$. There are two kinds of transitions, that is, a transition labeled with an event in Σ_τ or a time transition labeled with a number in $\mathbb{R}_+$. The operational semantics is defined by associating a set of firing rules with each and every process construct. The firing rules associated with the timed processes are present in Figure 5.7. The rest are available in [SLDZ09].

$$\frac{\epsilon \le d}{(V, Wait[d]) \xrightarrow{\epsilon} (V, Wait[d-\epsilon])}\ [\ wait1\ ] \qquad \frac{}{(V, Wait[0]) \xrightarrow{\tau} (V, Skip)}\ [\ wait2\ ]$$

$$\frac{(V,P) \xrightarrow{e} (V', P')}{(V, P\ timeout[d]\ Q) \xrightarrow{e} (V', P')}\ [\ to1\ ] \qquad \frac{}{(V, P\ timeout[0]\ Q) \xrightarrow{\tau} (V, Q)}\ [\ to2\ ]$$

$$\frac{(V,P) \xrightarrow{\tau} (V', P')}{(V, P\ timeout[d]\ Q) \xrightarrow{\tau} (V', P'\ timeout[d]\ Q)}\ [\ to3\ ]$$

$$\frac{(V,P) \xrightarrow{\epsilon} (V, P'), \epsilon \le d}{(V, P\ timeout[d]\ Q) \xrightarrow{\epsilon} (V, P'\ timeout[d-\epsilon]\ Q)}\ [\ to4\ ]$$

$$\frac{(V,P) \xrightarrow{a} (V', P')}{(V, P\ interrupt[d]\ Q) \xrightarrow{a} (V', P'\ interrupt[d]\ Q)}\ [\ ti1\ ]$$

$$\frac{(V,P) \xrightarrow{\epsilon} (V, P'), \epsilon < d}{(V, P\ interrupt[d]\ Q) \xrightarrow{e} (V, P'\ interrupt[d-\epsilon]\ Q)}\ [\ ti2\ ]$$

$$\frac{}{(V, P\ interrupt[0]\ Q) \xrightarrow{\tau} (V, Q)}\ [\ ti3\ ] \qquad \frac{(V,P) \xrightarrow{\tau} (V', P')}{(V, P\ within[d]) \xrightarrow{\tau} (V', P\ within[d])}\ [\ wi2\ ]$$

$$\frac{(V,P) \xrightarrow{e} (V', P')}{(V, P\ within[d]) \xrightarrow{e} (V', P')}\ [\ wi1\ ] \qquad \frac{(V,P) \xrightarrow{\epsilon} (V, P'), \epsilon < d}{(V, P\ within[d]) \xrightarrow{\epsilon} (V, P'\ within[d-\epsilon])}\ [\ wi3\ ]$$

$$\frac{(V,P) \xrightarrow{a} (V', P')}{(V, P\ deadline[d]) \xrightarrow{a} (V', P'\ deadline[d])}\ [\ dl1\ ]$$

$$\frac{(V,P) \xrightarrow{\checkmark} (V', P')}{(V, P\ deadline[d]) \xrightarrow{\checkmark} (V', P')}\ [\ dl2\ ]$$

$$\frac{(V,P) \xrightarrow{\epsilon} (V, P'), \epsilon < d}{(V, P\ deadline[d]) \xrightarrow{\epsilon} (V, P'\ deadline[d-\epsilon])}\ [\ dl2\ ]$$

FIGURE 5.7 Concrete firing rules.

- The semantics of $Wait[d]$ is captured by rule $wait1$ and $wait2$. Rule $wait1$ states that the process may idle for an arbitrary amount of time ϵ such that $\epsilon \leq d$. Afterward, $Wait[d]$ becomes $Wait[d - \epsilon]$. The valuation of the variables is unchanged. Rule $wait2$ states that the process becomes *Skip* via a τ-transition whenever d is 0.
- The semantics of P *timeout*$[d]$ Q is captured by rule $to1$ to $to4$. Rule $to1$ states that if an observable event e can be engaged by P, changing (V, P) to (V', P'), then $(V, P$ *timeout*$[d]$ $Q)$ becomes (V', P') so that Q is discharged. That is, P has performed an observable event before timeout occurs. Rule $to2$ states that if P instead performs a τ-transition, then Q and *timeout* operator remain (since an observable event is yet to be performed). Rule $to3$ states that if P may idle for less than or equal to d time units, so does P *timeout*$[d]$ Q. Rule $to4$ states that if d is 0, Q takes over control by a τ-transition.
- The semantics of P *interrupt*$[d]$ Q is defined by rule $ti1$ to $ti3$. Rule $ti1$ states that if event a (which may be observable or τ) can be engaged by P, changing (V, P) to (V', P'), then $(V, P$ *interrupt*$[d]$ $Q)$ can perform a as well. In contrast to rule $to1$, the *interrupt* operator remains. Intuitively, it states that before P is interrupted, P can do whatever it can. Rule $ti2$ states that if P may idle for less than or equal to d time units, so does P *interrupt*$[d]$ Q. Rule $ti3$ states that if d is 0, Q takes over control by a τ-transition.
- The semantics of P *within*$[d]$ is defined by rule $wi1$ to rule $wi3$. Rule $wi1$ states that if an observable event e occurs, then *within* is discharged, as the requirement is fulfilled. In contrast, rule $wi2$ states that if instead event τ occurs, then *within* remains. Rule $wi3$ state if P can idle for ϵ time units, then so does P *within*$[d]$ as long as $\epsilon \leq d$.
- Rule $dl1$, $dl2$, and $dl3$ define the semantics of P *deadline*$[d]$. Different from P *within*$[d]$, P *deadline*$[d]$ requires P to terminate (marked by a special event $\checkmark$) before d time units. Rule $dl1$ states that if P behaves as usual before the deadline is expired. Rule $dl2$ states that if P terminates, then *deadline* is discharged. Rule $dl3$ states if P can idle for ϵ time units, so does P *deadline*$[d]$ as long as $\epsilon \leq d$.

Definition 5.14: (Semantics of Stateful Timed CSP)

Let $\mathcal{S} = (Var, init_G, P)$ be a model. The concrete semantics of $\mathcal{S}$, denoted as $\mathcal{T}_{\mathcal{S}}$, is a transition system $(S, init, \Sigma \cup \mathbb{R}_+, T)$ such that S is a set of reachable concrete system configurations; $init = (init_G, P)$ is the initial configuration; and T satisfies: $((V, P), x, (V', P')) \in T$ if and only if $(V, P) \xrightarrow{x} (V', P')$. □

5.5.3 Verifying Stateful Timed CSP

Similar to Timed Automata, Stateful Timed CSP models also have infinite number of states because there may be infinitely many transitions between any two time points. Hence, abstraction is required if we want to apply model checking techniques. Different from Timed Automata, the timed process constructs rely on *implicit* clocks. For instance, process P *deadline*$[d]$ intuitively denotes that process P is constrained to terminate before d time units. That is, an implicit clock starts ticking when the process is *activated** and P must terminate no later when its reading is d. As a result, abstraction and verification techniques designed for Timed Automata, for example, zone abstraction, are not directly applicable. In order to build a finite abstraction from Stateful Timed CSP models, *dynamic zone abstraction* is proposed [SLDZ09]. The idea is to dynamically create clocks (only if necessary) to capture constraints introduced by the timed process constructs. In order to minimize the number of clocks, clocks are often shared by multiple timed process constructs. Further, clocks are pruned as soon as they become irrelevant. [SLDZ09] proves that dynamic

* Informally speaking, a process is activated if it has the control.

zone abstraction produces an abstract model, that is, a zone graph, which is both finite state and property preserving so that it is subject to model checking.

Given a model $\mathcal{S}$, by definition, $\mathcal{T}_{\mathcal{S}}$ is infinite if and only if either there are infinitely many configurations, which implies there are infinitely many values for V or P. In this chapter, *we assume that all variables have finite domains and all reachable processes are finite.** There can still be infinitely many values for P because parameters of timed process constructs in P (e.g., d in $Wait[d]$) can take infinitely many different values. Intuitively, the constants capture the reading of the implicit clocks associated with the processes. In the following, we abstract the constants by dynamic zone abstraction.

In Stateful Timed CSP, clocks are implicitly associated with timed processes. It starts ticking once the process becomes activated. Before applying zone abstraction, we associate clocks with time processes explicitly so as to differentiate parameters associated with different timed process constructs. In theory, each timed process construct is associated with a unique clock. Nonetheless, multiple timed processes can be activated at the same time during system execution and, therefore, the associated clocks are always of the same reading. For instance, assume a process P defined as follows.

$$(Wait[5]; Wait[3])\ interrupt[6]\ Q$$

There are three implicit clocks, one associated with $Wait[5]$ (say t_1), one with $Wait[3]$ (say t_2), and one with P (because of $interrupt[6]$, say t_3). Because $Wait[5]$ and P are activated, clock t_1 and t_3 have started. In contrast, clock t_2 starts only when $Wait[5]$ terminates. It can be shown that t_1 and t_3 always have the same reading and thus one clock is sufficient. It is known that the less clocks, the more efficient real-time model checking is [BY03]. In order to minimize the number of clocks, we introduce clocks at runtime so that timed processes which are activated at the same time share the same clock. Intuitively, a clock is introduced if and only if one or more timed processes have just become activated.

In the following, we show how to systematically associate clocks with timed processes. To distinguish from ordinary Stateful Timed CSP processes, let $\mathcal{Q}$ denotes the set of processes associated with explicit clocks. For simplicity, we write $Wait[d]_t$ (or $P\ timeout[d]_t\ Q$, $P\ interrupt[d]_t\ Q$, $P\ within[d]_t$, and $P\ deadline[d]_t$) to denote that the process is associated with clock t. Given a process P and a clock t, we define function $\mathcal{A}$ to return the corresponding process in $\mathcal{Q}$. Figure 5.8 presents the detailed definition. Intuitively speaking, A1 to A5 state that if a process is un-timed and none of its subprocesses are activated, then it is unchanged. A6 to A10 state that if a process has been associated with clocks (i.e., it is in $\mathcal{Q}$), then function $\mathcal{A}$ is applied to its activated subprocesses only. Note that $Wait[d]_{t'}$ denotes that it has already been associated with clock t'. A11 to A14 state that if the process itself is un-timed, then function $\mathcal{A}$ is applied to its activated subprocesses. A16 to A19 state that if the process is timed, then it is associated with t and function $\mathcal{A}$ is applied to its activated subprocesses. Given a process $P \in \mathcal{Q}$, we can obtain the set of clocks associated with P or any subprocess of P. The set, written as $cl(P)$, is referred to as the *active clocks* of P.

Example 5.5:

In the Fischer's mutual exclusion example, assume there are three processes. The first and second process have evaluated the condition **if** $(turn = -1)$ and become $Active(0)$ and $Active(1)$, whereas the third process has not made some move. So that the current process is

$$Active(0) \parallel Active(1) \parallel Proc(2)$$

* Stateful Timed CSP is expressive enough to capture irregular or even non-context-free languages [Hoa85].

$$
\begin{array}{lll}
\mathcal{A}(Stop, t) & = Stop & \text{- A1} \\
\mathcal{A}(Skip, t) & = Skip & \text{- A2} \\
\mathcal{A}(e \rightarrow P, t) & = e \rightarrow P & \text{- A3} \\
\mathcal{A}(a\{program\} \rightarrow P, t) & = a\{program\} \rightarrow P & \text{- A4} \\
\mathcal{A}(\textbf{if}\ (b)\ \{P\}\ \textbf{else}\ \{Q\}, t) & = \textbf{if}\ (b)\ \{P\}\ \textbf{else}\ \{Q\} & \text{- A5} \\
\mathcal{A}(Wait[d]_{t'}, t) & = Wait[d]_{t'} & \text{- A6} \\
\mathcal{A}(P\ timeout[d]_{t'}\ Q, t) & = \mathcal{A}(P, t)\ timeout[d]_{t'}\ Q & \text{- A7} \\
\mathcal{A}(P\ interrupt[d]_{t'}\ Q, t) & = \mathcal{A}(P, t)\ interrupt[d]_{t'}\ Q & \text{- A8} \\
\mathcal{A}(P\ within[d]_{t'}, t) & = \mathcal{A}(P, t)\ within[d]_{t'} & \text{- A9} \\
\mathcal{A}(P\ deadline[d]_{t'}, t) & = \mathcal{A}(P, t)\ deadline[d]_{t'} & \text{- A10} \\
\mathcal{A}(P|Q, t) & = \mathcal{A}(P, t)|\mathcal{A}(Q, t) & \text{- A11} \\
\mathcal{A}(P \setminus X, t) & = \mathcal{A}(P, t) \setminus X & \text{- A12} \\
\mathcal{A}(P; Q, t) & = \mathcal{A}(P, t); Q & \text{- A13} \\
\mathcal{A}(P \parallel Q, t) & = \mathcal{A}(P, t) \parallel \mathcal{A}(Q, t) & \text{- A14} \\
\mathcal{A}(Wait[d], t) & = Wait[d]_t & \text{- A15} \\
\mathcal{A}(P\ timeout[d]\ Q, t) & = \mathcal{A}(P, t)\ timeout[d]_t\ \mathcal{A}(Q, t) & \text{- A16} \\
\mathcal{A}(P\ interrupt[d]\ Q, t) & = \mathcal{A}(P, t)\ interrupt[d]_t\ \mathcal{A}(Q, t) & \text{- A17} \\
\mathcal{A}(P\ within[d], t) & = \mathcal{A}(P, t)\ within[d]_t & \text{- A18} \\
\mathcal{A}(P\ deadline[d], t) & = \mathcal{A}(P, t)\ deadline[d]_t & \text{- A19} \\
\mathcal{A}(P, t) & = \mathcal{A}(Q, t) \quad \text{if } P \widehat{=} Q & \text{- A20}
\end{array}
$$

FIGURE 5.8 Clock activation.

Assume t is a fresh clock, applying function $\mathcal{A}$ with t returns

$$
\begin{array}{l}
(update.0\{turn := 0\} \rightarrow Wait[\epsilon])\ within_t[\delta]; \cdots \\
\parallel (update.1\{turn := 1\} \rightarrow Wait[\epsilon])\ within_t[\delta]; \cdots \\
\parallel \textbf{if}\ (turn = -1)\ \{\ Active(2)\ \}\ \textbf{else}\ \{\ Proc(2)\ \}
\end{array}
$$

t is associated with the first process and the second process, but not the third process. Note that $Wait[\epsilon]$ has not been activated yet. □

5.5.4 Zones

The concrete firing rules presented in Figure 5.7 capture quantitative timing through parameters of the timed processes. Inevitably, there can be infinitely many constant values. Due to the difference between Stateful Timed CSP and Timed Automata, zone abstraction techniques developed for Timed Automata must be amended. In the following, we introduce the additional zone operations before presenting how to apply zone abstraction to Stateful Timed CSP models.

- *Add clocks:* Clocks may be introduced during system exploration as we have shown in Section 5.5.3. Assume the clock to be added is t_k and the given DBM is in its canonical form. Figure 5.9 shows how the DBM is updated with entries for t_k. For all i, D_k^i is set to be D_0^i and D_i^k is set to be D_i^0. Because t_k is a newly introduced clock, it must be equivalent to the dummy clock t_0. By a simple argument, it can be shown the resultant DBM is canonical if the given DBM is.
- *Prune clocks:* In our setting, clocks may be pruned. Because entries in a canonical DBM represent the tightest bounds on clock differences, pruning a clock t_i is simply to remove the i-row and i-column in the matrix. The remaining DBM is canonical, that is, the bounds can not be possibly tightened with less constraints. Given a DBM D and a set of clocks C, we write $D[C]$ to denote the

	t_0	t_1	$\cdots$	t_i	$\cdots$	t_{k-1}	$\mathbf{t_k}$
t_0	0	d_1^0	$\cdots$	d_i^0	$\cdots$	d_{k-1}^0	**0**
t_1	d_0^1	$*$	$\cdots$	$*$	$\cdots$	$*$	$\mathbf{d_0^1}$
$\cdots$	$\cdots$	$\cdots$	$\cdots$	$\cdots$	$\cdots$	$\cdots$	$\cdots$
t_i	d_0^i	$*$	$\cdots$	$*$	$\cdots$	$*$	$\mathbf{d_0^i}$
$\cdots$	$\cdots$	$\cdots$	$\cdots$	$\cdots$	$\cdots$	$\cdots$	$\cdots$
t_{k-1}	d_0^{k-1}	$*$	$\cdots$	$*$	$\cdots$	$*$	$\mathbf{d_0^{k-1}}$
$\mathbf{t_k}$	**0**	$\mathbf{d_1^0}$	$\cdots$	$\mathbf{d_i^0}$	$\cdots$	$\mathbf{d_{k-1}^0}$	**0**

FIGURE 5.9 Add clock.

DBM obtained by pruning all clocks other than those in C. In an abuse of notation, we write $D[t]$ to denote the constraint on t.

We omit discussion on rest of the zone operations, for example, adding a constraint. Interested readers are referred to [BY03] for details. We assume zone normalization [BY03] based on the clock ceiling is always applied. In the following, we present dynamic zone abstraction for Stateful Timed CSP. First, we define the notion of abstract system configurations as follows. An abstract system configuration is a triple (V, P, D), where V is a variable valuation; P is a process; and D is a zone. An abstract configuration is valid if and only if D is not empty. Intuitively speaking, an abstract configuration (V_a, P_a, D) abstracts a concrete configuration (V_c, P_c) if and only if $V_a = V_c$ and the following are satisfied:

- P_c differs from P_a only by the parameters of the timed process constructs and the fact that P_a is associated with clocks, whereas P_c is not.
- For every timed process construct of P_c, let d be the constant associated with it; let d' be the constant associated with the corresponding construct in P_a. If the construct is not associated with a clock in P_a, then $d = d'$. If the construct is associated with clock t in P_a, then $t = d$ satisfies $D[t]$.

For instance, if $P = Wait[3]; Wait[5]$ and $P' = Wait[4]_t; Wait[5]$, P' with zone $t \leq 4$ abstracts P. We write $(V, P_c) \in (V, P_a, D)$ to denote that (V, P_a, D) abstracts (V, P_c). Or equivalently, (V_c, P_c) is an instance of (V_a, P_a, D).

We define a function *idle* which, given a process in $\mathcal{Q}$, calculates exactly how long the process can idle. The result is in the form of a predicate over the clocks, that is, a zone. Figure 5.10 shows the detailed definition. Rule *idle*1 to *idle*5 state that if the process is un-timed and none of its subprocesses is activated, then the function returns true. Intuitively, it means that the process may idle for arbitrary amount of time. Rule *idle*6 to *rule*9 state that if subprocesses of the process are activated, then function *idle* is applied to the subprocesses. For instance, if the process is a choice (rule *idle*6) or a parallel composition (rule *idle*9) of P and Q, then the result is $idle(P) \wedge idle(Q)$. Intuitively, this means that process $P|Q$ (or $P \parallel Q$) may idle as long as both P and Q can idle. Rule *idle*10 to *idle*14 define the cases when the process is timed. For instance, process $Wait[d]_t$ may idle as long as t is less or equal to d.

In order to systematically apply zone abstraction, we define a set of *abstract* firing rules. The abstract firing rules eliminate concrete ϵ-transitions all together and use zones to ensure a process behaves correctly with respect to timing requirements. To distinguish from concrete firing rules, an abstract firing rule is written in the the form of $(V, P, D) \overset{x}{\rightsquigarrow} (V', P', D')$, where $x \in \Sigma_\tau$. Figure 5.11 presents the abstract firing rules for the timed processes.

- Rule *await* defines the abstract semantics of $Wait[d]$. In contrast to the concrete semantics, there is only one abstract rule. It states that a τ-transition occurs exactly when clock t reads d. Intuitively, $D^{\uparrow} \wedge t = d$ denotes the time when d time units elapsed since t starts. Afterward, the process becomes *Skip*.

$idle(Stop)$	$= true$	– rule $idle1$
$idle(Skip)$	$= true$	– rule $idle2$
$idle(e \rightarrow P)$	$= true$	– rule $idle3$
$idle(a\{program\} \rightarrow P)$	$= true$	– rule $idle4$
$idle(\mathbf{if}\ (b)\ \{P\}\ \mathbf{else}\ \{Q\})$	$= true$	– rule $idle5$
$idle(P \mid Q)$	$= idle(P) \wedge idle(Q)$	– rule $idle6$
$idle(P \setminus X)$	$= idle(P)$	– rule $idle7$
$idle(P; Q)$	$= idle(P); Q$	– rule $idle8$
$idle(P \parallel Q)$	$= idle(P) \wedge idle(Q)$	– rule $idle9$
$idle(Wait[d]_t)$	$= t \leq d$	– rule $idle10$
$idle(P\ timeout[d]_t\ Q)$	$= t \leq d \wedge idle(P)$	– rule $idle11$
$idle(P\ interrupt[d]_t\ Q)$	$= t \leq d \wedge idle(P)$	– rule $idle12$
$idle(P\ within[d]_t)$	$= t \leq d \wedge idle(P)$	– rule $idle13$
$idle(P\ deadline[d]_t)$	$= t \leq d \wedge idle(P)$	– rule $idle14$
$idle(P)$	$= idle(Q)$ if $P \widehat{=} Q$	– rule $idle15$

FIGURE 5.10 Idling calculation.

- Rule $ato1$, $ato2$, and $ato3$ define the abstract semantics of $P\ timeout[d]\ Q$. Rule $ato1$ states that if a τ-transition transforms (V, P, D) to (V', P', D'), then a τ-transition may occur given $(V, P\ timeout[d]_t\ Q, D)$ if zone $D' \wedge t \leq d$ is not empty. Intuitively, this means that the τ-transition must occur before d time units since t starts. Similarly, rule $ato2$ ensures that the occurrence of an observable event e from process P may occur only if $t \leq d$, that is, before timeout occurs. Rule $ato3$ states that timeout results in a τ-transition when the reading of t is exactly d. The constraint $t = d \wedge idle(P)$ ensures that process P may idle all the way until timeout occurs.
- Rule $ait1$ and $ait2$ define the abstract semantics of $P\ interrupt[d]\ Q$. Rule $ait1$ states that a transition originated from P may occur only if $t \leq d$, that is, before interrupt occurs. Rule $ait2$ states that interrupt results in a τ-transition when the reading of t is exactly d. The constraint $t = d \wedge idle(P)$ ensures that process P may idle until interrupt occurs.
- Rule $awi1$ and $awi2$ define the abstract semantics of $P\ within[d]$. Rule $awi1$ states that if a τ-transition occurs within d time units, then the resultant process is of the form $P'\ within[d]$, which means that it remains to perform some observable event before d time units. Rule $awi2$ states that once an observable event occurs, the *within* construct is removed.
- Rule $adl1$ and $adl2$ define the abstract semantics of $P\ deadline[d]$. Rule $adl1$ constraints that all transitions of P must occur within d time units. Rule $adl2$ states that if P terminates (by engaging in $\checkmark$), then *deadline* is removed.

The rest of the rules are similarly defined. Using the abstract firing rules, we can generate an abstract transition system which captures the abstract semantics of a model. Let $\langle t_0, t_1, \cdots \rangle$ be a sequence of clocks.

Definition 5.15: (Zone Graph of Stateful Timed CSP)

Let $\mathcal{S} = (Var, init_G, P)$ be a model. The time-abstract semantics of $\mathcal{S}$, denoted as $\mathcal{L}_\mathcal{S}$, is a transition system $(S, init, \Sigma_\tau, T)$ such that S is a set of valid abstract system configurations, $init = (init_G, P, true)$ is the initial abstract configuration and T is the smallest transition relation satisfying the following. For all $(V, P, D) \in S$, if t is the first clock in the sequence which is not in $cl(P)$, and

$$(V, \mathcal{A}(P, t), D \wedge t = 0) \overset{a}{\rightsquigarrow} (V', P', D')$$

then $((V, P, D), a, (V', P', D'[cl(P')])) \in T$.

$$\frac{}{(V, Wait[d]_t, D) \overset{\tau}{\rightsquigarrow} (V, Skip, D^{\uparrow} \wedge t = d)} \; [\,await\,]$$

$$\frac{(V, P, D) \overset{\tau}{\rightsquigarrow} (V', P', D')}{(V, P\ timeout[d]_t\ Q, D) \overset{\tau}{\rightsquigarrow} (V', P'\ timeout[d]_t\ Q, D' \wedge t \leq d)} \; [\,ato1\,]$$

$$\frac{(V, P, D) \overset{e}{\rightsquigarrow} (V', P', D')}{(V, P\ timeout[d]_t\ Q, D) \overset{e}{\rightsquigarrow} (V', P', D' \wedge t \leq d)} \; [\,ato2\,]$$

$$\frac{}{(V, P\ timeout[d]_t\ Q, D) \overset{\tau}{\rightsquigarrow} (V, Q, t = d \wedge idle(P))} \; [\,ato3\,]$$

$$\frac{(V, P, D) \overset{a}{\rightsquigarrow} (V', P', D')}{(V, P\ interrupt[d]_t\ Q, D) \overset{a}{\rightsquigarrow} (V', P'\ interrupt[d]_t\ Q, D' \wedge t \leq d)} \; [\,ait1\,]$$

$$\frac{}{(V, P\ interrupt[d]_t\ Q, D) \overset{\tau}{\rightsquigarrow} (V, Q, t = d \wedge idle(P))} \; [\,ait2\,]$$

$$\frac{(V, P, D) \overset{\tau}{\rightsquigarrow} (V', P', D')}{(V, P\ within[d]_t, D) \overset{\tau}{\rightsquigarrow} (V', P'\ within[d]_t, D' \wedge t \leq d)} \; [\,awi1\,]$$

$$\frac{(V, P, D) \overset{e}{\rightsquigarrow} (V', P', D')}{(V, P\ within[d]_t, D) \overset{e}{\rightsquigarrow} (V', P', D' \wedge t \leq d)} \; [\,awi2\,]$$

$$\frac{(V, P, D) \overset{a}{\rightsquigarrow} (V', P', D'), a \neq \checkmark}{(V, P\ deadline[d]_t, D) \overset{a}{\rightsquigarrow} (V', P'\ deadline[d]_t, D' \wedge t \leq d)} \; [\,adl1\,]$$

$$\frac{(V, P, D) \overset{\checkmark}{\rightsquigarrow} (V', P', D')}{(V, P\ deadline[d]_t, D) \overset{\checkmark}{\rightsquigarrow} (V', P', D' \wedge t \leq d)} \; [\,adl2\,]$$

FIGURE 5.11 Abstract firing rules.

Informally, given an abstract configuration (V, P, D), first, a clock t which is not currently associated with P is picked. The abstract configuration (V, P, D) is transformed to $(V, \mathcal{A}(P, t), D \wedge t = 0)$, that is, timed processes which just become activated are associated with t and D is conjuncted with $t = 0$. Then, an abstract firing rule is applied to get a target configuration (V', P', D') such that D' must not be empty (otherwise, the transition is infeasible). Lastly, clocks which are not in $cl(P')$ are pruned from D'. Note that for all $(V, P, D) \in S$, $cl(P) = cl(D)$.

Example 5.6:

Assume a model with no variables and process P defined as follows:

$$P \widehat{=} (a \rightarrow Wait[5]; b \rightarrow Stop)\ interrupt[3]\ c \rightarrow P$$

Intuitively, event b never occurs because interrupt always occurs first. The abstract LTS is shown in Figure 5.12, where transitions are labeled with the introduced clock, event, and pruned clocks. Let

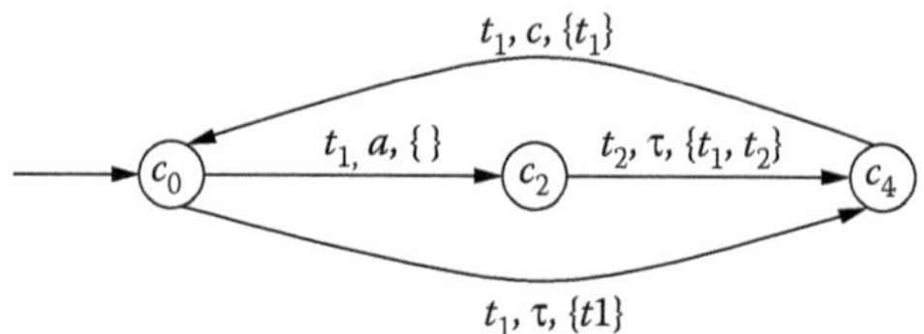

FIGURE 5.12 An abstract LTS.

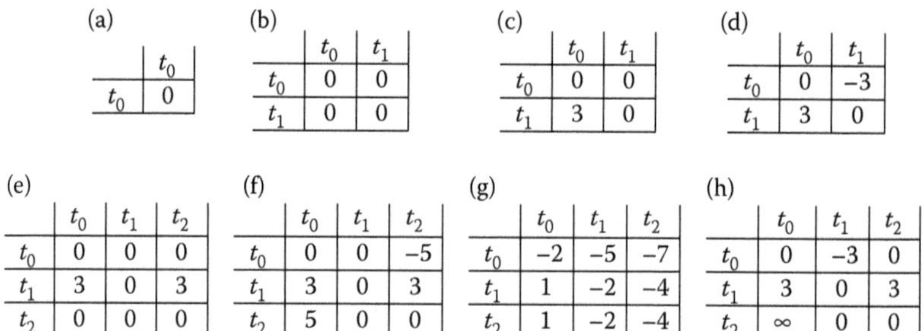

(a)

	t_0
t_0	0

(b)

	t_0	t_1
t_0	0	0
t_1	0	0

(c)

	t_0	t_1
t_0	0	0
t_1	3	0

(d)

	t_0	t_1
t_0	0	−3
t_1	3	0

(e)

	t_0	t_1	t_2
t_0	0	0	0
t_1	3	0	3
t_2	0	0	0

(f)

	t_0	t_1	t_2
t_0	0	0	−5
t_1	3	0	3
t_2	5	0	0

(g)

	t_0	t_1	t_2
t_0	−2	−5	−7
t_1	1	−2	−4
t_2	1	−2	−4

(h)

	t_0	t_1	t_2
t_0	0	−3	0
t_1	3	0	3
t_2	∞	0	0

FIGURE 5.13 DBM transformation.

$\langle t_1, t_2, \cdots \rangle$ be a sequence of clocks. The initial configuration is $c_0 = (\emptyset, P, true)$. For reference, the DBM representing *true* is presented in Figure 5.13a.

- Starting with c_0, we apply $\mathcal{A}$ to P with t_1 to get

$$c_1 = (\emptyset, (a \rightarrow Wait[5]; b \rightarrow Stop)\ interrupt[3]_{t_1}\ c \rightarrow P, t_1 = 0)$$

  The DBM representing $t_1 = 0$ is presented in Figure 5.13b. Next, we can apply either rule $ait1$ or $ait2$. Apply rule $ait1$, we get

$$c_2 = (\emptyset, (Wait[5]; b \rightarrow Stop)\ interrupt[3]_{t_1}\ c \rightarrow P, 0 \leq t_1 \leq 3)$$

  The DBM representing $0 \leq t_1 \leq 3$ is presented in Figure 5.13c. Apply rule $ait2$ to c_1, we get $c_3 = (\emptyset, c \rightarrow P, t_1 \geq 0 \wedge t_1 = 3)$. The DBM representing $t_1 \geq 0 \wedge t_1 = 3$ is presented in Figure 5.13d. Note that clock t_1 becomes irrelevant after the transition. After pruning t_1, we get $c_4 = (\emptyset, c \rightarrow P, true)$.
- Starting with c_2, we apply $\mathcal{A}$ to $(Wait[5]; b \rightarrow Stop)\ interrupt[3]_{t_1}\ c \rightarrow P$ with t_2 to get

$$c_5 = (\emptyset, (Wait[5]_{t_2}; b \rightarrow Stop)\ interrupt[3]_{t_1}\ c \rightarrow P, 0 \leq t_1 \leq 3 \wedge t_2 = 0)$$

  The DBM is represented in Figure 5.13e. Next, we can apply rule $ait1$ or $ait2$. Apply rule $ait1$ to c_5, we get

$$c_6 = (\emptyset, (Skip; b \rightarrow Stop)\ interrupt[3]_{t_1}\ c \rightarrow P, 0 \leq t_1 \leq 3 \wedge t_2 = 5)$$

  The DBM is presented in Figure 5.13f, whose canonical form is shown as Figure 5.13g. Notice that the zone is empty as $D_0^0 < 0$ and, therefore, this transition is invalid. Apply rule $ait2$ to c_5, we get

$$c_7 = (\emptyset, c \rightarrow P, t_1 \geq 0 \wedge t_2 \geq 0 \wedge t_2 \leq 5 \wedge t_1 = 3)$$

  Note that both clocks are irrelevant and, therefore, can be pruned. The resultant configuration is c_4.
- Starting with c_4, we have the transition to c_0. Note that t_1 is available and thus reused. □

It can be shown that $\mathcal{L}_S$ and $\mathcal{T}_S$ are time abstract equivalent [LY97] and, therefore, verification results based on $\mathcal{L}_S$ apply to the original Stateful Timed CSP models. In order to apply model checking techniques, we further establish that $\mathcal{L}_S$ is finite given any model $\mathcal{S}$.

Theorem 5.3:

$\mathcal{L}_S$ is finite for any model $\mathcal{S}$. □

By definition, the size of $\mathcal{L}_S$ is bounded by $\#V \times \#P \times \#D$, where $\#V$ denotes the number of variable valuations; $\#P$ denotes the number of processes; and $\#D$ denotes the number of zones. By assumption, all variables have finite domains and, therefore, $\#V$ is finite. By assumption, all reachable processes are finite. Thus, $\#P$ is finite if and only if values for parameters of the timed process constructs are finite and the number of clocks in $cl(P)$ is finite. It can be shown that all abstract firing rules preserve the parameters and, therefore, values for parameters is finite. By assumption, there can be only finitely many process constructs in any reachable process. Thus, $cl(P)$ is always finite. Therefore, $\#P$ is finite. Lastly, it is known that with zone normalization, the number of zones is finite given finitely many clocks and only integer constants [Dil89]. As a result, we conclude that $\mathcal{L}_S$ is finite, which implies that it is subject to model checking.

5.6 Summary

Real-time system modeling is of great importance and highly nontrivial. This chapter introduced the modeling languages and verification techniques for real-time system analysis.

In 1994, Alur and Dill proposed Timed Automata [AD94], which has become the de facto modeling language for real-time system with wide applications. This chapter introduced the syntax and semantics of Timed Automata with some examples. To perform automatic verification for Timed Automata, effective abstraction techniques are developed to reduce the infinite system state to finite ones, which make the system subject to model checking. This chapter covered two abstraction techniques (i.e., region abstraction and zone abstraction). A number of automatic verification support for Timed Automata based models have proven to be successful, for example, Cospan, Kronos, and Uppaal.

Models based on Timed Automata often have a simple structure. The advantage of a simple structure is that efficient model checking is made feasible. Nonetheless, designing and verifying compositional real-time systems are becoming an increasingly important task due to the widespread applications and increasing complexity of real-time systems. This chapter introduced two different approaches to model hierarchical systems with high-level timing requirements. One way to use the timed patterns to model fully hierarchical systems. Alternatively, hierarchical modeling languages like Stateful Timed CSP can be used to provide direct modeling and verification for hierarchical real-time systems.

References

[ABBL03] L. Aceto, P. Bouyer, A. Burgueño, and K. G. Larsen. The power of reachability testing for timed automata. *Theoretical Computer Science*, 300(1-3):411–475, 2003.

[AD94] R. Alur and D. L. Dill. A theory of timed automata. *Theoretical Computer Science*, 126: 183–235, 1994.

[BBBB09] C. Baier, N. Bertrand, P. Bouyer, and T. Brihaye. When are timed automata determinizable? In *36th International Colloquium on Automata, Languages and Programming*, volume 5556 of *Lecture Notes in Computer Science*, Greece, pages 43–54. Springer, 2009.

[BDM+98] M. Bozga, C. Daws, O. Maler, A. Olivero, S. Tripakis, and S. Yovine. Kronos: A model-checking tool for real-time systems. In *10th International Conference on Computer Aided Verification (CAV)*, volume 1427 of *Lecture Notes in Computer Science*, Canada, pages 546–550. Springer, 1998.

[BG06] H. Bowman and R. Gómez. How to stop time stopping. *Formal Aspects of Computing*, 18(4):459–493, 2006.

[BLP+99] G. Behrmann, K. G. Larsen, J. Pearson, C. Weise, and Wang Yi. Efficient timed reachability analysis using clock difference diagrams. In *11th International Conference on Computer Aided Verification (CAV)*, volume 1633 of *Lecture Notes in Computer Science*, Italy, pages 341–353. Springer, 1999.

[Bow99] H. Bowman. Modelling timeouts without timelocks. In *Proceedings of the 5th International AMAST Workshop on Formal Methods for Real-Time and Probabilistic Systems*, UK, pages 334–353, 1999.

[BY03] J. Bengtsson and Wang Yi. Timed automata: Semantics, algorithms and tools. In *Lectures on Concurrency and Petri Nets*, volume 3098 of *Lecture Notes in Computer Science*, Germany, pages 87–124. Springer, 2003.

[Cer92] K. Cerans. Decidability of bisimulation equivalences for parallel timer processes. In *Fourth International Workshop on Computer Aided Verification*, volume 663 of *Lecture Notes in Computer Science*, Canada, pages 302–315. Springer, 1992.

[Dav93] J. Davies. *Specification and Proof in Real-Time CSP*. Cambridge University Press, 1993.

[DHQ+08] J. S. Dong, P. Hao, S. C. Qin, J. Sun, and Wang Yi. Timed automata patterns. *IEEE Transactions on Software Engineering*, 34(6):844–859, 2008.

[Dil89] D. L. Dill. Timing assumptions and verification of finite-state concurrent systems. In *Automatic Verification Methods for Finite State Systems*, volume 407 of *Lecture Notes in Computer Science*, pages 197–212. Springer, New York, 1989.

[DMF99] J. S. Dong, B. P. Mahony, and N. Fulton. Modeling aircraft mission computer task rates. In *Formal Method (FM)*, volume 1708 of *Lecture Notes in Computer Science*, France, page 1855. Springer, 1999.

[Flo62] R. W. Floyd. Algorithm 97: Shortest path. *Communications of ACM*, 5(6):345, 1962.

[Gro] Object Management Group. UML resource page. Available at: http://www.omg.org/uml/.

[HG97] D. Harel and E. Gery. Executable object modeling with statecharts. *IEEE Computer*, 30(7): 31–42, 1997.

[HNSY94] T. A. Henzinger, X. Nicollin, J. Sifakis, and S. Yovine. Symbolic model checking for real-time systems. *Information and Computation*, 111(2):193–244, 1994.

[Hoa85] C. A. R. Hoare. *Communicating Sequential Processes*. International Series in Computer Science. Prentice-Hall, New Jersey, USA, 1985.

[HSLL97] K. Havelund, A. Skou, K. G. Larsen, and K. Lund. Formal modeling and analysis of an audio/video protocol: An industrial case study using UPPAAL. In *18th IEEE Real-Time Systems Symposium (RTSS)*, California, USA, pages 2–13. IEEE Computer Society, 1997.

[LMNS05] K. G. Larsen, M. Mikucionis, B. Nielsen, and A. Skou. Testing real-time embedded software using UPPAAL-TRON: An industrial case study. In *5th ACM International Conference on Embedded Software (EMSOFT)*, New Jersey, USA, pages 299–306. ACM, 2005.

[LP85] O. Lichtenstein and A. Pnueli. Checking that finite state concurrent programs satisfy their linear specification. In *Proceedings of POPL'85*, Louisiana, USA, pages 97–107. ACM, 1985.

[LPW97] K. G. Larsen, P. Pettersson, and Y. Wang. Uppaal in a nutshell. *International Journal on Software Tools for Technology Transfer*, 1(1-2):134–152, 1997.

[LPW01] M. Lindahl, P. Pettersson, and Y. Wang. Formal design and analysis of a gearbox controller. *International Journal on Software Tools for Technology Transfer*, 3(3):353–368, 2001.

[LV96] N. A. Lynch and F. W. Vaandrager. Action transducers and timed automata. *Formal Aspects of Computing*, 8(5):499–538, 1996.

[LW97] L. M. Lai and P. Watson. A case study in timed CSP: The railroad crossing problem. In *International Workshop on Hybrid and Real-Time Systems (HART)*, volume 1201 of *Lecture Notes in Computer Science*, France, pages 69–74. Springer, 1997.

[LY97] K. G. Larsen and Wang Yi. Time-abstracted bisimulation: Implicit specifications and decidability. *Information and Computation*, 134(2):75–101, 1997.

[Mil97] R. Milner. *Commnication and Concurrency*. Prentice Hall, New Jersey, USA, 1997.

[NS94] X. Nicollin and J. Sifakis. The algebra of timed processes, ATP: Theory and application. *Information and Computation*, 114(1):131–178, 1994.

[OW04] J. Ouaknine and J. Worrell. On the language inclusion problem for timed automata: Closing a decidability gap. In *19th IEEE Symposium on Logic in Computer Science (LICS 2004)*, Finland, pages 54–63. IEEE Computer Society, 2004.

[Pet99] P. Petterson. Modeling and verification of real-time systems using timed automata: Theory and practice, PhD thesis, Uppsala University, 1999.

[Rok93] T. G. Rokicki. Representing and modeling digital circuits, PhD thesis, Stanford University, 1993.

[Sch00] S. Schneider. *Concurrent and Real-Time Systems*. John Wiley and Sons, Chichester, UK, 2000.

[SLDP09] J. Sun, Y. Liu, J. S. Dong, and J. Pang. PAT: Towards flexible verification under fairness. In *20th International Conference on Computer Aided Verification (CAV)*, volume 5643 of *Lecture Notes in Computer Science*. France, Springer, 2009.

[SLDZ09] J. Sun, Y. Liu, J. S. Dong, and X. Zhang. Verifying stateful timed CSP using implicit clocks and zone abstraction. In *11th IEEEInternational Conference on Formal Engineering Methods (ICFEM)*, volume 5885 of *Lecture Notes in Computer Science*, Brazil, pages 581–600. Springer, 2009.

[TAKB96] S. Tasiran, R. Alur, R. P. Kurshan, and R. K. Brayton. Verifying abstractions of timed systems. In *7th International Conference on Concurrency Theory (CONCUR)*, volume 1119 of *Lecture Notes in Computer Science*, Italy, pages 546–562. Springer, 1996.

[Yi91] Wang Yi. CCS + time = An interleaving model for real time systems. In *18th International Colloquium on Automata, Languages and Programming (ICALP)*, volume 510 of *Lecture Notes in Computer Science*, Spain, pages 217–228. Springer, 1991.

6

Quantum Finite Automata

Daowen Qiu
Sun Yat-sen University
SQIG—Instituto de Telecomunicações
Instituto Superior Técnico
Chinese Academy of Sciences

Lvzhou Li
Sun Yat-sen University

Paulo Mateus
SQIG—Instituto de Telecomunicações
Instituto Superior Técnico

Jozef Gruska
Masaryk University

6.1 Introduction

Quantum information processing, communication, and security in general, and quantum computing in particular, have attracted large and growing attention of the academic community, and also outside of it, over the last two decades [1,2]. Since, the field can be seen as being a merge of arguably two of the most important areas of science and technology of the twentieth century, quantum physics and modern information processing and computing, it is expected to play a very important role in this century with broad implications.

Let us briefly mention some of the milestones in the development of quantum information processing. Landauer [3] has shown in 1961 that reversibility, an important characteristic of quantum evolution, is also of key importance for low-energy computation and also for supercomputing. Bennett [4] in 1973 made an additional important step by showing that reversibility, that used to be seen as a strange feature of quantum evolution, is not so alien for computation by showing that each Turing machine can be efficiently simulated by a reversible one. In 1980, Benioff [5] has demonstrated that, from the point of view of the information processing, quantum processes are as powerful as classical ones. Soon after that, Feynman [6] has pointed out in 1981 that there seems to be no efficient way to simulate all quantum processes on classical computers and, moreover, he indicated that to do that a new type of computer will be needed that will make use of the power of quantum phenomena and resources. All that made Deutsch [7] to reexamined the Church–Turing principle, to introduce quantum Turing machines (QTMs) and to show the existence of universal QTMs. Another important step was done by Yao [8] by showing that QTMs and quantum circuits are equivalent models of computation. All that created a basis for the development of theory of quantum computation. On a more practical side, the results by Shor [9] in 1994 and by Grover [10] in 1996 were very important. Shor proposed a quantum polynomial-time factorization algorithm and Grover found a way to speed up quadratically a variety of search problems over unstructured sets. These results had far-reaching impacts and made quantum information processing a very hot research topic for both theoreticians and experimentalists.

Models of automata used to play a very important role in (classical) theoretical computer science [11]. They were very helpful in finding power and limitations of various computation features and in developing a very powerful theory of computational complexity. Models of finite automata, as the key model behind the intuitive idea of real-time and finite memory computation, also helped to develop various design techniques for classical processors. All that is already a sufficient reason to assume that quantum versions of them can be expected to play an important role in the theory of quantum information processing. In addition, a vision of the most simple quantum processor as the one performing quantum actions on classical inputs leads also naturally to the models of quantum finite automata. The final reason for assuming that quantum finite automata could be of large importance is the fact that quantum memory seems to be very expensive and it is, therefore, of paramount importance to know what can be achieved with limited amounts of quantum resources.

A variety of models of quantum automata has already been introduced and explored to various degrees, similarly to the case of classical automata. They include quantum finite automata [12–34], quantum sequential machines (QSM) [35–37], quantum pushdown automata (QPDA) [38–40], QTM [41–47], quantum multicounter machines (QMCM) [47], quantum multistack machines (QMSM) [47], quantum cellular automata (QCA) [1,48,49], and orthomodular lattice-valued automata (l-VFA) [50–55] as well as other models. In this chapter, we discuss in detail only the basic versions of the following types of quantum automata: QFA, QSM, QPDA, QTM, and l-VFA. Several other versions of these models are mentioned only briefly because of space limitations. We are also not discussing here quantum cellular automata, which are important models of computation. There is a huge variety of approaches to QCA and there are many papers dealing with various attempts to introduce quantum features to classical cellular automata. Quantum cellular automata would require a special extensive paper or even a monograph to deal with. Actually, even for all models of quantum automata that we will deal with in some details, only main results and approaches could be presented—a monograph would be needed to cover all models and results in sufficient details.

There are various ways to see quantum automata as a modification of classical automata. One way is through probabilistic versions of the classical automata—in quantum case to each transition a probability amplitude is assigned instead of a probability. Another way is to see a configuration of a quantum automaton as a weighted superposition of the configurations of the corresponding classical automaton—with weights being probability amplitudes of configuration transitions. However, and this is a key point that brings another dimension into the actions of quantum automata, all that has to be done in such a way that evolution, a computation step, has to be governed by a unitary transformation. Moreover, for quantum automata to be useful, the outcome of their evolution has to be classical, be it acceptance or rejection or something else. This can be achieved only by a quantum measurement and various versions of quantum automata can be obtained by various policies when to perform a measurement and what kind of measurement to use.*

The remainder of the chapter is organized as follows. Section 6.2 is devoted to preliminaries. It introduces the very basic concepts from linear algebra, quantum mechanics, as well as from the formal language theory and probabilistic finite automata. All of them are concepts that will be needed in the rest of the chapter.

* A slightly different situation is in the case of the so-called orthomodular lattice-valued finite automata, where to each transition an element of an orthomodular lattice is assigned.

In Section 6.3, several models of QFA are presented that have been intensively explored in the literature: measure-once one-way quantum automata introduced by Moore and Crutchfiled [12], measure-many one-way quantum automata introduced by Kondacs and Watrous [13], one-way quantum finite automata together with classical states (1QFAC) [56], multiletter QFA [57,58], two-way QFA [13,19], generalized QFA [17,20], 1.5-way QFA [15], quantum automata with control language [23], and l-VFA [50–55]. In addition, we deal briefly also with other results related to models of quantum finite automata, especially those from papers [14,20,26,54].

Section 6.4 is devoted to several other models of automata. First, quantum sequential machines QSM are considered. They were introduced and explored first by Gudder [35] and then by Qiu [36] and by Li and Qiu [37,59]. QSM may be viewed as a quantum variant of stochastic sequential machines (SSM). A variety of quantum versions of classical pushdown automata [38–40] is considered next. The lattice-valued automata are the last model introduced and its power is analyzed in Section 6.4.

In the last section, after some concluding remarks, a variety of research directions and open problems concerning quantum finite automata are presented and their merit is discussed.

In this chapter, we usually mean *bounded error* for a machine $\mathcal{M}$ accepting a language $\mathcal{L}$ if we do not indicate explicitly another accepting fashion. The definition of bounded error is referred to Section 6.2.3.

6.2 Preliminaries

In this section, those very basic concepts and notations concerning linear algebra, quantum theory, and formal language theory are introduced that will be needed in the rest of the chapter.

6.2.1 Preliminaries from Linear Algebra

Let **C** denotes the set of complex numbers. Unless explicitly mentioned otherwise, the field of complex numbers will be the one we will work in this chapter. $\mathbf{C}^n$ will denote an n-dimensional vector space over **C**. An inner product on $\mathbf{C}^n \times \mathbf{C}^n$ is a mapping $(,)$: $\mathbf{C}^n \times \mathbf{C}^n \to \mathbf{C}$ satisfying the following properties:

1. $(x,x) \geq 0$ with equality if and only if $x = 0$
2. $(x,y) = (y,x)^*$
3. $(x, \lambda_1 y_1 + \lambda_2 y_2) = \lambda_1 (x, y_1) + \lambda_2 (x, y_2)$

where λ^* denotes the conjugate of complex number λ.

The inner product used in this chapter is the Euclidean inner product defined as

$$(x,y) = \sum_{i=1}^{n} x_i^* y_i. \tag{6.1}$$

This inner product introduces on $\mathbf{C}^n$ a norm as follows:

$$\|x\| = \sqrt{(x,x)} = \left(\sum_{i=1}^{n} |x_i|^2\right)^{\frac{1}{2}}. \tag{6.2}$$

A linear space with an inner product is called an *inner product space* and if it is complete, in a usual way, then it is called *Hilbert space.* A finite-dimensional inner product space is always a Hilbert space and only such will be used in the rest of the chapter.

The so-called Dirac bra-ket notation is usually used in quantum theory and we will use it also. For a column vector

$$v = \begin{bmatrix} v_1 \\ v_2 \\ \vdots \\ v_n \end{bmatrix} \tag{6.3}$$

notation $|v\rangle$ (ket-v) is mostly used and then $\langle v|$ (bra-v) denotes the row vector that is the conjugate transpose of v, that is $\langle v| = |v\rangle^\dagger$. Moreover, $\langle v|v'\rangle$ denotes the inner product of vectors $v = |v\rangle$ and $v' = |v'\rangle$ and $|v\rangle\langle v'|$ is then the matrix obtained when column vector $|v\rangle$ is multiplied by a row vector $\langle v'|$.

Two states v and v' of an inner product space are called *orthogonal* if $\langle v|v'\rangle = 0$.

For a matrix A, let A^*, A^T, and $A^\dagger$ denote conjugate, transpose, and conjugate transpose of the matrix A, respectively. I will denote the unitary matrix.

A matrix $A \in \mathbf{C}^{n\times n}$ is called *unitary* if $A \cdot A^\dagger = A^\dagger \cdot A = I$. If $x^\dagger A x \geq 0$ for any $x \in \mathbf{C}^n$, then the matrix A is called positive and for that the notation $A \geq 0$ is used.

6.2.2 Preliminaries from Quantum Theory

According to a quantum mechanical principle, to each closed quantum system* $\mathcal{S}$, a Hilbert space $\mathcal{H}_\mathcal{S}$ is associated and states of $\mathcal{S}$ correspond to the vectors of the norm one of $\mathcal{H}_\mathcal{S}$. In case $\mathcal{H}_\mathcal{S}$ is an n-dimensional vector space, then it has a basis consisting of n mutually orthogonal vectors and we will usually denote these vectors and the bases by

$$\{|i\rangle\}_{i=1}^n$$

and in such a case any vector of $\mathcal{H}_\mathcal{S}$ can be uniquely expressed as the following superposition:

$$|\psi\rangle = \sum_{i=1}^{n} \alpha_i |i\rangle \tag{6.4}$$

where α_i's are complex numbers, called amplitudes, that satisfy the following so-called normalization condition $\sum_{i=1}^n |\alpha_i|^2 = 1$. If such a state is measured with respect to the above bases, then the state collapses to one of the states $|i\rangle$ and to a particular state $|i_0\rangle$ with the probability $|\alpha_{i_0}|^2$ and i_0 is the outcome we receive into the classical world.

Each evolution of a finite n-dimensional quantum system is specified by a unitary $n \times n$ matrix U and each (discrete) step of the evolution changes any state $|\phi\rangle$ into the state $U|\phi\rangle$.

To extract some information from a quantum state $|\psi\rangle$, a measurement has to be performed. We will consider here only projective measurements that are defined by a set $\{P_m\}$ of the so-called projective operators/matrices, where indices m refer to the potential classical outcomes of measurements, with the property:

$$P_iP_j = \begin{cases} P_i & i = j, \\ 0 & i \neq j, \end{cases} \tag{6.5}$$

that, in addition, satisfies the following completeness condition:

$$\sum_{i=1}^{n} P_i = I. \tag{6.6}$$

* That is to a quantum system that has no interaction with an environment.

In case a state $|\psi\rangle$ is measured with respect to the set of projective operators $\{P_m\}$, the classical outcome m is obtained with the probability

$$p(m) = \|P_m|\psi\rangle\|^2, \tag{6.7}$$

and the state $|\psi\rangle$ "collapse" into the state

$$\frac{P_m|\psi\rangle}{\sqrt{p(m)}}. \tag{6.8}$$

A projective measurement $\{P_m\}$ is usually specified by an *observable* M, a Hermitian matrix that has a so-called spectral decomposition

$$M = \sum_m mP_m \tag{6.9}$$

where m are mutually different eigenvalues of M and P_m is a projector into the space of eigenvectors associated to the eigenvalue m.

6.2.3 Preliminaries on Probabilistic Finite Automata and Accepting Languages

Models of quantum automata have in principle a probabilistic behavior. In classical computing, we have also probabilistic models of computation. Of a special interest and importance for us are probabilistic finite automata (PFA) [60,61] that have close relations to quantum finite automata discussed in this chapter. Therefore, we first introduce basic concepts related to PFA.

Definition 6.1:

A probabilistic finite automaton (PFA) over the alphabet Σ is a four-tuple

$$\mathcal{M} = (S, \pi, \{M(\sigma)\}_{\sigma\in\Sigma}, \eta),$$

where

- $S = \{s_1, s_2, \cdots, s_n\}$ is a finite state set
- π is an n-dimensional stochastic row vector denoting the initial distribution over S
- $M(\sigma)$ is an $n \times n$ stochastic matrix (i.e., every row of $M(\sigma)$ is a stochastic vector) for every $\sigma \in \Sigma$, that is, $M(\sigma)(i, j)$ is the value of the transition probability from a state s_i to a state s_j when reading symbol σ
- η is an n-dimensional column vector with entries being 0's and 1's; $\eta_i = 1$ means s_i is an accepting state

---

Note that in the above definition, if it is further required that the stochastic matrix $M(\sigma)$ has only 1's and 0's as entries, then M reduces to a deterministic finite automaton (DFA). Given an input string $x = x_1x_2\cdots x_n \in \Sigma^*$, $\mathcal{M}$ has an accepting probability given by

$$P_{\mathcal{M}}(x) = \pi M(x_1)M(x_2)\cdots M(x_n)\eta. \tag{6.10}$$

A language $\mathcal{L}$ is said to be accepted (recognized) by PFA $\mathcal{M}$ *with a cut-point* $\lambda \in [0, 1)$, if for every $x \in \mathcal{L}$ it holds that $P_{\mathcal{M}}(x) > \lambda$, and for every $x \notin \mathcal{L}$ it holds that $P_{\mathcal{M}}(x) \leq \lambda$. In this case, we also say that $\mathcal{M}$ accepts the language $\mathcal{L}$ *with an unbounded error*. Languages accepted by PFA with a cut-point form the class of *stochastic languages*.

A language $\mathcal{L}$ is said to be accepted (recognized) by a PFA $\mathcal{M}$ *with an isolated cut-point* $\lambda \in [0, 1)$, if there exists $\epsilon > 0$ such that for every $x \in \mathcal{L}$ it holds that $P_{\mathcal{M}}(x) \geq \lambda + \epsilon$, and for every $x \notin \mathcal{L}$ it holds that $P_{\mathcal{M}}(x) \leq \lambda - \epsilon$. In this case, we also say that $\mathcal{M}$ accepts the language $\mathcal{L}$ *with bounded error.*

In the following sections, when discussing quantum automata, we will consider acceptance of languages in a similar way as defined above and we will not define it explicitly again. Moreover, unless specified otherwise, by acceptance we will always mean the acceptance with respect to bounded error.

6.3 Quantum Finite Automata

There is a variety of models of quantum finite automata (QFA). All of them can be seen as models of quantum processors in a similar way as various models of classical finite automata serve as models of classical processors that always have finite memory [11]. The first two main models of QFA were introduced independently by Moore and Crutchfield [12], as well as by Kondacs and Watrous [13] with different fashions, and then they were intensively investigated by others.

There are several ways to divide the QFA model. The two main ones are according to "head moves" and according to quantum resources and tools they operate with. According to the head moves, the main models are those whose heads always move and always in the same direction (so-called one-way QFA). As natural modifications of that are models whose heads can move in both directions (so called two-way QFA), or in one-direction only but sometimes they may be stationary (so-called 1.5 way QFA introduced by Amano and Iwama [15])—in the last two cases, however, models are not really with finite memory. Another distinction is whether models use only pure states, unitary operations, and projective measurements or work with a full power of quantum resources and operations—with mixed states, general quantum operations, and POVM measurements.

6.3.1 One-Way Quantum Finite Automata

All models of one-way quantum automata (1QFA) share the properties that their "heads" move in each step in the same direction (and create a superposition of "new heads")—to process an input word from left to right, symbol by symbol. 1QFA can further be divided into some subclasses such as MO-1QFA, MM-1QFA, CL-1QFA, *Latvian* QFA (LQFA), and generalized quantum finite automata (GQFA), which will be introduced in the following.

6.3.1.1 Measure-Once Quantum Finite Automata

MO-1QFA are the simplest model of quantum finite automata. They were first introduced by Moore and Crutchfield [12], and then studied by Brodsky and Pippenger [16] and others. MO-1QFA with bounded error acceptance mode accept exactly the so-called group languages, a proper subset of regular languages, but with unbounded error acceptance mode they can also accept some nonregular and even non-context-free languages. No matter what acceptance mode is considered, the language acceptance power of MO-1QFA does not surpass that of their classical counterparts—probabilistic finite automata. However, in some other aspects, such as state complexity, MO-1QFA show some advantages over DFA and PFA. We will introduce these results in more details below. First, let as define formally MO-1QFA.

Definition 6.2:

An MO-1QFA is a 5-tuple $\mathcal{M} = (Q, \Sigma, \delta, q_0, F)$, where

- $Q = \{q_0, q_1, \cdots, q_{n-1}\}$ is a finite set of states.
- Σ is a finite input alphabet.

- δ is a transition function $\delta : Q \times \Sigma \times Q \longrightarrow \mathbf{C}$ that satisfies the unitary condition

$$\sum_{p \in Q} \overline{\delta(q_1, \sigma, p)} \delta(q_2, \sigma, p) = \begin{cases} 1 & q_1 = q_2, \\ 0 & q_1 \neq q_2 \end{cases} \tag{6.11}$$

  for all states $q_1, q_2 \in Q$ and symbols $\sigma \in \Sigma$, where $\delta(p, \sigma, q)$ (resp. $|\delta(p, \sigma, q)|^2$) denotes the amplitude (resp. probability) of $\mathcal{M}$ going from the state p to the state q, upon reading symbol σ; $\overline{\delta(q_1, \sigma, p)}$ is the complex conjugate of $\delta(q_1, \sigma, p)$.
- q_0 denotes the initial state, and $F \subseteq Q$ denotes the set of accepting states.

The quantum state space of this model is a $|Q|$-dimensional Hilbert space denoted by H_Q. Each classical state $q_i \in Q = \{q_0, \ldots, q_{|Q|-1}\}$ corresponds to a basis state $|q_i\rangle$ in H_Q represented by a column vector with the $(i+1)$th entry being 1 and others being 0's. With this notational convenience, we can describe the above model as follows:

- The initial state q_0 can be represented as $|q_0\rangle = (1, 0, \cdots, 0)^{\mathrm{T}}$.
- For each $\sigma \in \Sigma$, the transition function δ can be represented by a unitary matrix $U(\sigma)$ with $U(\sigma)(j, i) = \delta(q_i, \sigma, q_j)$.
- The accepting set F corresponds to a projective operator $P_{\text{acc}} = \sum_{i:q_i \in F} |q_i\rangle\langle q_i|$.

The computing process of an MO-1QFA $\mathcal{M}$ on an input string $x = \sigma_1\sigma_2 \cdots \sigma_n \in \Sigma^*$ goes as follows: The tape head of $\mathcal{M}$ reads the input string from the left to right symbol by symbol, and meanwhile the unitary matrices $U(\sigma_1), U(\sigma_2), \ldots, U(\sigma_n)$ are applied, one by one, on the current state, starting with $|q_0\rangle$ as the initial state. Finally, a projective measurement $\{P_{\text{acc}}, I - P_{\text{acc}}\}$ is performed on the final state, in order to accept the input or not. Therefore, for the input string x, $\mathcal{M}$ has an accepting probability given by

$$P_{\mathcal{M}}(x) = \|P_{\text{acc}} U(\sigma_n) \cdots U(\sigma_2) U(\sigma_1) |q_0\rangle\|^2. \tag{6.12}$$

In the following, we give an example of MO-1QFA.

Example 6.1:

Let MO-1QFA $\mathcal{M} = (Q, \Sigma, \delta, q_0, F)$, where $Q = \{q_0, q_1\}$, $F = \{q_1\}$, $\Sigma = \{a\}$, q_0 is the initial state, and the transition function δ is as follows:

$$\delta(q_0, a, q_0) = \frac{1}{\sqrt{2}}, \quad \delta(q_0, a, q_1) = \frac{1}{\sqrt{2}},$$
$$\delta(q_1, a, q_0) = \frac{1}{\sqrt{2}}, \quad \delta(q_1, a, q_1) = -\frac{1}{\sqrt{2}}.$$

Then, the transition function can also be described by the following matrix:

$$U(a) = \begin{pmatrix} \frac{1}{\sqrt{2}} & \frac{1}{\sqrt{2}} \\ \frac{1}{\sqrt{2}} & -\frac{1}{\sqrt{2}} \end{pmatrix},$$

states q_0 and q_1 can be described by two vectors:

$$|q_0\rangle = \begin{pmatrix} 1 \\ 0 \end{pmatrix}, \quad |q_1\rangle = \begin{pmatrix} 0 \\ 1 \end{pmatrix},$$

and the accepting set F can be described by

$$P_{acc} = |q_1\rangle\langle q_1| = \begin{pmatrix} 0 & 0 \\ 0 & 1 \end{pmatrix}.$$

Next, we show how this automaton works on the input word aa.

1. The automaton starts in $|q_0\rangle$. After the first a has been fed, the state changes to $|\psi\rangle = \frac{1}{\sqrt{2}}|q_0\rangle + \frac{1}{\sqrt{2}}|q_1\rangle$, by applying $U(a)$ on $|q_0\rangle$.
2. After the second a has been fed, the state changes to $|q_0\rangle$, by applying $U(a)$ on $|\psi\rangle$. Then, P_{acc} is performed on the final state, and we obtain that the probability of $\mathcal{M}$ accepting aa is zero.

Moore and Crutchfield [12] first defined MO-1QFA. Gudder [62] considered the above definition without initial and accepting states. Moore and Crutchfield [12] presented some closure properties of MO-1QFA, and they also proved a Pumping lemma for this model (a different proof was given in Ref. [21]).

Theorem 6.1: ([12], Pumping lemma)

Let $\mathcal{M}$ be an MO-1QFA, and $P_{\mathcal{M}}(x)$ be the probability of $\mathcal{M}$ accepting x. Then, for any word w and any $\epsilon > 0$, there exists an integer k such that $|P_{\mathcal{M}}(uw^k v) - P_{\mathcal{M}}(uv)| \leq \epsilon$ for any words u, v. Moreover, if $\mathcal{M}$ is n-dimensional, then there is a constant c such that $k \leq (c\epsilon)^{-n}$.

Additionally, Moore and Crutchfield [12] derived a method for bilinearizing the representation of an MO-1QFA that can transform an MO-1QFA into the form of a generalized stochastic system, which enables one to study some related problems on MO-1QFA by using the classical methods.

Due to close connections between languages accepted by MO-1QFA and by group automata and reversible finite automata (RFA), we recall the last two models now (their relationships were dealt with in [54]) as follows.

Proposition 6.1:

Let $M = (Q, \Sigma, \delta)$ be a deterministic finite automaton (DFA). Then the following three statements are equivalent:

i. M is a *group automaton*, that is, for any $q \in Q$ and any $\sigma \in \Sigma$, there exists a unique $p \in Q$ such that $\delta(p, \sigma) = q$.
ii. M is a *BM-reversible automaton*, that is, for any $\sigma \in \Sigma$, there exists a string $x \in \Sigma^*$ such that for any $q \in Q$, $\delta(q, \sigma x) = q$.
iii. M is an AF-reversible automaton, that is, for any $q \in Q$ and any $\sigma \in \Sigma$, there is at most one $p \in Q$ such that $\delta(p, \sigma) = q$.

Bertoni and Carpentieri [21] studied some analogies and differences between MO-1QFA and probabilistic finite automata. In another work [22], they proved that the class of languages accepted by MO-1QFA with bounded error (i.e., with an isolated cut-point) is the class of reversible regular languages, where the class of reversible regular languages is equivalent to the class of languages accepted by AF-reversible automata mentioned in Proposition 6.1.

Brodsky and Pippenger [16] introduced the language classes RMO and UMO as the sets of languages recognized by MO-1QFA with bounded error and with unbounded error (i.e., with cut-point), respectively. They also proved that RMO is exactly the set of languages recognized by group automata, which attains the same result as the one in Ref. [22] mentioned above. Therefore, RMO is a proper subclass of regular languages. In addition, they showed that UMO is a subclass of stochastic languages (i.e., the class of languages recognized by probabilistic finite automata with a cut-point) ([16], Theorem 3.6). More exactly, UMO is a proper subclass of stochastic languages, since in Ref. [21] it was shown that UMO does not contain any finite language.

Blondel et al. [25] discussed the following decision problem regarding MO-1QFA: is the language recognized by an MO-1QFA $\mathcal{M}$ with (strict or nonstrict*) cut-point λ empty? Results concerning this decision problem show some interesting difference between MO-1QFA and probabilistic finite automata. For example, the problem whether the language $L_{>}$, defined in the footnote, is empty is decidable for MO-1QFA, while it is undecidable for probabilistic finite automata [25]. Hirvensalo [63] also considered this problem and obtained some improved results.

To determine whether two DFA or two probabilistic automata are equivalent or not is an important problem in the classical automata theory [11,60,61]. Thus, it is natural to consider the equivalence problem also for QFA. Two QFA are said to be equivalent (resp. k-equivalent) if they have the same accepting probability for every input string (resp. for every string with the length not longer than k). For the equivalence problem of MO-1QFA, Brodsky and Pippenger [16] sketched an algorithm to decide the equivalence between two MO-1QFA. Later, following the idea pointed out by Brodsky and Pippenger, Koshiba [64] further detailed a polynomial-time algorithm to deal with the equivalence of MO-1QFA. Li and Qiu [59] used a slightly different method to deal with this problem. We summarize the above result as follows.

Theorem 6.2: ([59])

Two MO-1QFA $\mathcal{A}_1$ and $\mathcal{A}_2$ are equivalent if and only if they are $(n_1+n_2)^2$-equivalent, where n_1 and n_2 are the numbers of states in $\mathcal{A}_1$ and $\mathcal{A}_2$, respectively. There is a polynomial-time algorithm in n_1, n_2 deciding whether $\mathcal{A}_1$ and $\mathcal{A}_2$ are equivalent or not.

The minimization problem for classical DFA and PFA has also been investigated in depth [11,60,61]. This problem has been solved in an elegant way for DFA. However, a different situation is in the case of PFA. In this case, research has focused mainly on methods to reduce the number of states and on verification of the minimality, especially when only elementary linear algebra is used (see Ref. [61])). As for quantum automata, we restate the minimization problem for MO-1QFA proposed by Moore and Crutchfield ([12], p. 304, Problem 5):

Is each quantum regular language (QRL) recognized by a unique QFA (up to isomorphism) with the minimal dimension? Here QFA means MO-1QFA, and the dimension is the number of basis states of QFA. Therefore, this problem is essentially about minimization of the number of states of MO-1QFA.

Recently, Mateus and Qiu [65] addressed this problem for semialgebraic families of bilinear machines and gave an EXPSPACE algorithm to find a minimal automaton in the family of automata equivalent to the given one. Therefore, they have solved the minimization problem for these cases (including the minimization of probabilistic automata and MO-1QFA). Moreover, they have solved an open problem concerning the covering minimization of stochastic sequential machines as proposed by Paz [61]. Indeed, with their method and technique one can solve the state minimization for all models of computation whose equivalence is known to be decidable.

More formally, the following result has been proved.

Theorem 6.3: ([65])

The state minimization of MO-1QFA is decidable in EXPSPACE.

---

* The language recognized with a strict cut-point λ is denoted by $L_{>} = \{w : P_{\mathcal{M}}(x) > x\}$; the language recognized with a nonstrict cut-point λ is denoted by $L_{\geq} = \{w : P_{\mathcal{M}}(x) \geq x\}$.

6.3.1.2 One-Way Quantum Finite Automata with Classical States

A special model of quantum automata called one-way quantum finite automata with classical states (1QFAC) was introduced in Ref. [56]. In this new model, a DFA is used to determine the choice of a unitary transformation to be applied for each input symbol. We now describe informally how an 1QFAC processes an input string. At the beginning, the automaton $\mathcal{A}$ is in the initial classical state and in the initial quantum state. By reading the first input symbol, the classical state transition results in a new classical state as the current classical state, and, the initial classical state together with the current input symbol determines a unitary transformation to process the initial quantum state, leading to a new quantum state as the current quantum state. Afterward, the machine reads the next input symbol, and similarly to the above process, its classical state will be updated by reading the current input symbol and, at the same time, with the current classical state and input symbol, a new unitary transformation is assigned to execute the current quantum state. Subsequently, the automaton $\mathcal{A}$ keeps processing input symbols, one by one, in a similar way until the last input symbol is processed. Afterward, a measurement determined by the last classical state is performed on the final quantum state in order to determine the acceptance or rejection.

Given a finite set $B \subseteq Q$, we denote by $\mathcal{H}(B)$ the subspace of the Hilbert space $\mathcal{H}(Q)$ generated by the basis states $\{|q\rangle : q \in B\}$. Furthermore, we denote by I and O the identity operator and zero operator on $\mathcal{H}(Q)$, respectively. The formal definition of 1QFAC is as follows:

Definition 6.3:

A 1QFAC $\mathcal{A}$ is defined by a 9-tuple

$$\mathcal{A} = (S, Q, \Sigma, \Gamma, s_0, |\psi_0\rangle, \delta, \mathbb{U}, \mathcal{M})$$

where:

- Σ is a finite set (the input alphabet)
- Γ is a finite set (the output alphabet)
- S is a finite set (the set of classical states)
- Q is a finite set (a finite classical set specifying basis of the underlying Hilbert space)
- s_0 is an element of S (the initial classical state)
- $|\psi_0\rangle$ is a unit vector in the Hilbert space $\mathcal{H}(Q)$ (the initial quantum state)
- $\delta : S \times \Sigma \to S$ is a map (the classical transition map)
- $\mathbb{U} = \{U_{s\sigma}\}_{s\in S,\sigma\in\Sigma}$ where $U_{s\sigma} : \mathcal{H}(Q) \to \mathcal{H}(Q)$ is a unitary operator for each s and σ (the *quantum transition operator* at s and σ)
- $\mathcal{M} = \{\mathcal{M}_s\}_{s\in S}$ where each $\mathcal{M}_s$ is a projective measurement over $\mathcal{H}(Q)$ with outcomes in Γ (the *measurement operator at* s)

Hence, each $\mathcal{M}_s = \{P_{s,\gamma}\}_{\gamma\in\Gamma}$ such that $\sum_{\gamma\in\Gamma} P_{s,j} = I$ and $P_{s,\gamma}P_{s,\gamma'} = \begin{cases} P_{s,\gamma}, & \gamma = \gamma', \\ O, & \gamma \neq \gamma'. \end{cases}$ Furthermore, if the automaton is in a classical state s and a quantum state $|\psi\rangle$ after reading the input string, then $\|P_{s,\gamma}|\psi\rangle\|^2$ is the probability of the machine producing the outcome γ on that input string.

Remark 6.1:

Map δ can be extended to a map $\delta^* : \Sigma^* \to S$ as usual. That is, $\delta^*(s, \epsilon) = s$; for any string $x \in \Sigma^*$ and any $\sigma \in \Sigma$, $\delta^*(s, \sigma x) = \delta^*(\delta(s, \sigma), x)$.

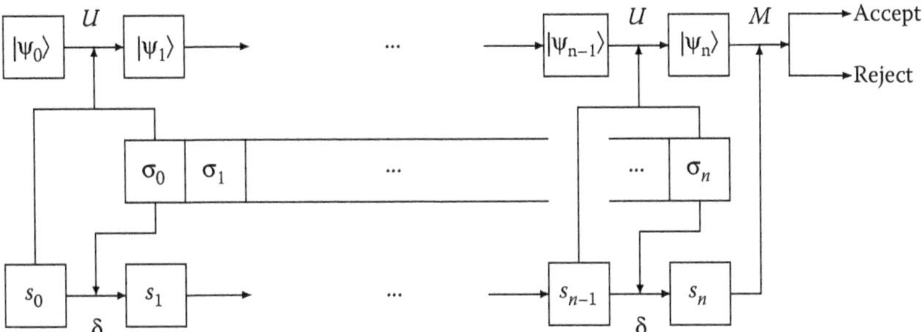

FIGURE 6.1 1QFAC dynamics as an acceptor of language.

Remark 6.2:

A specially interesting case of the above definition is when $\Gamma = \{a, r\}$, where a denotes *acceptance* and r denotes *rejection*. Then, $\mathcal{M} = \{\{P_{s,a}, P_{s,r}\} : s \in S\}$ and, for each $s \in S$, $P_{s,a}$ and $P_{s,r}$ are two projectors such that $P_{s,a} + P_{s,r} = I$ and $P_{s,a}P_{s,r} = O$. In this case, $\mathcal{A}$ is an acceptor of languages over Σ.

For intuition, we depict an acceptor 1QFAC in Figure 6.1.

In Ref. [56], the following results have been obtained. It was proved that the set of languages accepted by 1QFAC with bounded error consists precisely of all regular languages. For any prime number $m \geq 2$, there exists a regular language $L^0(m)$ whose minimal DFA needs $O(m)$ states and that cannot be accepted by the 1QFA proposed by Moore and Crutchfield and by Kondacs and Watrous, but there exists 1QFAC accepting $L^0(m)$ with only constant number of the classical states and $O(\log(m))$ of quantum basis states. By a bilinearization technique it was proved that any two 1QFAC $\mathcal{A}_1$ and $\mathcal{A}_2$ are equivalent if and only if they are $(k_1n_1)^2 + (k_2n_2)^2 - 1$-equivalent, and that there exists a polynomial-time $O([(k_1n_1)^2 + (k_2n_2)^2]^4)$ algorithm for determining their equivalence, where k_1 and k_2 are the numbers of classical states of $\mathcal{A}_1$ and $\mathcal{A}_2$, as well as n_1 and n_2 are the numbers of quantum basis states of $\mathcal{A}_1$ and $\mathcal{A}_2$, respectively. In summary, we describe them using the following results:

Proposition 6.2:

For any prime number $m \geq 2$, there exists a regular language $L^0(m)$ satisfying: (1) neither MO-1QFA nor MM-1QFA can accept $L^0(m)$; (2) the number of states in the minimal DFA accepting $L^0(m)$ is $m+1$; (3) for any $\varepsilon > 0$, there exists a 1QFAC with 2 classical states and $O(\log(m))$ quantum basis states such that for any $x \in L^0(m)$, x is accepted with no error, and the probability for accepting $x \notin L^0(m)$ is smaller than ε.

Theorem 6.4:

Two 1QFAC $\mathcal{A}_1$ and $\mathcal{A}_2$ are equivalent if and only if they are $(k_1n_1)^2 + (k_2n_2)^2 - 1$-equivalent. Furthermore, there exists a polynomial-time algorithm running in time $O([(k_1n_1)^2 + (k_2n_2)^2]^4)$ that takes as input two 1QFAC $\mathcal{A}_1$ and $\mathcal{A}_2$ and determines whether $\mathcal{A}_1$ and $\mathcal{A}_2$ are equivalent, where k_i and n_i are the numbers of classical and quantum basis states of $\mathcal{A}_i$, respectively, $i = 1, 2$.

Remark 6.3:

Using the ideas from Ref. [65], the minimization of quantum basis states for 1QFAC has been showed to be decidable in EXPSPACE [56].

6.3.1.3 Measure-Many Quantum Finite Automata

MM-1QFA were first proposed by Kondacs and Watrous [13], and then studied by Ambainis and Freivalds [14], by Brodsky and Pippenger [16], and by others. MM-1QFA are strictly more powerful than MO-1QFA, but still recognize with bounded error only a proper subclass of regular languages. The behavior of MM-1QFA is more complicated than that of MO-1QFA, and thus a complete characterization of the languages recognized by MM-1QFA is still not clear. The definition of MM-1QFA is similar to that of MO-1QFA. The essential difference between them is that in an MO-1QFA only one measurement is allowed at the end of the input string, but in an MM-1QFA, the measurement is allowed after each symbol has been read.

In the following, after giving the formal definition of MM-1QFA, we will present some main results regarding this model.

Definition 6.4:

An MM-1QFA is a 6-tuple $\mathcal{A} = (Q, \Sigma, \delta, q_0, Q_{acc}, Q_{rej})$, where Q is a finite set of states, Σ is a finite input alphabet, equipped with an end-marker symbol $\$ \notin \Sigma$ (denote $\Gamma = \Sigma \cup \{\$\}$), $\delta : Q \times \Gamma \times Q \to \mathbf{C}$ is a transition function satisfying Equation 6.11, and state q_0 is the initial state of $\mathcal{A}$. The set Q is partitioned into three subsets: Q_{acc} is the set of accepting states, Q_{rej} is the set of rejecting states, and Q_{non} is the set of nonhalting states.

Remark 6.4:

(i) Note that in the original definition of MM-1QFA given by Kondacs and Watrous [13], the input alphabet Σ had two end-markers at the start and the end of the input string. However, Brodsky and Pippenger [16] showed that if only one end-maker at the end of the tape is used, then this does not affect the acceptance power of MM-1QFA.

Similarly to the definition of MO-1QFA, we can also describe MM-1QFA using the language of matrices. Then, the initial state q_0 of MM-1QFA $\mathcal{A}$ can be represented by a unit vector $|q_0\rangle$ in the Hilbert space $\mathcal{H}_Q$; for any $\sigma \in \Gamma$, the transition function δ can be represented by a $|Q| \times |Q|$ unitary matrix $U(\sigma)$ with $U(\sigma)(j, i) = \delta(q_i, \sigma, q_j)$. The whole space $\mathcal{H}_Q$ is divided into three subspaces: (i) $E_{non} = \text{span}\{|q\rangle : q \in Q_{non}\}$, (ii) $E_{acc} = \text{span}\{|q\rangle : q \in Q_{acc}\}$, (iii) $E_{rej} = \text{span}\{|q\rangle : q \in Q_{rej}\}$, and correspondingly there are three projectors $P_{non}, P_{acc}, P_{rej}$ onto the three subspaces, respectively. Thus, $M = \{P_{non}, P_{acc}, P_{rej}\}$ is a projective measurement on $\mathcal{H}_Q$.

The computing process of MM-1QFA is similar to that of MO-1QFA except that after each input symbol σ is read and the corresponding unitary operator $U(\sigma)$ is applied to the current quantum state of the system, the projection measurement M is applied on the current state. If the measurement result is "*non*," then the computation continues; if the result is "*acc*," then $\mathcal{A}$ accepts, otherwise it rejects. After every measurement, the state collapses into the subspace specified by the projection operator that has been applied and the state has been renormalized.

Since there is a nonzero probability that the automaton A halts partway through the computation, it is useful to keep track of the cumulative accepting and rejecting probabilities. Thus, we can represent the current state of $\mathcal{A}$ as a triple $(|\psi\rangle, p_{acc}, p_{rej})$, where p_{acc}, p_{rej} are, respectively, the cumulative probabilities of accepting and rejecting. Then the initial state of $\mathcal{A}$ can be represented by $(|q_0\rangle, 0, 0)$, and the evolution of $\mathcal{A}$ on reading a symbol σ is denoted by

$$(|\psi\rangle, p_{acc}, p_{rej}) \mapsto (P_{non}|\psi'\rangle, p_{acc} + \|P_{acc}|\psi'\rangle\|^2, p_{rej}p_{acc} + \|P_{rej}|\psi'\rangle\|^2) \tag{6.13}$$

where $|\psi'\rangle = U(\sigma)|\psi\rangle$. On an input string $\sigma_1\sigma_2\cdots\sigma_n\$$, the accepting probability of $\mathcal{A}$ is given by

$$P_{\mathcal{A}}(\sigma_1\ldots\sigma_n) = \sum_{k=1}^{n+1} \|P_{\text{acc}}U(\sigma_k)\prod_{i=1}^{k-1}(P_{\text{non}}U(\sigma_i))|q_0\rangle\|^2, \tag{6.14}$$

where let $\$ = \sigma_{n+1}$, and we define $\prod_{i=1}^{n} A_i = A_nA_{n-1}\cdots A_1$, instead of $A_1A_2\cdots A_n$.

In the following, we give an example of MM-1QFA.

Example 6.2:

Let MM-1QFA $\mathcal{M} = (Q, \Sigma, \{U(\sigma)\}_{\sigma\in\Sigma\cup\{\$\}}, q_0, Q_{\text{acc}}, Q_{\text{rej}})$, where $Q = \{q_0, q_1, q_{\text{acc}}, q_{\text{rej}}\}$ with the set of accepting states $Q_{\text{acc}} = \{q_{\text{acc}}\}$ and the set of rejecting states $Q_{\text{rej}} = \{q_{\text{rej}}\}$; $\Sigma = \{a\}$; q_0 is the initial state; $\{U(\sigma)\}_{\sigma\in\Sigma\cup\{\$\}}$ are described below:

$$U(a)|q_0\rangle = \frac{1}{2}|q_0\rangle + \frac{1}{\sqrt{2}}|q_1\rangle + \frac{1}{2}|q_{\text{acc}}\rangle,$$

$$U(a)|q_1\rangle = \frac{1}{2}|q_0\rangle - \frac{1}{\sqrt{2}}|q_1\rangle + \frac{1}{2}|q_{\text{acc}}\rangle,$$

$$U(\$)|q_0\rangle = |q_{\text{acc}}\rangle, \quad U(\$)|q_1\rangle = |q_{\text{rej}}\rangle.$$

The transition function can also be defined by describing the mapping δ. For example, $U(a)|q_0\rangle = \frac{1}{2}|q_0\rangle + \frac{1}{\sqrt{2}}|q_1\rangle + \frac{1}{2}|q_{\text{acc}}\rangle$ would be

$$\delta(q_0, a, q_0) = \frac{1}{2}, \quad \delta(q_0, a, q_1) = \frac{1}{\sqrt{2}},$$

$$\delta(q_0, a, q_{\text{acc}}) = \frac{1}{2}, \quad \delta(q_0, a, q_{\text{rej}}) = 0.$$

There are some transitions that we have not described. For example, $U(a)|q_{\text{acc}}\rangle$ has not been specified. These values are not important and can be arbitrary. We need them to be such that $U(a)$ is unitary but this is not difficult. As long as all specified $U(a)|q_i\rangle$ are orthogonal, the remaining values can be assigned so that the whole $U(a)$ is unitary.

Next, we show how this automaton works on the input word $aa\$$.

1. The automaton starts in $|q_0\rangle$. Then $U(a)$ is applied, giving $\frac{1}{2}|q_0\rangle + \frac{1}{\sqrt{2}}|q_1\rangle + \frac{1}{2}|q_{\text{acc}}\rangle$. This state is observed with two possible results. With probability $\left(\frac{1}{2}\right)^2$, the accepting state is observed and then the computation terminates. Otherwise, a nonhalting state $\frac{1}{2}|q_0\rangle + \frac{1}{\sqrt{2}}|q_1\rangle$ (unnormalized) is observed and then the computation continues.
2. After the second a is fed, the state $\frac{1}{2}|q_0\rangle + \frac{1}{\sqrt{2}}|q_1\rangle$ is mapped to $\frac{1}{2}\left(\frac{1}{2} + \frac{1}{\sqrt{2}}\right)|q_0\rangle + \frac{1}{\sqrt{2}}\left(\frac{1}{2} - \frac{1}{\sqrt{2}}\right)|q_1\rangle + \frac{1}{2}\left(\frac{1}{2} + \frac{1}{\sqrt{2}}\right)|q_{\text{acc}}\rangle$. This is observed with two possible results. With probability $\left[\frac{1}{2}\left(\frac{1}{2} + \frac{1}{\sqrt{2}}\right)\right]^2$, the computation terminates in the accepting state q_{acc}. Otherwise, the computation continues with a new no-halting state $\frac{1}{2}\left(\frac{1}{2} + \frac{1}{\sqrt{2}}\right)|q_0\rangle + \frac{1}{\sqrt{2}}\left(\frac{1}{2} - \frac{1}{\sqrt{2}}\right)|q_1\rangle$ (unnormalized).
3. After the last symbol \$ is fed, the automaton's state turns to $\frac{1}{2}\left(\frac{1}{2} + \frac{1}{\sqrt{2}}\right)|q_{\text{acc}}\rangle + \frac{1}{\sqrt{2}}\left(\frac{1}{2} - \frac{1}{\sqrt{2}}\right)|q_{\text{rej}}\rangle$. This is observed. The computation terminates in the accepting state $|q_{\text{acc}}\rangle$ with probability $\left[\frac{1}{2}\left(\frac{1}{2} + \frac{1}{\sqrt{2}}\right)\right]^2$ or in the rejecting state $|q_{\text{rej}}\rangle$ with probability $\left[\frac{1}{\sqrt{2}}\left(\frac{1}{2} - \frac{1}{\sqrt{2}}\right)\right]^2$.

The total accepting probability is $(\frac{1}{2})^2 + [\frac{1}{2}(\frac{1}{2} + \frac{1}{\sqrt{2}})]^2 + [\frac{1}{2}(\frac{1}{2} + \frac{1}{\sqrt{2}})]^2 = \frac{5}{8} + \frac{1}{2\sqrt{2}}$.

Most of the papers exploring MM-1QFA discuss the open problem of a full characterization of the class of languages accepted by MM-1QFA. The main, but only partial, result along these lines is due to Kondacs and Watrous [13]. They proved that the class of languages accepted by MM-1QFA with bounded error is a proper subset of regular languages and that the regular language $L = \{a, b\}^*a$ cannot be accepted by MM-1QFA with bounded error.

Brodsky and Pippenger [16] further dealt with the problem to characterize the class of languages accepted by MM-1QFA. In a similar way, as classes RMO and UMO were defined, they [16] introduced classes RMM and UMM as classes of languages accepted by MM-1QFA with bound error and with unbounded error (i.e., with a cut-point), respectively, and they studied the closure properties of them. Specifically, they showed that the classes RMM and UMM are closed under complement, inverse homomorphism, and word quotient, but are not closed under homomorphisms. As an open problem they left the problem whether the classes RMM and UMM are closed under Boolean operations.

A negative answer to this problem was given by Ambainis at al. [30]. In this paper, they also gave a variety of necessary and sufficient conditions for some special regular languages to be accepted by MM-1QFA.

Recently, Yakaryilmaz and Cem Say [66,67] proved that UMM equals to the class of stochastic languages. Thus, with respect to unbounded error acceptance, MM-1QFA and probabilistic finite automata (PFA) have the same language recognition power. However, MM-1QFA may have some superiority over PFA in other aspects. For example, Yakaryilmaz and Cem Say [68] proved that the class of languages accepted by MM-1QFA with the cut-point 0 properly contains regular languages. As we know, the class of languages accepted by PFA with the cut-point 0 equals to the class of regular languages [61]. Thus, MM-1QFA are more powerful than PFA if the language acceptance fashion is with the cut-point 0. In the following, we will see that MM-1QFA also show superiority in other aspects.

An important result concerning MM-1QFA, proved by Ambainis and Freivalds [14], is that an MM-1QFA could accept a language with probability higher than $\frac{7}{9}$ if and only if the language could be accepted by an AF-reversible automaton. In [29], Ambainis, Bonner, Freivalds, and Kikusts constructed an infinite hierarchy of regular languages such that the current language in the hierarchy can be accepted by MM-1QFA with a probability smaller than the corresponding probability for accepting the preceding language in the hierarchy. Moreover, these probabilities converge to $\frac{1}{2}$. Ambainis and Kikusts [24] determined the maximum probabilities achieved by MM-1QFA for several languages, and they showed that any language that is not accepted by AF-reversible automata can only be accepted by an MM-1QFA with probability not more than 0.7726.

Golovkins et al. [69,70] studied the model of probabilistic reversible automata and its relation to MM-1QFA and MO-1QFA. Dzelme-Bērzina [32] discussed the connection between the formulas of logic and QFA (MM-1QFA and MO-1QFA) in respect to the language recognition of QFA. These two papers bring a new approach to the problem of characterization of the language recognition power of QFA.

Another interesting area of research concerns the relation between 1QFA and other models of automata with respect to the number of states needed to recognize the same language. A first result is from Kikusts [71] who constructed an MM-1QFA that is quadratically smaller than any equivalent DFA. A much stronger result was shown by Ambainis and Freivalds [14]. They showed that for the language $L_p = \{a^i : i \text{ is divisible by } p \text{ and } p \text{ is a prime}\}$, there exists an MM-1QFA accepting it with probability arbitrarily close to 1, which is exponentially smaller that any equivalent PFA. In fact, this result also holds for MO-1QFA, since the QFA constructed in Ref. [14] is exactly an MO-1QFA. Recently, Ambainis and Nahimovs [28] presented an improved construction of MO-1QFA which achieve an exponential advantage over classical finite automata.

In contrast, Ambainis et al. [18] showed for the language $L_m = \{w1 | w \in \{0, 1\}^*, |w| \leq m\}$, that any MM-1QFA accepting this language has exponentially more states than the minimal DFA accepting this language. Papers [72–77] contain a variety of other results comparing state complexity of MO-1QA and MM-1QA or various special languages.

In [78] the following open problem was formulated: Is it decidable whether two MM-QFA are equivalent? Then Li and Qiu [27] answered this question affirmatively. We present their result in the following form.

Theorem 6.5: ([27])

Two MM-1QFA $\mathcal{A}_1$ and $\mathcal{A}_2$ are equivalent if and only if they are $3n_1^2 + 3n_2^2 - 1$-equivalent, where n_1 and n_2 are numbers of states in $\mathcal{A}_1$ and $\mathcal{A}_2$, respectively. There is a polynomial-time algorithm in n_1, n_2 deciding whether $\mathcal{A}_1$ and $\mathcal{A}_2$ are equivalent or not.

Recently, Li and Qiu [79] mentioned that the above equivalence criterion can be slightly improved to $n_1^2 + n_2^2 - 1$.

Remark 6.5:

Since the equivalence problem of MM-1QFA has been solved, by means of the idea from Ref. [65] the state minimization of MM-1QFA can also be proved to be decidable in EXPSPACE.

6.3.1.4 Quantum Finite Automata with Control Languages

Bertoni et al. [23] introduced a quantum computing model called *quantum finite automata with control languages* (CL-1QFA). This model can be seen as belonging to the class of 1QFA, since its tape head is allowed to move only toward right at each step. Similarly as MM-1QFA, this model also requires a measurement after reading each symbol. However, this model differs much from MO-1QFA and MM-1QFA as illustrated in the following two observations:

i. CL-1QFA allow more general measurements to be used. They allow any projective measurement to be used after each input symbol is processed.
ii. The accepting condition of this model is also different from that of the previous models. Suppose that measurement results of a CL-1QFA M are from a set of possible results $\mathcal{C} = \{c_1, \ldots, c_n\}$. Given an input word x, let measurements during processing of x by a CL-1QFA provide an $y \in \mathcal{C}^*$ with probability $p(y|x)$. It is defined that M accepts x if and only if y belongs to a beforehand given regular language $\mathcal{L} \subseteq \mathcal{C}^*$. Intuitively, $\mathcal{L}$ acts here as a controller, which is reflected in the name of this model.

More formally, the definition of CL-1QFA is as follows.

Definition 6.5:

Given an input alphabet Σ and the end-marker symbol $\$ \notin \Sigma$, a CL-1QFA over the working alphabet $\Gamma = \Sigma \cup \{\$\}$ is a five-tuple $\mathcal{M} = (Q, |\psi_0\rangle, \{U(\sigma)\}_{\sigma\in\Gamma}, \mathcal{O}, \mathcal{L})$, where

- $Q = \{q_1, \ldots, q_n\}$ is the basic state set; at any time, the configuration of $\mathcal{M}$ is a superposition of these basic states
- $|\psi_0\rangle$ is a unit vector in Hilbert space $\mathbf{C}^n$, denoting the initial configuration of $\mathcal{M}$
- for any $\sigma \in \Gamma$, $U(\sigma)$ is an $n \times n$ unitary matrix
- $\mathcal{O}$ is an observable with the set of possible results $\mathcal{C} = \{c_1, \ldots, c_s\}$ and with the projectors set $\{P(c_i) : i = 1, \ldots, s\}$, where $P(c_i)$ denotes the projector onto the eigenspace corresponding to c_i;
- $\mathcal{L} \subseteq \mathcal{C}^*$ is a regular language, called as a control language.

A configuration of $\mathcal{M}$ is represented by a unit vector $|\psi\rangle = \sum_i \alpha_i |q_i\rangle$ with $q_i \in Q$. Any input string w to CL-1QFA $\mathcal{M}$ has the form $w \in \Sigma^*\$$, with symbol \$ denoting the end of the input. Now, we define the behavior of $\mathcal{M}$ on an input $x_1 \ldots x_n\$$. The computation starts in the initial state $|\psi_0\rangle$, and then the transformations associated with the symbols in string $x_1 \ldots x_n\$$ are performed in succession. The transformation associated with any symbol $\sigma \in \Gamma$ consists of two steps:

1. $U(\sigma)$ is applied to the current state $|\phi\rangle$ of $\mathcal{M}$, yielding the new state $|\phi'\rangle = U(\sigma)|\phi\rangle$.
2. The observable $\mathcal{O}$ is measured on $|\phi'\rangle$. According to quantum mechanics principle, this measurement yields a result c_k with probability $p_k = ||P(c_k)|\phi'\rangle||^2$, and the state of $\mathcal{M}$ collapses to $P(c_k)|\phi'\rangle/\sqrt{p_k}$.

Thus, the computation on string $x_1 \ldots x_n\$$ leads to a sequence $y_1 \ldots y_{n+1} \in C^*$ with probability $p(y_1 \ldots y_{n+1}|x_1 \ldots x_n\$)$ given by

$$p(y_1 \ldots y_{n+1}|x_1 \ldots x_n\$) = \left\| \prod_{i=1}^{n+1} (P(y_i)U(x_i))|\psi_0\rangle \right\|^2, \tag{6.15}$$

where we let $x_{n+1} = \$$, and $\prod_{i=1}^{n} A_i = A_n \cdots A_2 A_1$. Thus, the probability of $\mathcal{M}$ accepting x is

$$P_{\mathcal{M}}(x) = \sum_{y_1 \ldots y_{n+1} \in \mathcal{L}} p(y_1 \ldots y_{n+1}|x_1 \ldots x_n\$). \tag{6.16}$$

Bertoni et al. [23] defined this model, and they studied some closure properties of the languages accepted by CL-1QFA. They showed that the class of languages accepted by CL-1QFA with bounded error is closed under Boolean operations, which is contrary to the class accepted by MM-1QFA which is not closed under Boolean operations [30]. Also, they showed that the languages accepted with bound error by CL-1QFA are still regular languages. The following inclusions were presented in Ref. [23]:

$$\text{RMO} \subset \text{RMM} \subset \text{RQC}, \tag{6.17}$$

where RMO, RMM, RQC denote the classes of languages accepted with bounded error by MO-1QFA, MM-1QFA, and CL-1QFA, respectively. Mereghetti and Palano [80] further showed that RQC is exactly the class of regular languages.

Li and Qiu [27] showed that two CL-1QFA $\mathcal{A}_1$ and $\mathcal{A}_2$ with control languages (regular languages) $\mathcal{L}_1$ and $\mathcal{L}_2$, respectively, are equivalent if and only if they are $(c_1 n_1^2 + c_2 n_2^2 - 1)$-equivalent, where n_1 and n_2 are numbers of states in $\mathcal{A}_1$ and $\mathcal{A}_2$, respectively, and c_1 and c_2 are the numbers of states in the minimal DFA that recognize $\mathcal{L}_1$ and $\mathcal{L}_2$, respectively. Furthermore, when the control languages are given in the form of DFA, there exists a polynomial-time algorithm to determine whether the two CL-1QFA are equivalent or not.

Remark 6.6:

The equivalence problem of CL-1QFA has been solved in Ref. [27]. By means of the idea from Ref. [65], the minimization of quantum basis states for CL-1QFA can be proved to be decidable in EXPSPACE.

6.3.1.5 Multiletter Quantum Finite Automata

Belovs et al. [57] proposed a new one-way QFA model, namely, the so-called multiletter QFA, that should be thought of as a quantum counterpart of the quite restricted classical model of finite automata—*one-way multihead finite automata* [81]. Roughly speaking, a k-letter QFA is not limited to see only one, the just-incoming input letter, but can see several of the earlier received letters as well. Therefore, the quantum state transition which such a k-letter automaton performs at each step depends on the last k letters received. Otherwise such automata work as MO-1QFA already discussed.

At first we introduce classical k-letter deterministic finite automata (k-letter DFA).

Definition 6.6: ([57])

A k-letter *deterministic finite automaton* (k-letter DFA) is defined by a quintuple $(Q, Q_{acc}, q_0, \Sigma, \gamma)$, where Q is a finite set of states, $Q_{acc} \subseteq Q$ is the set of accepting states, $q_0 \in Q$ is the initial state, Σ is a finite input alphabet, and γ is a transition function that maps $Q \times T^k$ to Q, where $T = \{\Lambda\} \bigcup \Sigma$ and letter $\Lambda \notin \Sigma$ denotes the empty letter, and $T^k \subset T^*$ consists of all strings of length k.

We describe now the computing process of a k-letter DFA on an input string x in Σ^*, where $x = \sigma_1\sigma_2 \cdots \sigma_n$. A k-letter DFA has a tape which contains the letter Λ in its first $k-1$ positions followed by the input string x. The automaton starts in the initial state q_0 and has k reading heads which initially are on the first k positions of the tape (clearly, the kth head reads σ_1 and the other heads read Λ). In each step, symbols read by k heads and the current internal state determine a new internal state and all reading heads move in parallel by one symbol to the right. This process ends when the last input symbol is read by a head. The input string x is accepted if and only if at the end of the computation automaton enters an accepting state.

Clearly, k-letter DFA are not more powerful than DFA. Indeed, the family of languages accepted by k-letter DFA for $k \geq 1$ is exactly that of regular languages.

Now we recall the definition of multiletter QFA [57].

Definition 6.7: ([57])

A k-letter QFA $\mathcal{A}$ is defined by a quintuple $\mathcal{A} = (Q, Q_{acc}, |\psi_0\rangle, \Sigma, \mu)$, where Q is a set of states, $Q_{acc} \subseteq Q$ is the set of accepting states, $|\psi_0\rangle$ is the initial quantum state from $\mathcal{H}_Q$, Σ is a finite input alphabet, and μ is a function that assigns a unitary transition matrix U_w on $\mathbf{C}^{|Q|}$ to each string $w \in (\{\Lambda\} \cup \Sigma)^k$.

In the following, we give an example of 2-letter QFA.

Example 6.3:

A 2-letter QFA $\mathcal{A} = (Q, Q_{acc}, |\psi_0\rangle, \Sigma, \mu)$ is given as $Q = \{q_0, q_1\}, Q_{acc} = \{q_1\}, |\psi_0\rangle = |q_0\rangle = \begin{pmatrix} 1 \\ 0 \end{pmatrix}, \Sigma = \{a, b\}, \mu(\Lambda b) = \mu(bb) = \begin{pmatrix} 1 & 0 \\ 0 & 1 \end{pmatrix}, \mu(\Lambda a) = \mu(ab) = \mu(ba) = \begin{pmatrix} 0 & 1 \\ 1 & 0 \end{pmatrix}$. Note that this automata was given in Ref. [57] to accept the language $(a+b)^*b$.

A k-letter QFA $\mathcal{A}$ works in the same way as an MO-1QFA [12,16], except that it applies unitary transformations corresponding not only to the last letter but to the last k letters received (like a k-letter DFA). When $k = 1$, it is exactly an MO-1QFA. According to Ref. [57], all languages accepted by k-letter QFAs with bounded error are regular languages for any k.

Let us now determine the probability $P_{\mathcal{A}}(x)$ that a k-letter QFA $\mathcal{A} = (Q, Q_{acc}, |\psi_0\rangle, \Sigma, \mu)$ accepts an input string $x = \sigma_1\sigma_2 \cdots \sigma_m$. From the definition of k-letter QFA it follows that for any $w \in (\{\Lambda\} \cup \Sigma)^k$, $\mu(w)$ is a unitary matrix. The definition of the mapping μ can be used to assign, in a usual way, a unitary matrix to any string $x = \sigma_1\sigma_2 \cdots \sigma_m \in \Sigma^*$. By $\overline{\mu}$ (induced by μ) we will denote such a map from Σ^* to the set of all $|Q| \times |Q|$ unitary matrices. More specifically, $\overline{\mu}$ is induced by μ in the following way: For

$x = \sigma_1\sigma_2\cdots\sigma_m \in \Sigma^*$,

$$\overline{\mu}(x) = \begin{cases} \mu(\Lambda^{k-1}\sigma_1)\mu(\Lambda^{k-2}\sigma_1\sigma_2)\cdots\mu(\Lambda^{k-m}x), & \text{if } m < k, \\ \mu(\Lambda^{k-1}\sigma_1)\mu(\Lambda^{k-2}\sigma_1\sigma_2)\cdots\mu(\sigma_{m-k+1}\sigma_{m-k+2}\cdots\sigma_m), & \text{if } m \geq k, \end{cases} \tag{6.18}$$

which specifies how $\mathcal{A}$ process an input string x.

Let P_{acc} denotes the projection operator on the subspace spanned by Q_{acc}. Using the above notations it holds:

$$P_{\mathcal{A}}(x) = \|\langle\psi_0|\overline{\mu}(x)P_{\text{acc}}\|^2. \tag{6.19}$$

Belovs et al. [57] showed that the language $(a+b)^*b$ can be accepted by a 2-letter QFA but, as proved in Ref. [13], cannot be accepted by any MM-1QFA with bounded error. Therefore, multiletter QFA can accept with bounded error some regular languages that cannot be accepted by any MM-1QFA and MO-1QFA, but still they cannot accept the whole class of regular languages.

Let us denote by $\mathcal{L}(QFA_k)$ the class of languages accepted with bounded error by k-letter QFAs. Any given k-letter QFA can be simulated by some $(k+1)$-letter QFA. Qiu and Yu [58] proved that $\mathcal{L}(QFA_k) \subset \mathcal{L}(QFA_{k+1})$ for $k = 1, 2, \ldots$, where the inclusion $\subset$ is proper. Therefore, $(k+1)$-letter QFA are computationally more powerful than k-letter QFA. Note that such a hierarchy does not hold in the case of multiletter DFA. Indeed, as pointed out before, k-letter DFA accept exactly the class of regular languages for any $k \geq 1$.

Concerning the equivalence problem of multiletter QFA, Qiu and Yu [58] proved that any two such automata, a k_1-letter QFA $\mathcal{A}_1$ and a k_2-letter QFA $\mathcal{A}_2$ over the same single-letter input alphabet $\Sigma = \{\sigma\}$ are equivalent if and only if they are $((n_1+n_2)^4 + k - 1)$-equivalent, where n_1 and n_2 are the numbers of states of $\mathcal{A}_1$ and $\mathcal{A}_2$, respectively, and $k = \max(k_1, k_2)$. Recently, Qiu et al. [34] further consider the equivalence problem of multiletter QFA for an arbitrarily finite alphabet, and they proved that a k_1-letter QFA $\mathcal{A}_1$ and a k_2-letter QFA $\mathcal{A}_2$ over the input alphabet $\Sigma = \{\sigma_1, \sigma_2, \ldots, \sigma_m\}$ are equivalent if and only if they are $((n_1+n_2)^2 m^{k-1} - m^{k-1} + k)$-equivalent, where n_1 and n_2 are the numbers of states of $\mathcal{A}_1$ and $\mathcal{A}_2$, respectively, and $k = \max(k_1, k_2)$. When $k = 1$, the result reduces to the one for MO-1QFA, and when $m = 1$, the result reduces to the one given in Ref. [58], as mentioned above. Note that the equivalence problem for the case of arbitrary alphabet is more complex than the one for the case of single-letter alphabet. In summary, these results are described as follows.

Theorem 6.6:

$\mathcal{A}_1$ and $\mathcal{A}_2$ above are equivalent if and only if they are $(n^2 m^{k-1} - m^{k-1} + k)$-equivalent, where $k = \max(k_1, k_2)$ and $n = n_1 + n_2$, with n_i being the number of states in Q_i, $i = 1, 2$.

Theorem 6.7:

Let Σ, k_1-letter QFA $\mathcal{A}_1$, and k_2-letter QFA $\mathcal{A}_2$ be the same as above. Then, there exists a polynomial-time $O(m^{2k-1}n^8 + km^k n^6)$ algorithm to determine whether or not $\mathcal{A}_1$ and $\mathcal{A}_2$ are equivalent.

Remark 6.7:

Using the idea from Ref. [65] and the technique developed for showing decidability of the equivalence problem for multiletter QFA [34], Qiu et al. [34] have proved that the minimization of states for multiletter QFA is decidable in EXPSPACE.

Theorem 6.8:

The state minimization problem of multiletter QFAs is decidable in EXPSPACE.

Concerning the state complexity compared with 1QFAC, we have:

Proposition 6.3:

For any $m \geq 2$ and any string z with $|z| = n \geq 1$, there exists a regular language $L_z(m)$ that cannot be accepted by any multiletter QFA, but there exists a 1QFAC $\mathcal{A}_m$ accepting $L_z(m)$ with $n+1$ classical states (independent of m) and 2 quantum basis states. In contrast, the minimal DFA accepting $L_z(m)$ has $m(n+1)$ states.

6.3.1.6 Latvian Quantum Finite Automata

Ambainis et al. [26] defined a QFA model, called Latvian QFA (LQFA), which can be regarded as a generalized version of MO-1QFA. The definition of LQFA is almost the same as the one of MO-1QFA except that for any symbol σ in the input alphabet Σ, the transition function is a combination of a unitary matrix and a projective measurement, instead of a unitary matrix only. This change enhances the language recognition power of LQFA.

As shown in Ref. [26], LQFA are closely related with the probabilistic model PRA-C [69] (a special probabilistic finite automata), and from the results in Ref. [26] it follows that the two models recognize the same class of languages, that is, the languages whose syntactic monoid is a block group [26]. More specifically, a language is recognized by LQFA with bounded error if and only if the language is a Boolean combination of languages of the form $L_0 a_1 L_1 \ldots a_k L_k$, where the a_i's are letters and the L_i's are group languages (i.e., the languages accepted by group automata defined in Proposition 6.1). Hence, LQFA can recognize with bounded error only a proper subset of regular languages. For example, LQFA cannot recognize regular languages $\Sigma^* a$ and $a\Sigma^*$ [26].

The class of languages recognized with bounded error by LQFA is a proper subset of the languages recognized by MM-1QFA. Indeed, on the one hand MM-1QFA can recognize all languages recognized by LQFA [26], and on the other hand, MM-1QFA can recognize language $a\Sigma^*$ that cannot be recognized by LQFA. For example, to recognize $a\{a,b\}^*$, we construct an MM-1QFA with an initial state $|q_0\rangle$, a rejecting state $|q_{\text{rej}}\rangle$ and an accepting state $|q_{\text{acc}}\rangle$, and the transition function is described by

$$U(a)|q_0\rangle = \frac{1}{\sqrt{2}}|q_1\rangle + \frac{1}{\sqrt{2}}|q_{\text{acc}}\rangle, \quad U(a)|q_1\rangle = \frac{1}{\sqrt{2}}|q_1\rangle - \frac{1}{\sqrt{2}}|q_{\text{acc}}\rangle,$$
$$U(b)|q_1\rangle = \frac{1}{\sqrt{2}}|q_1\rangle + \frac{1}{\sqrt{2}}|q_{\text{acc}}\rangle, \quad U(b)|q_0\rangle = |q_{\text{rej}}\rangle, \quad U(\$)|q_0\rangle = |q_{\text{acc}}\rangle, \quad U(\$)|q_1\rangle = |q_{\text{rej}}\rangle,$$

where those transitions that are not important have not been described.

6.3.1.7 General Quantum Finite Automata

The model of GQFA was firstly studied by Nayak [17] and Ambainis et al. [20]. The model can be seen as a generalized version of MM-1QFA, or an measure-many version of LQFA. The definition of GQFA is almost the same as the one of MM-1QFA except that for any symbol σ in the input alphabet Σ, the transition function is a combination of a unitary matrix and a projective measurement,* instead of a unitary matrix only.

An interesting result about GQFA is that there exist languages $L_m = \{w0 \mid w \in \{0,1\}^*, |w| \leq m\}, m \geq 1$ for which GQFA need exponentially more states than any DFA accepting them [17,20]. From the results about LQFA stated above, it follows that GQFA are strictly powerful than LQFA. However, GQFA still cannot recognize with bounded error all regular languages; for example, GQFA cannot recognize the language $\{a, b\}^*a$ [17]. It is still not yet known whether GQFA can recognize with bounded error strictly more languages than MM-1QFA.

Concerning the unbounded error acceptance, Yakaryilmaz and Cem Say [83] proved that the class of languages recognized by GQFA with unbounded error equals the class of stochastic languages. Thus, the capability of performing additional intermediate projective measurements does not increase the recognition power of GQFA over MM-1QFA in this accepting mode.

We have presented so far a series of models of QFA in which more and more powerful operations were allowed. However, in quantum mechanics the most general operation is the completely positive, trace-preserving mapping (CPT) [2]. Hence, it is natural to consider an QFA model based on CPT operations. Indeed, Hirvensalo [84] proposed such a QFA in which the transition function associated with each input symbol is a CPT, and other elements are the same as those in MO-1QFA. We denote the model of Hirvensalo [84] by MO-1gQFA. A more systemic investigation on such models can be referred to Li and Qiu [33]. The main results of [33] are as follows:

i. Characterization of MO-1gQFA. It was proved that MO-1gQFA recognize with bounded error exactly the class of regular languages. Also, a sufficient and necessary condition was obtained for the equivalence problem of MO-1gQFA.
ii. Characterization of MM-1gQFA. Generally, the times of the measurement performed in the computation affect the computational power of 1QFA. For instance, MM-1QFA are strictly more powerful than MO-1QFA, and GQFA also recognize more languages than LQFA. Inspired by this one, the authors of Ref. [33] defined a measure-many version of MO-1gQFA, denoted by MM-1gQFA, of which the elements are almost the same as those of MM-1QFA except that there is a CPT associated with each input symbol. In Ref. [33], it was proved that MM-1gQFA also recognize only regular languages with bounded error. Thus, MM-1gQFA and MO-1gQFA have the same language recognition power, which is greatly different from the conventional case in which the times of the measurement performed in the computation generally affect the language recognition power of one-way QFA. In addition, a sufficient and necessary condition was given for two MM-1gQFA to be equivalent.

There are another two 1QFA models, defined by Paschen [85] and by Ciamarra [86]. As shown in Ref. [33], they can be seen as special cases of the QFA model based on the CPT operations.

Qiu et al. [56] recently proposed another model of 1QFA named *one-way quantum finite automata together with classical states* (1QFAC) where quantum states and classical states exist in parallel, accompanied with a quantum transition function and a classical one.

Remark 6.8:

Since the equivalence problem of MO-1gQFA and MM-1gQFA has been proved to be decidable in [33], one can prove that the states minimization problem for MO-1gQFA and MM-1gQFA is decidable in EXPSPACE by virtue of the idea from [65].

---

* Note that in the original definition [17], the transition function is a finite sequence of unitary transformations and projective measurements. As shown in Ref. [82], the two definitions are equivalent.

6.3.2 Two-Way Quantum Finite Automata

2QFA were first defined by Kondacs and Watrous [13]. 2QFA can be seen as a quantum variant of classical two-way finite automata [11], but they are more powerful than the classical counterparts.

In a way, the definition of 2QFA is somewhat similar to that of MM-1QFA, but the tape heads are allowed to move toward right or left, or to be stationary (the tape head of MM-1QFA always moves toward right). It is this difference that is behind the fact that in this case conditions that a transition mapping has to satisfy are much more complex than in the case of MM-1QFA. Also, it is exactly this difference that makes 2QFA strictly more powerful than MM-1QFA. Indeed, 2QFA cannot only accept all regular languages, but can also accept, with one-side error, some context-free languages, and even non-context-free languages in linear time [13].*

The formal definition of 2QFA is as follows.

Definition 6.8:

A 2QFA $\mathcal{A}$ is a 6-tuple $(\Sigma, Q, q_0, Q_{\text{acc}}, Q_{\text{rej}}, \delta)$ and it is specified by the finite input alphabet Σ, the finite set of states Q, the initial state q_0, the sets $Q_{\text{acc}} \subset Q$ and $Q_{\text{rej}} \subset Q$ of accepting and rejecting states, respectively, with $Q_a \cap Q_r = \emptyset$, and the transition function

$$\delta : Q \times \Gamma \times Q \times \{\leftarrow, \downarrow, \rightarrow\} \longrightarrow \mathbf{C}$$

where $\Gamma = \Sigma \cup \{\#, \$\}$ is the tape alphabet of $\mathcal{A}$, and # and \$ are endmarkers not in Σ. The transition function should satisfy the following conditions (of well-formedness) for any $q_1, q_2 \in Q$, $\sigma, \sigma_1, \sigma_2 \in \Gamma$, $d \in \{\leftarrow, \downarrow, \rightarrow\}$:

1. Local probability and orthogonality condition:

$$\sum_{q',d} \overline{\delta(q_1, \sigma, q', d)}\delta(q_2, \sigma, q', d) = \begin{cases} 1 & q_1 = q_2; \\ 0 & q_1 \neq q_2. \end{cases} \tag{6.20}$$

2. Separability condition I:

$$\sum_{q'} \left(\overline{\delta(q_1, \sigma_1, q', \rightarrow)}\delta(q_2, \sigma_2, q', \downarrow) + \overline{\delta(q_1, \sigma_1, q', \downarrow)}\delta(q_2, \sigma_2, q', \leftarrow) \right) = 0. \tag{6.21}$$

3. Separability condition II:

$$\sum_{q'} \overline{\delta(q_1, \sigma_1, q', \rightarrow)}\delta(q_2, \sigma_2, q', \leftarrow) = 0. \tag{6.22}$$

Remark 6.9:

The above three conditions are equivalent to the requirement that evolution of $\mathcal{A}$ is unitary. Notably, there are two endmarkers # and \$ in the definition, and, as far as we are aware, no proof shows that the model with one endmarker is equivalent to the one with two.

---

* Acceptance of a language L with one-side error means that if x is in L, the machine accepts it with probability 1; if x is not in L, the machine rejects it with probability at least $1 - \epsilon$.

Kondacs and Watrous [13] proved that every regular language is accepted by a 2QFA. It is well known that any DFA can be simulated by a two-way reversible finite automaton (i.e., by a 2QFA, amplitudes of which are only 0 and 1).

It is well known that the language $L_{eq} = \{a^n b^n | n \in \mathbf{N}\}$, where $\mathbf{N}$ is the set of natural numbers, is nonregular. Kondacs and Watrous [13] proved the following theorem.

Theorem 6.9:

The language L_{eq} can be recognized by a 2QFA with one-sided error in linear time.

Yakaryilmaz and Cem Say [31] gave an improved construction of a 2QFA recognizing the above language. Note that Freivalds [87] proved that *two-way probabilistic finite automata* (2PFA) can recognize $L_{eq} = \{a^n b^n | n \in \mathbf{N}\}$ with arbitrarily small error, but in exponential expected time [88]. It has been actually proven that any 2PFA recognizing a nonregular language with bounded error needs take exponential expected time [89,90].

In the theory of classical computation, Dwork and Stockmeyer [91] introduced and explored interactive proof systems (IP) based on *two-way probabilistic finite automata*. Inspired by this work, Nishimura and Yamakami [92] presented an application of quantum finite automata (2QFA and 1QFA) to interactive proof systems.

Remark 6.10:

Ambainis and Watrous[19] proposed a different two-way quantum finite automata model—*two-way finite automata with quantum and classical states* (2QCFA). In this model, there are both quantum and classical states, and correspondingly to them two transfer functions: (i) one specifies unitary operator or measurement for the evolution of quantum states and (ii) the other describes the evolution of classical part of the machine, including the classical internal states and the tape head. This quantum computing device may be more easy to implement than ordinary 2QFA, since the moves of tape heads of 2QCFA are classical. In spite of this restriction, 2QCFA have more power than 2PFA. Indeed, as Ambainis and Watrous [19] pointed out, 2QCFA can clearly recognize all regular languages with certainty, and particularly, they [19] proved that this model can also recognize nonregular languages $L_{\rm eq}$ and palindromes $L_{\rm pal} = \{x \in \{a, b\}^* | x = x^R\}$, where notably the complexity for recognizing $L_{\rm eq}$ is polynomial time in case of one-sided error. As it is known, no 2PFA can recognize $L_{\rm pal}$ with bounded error [91]. Therefore, this is also an interesting model of quantum computation.

Remark 6.11:

Closure properties of languages accepted by 2QCFA were explored by Qiu [93]. It was shown, for example, that this class is closed under all Boolean operations as well as on reversal operation. As a corollary, it was shown closure on the intersection, complement, and reversal operations of the class of languages recognized by 2QCFA with one-sided error probabilities. Furthermore, it was verified that the class of languages recognized by 2QCFA with error probabilities is closed under catenation under certain restricted condition (this result also holds for the case of one-sided error probabilities) [93]. Moreover, in Ref. [93] it has been investigated how state complexity, with respect to the 2QCFA, of results of some language operations depend on state complexity of the languages operations are applied on. Some examples were included for an application of the derived results. For instance, the language $\{xx^R | x \in \{a, b\}^*, \#_x(a) = \#_x(b)\}$ was shown to be recognizable by 2qcfa with one-sided

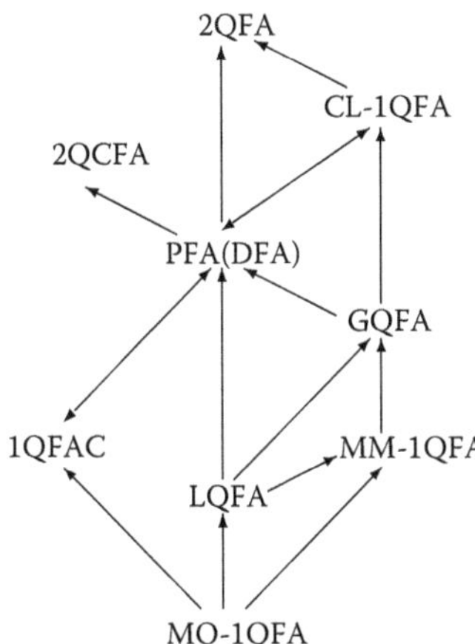

FIGURE 6.2 A diagram illustrating known inclusions among the families of languages recognized with bounded error by most of the currently known 1QFA. One-directional arrows denote containment relation; bidirectional equivalence.

error probability $0 \leq \epsilon < \frac{1}{2}$ in polynomial time, where $\#_x(a)$ denotes the number of a appearing in the word x.

Remark 6.12:

It is worth pointing out that we do not know whether or not the closure properties presented in Ref. [93] apply also to 2QFA [13], because the evolution has to be unitary in 2QFA. Therefore, either more technical proof methods are needed or some restrictions are needed such that closure properties can be shown (e.g., we may restrict the initial state not to be entered again).

Finally, we would like to mention another interesting quantum finite automata model introduced by Amano and Iwama [15]. This model is denoted as 1.5QFA and its tape heads are allowed to move right or to be stationary, but not toward left. As an important result, Amano and Iwama [15] showed that the emptiness problem for 1.5QFA is undecidable.

6.3.3 Language Family Inclusions

Figure 6.2 depicts the inclusion relations among the families of languages recognized by most of the currently known 1QFA. For all types of 1QFA depicted in the Figure 6.1 only language families with respect to bounded error acceptance are considered in the figure. Most of the inclusion relations depicted in Figure 6.2 are proper, except for the following three cases: (i) it is still not known whether GQFA can recognize any language not recognized by MM-1QFA; (ii) CL-1QFA, 1QFAC, and PFA recognize the same language class—regular languages; and (iii) the relationship between 2QCFA and 2QFA is still not clear.

6.4 More Powerful Quantum Automata

There is a variety of quantum automata models that have an intermediate power: they are more powerful than finite ones, but less powerful as such universal ones as QTMs. In the following, we present a description of some of them.

6.4.1 Quantum Sequential Machines

Sequential quantum machines (SQM), as a quantum variant of *stochastic sequential machines* (SSM) [61], were introduced by Gudder [35], and an equivalent version, called *quantum sequential machines* (QSM), was proposed by Qiu [36]. The definition of QSM is a natural generalization of SSM [61].

It is known (see [61]) that a very crucial result on SSM is that two SSM, with n and n' states, respectively, and the same input and output alphabets are equivalent if and only if they are $(n+n'-1)$-equivalent (Theorem 2.7 in [61]). (Equivalence of two SSM means that their accepting probabilities are the same for any input, while $(n+n'-1)$-equivalent means that their accepting probabilities are the same for any input string whose length is not more than $(n+n'-1)$ are equivalent.)

Motivated by the above result, Gudder in Ref. [35] proposed as an open problem whether the above result holds for SQM. More specifically, let $\mathcal{M}$ and $\mathcal{M}'$ be SQM with n and n' states, respectively, and the same input and output alphabets. Is it true that $\mathcal{M}$ and $\mathcal{M}'$ are equivalent (i.e., $p_{\mathcal{M}}(v|u)=p_{\mathcal{M}'}(v|u)$ for all words u, v, where $p_{\mathcal{M}}(v|u)$ is the probability that the machine $\mathcal{M}$ prints the word v having been fed with the word u that has the same length as v) if and only if $p_{\mathcal{M}}(v|u)=p_{\mathcal{M}'}(v|u)$ for all words u, v with length not larger than $n+n'-1$? (see [35], p. 2159). Qiu [36] gave a negative answer to this question by constructing a counterexample. However, as Qiu [36] mentioned, no sufficient and necessary condition for the equivalence between two QSM had been discovered by that time.

In 2006, Li and Qiu [37] showed that if the condition for $(n+n'-1)$-equivalence is appropriately relaxed, then there exists a sufficient and necessary condition for the equivalence between two QSM. More precisely, Li and Qiu [37] showed that two QSM $\mathcal{M}$ and $\mathcal{M}'$ are equivalent if and only if they are $(n+n')^2$-equivalent, where n and n' are numbers of states in $\mathcal{M}$ and $\mathcal{M}'$, respectively. This means that the open problem suggested by Gudder [35] has been solved.

However, the just mentioned method to check the equivalence likely needs exponential expected time. Therefore, Li and Qiu [59] started to explore the time complexity of the equivalence problem between QSM and also related equivalence problems. In Ref. [59], the authors presented a polynomial-time algorithm for deciding the equivalence between QSM.

As a consequence, the equivalence problem between QSM has been solved completely.

Remark 6.13:

Since the equivalence problem of QSM has been solved in Ref. [59], one can prove that the minimization of states for QSM is decidable in EXPSPACE by using the idea of Ref. [65].

6.4.2 Quantum Pushdown Automata

6.4.2.1 MC-QPDA

Quantum pushdown automata (QPDA) were first suggested by Moore and Crutchfield [12], but they actually concentrated more on the so-called generalized quantum pushdown automata, in which the evolution operators are not required to be unitary and not even isometric. We denote quantum pushdown automata of [12] by MC-QPDA. However, we will not deal with them in more detail because they can hardly be seen as being quantum in a reasonable sense.

6.4.2.2 Go-QPDA

Motivated by the definition of quantum finite automata in Ref. [13], Golovkins [38] introduced another type, much more proper, of quantum pushdown automata. The quantum pushdown automata of Golovkins [38], which we denote by Go-QPDA, satisfy the unitary condition for evolution; that is to say, the transition function is required to fulfil the so-called well-formedness conditions that are quite complex. In particular, Golovkins [38] verified that the language $\{0,1\}^*1$ can be recognized by Go-QPDA (more exactly, by Go-QPDA with transition restricted in $\{0,1\}$), but it was proved by Kondacs and Watrous [13,35] that the language $\{0,1\}^*1$ cannot be accepted by MM-1QFA. Furthermore, Golovkins [38] showed that Go-QPDA with transition restricted in $\{0,1\}$ can recognize nonregular language $\{w \in \{a,b\}^* : |w|_a = |w|_b\}$. Notably, Golovkins [38] has showed that Go-QPDA can recognize some languages that cannot be recognized by deterministic pushdown automata (that are not recognized even by probabilistic pushdown automata either).

6.4.2.3 QY-QPDA

Qiu and Ying [39] introduced another quantum pushdown automata which was called q quantum pushdown automata (qQPDA) in Ref. [39], named after q quantum automata introduced by Gudder [62], but here we call them QY-quantum pushdown automata (QY-QPDA). To a certain extent, QY-QPDA have some advantages than other quantum pushdown automata. Indeed, the languages recognized by MC-QPDA with the empty stack are also accepted by some QY-QPDA; compared with Go-QPDA of Golovkins [38], the conditions for transition function δ on qQPDA are relatively simple. In Ref. [39], various properties of the languages accepted by QY-QPDA were discussed. In particular, it was proved that every η-q quantum context-free language is also η'-q quantum context-free language for any $\eta \in [0,1)$ and $\eta' \in (0,1)$, which is interesting and corresponds to the similar property of quantum regular languages investigated by Gudder in [62]. Here, η'-q quantum context-free languages are indeed the usual cup-point languages accepted by QY-QPDA, similar to those defined in probabilistic automata [61].

However, to preserve evolution to be unitary, the conditions of the transition functions defined in QY-QPDA [39] lose local property. Therefore, how to improve this point is also an unsolved problem.

6.4.2.4 Gu-QPDA

Gudder in Ref. [35] also presented a definition of quantum pushdown automata by directly generalizing the classical deterministic pushdown automata [11]. We call them Gu-QPDA here. However, the transition operators defined in Ref. [35] are just isometric but not unitary. It is worth mentioning that the conditions of the transition functions in Gu-QPDA are simpler than those in Go-QPDA, and isometry still has reasonable physical meaning. Gu-QPDA seem to be, therefore, worth of further investigations.

6.4.2.5 Na-QPDA

Another kind of interesting quantum pushdown automata was proposed by Nakanishi [40]. They are actually quantum pushdown automata with classical stack operations. In brief, we denote them Na-QPDA here. In an Na-QPFA, the stack is still classical. Notably, Nakanishi [40] showed that Na-QPDA can recognize some non-context-free language L with one-side error (i.e., there exists an $\epsilon \in [0,1/2)$ such that the accepting probability for $x \in L$ is greater than $1-\epsilon$ and, the rejecting probability for $x \notin L$ is one, or vice versa).

6.4.2.6 *l*-VPDA

As already pointed out before, one of the fundamental issues in quantum computing is to deal with automata theory based on quantum logic, that is, orthomodular lattice-valued automata. In Ref. [54], Qiu established a basic framework of orthomodular lattice-valued pushdown automata (l-VPDA), and showed that the class of the languages accepted by l-VPDAs *by empty stack* coincides with that accepted by l-VPDAs *by final state.*

6.4.3 Lattice-Valued Automata

Another approach to the study of quantum computing models can be through automata theory based on quantum logic, that is orthomodular lattice-valued automata [50–55].

Quantum logic was suggested by Birkhoff and Neumann [94] in 1936 to study the logical basic of quantum mechanics, and it originated from the Hilbert space's formalization of quantum mechanics. Since a state of a quantum system can be described by a closed subspace of a Hilbert space, while all the closed subspaces of a Hilbert space are endowed with the algebraic structure of orthomodular lattices, it was proposed that orthomodular lattices could be the algebraic version of quantum logic. Actually, orthomodular lattices were sometimes considered directly as quantum logic [95]. Thus, investigating orthomodular lattice-valued automata (l-VFA) may be considered to be a different approach of the logical basic of quantum computing. Ying [50,51] first defined orthomodular lattice-valued finite automata and studied some of the related properties, and then Qiu [52–55] dealt with l-VFA from different aspects. In Ref. [54], Qiu further investigated orthomodular lattice-valued pushdown automata (l-VPDA). Afterward, there is a variety of other recent papers to deal with this model of computation.

The definition of l-VFA is similar to that of classical finite automata [11], except that to each transition is assigned an element of an orthomodular lattice. One of the interesting results, in Ref. [53], is that some properties of l-VFA hold if and only if the truth-value set of logic satisfies the distributivity condition that is equivalent to the orthomodular lattice underlying logic reducing to a Boolean algebra, while some properties in l-VFA hold if and only if the truth-value set of logic satisfies a strictly weaker distributivity.

Finally, we would like to mention *automata theory based on residuated lattice-valued logic* (called residuated lattice-valued automata) established by Qiu [96,97], since residuated lattices, as a very generic algebra, have been extensively studied as the algebraic structures for fuzzy logics and other nonclassical logics, for example, for MV-algebras. Indeed, some scholars have already considered the relation between MV-algebras and quantum computation [98].

6.5 Concluding Remarks and Further Work

QFA are one of the crucial quantum computing models. To clarify their power, limitations, and advantages is so important for our understanding of the power and limitations of quantum computing, as conventional automata have played in classical theory of computation [11]. In this chapter, we have loosely reviewed some fundamental quantum computing models. We have partly outlined their definitions, basic properties, power of computation, their strengths and weaknesses, as well as their mutual relationships. The chapter should be seen only as a brief introduction into this interesting area of research and it is hoped that it could stimulate further research in this field.

It is clear that in this field there are still many interesting and important questions and problems that have not been answered and solved yet. Actually, some very basic questions are still waiting for solutions. In the following, we will try to list a number of problems related to QFA, and we are confident that to solve anyone of these is of importance and significance.

- *Decidability problems*
  - Is it decidable for a given 2QFA (or 2QCFA) $\mathcal{M}$ and a threshold λ, if there exists an input word x for which $p_{\mathcal{M}}(x) > \lambda$ and if there exists an input word x for which $p_{\mathcal{M}}(x) \geq \lambda$? (Here, $p_{\mathcal{M}}(x)$ denotes the probability of $\mathcal{M}$ accepting x.)
  - For any two given Go-QPDA $\mathcal{M}_1$ and $\mathcal{M}_2$, whether intersection of languages accepted of these two automata, with respect to a bounded error, is empty or not?
- *Equivalence problems*
  - How to determine whether given two 2QFA are equivalent? Does there exist a polynomial-time algorithm to determine the equivalence between them?
  - Similarly, how to determine whether any two 2QCFA are equivalent? Does there exist a polynomial-time algorithm to determine the equivalence between them?
  - Can we solve the equivalence problem for other quantum computing models? Indeed, we have solved them for QSM, MO-1QFA, MM-1QFA, CL-1QFA, and multiletter QFA.

- *States complexity* (we may refer to [99] concerning the states complexity of classical finite automata)
  - State complexity of language with respect to a fixed QFA (such as MO-1QFA, or MM-1QFA, or 2QFA). That is, given a language $\mathcal{L}$ accepted with bounded error by a QFA $\mathcal{M}$, how to find a minimal state QFA of the same type to accept $\mathcal{L}$?
  - State complexity of transformation of one QFA to another. For example, suppose a language $\mathcal{L}$ is accepted with bounded error by an MO-1QFA, MM-1QFA, 1QFAC, or 2QFA. If $\mathcal{L}$ is accepted by a 2QFA with the minimal number of states n_1, then how many states has a minimal MM-1QFA accepting $\mathcal{L}$? Similarly, how many is the number of the minimal states of an MO-1QFA accepting $\mathcal{L}$? How about 1QFAC?
  - State complexity of operations on QFA. In other words, if the two languages $\mathcal{L}_1$ and $\mathcal{L}_2$ are accepted by two MM-1QFA (MO-1QFA, or 1QFAC, or 2QFA) $\mathcal{M}_1$ and $\mathcal{M}_2$, with the minimal numbers of states, say, n_1 and n_2, respectively, then is it true that intersection and union of $\mathcal{L}_1$ and $\mathcal{L}_2$ can be accepted by some MM-1QFA (MO-1QFA, or 1QFAC, or 2QFA)? What is the relation between the number of states of such an automaton of MM-1QFA (MO-1QFA, or 1QFAC, or 2QFA) to n_1 and n_2?
  - Find another interesting examples of languages accepted by 1QFA that need exponentially less states when accepted by a 1QFA than when accepted by a DFA.
- *Closure properties*
  - Closure properties of the languages accepted by 2QCFA have been investigated in [93]. What about the closure properties for languages accepted by 2QFA with bounded error or with cut-point?
  - What about the closure properties for languages accepted by Go-QPDA with bounded error or with cut-point?
  - What about the closure properties for languages accepted by 2QFA and MO-1QFA as well as MM-1QFA with bounded error or with cut-point?
- *Language classes.* How to characterize classes of languages accepted by 2QFA and Go-QPDA? Moreover, what is the relation between classes of languages accepted by 2QFA and Go-QPDA, respectively?
- *Restriction of amplitudes.* How can we restrict amplitudes of QFA, and what would be the price to pay for that? This problem has already been investigated for QTMs by showing that restricting QTM to the rational amplitudes does not reduce their computational power.
- *Entanglement for QFA—What is the role of entanglement for QFA?* As we know, entanglement often brings more power to quantum information processing and makes some classically impossible tasks possible. Entanglement so far played no role in QFA, are we missing some point?
- How to establish quantum interactive proof systems based on 2QFA, as a counterpart of the interactive proof systems based on 2PFA [91]? As a start up, see Ref. [92].
- How to introduce measurement-based QFA, and how about their computing power?

Acknowledgments

This work is supported in part by the National Natural Science Foundation (Nos. 60873055, 61073054, 61100001), the Natural Science Foundation of Guangdong Province of China (No. 10251027501000004), the Fundamental Research Funds for the Central Universities (Nos. 10lgzd12,11lgpy36), the Research Foundation for the Doctoral Program of Higher School of Ministry of Education (Nos. 20100171110042, 20100171120051), the Program for New Century Excellent Talents in University (NCET) of China, the China Postdoctoral Science Foundation project (Nos. 20090460808, 201003375), and the project (No. MSM00216233419) of Czech Republik, the project of SQIG at IT, funded by FCT and EU FEDER projects Quantlog POCI/MAT/55796/2004 and QSec PTDC/EIA/67661/2006, IT Project

QuantTel, NoE Euro-NF, the SQIG LAP initiative, and FCT project PTDC/EEA-TEL/103402/2008 QuantPrivTel.

References

1. Gruska, J. 1999. *Quantum Computing*. London: McGraw-Hill.
2. Nielsen, M. A. and I. L. Chuang. 2000. *Quantum Computation and Quantum Information*. Cambridge: Cambridge University Press.
3. Landauer, R. 1961. Irreversibility and heat generation in the computing process. *IBM Journal of Research and Development* 5: 183.
4. Bennett, C. 1973. Logical reversibility of computation. *IBM Journal of Research and Development* 17: 525–532.
5. Benioff, P. 1980. The computer as a physical system: A microscopic quantum mechanical Hamiltonian model of computers as represented by Turing machines. *Journal of Statistical Physics* 22: 563–591.
6. Feynman, R. P. 1982. Simulating physics with computers. *International Journal of Theoretical Physics* 21: 467–488.
7. Deutsch, D. 1985. Quantum theory, the Church-Turing principle and the universal quantum computer. *Proceedings of Royal Society of London A* 400: 97–117.
8. Yao, A. C. 1993. Quantum circuit complexity. Paper presented at Proceedings of 34th IEEE Annual Symposium Foundations of Computer Science, Palo Alto, California, pp. 352–361.
9. Shor, P. W. 1994. Algorithm for quantum computation: Discrete logarithms and factoring. Paper presented at Proceedings of 37th IEEE Annual Symposium on Foundations of Computer Science, Burlington, Vermont, pp. 124–134.
10. Grover, L. 1996. A fast quantum mechanical algorithms for datdbase search. Paper presented at Proceedings of 28th Annual ACM Symposium on Theory of Computing, Philadelphia, Pennsylvania, pp. 212–219.
11. Hopcroft, J. E. and J. D. Ullman. 1979. *Introduction to Automata Theory, Languages, and Computation*. New York: Addision-Wesley.
12. Moore, C. and J. P. Crutchfield. 2000. Quantum automata and quantum grammars. *Theoretical Computer Science* 237: 275–306. Also see quant-ph/9707031.
13. Kondacs, A. and J. Watrous. 1997. On the power of finite state automata. Paper presented at Proceedings of 38th IEEE Annual Symposium on Foundations of Computer Science, Miami Beach, Florida, pp. 66–75.
14. Ambainis, A. and R. Freivalds. 1998. One-way quantum finite automata: Strengths, weaknesses and generalizations. Paper presented at Proceedings of the 39th Annual Symposium on Foundations of Computer Science, Palo Alfo, California, pp. 332–341. Also see quant-ph/9802062.
15. Amano, M. and K. Iwama. 1999. Undecidability on quantum finite automata. Paper presented at Proceedings of the 31st Annual ACM Symposium on Theory of Computing, Atlanta, Georgia, pp. 368–375.
16. Brodsky, A. and N. Pippenger. 2002. Characterizations of 1-way quantum finite automata. *SIAM Journal on Computing* 31: 1456–1478. Also see quant-ph/9903014.
17. Nayak, A. 1999. Optimal lower bounds for quantum automata and random access codes. Paper presented at Proceedings of the 40th Annual Symposium on Foundations of Computer Science, New York, pp. 124–133.
18. Ambainis, A., A. Nayak, A. Ta-Shma, and U. Vazirani. 1999. Dense quantum coding and a lower bound for for 1-way quantum automata. Paper presented at Proceedings of the 31st Annual ACM Symposium on Theory of Computing, Atlanta, Georgia, pp. 376–383.
19. Ambainis, A. and J. Watrous. 2002. Two-way finite automata with quantum and classical states. *Theoretical Computer Science* 287: 299–311.

20. Ambainis, A., A. Nayak, A. Ta-Shma, et al. 2002. Dense quantum coding and quantum automata. *Journal of the ACM* 49(4): 496–511.
21. Bertoni, A. and M. Carpentieri. 2001. Analogies and differences between quantum and stochastic automata. *Theoretical Computer Science* 262: 69–81.
22. Bertoni, A. and M. Carpentieri. 2001. Regular languages accepted by quantum automata. *Information and Computation* 165: 174–182.
23. Bertoni, A., C. Mereghetti, and Palano B. 2003. Quantum computing: 1-Way quantum automata. Paper presented at Proceedings of the 9th International Conference on Developments in Language Theory (DLT 2003), Palermo, Italy, Lecture Notes in Computer Science, vol. 2710: 1–20.
24. Ambainisa, A. and A. Kikusts. 2003. Exact results for accepting probabilities of quantum automata. *Theoretical Computer Science* 295: 3–25.
25. Blondel, V. D., E. Jeandel, P. Koiran, et al. 2005. Decidable and undecidable problems about quantum automata. *SIAM Journal on Computing* 34(6): 1464–1473.
26. Ambainis, A., M. Beaudry, M. Golovkins, et al. 2006. Algebraic results on quantum automata. *Theory of Computing Systems* 39: 1654–188.
27. Li, L. Z. and D. W. Qiu. 2008. Determining the equivalence for 1-way quantum finite. *Theoretical Computer Science* 403: 42–51.
28. Ambainis, A. and N. Nahimovs. 2009. Improved constructions of quantum automata. *Theoretical Computer Science* 410: 1916–1922.
29. Ambainis, A., R. Bonner, R. Freivalds, et al. 1999. Probabilities to accept languages by quantum finite automata. Paper presented at Proceedings of the 5th Annual International Conference on Computation and Combinatorics(COCOON'99), Tokyo, Japan, Lecture Notes in Computer Science, vol. 1627: 174–183.
30. Ambainisa, A., A. Kikusts, and M. Valdats. 2000. On the class of languages recognizable by 1-way quantum finite automata. arXiv: quant-ph/0009004.
31. Yakaryilmaz, A. and A. C. Cem Say. 2009. Efficient probability amplification in two-way quantum finite automata. *Theoretical Computer Science* 410: 1932–1941.
32. Dzelme-Bērzina. 2009. Mathematical logic and quantum finite state automata. *Theoretical Computer Science* 410: 1952–1959.
33. Li, L. Z., D. W. Qiu, Z. F. Zou, L. J. Li, L. H. Wu., and P. Mateus. 2012. Characterizations of one-way general quantum finite automata. *Theoretical Computer Science* 419: 73–91.
34. Qiu, D. W., L. Z. Li, X. Zou, P. Mateus, and J. Gruska. 2011. Multi-letter quantum finite automata: decidability of the equivalence and minimization of states. *Acta Informatica* 48: 271–290.
35. Gudder, S. 2000. Quantum computers. *International Journal of Theoretical Physics* 39: 2151–2177.
36. Qiu, D. W. 2002. Characterization of sequential quantum machines. *International Journal of Theoretical Physics* 41: 811–822.
37. Li, L. Z. and D. W. Qiu. 2006. Determination of equivalence between quantum sequential machines. *Theoretical Computer Science* 358: 65–74.
38. Golovkins, M. 2000. Quantum pushdown automata. Paper presented at Proceedings of the 27th Conference on Current Trends in Theory and Practice of Informatics (SOFSEM'00), Milovy, Lecture Notes in Computer Science, vol. 1963: 336–346.
39. Qiu, D. W. and M. S. Ying. 2004. Characterizations of quantum automata. *Theoretical Computer Science* 312: 479–489.
40. Nakanishi, M. 2004. On the power of one-sided error quantum pushdown automata with classical stack operations. Paper presented at Proceedings of 10th Annual International Computing and Combinatorics Conference (COCOON 2004), Jeju Island, Korea, Lecture Notes in Computer Science, vol. 3106: 179–187.
41. Adleman, L., J. DeMarrais, and H. Huang. 1997. Quantum computability. *SIAM Journal on Computing* 26(5): 1524–1540.

42. Bernstein, E. and U. Vazirani. 1997. Quantum complexity theory. *SIAM Journal on Computing* 26(5): 1411–1473.
43. Nishimura, H. and M. Ozawa. 2002. Computational complexity of uniform quantum circuit families and quantum Turing machines. *Theoretical Computer Science* 276: 147–181.
44. Watrous, J. 1999. Space-bounded quantum complexity. *Journal of Computer and System Sciences* 59(2): 281–326.
45. Hirvensalo, M. 1997. On quantum computation. PhD dissertation, Turku Center for Computer Science.
46. Müller, M. 2008. Strongly universal quantum Turing machines and invariance of Kolmogorov complexity. *IEEE Transation on Information Theory* 54(2): 763–780.
47. Qiu, D. W. 2008. Simulations of quantum Turing machines by quantum multi-stack machines. Paper presented at Computability in Europe (CIE 2008), Athens, Greece, June 15–20. Also see quant-ph/0501176.
48. Watrous, J. 1995. On one-dimensional cellular automata. Paper presented at Proceedings of 36th IEEE Annual Symposium Foundations of Computer Science, Milwaukee, Wisconsin, pp. 528–537.
49. Dürr, C. and M. Santha. 2002. A decision procedure for unitary linear quantum cellular automata. *SIAM Journal on Computing* 31(4): 1076–1089.
50. Ying, M. S. 2000. Automata theory based on quantum logic I. *International Journal of Theoretical Physics* 39: 981–991.
51. Ying, M. S. 2002. Automata theory based on quantum logic II. *International Journal of Theoretical Physics* 39: 2545–2557.
52. Qiu, D. W. 2003. Automata and grammar theory based on quantum logic. *Journal of Software* (in Chinese) 14(1): 23–27.
53. Qiu, D. W. 2004. Automata theory based on quantum logic: Some characterizations. *Information and Compututation* 190: 179–195.
54. Qiu, D. W. 2007. Automata theory based on quantum logic: Reversibilities and pushdown automata. *Theoretical Computer Science* 386: 38–56.
55. Qiu, D. W. 2007. Notes on automata theory based on quantum logic. *Science in China (Series F: Information Sciences)* 50(2): 154–169.
56. Qiu, D. W., P. Mateus, and A. Sernadas. 2009. One-way quantum finite automata together with classical states. arXiv:0909.1428.
57. Belovs, A., A. Rosmanis, and J. Smotrovs. 2007. Multi-letter reversible and quantum finite automata. Paper presented at Proceedings of the 13th International Conference on Developments in Language Theory (DLT 2007), Stuttgart, Germany, Lecture Notes in Computer Science, vol. 4588: 60–71.
58. Qiu, D. W. and S. Yu. 2009. Hierarchy and equivalence of multiletter quantum finite automata. *Theoretical Computer Science* 410: 3006–3017. Also see arXiv: 0812.0852.
59. Li, L. Z. and D. W. Qiu. 2009. A note on quantum sequential machines. *Theoretical Computer Science* 410: 2529–2535.
60. Rabin, M. O. 1963. Probabilistic automata. *Information and Control* 6: 230–244.
61. Paz, A. 1971. *Introduction to Probabilistic Automata.* New York: Academic Press.
62. Gudder, S. 1999. Quantum automata: An overview. *International Journal of Theoretical Physics* 38: 2261–2282.
63. Hirvensalo, M. 2007. Improved undecidability results on the emptiness problem of probabilistic and quantum cut-point languages. Paper presented at Proceedings of SOFSEM'2007, Harrachov, Czech Republic, Lecture Notes in Computer Science, vol. 4362: 309–319.
64. Koshiba, T. 2001. Polynomial-time algorithms for the equivalence for one-way quantum finite automata. Paper presented at Proceedings of the 12th International Symposium on Algorithms and Computation (ISAAC 2001), Christchurch, New Zealand, Lecture Notes in Computer Science, vol. 2223: 268–278.

65. Mateus, P. and D. W. Qiu. 2010. On the complexity of minimizing probabilistic and quantum automata. Submitted to *Information and Computation.*
66. Yakaryilmaz, A. and A. C. Cem Say. 2009. Languages recognized with unbounded error by quantum finite automata. Paper presented at Proceedings of the 4th Computer Science Symposium in Russia, Novosibirsk Akademgorodok, Russia, Lecture Notes in Computer Science, vol. 5675: 356–367.
67. Yakaryilmaz, A. and A. C. Cem Say. 2010. Unbounded-error quantum computation with small space bounds. arXiv:1007.3624.
68. Yakaryilmaz, A. and A. C. Cem Say. 2010. Languages recognized by nondeterministic quantum finite automata. *Quantum Information and Computation* 10: 747–770. Also see arXiv:0902.2081.
69. Golovkins, M. and M. Kravtsev. 2002. Probabilistic reversible automata and quantum automata. Paper presented at Proceedings of the 8th Annual International Conference on Computing and Combinatorics (COCOON 2002), Singapore, Lecture Notes in Computer Science, vol. 2387: 574–583.
70. Golovkins, M. M. Kravtsev, and V. Kravcevs. 2009. On a class of languages recognizable by probabilistic reversible decide-and-halt automata. *Theoretical Computer Science* 410: 1942–1951.
71. Kikusts, A. 1998. A small 1-way quantum finite automaton. arXiv: quant-ph/9810065.
72. Ablayev, F. and A. Gainutdinova. 2000. On the lower bounds for one-way quantum automata. Paper presented at Proceedings of 25th International Symposium on Mathematical Foundations of Computer Science (MFCS 2000), Bratislava, Slovak Republic, Lecture Notes in Computer Science, vol. 1893: 132–140.
73. Bertoni, A., C. Mereghetti, and B. Palano. 2005. Small size quantum automata recognizing some regular languages. *Theoretical Computer Science* 340: 394–407.
74. Bertoni, A., C. Mereghetti, and B. Palano. 2003. Lower bounds on the size of quantum automata accepting unary languages. Paper presented at Proceedings of 8th Italian Conference on Theoretical Computer Science (ICTCS 2003), Bertinoro, Italy, Lecture Notes in Computer Science, vol. 2841: 86–96.
75. Mereghetti, C. and B. Palano. 2001. Upper bounds on the size of one-way quantum finite automata, Paper presented at Proceedings of the 7th Italian Conference on Theoretical Computer Science (ICTCS 2001), Torino, Italy, Lecture Notes in Computer Science, vol. 2202: 123–135.
76. Mereghetti, C. and B. Palano. 2002. On the size of one-way quantum finite automata with periodic behaviors. *Theoretical Informatics and Applications* 36: 277–291.
77. Mereghetti, C. and B. Palano. 2007. Quantum automata for some multiperiodic languages. *Theoretical Computer Science* 387: 177–186.
78. Gruska, J. 2000. Descriptional complexity issues in quantum computing. *Journal of Automata, Languages and Combinatorics* 5(3): 191–218.
79. Li, L. Z. and D. W. Qiu. 2010. Revisiting the power and equivalence of one-way quantum finite automata. Paper presented at 2010 International Conference on Intelligent Computing (ICIC 2010), Changsha, China, Lecture Notes in Artificial Intelligence, vol. 6216: 1–8.
80. Mereghetti, C. and B. Palano. Quantum finite automata with control language. *Theoretical Informatics and Applications* 40: 315–332.
81. Hromkovič, J. 1983. One-way multihead deterministic finite automata. *Acta Informatica* 19: 377–384.
82. Mercer, M. 2008. Lower bounds for generalized quantum finite automata. Paper presented at the 2nd International Conference on Language and Automata Theory and Applications (LATA 2008), Tarragona, Spain, Lecture Notes in Computer Science, vol. 5196: 373–384.
83. Yakaryilmaz, A. and A. C. Cem Say. 2009. Language recognition by generalized quantum finite automata with unbounded error. arXiv:0901.2703v2.
84. Hirvensalo, M. 2008. Various aspects of finite quantum automata. Paper presented at Developments in Language Theory: 12th International Conference (DLT 2008), Kyoto, Japan, Lecture Notes in Computer Science, vol. 5257: 21–33.

85. Paschen, K. 2000. Quantum finite automata using ancilla qubits. University of Karlsruhe, Technical report.
86. Ciamarra, M. P. 2001. Quantum reversibility and a new model of quantum automaton, Paper presented at Proceeding of 13th International Symposium on Fundamentals of Computation Theory (FCT 2001), Riga, Latvia, Lecture Notes in Computer Science, vol. 2138: 376–379.
87. Freivalds, R. 1981. Probabilistic two-way machines. Paper presented at Proceedings on Mathematical Foundations of Computer Science, Strbske Pleso, Czechoslovakia, Lecture Notes in Computer Science, vol. 188: 33–45.
88. Greenberg, A. and A. Weiss. 1986. A lower bound for probabilistic algorithms for finite state machines. *Jounal of Computer and System Sciences* 33(1): 88–105.
89. Dwork, C. and L. Stockmeyer. 1990. A time-complexity gap for two-way probabilistic finite state automata. *SIAM Jounal on Computing* 19: 1011–1023.
90. Kaneps, J. and R. Freivalds. 1991. Running time to recognize nonregular languages by 2-way probabilistic automata. Paper presented at Proceedings of the 18th International Colloquium on Automata, Languages and Programming (ICALP 1991), Madrid, Spain, Lecture Notes in Computer Science, vol. 510: 174–185.
91. Dwork, C. and L. Stockmeyer. 1992. Finite state verifier I: The power of interaction. *Journal of the ACM* 39(4): 800–828.
92. Nishimura, H. and T. Yamakami. 2009. An application of quantum finite automata to interactive proof systems. *Journal of Computer and System Sciences* 75: 255–269.
93. Qiu, D. W. 2008. Some observations on two-way finite automata with quantum and classical states. Paper presented at 2008 International Conference on Intelligent Computing (ICIC 2008), Shanghai, China, Lecture Notes in Computer Science, vol. 5226: 1–8. Also, see quant-ph/0701187.
94. Birkhoff, G. and J. von Neumann. 1936. The logic of quantum mechanics. *Annals of Mathematics* 37: 823–843.
95. Pták, P. and S. Pulmannová. 1991. *Orthomodular Structures as Quantum Logics*. Dordrecht: Kluwer.
96. Qiu, D. W. 2001. Automata theory based on complete residuated lattice-valued logic. *Science in China (Series F: Information Sciences)* 44(6): 419–429.
97. Qiu, D. W. 2002. Automata theory based on complete residuated lattice-valued logic (II). *Science in China (Series F: Information Sciences)* 45(6): 442–452.
98. Ledda, A., M. Konig, F. Paoli, *et al.* 2006. MV-algebras and quantum computation. *Studia Logica* 82: 245–270.
99. Yu, S. 1998. Regular languages. In *Handbook of Formal Languages*, edited by G. Rozenberg, and A. Salomaa, 41–110. Berlin: Springer-Verlag.

7

Finite Automata Minimization

Marco Almeida, Nelma Moreira, and Rogério Reis
University of Porto

7.1 Introduction

The problem of finding the minimal DFA equivalent to a given automaton can be traced back to the 1950s with the works of Huffman (1954, 1955) and Moore (1956). Having applications on compiler construction, pattern matching, hardware circuit minimization, and XML processing to name a few, over the years several alternative (and increasingly efficient) algorithms have been proposed. Although they all depend upon computing an equivalence relation on the set of states, several approaches are possible: explicitly computing the equivalence relation, computing the partition (of states) that it induces, or computing the complement of the equivalence relation (finding all pairs of distinguishable states).

Aho et al. (1974, pp. 157–162), for example, present the minimization of finite automata as an application of a partitioning algorithm.

The algorithm which authors typically present as "Moore's algorithm," based on the works of Huffman and Moore, finds the pairs of distinguishable states and uses this information to create the equivalence classes.

Hopcroft's algorithm presents a more efficient approach to the set partitioning problem, resulting in the fastest (in terms of worst-case running time complexity) known DFA minimization algorithm.

Brzozowski's simple and elegant algorithm computes the equivalence classes in a somewhat mysterious manner.

There have also been some recent developments in incremental minimization algorithms. These may be halted at any time, returning an equivalent, partially minimized DFA, that can later be used to resume the minimization process.

7.2 Preliminaries

An *alphabet* is a (nonempty) finite set of symbols, usually denoted by Σ. A *word* is a finite, possibly empty, sequence of symbols. We denote the *empty word* by ϵ, and the set of all the words over the alphabet Σ by $\Sigma^\star$. A *language* is any subset of $\Sigma^\star$.

7.2.1 Deterministic Finite Automata

Formally, a *deterministic finite automaton* (DFA) is a quintuple $D = (Q, \Sigma, \delta, q_0, F)$, where

- Q is a finite set of *states*;
- Σ is a finite, nonempty, set of *symbols*;
- $\delta : Q \times \Sigma \to Q$ is the *transition function*;
- $q_0 \in Q$ is the *initial state* (or *start state*);
- $F \subseteq Q$ is the set of *final states* (or *accepting states*).

To simplify the notation, unless where otherwise noted, given an arbitrary DFA, n denotes its number of states and k the size of the alphabet.

A DFA is graphically represented by a *transition diagram*. A transition diagram is a directed graph with labeled arcs where: each node corresponds to a state; each transition such as $\delta(p, a) = q$ is represented by an arc from node p to node q, labeled by a; the initial state is signaled by an unlabeled incoming arrow; and the final states are represented by a doubled border.

When the transition function δ is total, we say that the DFA is *complete*. Note, however, that any incomplete DFA can easily be transformed into a complete one with a process that runs in $O(kn)$. The DFA of Figure 7.1a, for example, is missing transitions by b from states q_1 and q_2. In order to make it complete, we simply add a new nonfinal state—called a *sink state* or *dead state*—with a self-loop for every symbol, and add the missing transitions pointing there. The DFA on Figure 7.1b results of this construction.

As previously defined, the transition function accepts only single symbols as input. In order to describe what happens when we process a word we need to define an *extended transition function*. This function, denoted by $\hat{\delta}$, returns the state an automaton reaches after processing some sequence of symbols. It is

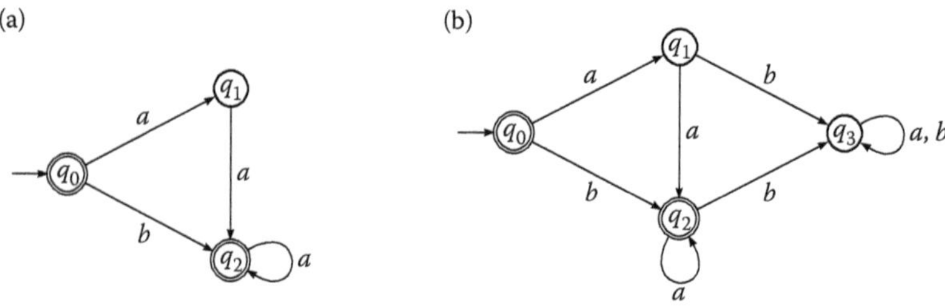

FIGURE 7.1 (a) Transition diagram of an incomplete DFA. (b) Transition diagram of a complete DFA.

defined inductively as follows:

$$\hat{\delta} : Q \times \Sigma^\star \longrightarrow Q$$
$$\hat{\delta}(q, \epsilon) = q,$$
$$\hat{\delta}(q, aw) = \hat{\delta}(\delta(q, a), w),$$

for all $q \in Q$, $a \in \Sigma$, and $w \in \Sigma^\star$.

We say that the words w such that $\hat{\delta}(q_0, w) \in F$ are *accepted* by the DFA. The language accepted by a DFA D is thus defined as

$$L(D) = \{w \in \Sigma^\star \mid \hat{\delta}(q_0, w) \in F\}.$$

Two DFAs $D_1 = (Q_1, \Sigma, \delta_1, q_1, F_1)$ and $D_2 = (Q_2, \Sigma, \delta_2, q_2, F_2)$ are *equivalent*, and we write $D_1 \sim D_2$, if they accept the same language, that is, $L(D_1) = L(D_2)$. We say that D_1 and D_2 are *isomorphic* if they are equal up to a renaming of its states, that is, if there exists a bijection $f : Q_1 \rightarrow Q_2$ such that $f(q_1) = q_2$, $f(\delta_1(q, a)) = \delta_2(f(q), a)$, and $f(F_1) = F_2$, for all $q \in Q_1$, $a \in \Sigma$.

Let $D = (Q, \Sigma, \delta, q_0, F)$ be a DFA. The *size* of D, denoted by $|D|$, is given by its number of states, hence $|D| = |Q|$. A state $q \in Q$ is said to be *initially connected* if there is a path from the initial state q_0 to q in the induced digraph, i.e., if there exists a word $w \in \Sigma^\star$ such that $\hat{\delta}(q_0, w) = q$. We say that D is initially connected if all its states are initially connected. Unless otherwise stated, we assume all DFAs to be initially connected. Given two distinct states q_1 and q_2 in Q, we say that q_1 is *distinguishable* from q_2, and write $q_1 \nsim q_2$, if there exists a word $w \in \Sigma^\star$ such that $\hat{\delta}(q_1, w) \in F$ and $\hat{\delta}(q_2, w) \notin F$, or vice versa. When no such word exists, the states q_1 and q_2 are *equivalent* (or *indistinguishable*) and we write $q_1 \sim q_2$. Formally, we define the *equivalence* relation $\sim$ on the set of states Q as the following set of pairs

$$\{(q_1, q_2) \in Q^2 \mid \forall w \in \Sigma^\star\ (\hat{\delta}(q_1, w) \in F \leftrightarrow \hat{\delta}(q_2, w) \in F)\}.$$

The relation $\sim$ is trivially an *equivalence relation.*

Lemma 7.1:

The relation $\sim$ is right-invariant, that is,

$$\forall q_1, q_2 \in Q\ \forall a \in \Sigma\ (q_1 \sim q_2 \rightarrow \delta(q_1, a) \sim \delta(q_2, a)).$$

Proof. Let $q_1, q_2 \in Q$

$$\begin{aligned} q_1 \sim q_2 &\Rightarrow \forall a \in \Sigma\ \forall w \in \Sigma^\star\ (\hat{\delta}(q_1, aw) \in F \leftrightarrow \hat{\delta}(q_2, aw) \in F) && \text{by definition of } \sim, \\ &\Leftrightarrow \forall a \in \Sigma\ \forall w \in \Sigma^\star\ (\hat{\delta}(\delta(q_1, a), w) \in F \leftrightarrow \hat{\delta}(\delta(q_2, a), w) \in F) && \text{by definition of } \hat{\delta}, \\ &\Leftrightarrow \forall a \in \Sigma\ \delta(q_1, a) \sim \delta(q_2, a) && \text{by definition of } \sim. \end{aligned}$$

□

Let us consider the *partition* of Q induced by this equivalence relation

$$Q/\!\sim\, = \{[\,q\,] \mid q \in Q\},$$

where $[\,q\,]$ is the *equivalence class* of state q defined as $[\,q\,] = \{q' \mid q' \sim q\}$.

For any DFA $D = (Q, \Sigma, \delta, q_0, F)$, indistinguishable states are redundant, and thus can be safely collapsed without changing the language accepted. In the resulting DFA, called the *quotient automaton* and denoted by $D/\sim$, each state corresponds to an equivalence class of $\sim$. The formal definition is as follows. Let

$$D/\sim = (Q/\sim, \Sigma, \tilde{\delta}, [\, q_0 \,], \tilde{F}),$$

where $\tilde{F} = \{[\, q \,] \mid q \in F\}$ and

$$\begin{aligned} \tilde{\delta} : Q/\sim \times \Sigma &\longrightarrow Q/\sim \\ \tilde{\delta}([\, q \,], a) &\longmapsto [\, \delta(q, a) \,]. \end{aligned}$$

Lemma 7.1 ensures that this definition is sound.

Lemma 7.2:

For all $q \in Q$ and $w \in \Sigma^\star$, $\hat{\tilde{\delta}}\left([q], w\right) = [\, \hat{\delta}(q, w) \,]$.

Proof. The proof follows by induction on the length of w. For $|w| = 0$, then $w = \epsilon$ and

$$\begin{aligned} \hat{\tilde{\delta}}([\, q \,], \epsilon) &= [\, q \,] && \text{by definition of } \hat{\tilde{\delta}}, \\ &= [\, \hat{\delta}(q, \epsilon) \,] && \text{by definition of } \hat{\delta}. \end{aligned}$$

Let us suppose that for all words w with $|w| = n - 1$, for all $q \in Q$ we have $\hat{\tilde{\delta}}([\, q \,], w) = [\, \hat{\delta}(q, w) \,]$. Any word of size n can be written as aw with $a \in \Sigma$ and w a word of size $n - 1$. Thus

$$\begin{aligned} \hat{\tilde{\delta}}\left([q], aw\right) &= \hat{\tilde{\delta}}\left(\tilde{\delta}([q], a), w\right) && \text{by definition of } \hat{\tilde{\delta}}, \\ &= \hat{\tilde{\delta}}([\, \delta(q, a) \,], w) && \text{by definition of } \tilde{\delta}, \\ &= [\, \hat{\delta}(\delta(q, a), w) \,] && \text{by induction hypothesis}, \\ &= [\, \hat{\delta}(q, aw) \,] && \text{by definition of } \hat{\delta}. \end{aligned}$$

□

Theorem 7.1:

A DFA D and its quotient $D/\sim$ are equivalent.

Proof. For any $w \in \Sigma^\star$,

$$\begin{aligned} w \in L(D/\sim) &\Leftrightarrow \hat{\tilde{\delta}}([\, q_0 \,], w) \in \tilde{F} && \text{by definition of } L(D/\sim), \\ &\Leftrightarrow [\, \hat{\delta}(q_0, w) \,] \in \tilde{F} && \text{by Lemma 7.2}, \\ &\Leftrightarrow \hat{\delta}(q_0, w) \in F && \text{by definition of } \tilde{F} \text{ and } \sim, \\ &\Leftrightarrow w \in L(D) && \text{by definition of } L(D). \end{aligned}$$

□

Lemma 7.3:

Given two equivalent DFAs, $D = (Q, \Sigma, \delta, q_0, F)$ and $D' = (Q', \Sigma, \delta', q_0', F')$, their respective quotient automata, $D/\sim$ and $D'/\sim$, are isomorphic.

Proof. Because both D and D' are initially connected, their respective quotient automata, trivially, are initially connected too. For each state q of $D/\sim$, let $r(q)$ represents the smallest word (in the lexicographical order) for which $\hat{\tilde{\delta}}([q_0], w) = q$. Because r is injective ($\tilde{\delta}$ is injective too), r^{-1}, its inverse function, exists and it is well defined for the appropriate domain. Let us call r' the corresponding function defined for $D'/\sim$. To see that the application

$$\begin{aligned} r^{-1} \circ r' : Q'/\sim &\longrightarrow Q/\sim \\ q &\longmapsto r^{-1}(r'(q)) \end{aligned}$$

is injective, let q_1 and q_2 be two states of $D'/\sim$. Then $r'(q_1) \neq r'(q_2)$, and we cannot have $r^{-1}(r'(q_1)) = r^{-1}(r'(q_2))$, because then, given the fact that we are dealing with a quotient automaton, there must exist a word w such that

$$\hat{\tilde{\delta}}'(q_1, w) \in \tilde{F}' \dot{\vee} \hat{\tilde{\delta}}'(q_2, w) \in \tilde{F}',$$

thus

$$r'(q_1)w \in L(D'/\sim) \dot{\vee} r'(q_2)w \in L(D'/\sim),$$

and, because

$$r^{-1}(r'(q_1)) = r^{-1}(r'(q_2))$$

we will have

$$r'(q_1)w \in L(D/\sim) \leftrightarrow r'(q_2)w \in L(D/\sim),$$

and this contradicts the supposition that those automata were equivalent. An entirely symmetric argument would show that $r'^{-1} \circ r$ is injective and thus $r^{-1} \circ r'$ is a bijection.

The rest of the proof that $r^{-1} \circ r'$ is an isomorphism between $D'/\sim$ and $D/\sim$ trivially results of its own definition. □

Theorem 7.2:

Given a DFA D, no DFA D' such that $D \sim D'$ has less states than the quotient automaton $D/\sim$.

Proof. By definition, the quotient automaton of a DFA D, $D/\sim$, cannot have more states than D. Let D' be a DFA equivalent to D. Then $D'/\sim$ is equivalent to D' (by Theorem 7.1) and is isomorphic to $D/\sim$ (by Lemma 7.3). Therefore, it cannot be smaller than $D/\sim$. □

Theorem 7.3:

A minimal DFA is unique up to isomorphism.

Proof. The proof follows directly from Theorem 7.1, Theorem 7.2, and Lemma 7.3. □

Corollary 7.1:

A regular language can be unequivocally identified by the minimal DFA that accepts it.

Proof. It is a direct consequence of Theorem 7.3. □

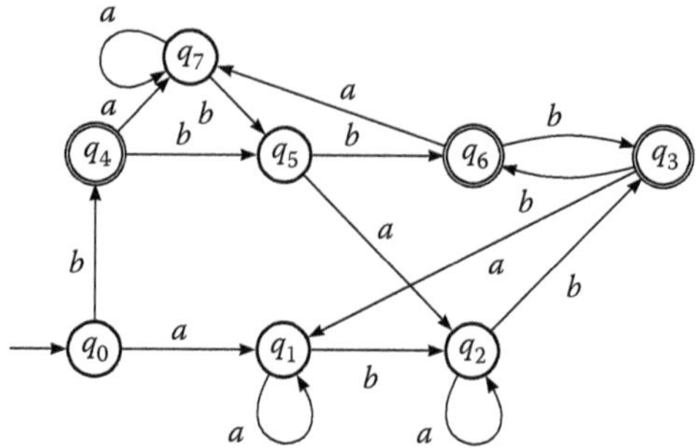

FIGURE 7.2 The transition diagram of DFA D.

Example 7.1:

Consider the DFA $D = (\{q_0, q_1, q_2, q_3, q_4, q_5, q_6, q_7\}, \{a.b\}, \delta, q_0, \{q_3, q_4 q_6\})$ of Figure 7.2. To compute the equivalence relation $\sim$ on the set of states, we incrementally compute the set of distinguishable pairs of states, N. The correctness of this method follows from Lemma 7.1 by contraposition. First, we consider the set N_0 of distinguishable pairs of states (q, q') such that one state is final and the other is not $(q \in F \dot{\vee} q' \notin F)$. The following table presents the set N_0 of such pairs, by noting that if $(q, q') \in N$ then $(q', q) \in N$:

q_1							
q_2							
q_3	×	×	×				
q_4	×	×	×				
q_5				×	×		
q_6	×	×	×			×	
q_7				×	×		×
	q_0	q_1	q_2	q_3	q_4	q_5	q_6

Let N_1 be the relation

$$N_1 = N_0 \cup \left\{(q, q') \in Q^2 \mid \exists \sigma \in \Sigma, (\delta(q, \sigma), \delta(q', \sigma)) \in N_0)\right\}.$$

Then,

$$\begin{aligned}
&(\delta(q_0, b), \delta(q_1, b)) = (q_4, q_2) \in N_0 \Rightarrow (q_0, q_1) \in N_1 \\
&(\delta(q_0, a), \delta(q_2, a)) = (q_1, q_2) \notin N_0 \wedge (\delta(q_0, b), \delta(q_2, b)) = (q_4, q_3) \notin N_0 \\
&(\delta(q_0, a), \delta(q_5, a)) = (q_1, q_2) \notin N_0 \wedge (\delta(q_0, b), \delta(q_5, b)) = (q_4, q_6) \notin N_0 \\
&(\delta(q_0, b), \delta(q_7, b)) = (q_4, q_5) \in N_0 \Rightarrow (q_0, q_7) \in N_1 \\
&(\delta(q_1, b), \delta(q_2, b)) = (q_2, q_3) \in N_0 \Rightarrow (q_1, q_2) \in N_1 \\
&(\delta(q_1, b), \delta(q_5, b)) = (q_2, q_6) \in N_0 \Rightarrow (q_1, q_5) \in N_1 \\
&(\delta(q_1, a), \delta(q_7, a)) = (q_1, q_7) \notin N_0 \wedge (\delta(q_1, b), \delta(q_7, b)) = (q_2, q_5) \notin N_0 \\
&(\delta(q_2, a), \delta(q_5, a)) = (q_2, q_2) \notin N_0 \wedge (\delta(q_2, b), \delta(q_5, b)) = (q_2, q_6) \notin N_0 \\
&(\delta(q_2, b), \delta(q_7, b)) = (q_3, q_5) \in N_0 \Rightarrow (q_2, q_7) \in N_1 \\
&(\delta(q_3, b), \delta(q_4, b)) = (q_6, q_5) \in N_0 \Rightarrow (q_3, q_4) \in N_1 \\
&(\delta(q_3, a), \delta(q_6, a)) = (q_1, q_7) \notin N_0 \wedge (\delta(q_3, b), \delta(q_6, b)) = (q_6, q_3) \notin N_0 \\
&(\delta(q_4, b), \delta(q_6, b)) = (q_5, q_3) \in N_0 \Rightarrow (q_4, q_6) \in N_1 \\
&(\delta(q_5, b), \delta(q_7, b)) = (q_6, q_5) \in N_0 \Rightarrow (q_5, q_7) \in N_1.
\end{aligned}$$

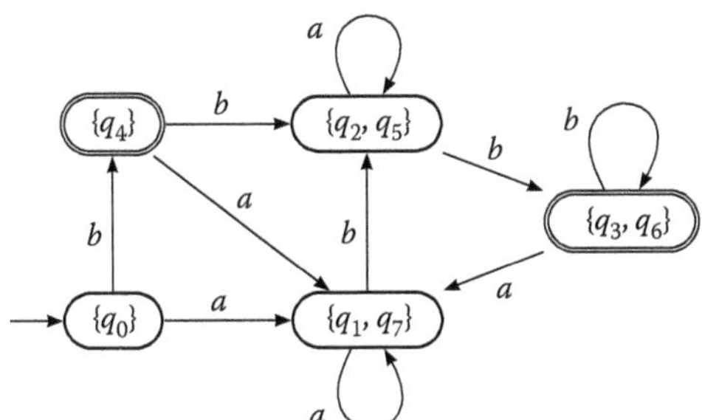

FIGURE 7.3 The transition diagram of minimal DFA equivalent to D.

The table for N_1 is then the following:

	q_0	q_1	q_2	q_3	q_4	q_5	q_6
q_1	×						
q_2		×					
q_3	×	×	×				
q_4	×	×	×	×			
q_5		×		×	×		
q_6	×	×	×		×	×	
q_7	×		×	×	×	×	×

We can repeat the procedure, by considering:

$$N_2 = N_1 \cup \left\{(q, q') \in Q^2 \mid \exists \sigma \in \Sigma,\ (\delta(q, \sigma), \delta(q', \sigma) \in N_1)\right\}.$$

In this case, the pairs of states to be added are

$$(\delta(q_0, b), \delta(q_2, b)) = (q_4, q_3) \in N_1 \Rightarrow (q_0, q_2) \in N_2$$
$$(\delta(q_0, a), \delta(q_5, a)) = (q_1, q_2) \in N_1 \Rightarrow (q_0, q_5) \in N_2.$$

And, the new table is

	q_0	q_1	q_2	q_3	q_4	q_5	q_6
q_1	×						
q_2	×	×					
q_3	×	×	×				
q_4	×	×	×	×			
q_5	×	×		×	×		
q_6	×	×	×		×	×	
q_7	×		×	×	×	×	×

The next iteration does not add more pairs of states, thus we have $N = N_2$. We conclude that the equivalence classes of the complementary relation, $\sim$, are

$$\left\{\{q_0\}, \{q_1, q_7\}, \{q_2, q_5\}, \{q_3, q_6\}, \{q_4\}\right\}.$$

The minimal DFA equivalent to D is presented in Figure 7.3.

7.2.2 Nondeterministic Finite Automata

A nondeterministic finite automaton (NFA) is a quintuple $N = (Q, \Sigma, \delta, q_0, F)$, where Q, Σ, q_0, and F are defined exactly the same way as for a DFA. The transition function is defined as $\delta : Q \times \Sigma \to 2^Q$, returning a set of states instead of a single state.

Just like in the deterministic case, we need to extend the transition function δ to a function $\hat{\delta}$ that describes the processing of a word. Given an NFA $N = (Q, \Sigma, \delta, q_0, F)$, we define $\hat{\delta}$ by

$$\hat{\delta} : Q \times \Sigma^\star \to 2^Q$$
$$\hat{\delta}(p, \epsilon) = \{ p \}$$
$$\hat{\delta}(p, w) = \bigcup_{p' \in \hat{\delta}(p,a)} \hat{\delta}(p', v).$$

where $w = av$ such that $v \in \Sigma^\star$ and $a \in \Sigma$.

Sometimes it is useful to extend the transition function even further so that it will take a set of states $P \subseteq Q$ and a symbol $a \in \Sigma$, and returns the set of all states accessible from each $p' \in P$ when reading a. We define this new transition function in the following way:

$$\Delta : 2^Q \times \Sigma \to 2^Q$$
$$\Delta(P, a) = \bigcup_{p' \in P} \delta(p', a).$$

An NFA *accepts* a word $w \in \Sigma^\star$ if, while reading the symbols of w, there is at least a sequence of choices which reaches a final state. Formally, we define the *language* accepted by an NFA $N = (Q, \Sigma, \delta, q_0, F)$, by

$$L(N) = \{ w \in \Sigma^\star \mid \hat{\delta}(q, w) \cap F \neq \emptyset \}.$$

Again, similarly to the deterministic case, we say that two NFAs N_1 and N_2 are *equivalent*, and write $N_1 \sim N_2$, if they accept the same language.

A common extension to NFAs is the use of transitions labeled by ϵ, the *empty* word. Formally, this special kind of automata (ϵ-NFA) is defined by the same quintuple $N_\epsilon = (Q, \Sigma, \delta, q_0, F)$ as the previous NFAs but the domain of the transition function, defined as $\delta : Q \times (\Sigma \cup \{ \epsilon \}) \to 2^Q$, is equipped with the extra symbol. This "feature" allows an NFA to make a transition spontaneously, without reading any input. Although it does not add any expressive power (a language is accepted by an NFA if and only if it is accepted by some ϵ-NFA), it does sometimes simplify the construction of an automaton.

Another generalization we may consider is a *non-deterministic initial state.* Such an NFA is defined with the same quintuple, except that there is a set of initial states rather than exactly one. Thus, the computation of the acceptance of a given word starts from a nondeterministically chosen initial state. We can trivially convert such an NFA to an ϵ-NFA by adding a new initial state i, creating an ϵ-transition from i to each of the other initial states, and then defining i as the only initial state.

7.2.2.1 Equivalence of Nondeterministic and Deterministic Finite Automata

Although, when defining some languages, an NFA is easier to construct than a DFA, the expressive power of both formalisms is equivalent and any regular language described by an NFA can also be described by some DFA. Moreover, there is an algorithm for making any NFA deterministic, that is, computing an equivalent DFA. This is achieved with the *subset construction* which, in the worst case, constructs all the subsets of the set of states of N. The notion of NFA, its equivalence to DFAs, and the subset construction are results due to Rabin and Scott (1959).

Given an NFA $N = (Q, \Sigma, \delta, q_0, F)$ such that $|Q| = n$, we can construct a DFA $D = (Q', \Sigma, \delta', q_0', F')$ in the following way:

$$
\begin{aligned}
Q' &= 2^Q; \\
\delta'(p, a) &= \Delta(p, a), \quad \text{for } p \in Q',\ a \in \Sigma; \\
q_0' &= \{\, q_0 \,\}; \\
F' &= \{\, p \in Q' \mid p \cap F \neq \emptyset \,\}.
\end{aligned}
$$

Notice that this construction implies that $|Q'| = 2^n$ even when the transition function is not total (Hopcroft et al. 2000). Consequently, the size of the DFA D will be exponentially large even when some (or most) of the states in Q' are not actually used/necessary. This results in a huge, possibly disconnected DFA. There is a more constructive approach, as demonstrated by the procedure Fa-Deterministic. The blowup in the number of states may sometimes be avoided by assuring that only reachable states are added to the set Q'. Such an implementation of the algorithm follows the transition function δ and incrementally builds the set of states Q'. This assures that a state q will be one of the states of the resulting DFA D if and only if it is used by the transition function δ'.

```
def FA-DETERMINISTIC(Q, Σ, δ, q0, F):
  Q' := ∅
  δ' := NIL
  V := ∅
  V := PUSH(V, {q0})
  while V ≠ ∅:
      p := POP(V)
      for a ∈ Σ:
          T := Δ(p, a)
          δ'(p, a) := T
          if T ∉ Q':
              V := PUSH(V, T)
          Q' := Q' ∪ {T}
  F' = ∅
  for p ∈ Q':
      if p ∩ F ≠ ∅:
          F' = F' ∪ {p}
  D := (Q', Σ, δ', {q0}, F')
  return D
```

7.3 Minimization

Over the last decades several different finite automata minimization algorithms have been developed and published. A taxonomy of several of those algorithms can be found in Bruce Watson's PhD thesis (1995). Although they all depend upon computing an equivalence relation on the set of states, several approaches are possible: explicitly computing the equivalence relation, computing the partition (of states) that it induces, or computing the complement of the equivalence relation (finding all pairs of distinguishable states).

In Aho et al. (1974, pp. 157–162), for example, the minimization of finite automata is presented as an application of a partitioning algorithm. The problem of minimizing states in a finite automaton $D = (Q, \Sigma, \delta, q_0, F)$ is equivalent to the problem of finding the coarsest (having fewest blocks) partition P of Q such that

- P is consistent with the initial partition $\{\, F, Q - F \,\}$, that is, each block in P is a subset either F or $Q - F$

- if the states p and q are in the same block, then the states $\delta(p,a)$ and $\delta(q,a)$ are also in the same block, for each $a \in \Sigma$

The algorithm which authors typically present as "Moore's algorithm," based on the works of Huffman and Moore, finds the pairs of distinguishable states and uses this information to create the equivalence classes. This is also the approach used in Example 7.1.

Hopcroft's algorithm presents a more efficient approach to the set partitioning problem, resulting in the fastest (in terms of worst-case running time complexity) known DFA minimization algorithm.

Brzozowski's simple and elegant algorithm computes the equivalence classes in a somewhat distinct manner. A result due to Champarnaud et al. (2002) shows that given a DFA $D = (Q, \Sigma, \delta, q_0, F)$, two states p and q are equivalent if and only if after the two first operations (reversal and subset construction) they both belong to the same set $S \subseteq 2^Q$.

In the following sections, we assume that the states of a given automaton are represented by integers, and thus it is possible to order them. This ordering is used to normalize pairs of states.

```
def NORMALISE-PAIR(p, q):
    if p < q:
        return (p, q)
    else:
        return (q, p)
```

The normalization step allows us to improve the behavior of some minimization algorithms by ensuring that only $(n^2 - n)/2$ pairs of states are considered.

7.4 Moore's Algorithm

One way to minimize a DFA, usually credited to Huffman (1954, 1955) and Moore (1956), is to determine the set of all distinguishable states. Having been found, a minimal equivalent automaton can be constructed by collapsing each set of mutually indistinguishable (equivalent) states into a single state.

Although neither Huffman nor Moore explicitly presents an algorithm for minimizing finite automata, they do define state equivalence, the distinguishability relation (and associated set partition), prove that a minimal finite automata has no equivalent states and that a minimal DFA is in fact unique.

Let $D = (Q, \Sigma, \delta, q_0, F)$ be a DFA. The following version of the algorithm, DFA-MINIMISE-MOORE, runs in $O\left(|\Sigma||Q|^2\right)$ time. It is based on the algorithms presented by Hopcroft and Ullman (1979, p. 70)—attributed to Huffman (1954) and Moore (1956)—and Shallit (2008, p. 87).

In their presentation, Hopcroft and Ullman use the variable L to map each pair of states to a list of related pairs of states. We implement it as an associative array (dictionary).

Given a DFA D, DFA-MINIMISE-MOORE returns the minimal DFA D' such that $D \sim D'$.

```
def DFA-MINIMISE-MOORE(Q, Σ, δ, q0, F):
  for p ∈ Q:
      for q ∈ {x | x ∈ Q, x > p}:
          L[(p, q)] := ∅
          if (p ∈ F) ∨̇ (q ∈ F):
              M[(p, q)] := TRUE
          else:
              M[(p, q)] := FALSE
  for p ∈ Q:
      for q ∈ {x | x ∈ Q, x > p}:
          if (p ∈ F) ⇔ (q ∈ F):
              m := FALSE
              for a ∈ Σ:
```

```
14                      x := NORMALISE-PAIR(δ(p, a), δ(q, a))
15                      if M[x]:
16                          m := TRUE
17                          break
18              if m:
19                  M[(p, q)] := TRUE
20                  MARK-SET(p, q)
21              else:
22                  for a ∈ Σ:
23                      if δ(p, a) ≠ δ(q, a):
24                          x := NORMALISE-PAIR(δ(p, a), δ(q, a))
25                          L[x] := L[x] ∪ { (p, q) }
26  D' := (Q, Σ, δ, q0, F)
27  for p ∈ Q:
28      for q ∈ { x | x ∈ Q, x > p }:
29          if M[(p, q)] = FALSE
30              JOIN-STATES(D', {{p, q}})
31  return D'
```

The algorithm DFA-MINIMISE-MOORE proceeds by finding the smallest word which distinguishes a pair of states. Every pair of states (p, q) is assumed to be equivalent until some word which distinguishes p from q is found. The two global variables, L and M, are used to keep a set associated to each pair of states, and mark a pair of states as either equivalent or distinguishable, respectively. They are global only to simplify the description of DFA-MINIMISE-MOORE and its integration with the auxiliary procedure MARK-SET. Given two states p and q, $L[(p, q)]$ contains a set of pairs of states which are distinguishable if and only if p is distinguishable from q. The states p and q are marked as distinguishable by making $M[(p, q)] = $ TRUE. Notice that we use M as an associative array only to make it similar to L and keep the notation simple, it could just as easily be implemented as a plain matrix.

The loop in lines 2–8 separates the trivially distinguishable pairs of states (those such that one state is final and the other is not) from the remaining pairs, assumed to be equivalent until proven otherwise. The loop in lines 9–24 iterates through all possible pairs of states. For each pair of states (p, q) not yet marked as distinguishable (line 11), if an already marked pair of states reachable from p and q is found (line 17), (p, q) and all pairs on its set are also marked as distinguishable (line 19). If, on the other hand, no two states directly accessible from the pair (p, q) is marked, it is added to the set of each pair directly accessible from it (lines 21–24). In lines 25–29 a copy of the original DFA is created, and each pair of equivalent states is merged by the call to JOIN-STATES. The minimal DFA D' is returned at line 30.

```
def MARK-SET(p, q):
    for (p', q') ∈ L[(p, q)]:
        if M[(p', q')] = FALSE:
            M[(p', q')] := TRUE
            MARK-SET(p', q')
```

The procedure MARK-SET recursively marks all unmarked pairs (p', q') on the list of (p, q), and all pairs on the list of (p', q'), and so on.

7.4.1 An Example

Consider the nonminimal DFA presented in Figure 7.4. After the initialization step (lines 2–8), variables L and M are as presented in Table 7.1.

During the execution of the main loop (lines 9–24), both variables L and M are updated to the values presented in Table 7.2.

All states p and q such that $M[(p, q)] = $ FALSE are equivalent. The loop at lines 26–29 merges these states on the corresponding equivalence class, updating the structure of a new DFA, constructed at line 25.

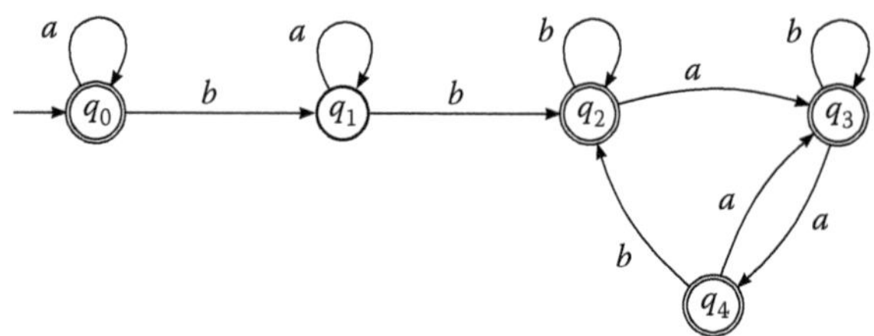

FIGURE 7.4 Example of a DFA that is not minimal.

TABLE 7.1 Value of the Variables L and M, after Initialized, in DFA-MINIMISE-MOORE

States	M	L	States	M	L
(q_0, q_1)	TRUE	$\emptyset$	(q_1, q_3)	TRUE	$\emptyset$
(q_0, q_2)	FALSE	$\emptyset$	(q_1, q_4)	TRUE	$\emptyset$
(q_0, q_3)	FALSE	$\emptyset$	(q_2, q_3)	FALSE	$\emptyset$
(q_0, q_4)	FALSE	$\emptyset$	(q_2, q_4)	FALSE	$\emptyset$
(q_1, q_2)	TRUE	$\emptyset$	(q_3, q_4)	FALSE	$\emptyset$

TABLE 7.2 Value of the Variables L and M after the Main Loop of DFA-MINIMISE-MOORE

States	M	L	States	M	L
(q_0, q_1)	TRUE	$\emptyset$	(q_1, q_3)	TRUE	$\emptyset$
(q_0, q_2)	TRUE	$\emptyset$	(q_1, q_4)	TRUE	$\emptyset$
(q_0, q_3)	TRUE	$\emptyset$	(q_2, q_3)	FALSE	$\{(q_2, q_3), (q_3, q_4)\}$
(q_0, q_4)	TRUE	$\emptyset$	(q_2, q_4)	FALSE	$\emptyset$
(q_1, q_2)	TRUE	$\emptyset$	(q_3, q_4)	FALSE	$\{(q_2, q_3), (q_3, q_4)\}$

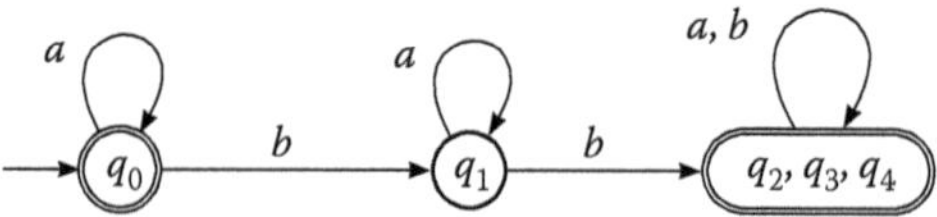

FIGURE 7.5 Minimal DFA equivalent to the one presented in Figure 7.4.

The equivalent minimal DFA, obtained after collapsing the non-distinguishable states, is presented in Figure 7.5.

7.5 Hopcroft's Algorithm

In terms of worst-case complexity analysis, Hopcroft's algorithm, presented in Hopcroft (1971), is the best known algorithm for deterministic finite automata minimization. It runs on $O(kn \log n)$ time when applied to a DFA with n states over an alphabet of k symbols and is presented as DFA-MINIMISE-HOPCROFT.

Although widely known and cited, the original algorithm's proofs of correctness and running time are quite difficult to understand. Detailed and easy to follow tutorial reconstructions of the original algorithm, including time complexity analysis and correctness proofs, are presented in Gries (1972) and Knuutila (2001).

Unlike Moore's algorithm (and several of its variations (Hopcroft et al. 2000, pp. 154–157)), DFA-MINIMISE-HOPCROFT does not identify pairs of distinguishable states. Instead, it proceeds by *refining* the coarsest partition that separates final and non-final states, until reaching a partition that is right-invariant. Let $D = (Q, \Sigma, \delta, q_0, F)$ be a DFA. The initial partition is $P = \{F, Q - F\}$. At each step of the algorithm, a block $B \in P$ and a symbol $a \in \Sigma$ are selected to refine the partition. This refinement process *splits* every other block $B' \in P$ according to whether a state of B', when consuming a symbol $a \in \Sigma$, reaches a state which is in B or not. Formally, we call this procedure SPLIT and define it in the following way.

Definition 7.1:

$$\text{SPLIT}(B', B, a) = \left(B' \cap \check{\delta}(B, a), B' \cap \overline{\check{\delta}(B, a)}\right)$$

where

$$\check{\delta}(Q', a) = \bigcup_{q \in Q'} \delta^{-1}(q, a), \quad Q' \subseteq Q.$$

DFA-MINIMISE-HOPCROFT terminates when there are no more blocks to refine. In the end, each block of the partition is a set of equivalent states and represents a single state of the minimal DFA.

```
1   def DFA-MINIMISE-HOPCROFT(Q, Σ, δ, q, F):
2     P := { Q − F, F }
3     L := PUSH(∅, F)
4     while L ≠ ∅:
5         C := POP(L)
6         for a ∈ Σ:
7             for B ∈ P:
8                 (B1, B2) = SPLIT(B, C, a)
9                 P := P − { B }
10                P := P ∪ { B1, B2 }
11                if |B1| < |B2|:
12                    L := PUSH(L, B1)
13                else:
14                    L := PUSH(L, B2)
15    D′ := D
16    JOIN-STATES(D′, P)
17    return D′
```

The data structure L, implemented as a stack, contains the blocks of P which are yet to be visited. At each iteration of the main loop (lines 9–19), one element (the *splitter*) is removed from L (line 10) and used in the splitting process (line 13). Although the choice of the element does not influence the correctness of the algorithm, Baclet and Pagetti (2006) present some experimental results stating that a *last-in-first-out* (stacked) policy yields better practical results.

Next, having selected a splitter set C, all elements of P (lines 12–19) are refined and used to update P and L. At this point, the original algorithm by Hopcroft applies the splitting process to and refines only blocks $B \in P$ such that there exists $t \in B$ with $\delta(t, a) \in \check{\delta}(C, a)$ for some $a \in \Sigma$. We omit this selection and iterate through all blocks in P because, for a given $B \in P$, if $\delta(t, a) \notin \check{\delta}(C, a)$ for all $t \in B$, the call SPLIT(B, C, a) will return $(\emptyset, B)$. This will change neither P nor L because

- B is removed from P at line 14, but $B_2 = B$ is added at line 15
- at line 16, $B_1 = \emptyset$; thus $|B_1| < |B_2|$ and $\emptyset$ is added to L at line 17

TABLE 7.3 Partition Refinement Process of DFA-MINIMISE-HOPCROFT, Applied to the DFA of Figure 7.4

C	a	B	B_1	B_2
$\{q_1\}$	a	$\{q_0, q_2, q_3, q_4\}$	$\emptyset$	$\{q_0, q_2, q_3, q_4\}$
$\{q_1\}$	a	$\{q_1\}$	$\{q_1\}$	$\emptyset$
$\{q_1\}$	b	$\{q_0, q_2, q_3, q_4\}$	$\{q_0\}$	$\{q_2, q_3, q_4\}$
$\{q_1\}$	b	$\{q_1\}$	$\emptyset$	$\{q_1\}$
$\{q_0\}$	a	$\{q_2, q_3, q_4\}$	$\emptyset$	$\{q_2, q_3, q_4\}$
$\{q_0\}$	a	$\{q_0\}$	$\{q_0\}$	$\emptyset$
$\{q_0\}$	a	$\{q_1\}$	$\emptyset$	$\{q_1\}$
$\{q_0\}$	b	$\{q_2, q_3, q_4\}$	$\emptyset$	$\{q_2, q_3, q_4\}$
$\{q_0\}$	b	$\{q_0\}$	$\emptyset$	$\{q_0\}$
$\{q_0\}$	b	$\{q_1\}$	$\emptyset$	$\{q_1\}$

```
1 def SPLIT(B, C, a):
2     X := ∅
3     for q ∈ C:
4         X := X ∪ { δ⁻¹(q, a) }
5     Y := Q − X
6     return (B ∩ X, B ∩ Y)
```

A call such as SPLIT(B, C, a) refines the set B into two smaller sets according to the splitter C and the transitions labeled by the symbol a (cf. Definition 7.1). The sets $\check{\delta}(C, a)$ and $\overline{\check{\delta}(C, a)}$ are computed at lines 2–4 and 5, respectively, and the refined classes, obtained from the intersection with B, are returned at line 6.

7.5.1 An Example

Recall the nonminimal DFA presented in Figure 7.4. Since the number of final states is greater than the number of nonfinal states, the variables P and L of DFA-MINIMISE-HOPCROFT are initialized (lines 2–8) as follows:

$$P = \{\{q_0, q_2, q_3, q_4\}, \{q_1\}\},$$
$$L = \{\{q_1\}\}.$$

At each step of the execution of the main loop (lines 9–19), the partition P is successively refined by the splitting process as presented in Table 7.3.

After the main loop is finished, each block of P represents an equivalence class, and, in the particular case of the example we are considering, is as follows:

$$P = \{\{q_0\}, \{q_1\}, \{q_2, q_3, q_4\}\}.$$

The equivalent minimal DFA, obtained by collapsing all elements of each equivalence class into a single state, is presented in Figure 7.5.

7.6 Brzozowski's Algorithm

Given a (possibly nondeterministic) finite automaton A, Brzozowski's algorithm builds the minimal DFA A_m, such that A_m is equivalent to A, simply by applying two successive reverse and subset construction operations. We must take into account that an NFA which results from the reversal operation may have more than one initial state.

We present the implementation of Brzozowski's minimization algorithm (Brzozowski 1962), specialized for the deterministic case, as DFA-MINIMISE-BRZOZOWSKI.

```
def DFA-MINIMISE-BRZOZOWSKI(D):
    D^R := FA-DETERMINISTIC(DFA-REVERSE(D))
    D_m := FA-DETERMINISTIC(DFA-REVERSE(D^R))
    return D_m
```

Let N be a nondeterministic finite automaton. The procedure NFA-REVERSE returns an NFA N_r such that the language recognized by N is the reversal of the one recognized by N_r, that is, $L(N_r) = L^R(N)$. DFA-REVERSE performs the equivalent transformation to DFAs.

The procedure FA-DETERMINISTIC takes an arbitrary NFA N as argument and returns an equivalent DFA D such that $L(N) = L(D)$. It simply applies the well-known subset construction method, for which we present an implementation in Section 7.2.2.

```
def DFA-REVERSE(Q, Σ, δ, q0, F):
    return NFA-REVERSE(Q, Σ, δ, {q0}, F)
```

```
def NFA-REVERSE(Q, Σ, δ, I, F):
  I_r := F
  F_r := I
  for q ∈ Q:
      for a ∈ Σ:
          δ_r(q, a) := ∅
  for q ∈ Q:
      for a ∈ Σ:
          for t ∈ δ(q, a):
              δ_r(t, a) := δ_r(t, a) ∪ { q }
    N_r := (Q, Σ, δ_r, I_r, F_r)
    return N_r
```

Because of the peculiar way by which Brzozowski's algorithm computes the minimal DFA, Watson, in his taxonomy (Watson 1995, p. 193), assumed it to be unique and placed it apart all other algorithms. Later, however, after analyzing how the sequential determinizations perform the minimization, Champarnaud et al. (2002) showed that DFA-MINIMISE-BRZOZOWSKI *does* compute state equivalences.

Since, it invokes the subset construction algorithm FA-DETERMINISTIC twice—in order to make the two reversed automata deterministic—DFA-MINIMISE-BRZOZOWSKI is exponential in the worst case. Moreover, (Nicaud 2000) has proved that, for the specific case of group automata, the algorithm is also exponential on the average case. Nonetheless, some authors (e.g., Watson (1995, p. 333)) have stated that it does present very good practical results, sometimes even outperforming Hopcroft's $O(kn \log n)$ algorithm.

7.6.1 An Example

Consider the nonminimal deterministic finite automaton D in Figure 7.6. The sequence of operations by which DFA-MINIMISE-BRZOZOWSKI operates to minimize this DFA is illustrated in Figure 7.7.

Figure 7.7a shows the result of the first reversal operation. Notice that the resulting DFA is constructed directly from D, using DFA-REVERSE, and presents only the following differences:

- The initial state of D is the only accepting state of the new DFA
- The set of initial states of the new DFA consists on all accepting states of D
- The direction of all arcs in D is reversed

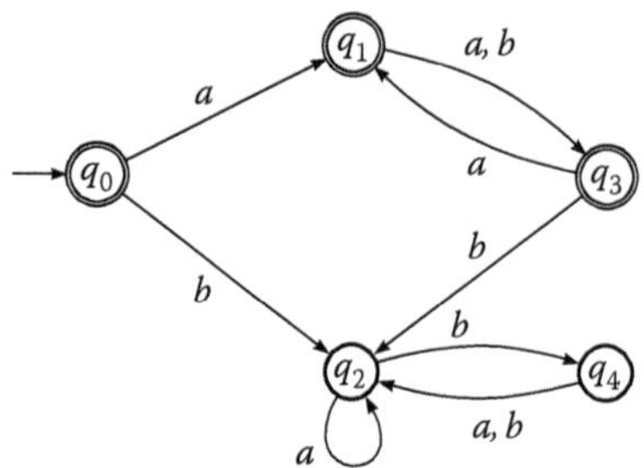

FIGURE 7.6 Example of a DFA that is not minimal.

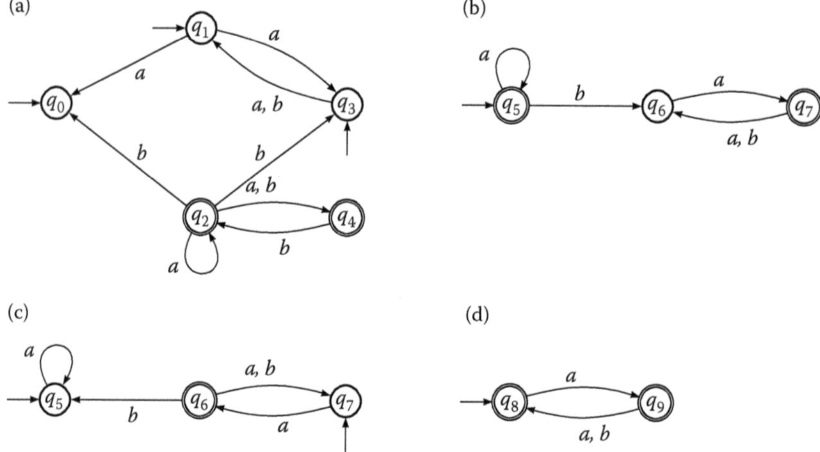

FIGURE 7.7 Minimizing a DFA with DFA-MINIMISE-BRZOZOWSKI. (a) The first reversal. (b) The first subset construction. (c) The second reversal. (d) Second subset construction: minimal DFA.

The result of the first application of the subset construction method, DFA D^R, is presented in Figure 7.7b.

Figure 7.7c shows the reversal of D^R, again, an NFA with multiple initial states. The equivalent minimal DFA, result of the second application of the subset construction method and final step of Brzozowski's algorithm, is presented in Figure 7.7d.

7.7 Incremental Minimization

Given an arbitrary DFA D as input, *incremental minimization* algorithms may be halted at any time returning a partially minimized DFA D' such that $|D'| \leq |D|$ and $D' \sim D$. Whenever the minimization process is interrupted, calling the incremental minimization algorithm with the output of the halted process resumes the minimization process. Incremental minimization algorithms also allow, for example, to minimize some automaton D at the same time as D is being used to process a word for acceptance.

In 2001, Watson presented the first (known) incremental DFA minimization algorithm (Watson 2001). In collaboration with Watson, Daciuk later proposed an improved version of the same algorithm (Watson & Daciuk 2003), this time making use of full memoization in order to obtain a lower worst-case running time. While Watson's original incremental minimization algorithm presents an exponential running time—$O(|\Sigma|^{\max(0,n-2)})$ in the worst case—the memoized version is $O(|\Sigma|n^2\alpha(n^2))$, where α

denotes an inverse of the Ackermann function. Since $\alpha(x) \leq 4$ for any $x \leq 16^{512}$, it can be considered a constant for all "practical" values of x, resulting in an almost quadratic algorithm. The published version of that algorithm, however, was found to be incorrect and, therefore, we are not able to include it in this chapter.

7.7.1 Watson's Incremental Algorithm

DFA-MINIMISE-WATSON is a straightforward implementation of Watson's original algorithm, specialized for initially connected DFAs. At each point of the main loop (lines 6–11), variables E and $\bar{E}$ maintain the sets of pairs of states already known to be equivalent and not equivalent, respectively. This loop can be interrupted at any time, using the partially computed set of equivalent pairs of states E to build a (possibly) smaller equivalent DFA.

Only for matters of efficiency, a variable S, containing a set of presumably equivalent pairs of states, is made global. Lines 12–15 are not part of the original algorithm, but we include them in order to make the construction of the minimal DFA explicit.

```
1  def DFA-MINIMISE-WATSON(Q, Σ, δ, q₀, F):
2    E := { (q,q) | q ∈ Q }
3    Ē := ((Q − F) × F) ∪ (F × (Q − F))
4    k := max(0, |Q| − 2)
5    S := ∅
6    for p ∈ Q:
7        for q ∈ Q:
8            if WATSON-EQUIV-P(p, q, k) :
9                E := E ∪ { (p,q), (q,p) }
10           else:
11               Ē := Ē ∪ { (p,q), (q,p) }
12   D′ := (Q, Σ, δ, q, F)
13   for (p,q) ∈ E:
14       JOIN-STATES(D′, (p,q))
15   return D′
```

DFA-MINIMISE-WATSON makes use of an auxiliary function, WATSON-EQUIV-P, which tests the equivalence of any two states, p and q. Variable l is used to limit the recursion depth of the auxiliary procedure WATSON-EQUIV-P, only for efficiency reasons. The proof that a word of size $|Q| - 2$ suffices to distinguish any pair of states may be found in Conway (1971, p. 11) or Wood (1987, p. 129).

```
def WATSON-EQUIV-P(p, q, l):
  if l = 0:
      return (p ∈ F ⇔ q ∈ F)
  elif (p,q) ∈ S:
      return TRUE
  else:
      eq := (p ∈ F ⇔ q ∈ F)
      S := S ∪ { (p,q) }
      for a ∈ Σ:
          if eq = FALSE:
              return FALSE
          eq := eq ∧ WATSON-EQUIV-P(δ(p,a), δ(q,a), l − 1)
      S := S − { (p,q) }
  return eq
```

WATSON-EQUIV-P follows the transition function and recursively calls itself with increasingly longer words (line 12) until one of the following two conditions occurs:

- The recursion limit was reached (line 2), which means that the two states used in the first call to WATSON-EQUIV-P (line 8 of DFA-MINIMISE-WATSON) are equivalent if and only if they are either both final or both not final, since the longest word necessary to distinguish them has been computed;
- A loop was found (line 4), and clearly the states are equivalent since no word that distinguishes the states has been, or will be, found.

The ability to interrupt the algorithm presents an excellent opportunity to save computational resources when, for example, given a DFA, the goal is not to obtain the equivalent minimal automaton, but solely to check if it is already minimal. The procedure DFA-MINIMAL-WATSON-P implements a simplified version of Watson's algorithm which halts, returning FALSE, when the first pair of equivalent states is found. Naturally, if no pair of states is found to be equivalent, the algorithm returns TRUE, indicating that the input DFA is in fact already minimal.

```
def DFA-MINIMAL-WATSON-P(Q, Σ, δ, q0, F):
  k := max(0, |Q| − 2)
  S := ∅
  for p ∈ Q:
      for q ∈ Q:
          if WATSON-EQUIV-P(p, q, k) :
              return FALSE
  return TRUE
```

7.7.2 A Quadratic Incremental Method

This algorithm (Almeida et al. 2010) uses a disjoint-set data structure to represent the DFA's states, and UNION-FIND to mark pairs of equivalent states and update the equivalence classes. This approach allows to maintain the transitive closure of the equivalence classes in a very concise and elegant manner. The pairs of states marked as distinguishable are stored in an auxiliary data structure in order to avoid repeated computations.

Unlike the usual techniques, which explicitly compute the equivalence classes of the set of states, this algorithm proceeds by testing the equivalence of pairs of states. The intermediate results are stored for speeding up future computations and assuring quadratic running time and memory usage.

The quadratic time bound of the minimization procedure DFA-MINIMISE-INCREMENTAL is achieved by testing each pair of states for equivalence exactly once. This is assured by storing the intermediate results of all calls to the pairwise equivalence-testing function INCREMENTAL-EQUIV-P.

```
def DFA-MINIMISE-INCREMENTAL(Q, Σ, δ, q0, F):
 for q ∈ Q:
     MAKE(q)
 Ē := {NORMALISE-PAIR(p, q) | p ∈ F, q ∈ Q − F}
 for p ∈ Q:
     for q ∈ {x | x ∈ Q, x > p}:
         if (p, q) ∈ Ē:
             continue
         if FIND(p) = FIND(q):
             continue
         E := SET-MAKE(|Q|²)
         H := SET-MAKE(|Q|²)
         if INCREMENTAL-EQUIV-P(p, q):
             for (p′, q′) ∈ SET-ELEMENTS(E):
```

```
15                  Union(p′, q′)
16         else:
17             for (p′, q′) ∈ Set-Elements(H):
18                  Ē := Ē ∪ {(p′, q′)}
19   C := {}
20   for p ∈ Q:
21       r := Find(p)
22       C[r] := C[r] ∪ {p}
23   D′ := D
24   Join-States(D′, C)
25   return D′
```

Dfa-Minimise-Incremental starts by creating the trivial equivalence classes (lines 2–3); these are singletons as no states are yet marked as equivalent. The global variable $\bar{E}$, used to store the distinguishable pairs of states, is also initialized (line 4) with the trivial identifications. Variables H and E, which are global and reset before each call to Incremental-Equiv-P, maintain the history of calls to the transition function and the set of potentially equivalent pairs of states, respectively.

The main loop of Dfa-Minimise-Incremental (lines 5–18) iterates through all the normalized pairs of states and, for those not yet known to be either distinguishable or equivalent, calls the pairwise equivalence test Incremental-Equiv-P. Every call to Incremental-Equiv-P is conclusive and the result is stored either by merging the corresponding equivalence classes (lines 13–15), or updating $\bar{E}$ (lines 16–18). Thus, each recursive call to Incremental-Equiv-P will avoid one iteration on the main loop of Dfa-Minimise-Incremental by skipping (lines 7–10) that pair of states.

Finally, at lines 19–22, the set partition of the corresponding equivalence classes is created. Next, the DFA D is copied to D' (line 23) and the equivalent states are merged by the call to Join-States. The last instruction, at line 25, returns the minimal DFA D' equivalent to D.

```
1  def Incremental-Equiv-P(p, q):
2    if (p, q) ∈ Ē:
3        return False
4    if Set-Search((p, q), H) ≠ Nil:
5        return True
6    H := Set-Insert((p, q), H)
7    for a ∈ Σ:
8        (p′, q′) := Normalise-Pair(Find(δ(p, a)), Find(δ(q, a)))
9        if p′ ≠ q′ and Set-Search((p′, q′), E) = Nil:
10           E := Set-Insert((p′, q′), E)
11           if not Incremental-Equiv-P(p′, q′):
12               return False
13           else:
14               H := Set-Remove((p′, q′), H)
15   E := Set-Insert((p, q), E)
16   return True
```

Incremental-Equiv-P is used to test the equivalence of the two states, p and q, passed as arguments. The global variables E and H are updated with the pair (p, q) during each nested recursive call. As there is no recursion limit, Incremental-Equiv-P will only return when p is distinguishable from q (line 3) or when a cycle is found (line 5). If a call to Incremental-Equiv-P returns False, then all pairs of states recursively tested are distinguishable and variable H—used to store the sequence of calls to the transition function—will contain a set of distinguishable pairs of states. If it returns True, no pair of distinguishable states was found within the cycle and variable E will contain a set of equivalent states. This is the strategy which assures that each pair of states is tested for equivalence exactly once: every call to Incremental-Equiv-P is conclusive and the result stored for future use. It does, however, lead to an increased usage of memory.

7.7.2.1 Efficient Set Implementation

Throughout the execution of DFA-MINIMISE-INCREMENTAL and INCREMENTAL-EQUIV-P, mainly due to the heavy usage of the variables E and H, several set-theoretical operations—SET-MAKE, SET-SEARCH, SET-INSERT, SET-REMOVE, and SET-ELEMENTS—are executed. In order to achieve the desired quadratic upper bound, all these operations must be performed in $O(1)$, which may be accomplished by using a data structure that combines a hash-table with a doubly linked list to represent and manipulate sets.

This is another space–time trade-off that assures the desired worst-case complexity on all operations. The hash-table maps a given value (state of the DFA) to the address on which it is stored in the linked list. Since we know the size of the hash-table in advance—n^2 elements for a DFA with n states—searching, inserting, and removing elements is $O(1)$. The linked list assures that, at lines 14–15 and 17–18 of DFA-MINIMISE-INCREMENTAL, the loop is repeated only on the elements that were actually used in the calls to INCREMENTAL-EQUIV-P, instead of iterating through the entire hash-table.

7.7.3 An Example

Let us walk through the minimization process of the DFA presented in Figure 7.8, exemplifying the incremental minimization algorithm in action. It is a simple complete, initially connected DFA, with 5 states over an alphabet of two symbols, a and b. The only final state (and also the initial state) is q_0.

After creating the UNION-FIND data structures in lines 2–3, DFA-MINIMISE-INCREMENTAL will initialize the set of distinguishable pairs of states (variable $\bar{E}$ at line 4) with the trivial identifications: $\{(q_0,q_1),(q_0,q_3),(q_0,q_2),(q_0,q_4)\}$.

Following the order induced by the states' names, the first pair to be tested for equivalence is (q_0,q_1). These states are already known to be distinguishable and will be ignored by lines 7–8. The same is true for the pairs (q_0,q_2), (q_0,q_3), and (q_0,q_4).

The next pair to be tested for equivalence is (q_1,q_2). Nothing is known about this pair so INCREMENTAL-EQUIV-P is called. Following the transitions by a, the pair (q_3,q_4) is reached. From (q_3,q_4), regardless of the symbol used, a loop is reached: (q_4,q_4) by a, and (q_1,q_1) by b. Another loop is found when trying to follow the transitions of the states (q_1,q_2) by the symbol b. This results on INCREMENTAL-EQUIV-P returning TRUE with $E=\{(q_1,q_2),(q_3,q_4)\}$, and DFA-MINIMISE-INCREMENTAL making state q_1 equivalent to state q_2, and state q_3 equivalent to state q_4 by merging the corresponding UNION-FIND sets at lines 13–15. The partially minimized DFA obtained at this point is presented in Figure 7.9.

Continuing with the minimization process, again, nothing is yet known about the equivalence of the states q_1 and q_3. The call to INCREMENTAL-EQUIV-P returns TRUE and sets the global variable $E=\{(q_1,q_3),(q_2,q_4)\}$. After merging the equivalence classes of the states (q_1,q_3) and (q_2,q_4), the remaining pairs of states—(q_1,q_4), (q_2,q_3), (q_2,q_4), and (q_3,q_4)—are known to be equivalent and are ignored at lines 9–10 of DFA-MINIMISE-INCREMENTAL.

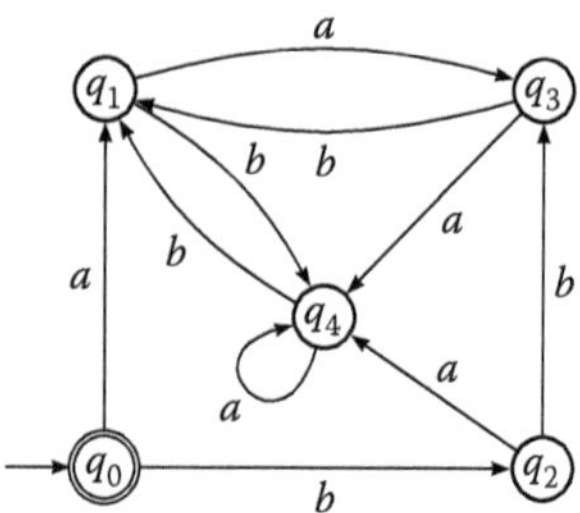

FIGURE 7.8 Example of a nonminimal DFA.

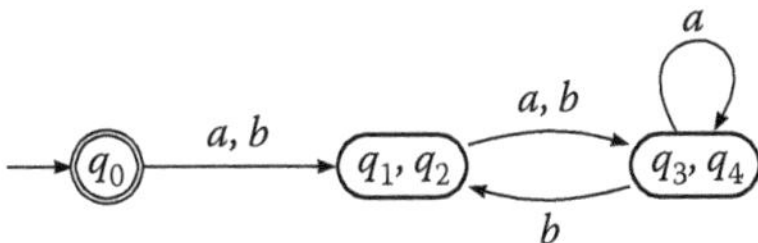

FIGURE 7.9 Intermediate DFA when applying the incremental minimization algorithm to the DFA in Figure 7.8.

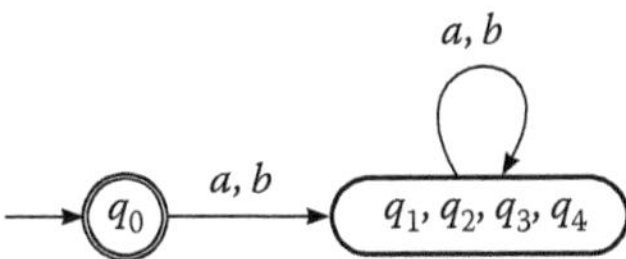

FIGURE 7.10 Minimal DFA equivalent to the one presented in Figure 7.8.

Since there are no more states to test, line 19 of DFA-MINIMISE-INCREMENTAL is reached. The set partition corresponding to the equivalence classes is created (lines 20–22), and the equivalent states are merged (line 24). The resulting minimal DFA is presented in Figure 7.10.

7.8 Experimental Results

Apart from the theoretical worst-case run-time analysis, which authors typically present, not much is known about the practical performance of finite automata minimization algorithms. Such information is essential, however, when, for example, one has to choose an appropriate algorithm for a given application. Some exceptions are Watson (1995) (although it uses rather small and biased samples), Tabakov and Vardi (2005) that compares Hopcroft's and Brzozowski's algorithms using samples obtained with an ad-hoc random automata generator, Baclet and Pagetti (2006) which analyzes different implementations of the equivalence class refinement process in Hopcroft's algorithm, Bassino et al. (2007) comparing Moore and Hopcroft's DFA minimization algorithms, Almeida et al. (2008), Almeida (2011) that experimentally compare the practical performance of several automata minimization algorithms, and, more recently, Bassino et al. (2009) which study the average complexity of Moore's minimization algorithm.

In this section, we present experimental comparative results on the performance of the algorithms described in the previous sections.

All benchmarks were performed with samples obtained from the uniform random generator described by Almeida et al. (2007) and FAdo (1999–2010). The results are thus conclusive, statistically significant, and reflect the trends of the population. Each process was allowed to run during a maximum period of 24 h and use up to 2 GB of RAM. This time limit was enforced by the `psmon` tool (Almeida 2009–2010). Running times were obtained using the GNU `time` utility (Free Software Foundation 1999) and consider only the CPU time, thus assuring that no other process skews an algorithms' performance data.

In order to evaluate the performance of a given algorithm, we must consider some *random sample* with *reasonable* size as input. Too large a sample implies a waste of the resources which we were trying to save in the first place; a sample too small diminishes the utility of the results. By fixing a confidence interval and a confidence level (margin of error), we can calculate the correct sample size and thus assure that the results are suitable for drawing statistically significant conclusions. According to Cochran (1977, p. 75), given a margin of error ϵ and an estimated proportion p, we may use the formula

$$N = \frac{z^2 p(1-p)}{\epsilon^2} \tag{7.1}$$

to determine the appropriate sample size N. By application of the central limit theorem, we may assume that the data is normally distributed and take z from the abscissa of the normal curve that cuts off an area of the confidence level γ at the tails, that is,

$$P(-z < Z < z) = \gamma.$$

As we know that the population is large (cf. Reis et al. (2005)) but do not have a proper estimate for the proportion p, we assume $p = \frac{1}{2}$ for maximum variability. We may then simplify Equation 7.1 in the following way:

$$N = \frac{z^2}{4\epsilon^2} = \left(\frac{z}{2\epsilon}\right)^2. \tag{7.2}$$

For benchmarking purposes, we produced samples of 20,000 random initially connected DFAs with the following characteristics:

- $n \in \{5, 10, 20, 30, 40, 50, 60, 70, 80, 90, 100\}$ states, each over an alphabet of $k \in \{x \mid 2 \leq x \leq 16\} \cup \{18, 20, 25, 30, 35, 40, 45, 50\}$ symbols;
- $n \in \{1000\}$ states, each over an alphabet of $k \in \{2, 3, 5\}$ symbols.

7.8.1 Analysis of the Results

The graphics in the following figures show the *performance* of each algorithm when minimizing the datasets of 20,000 elements, measured as *minimized DFAs per second*.

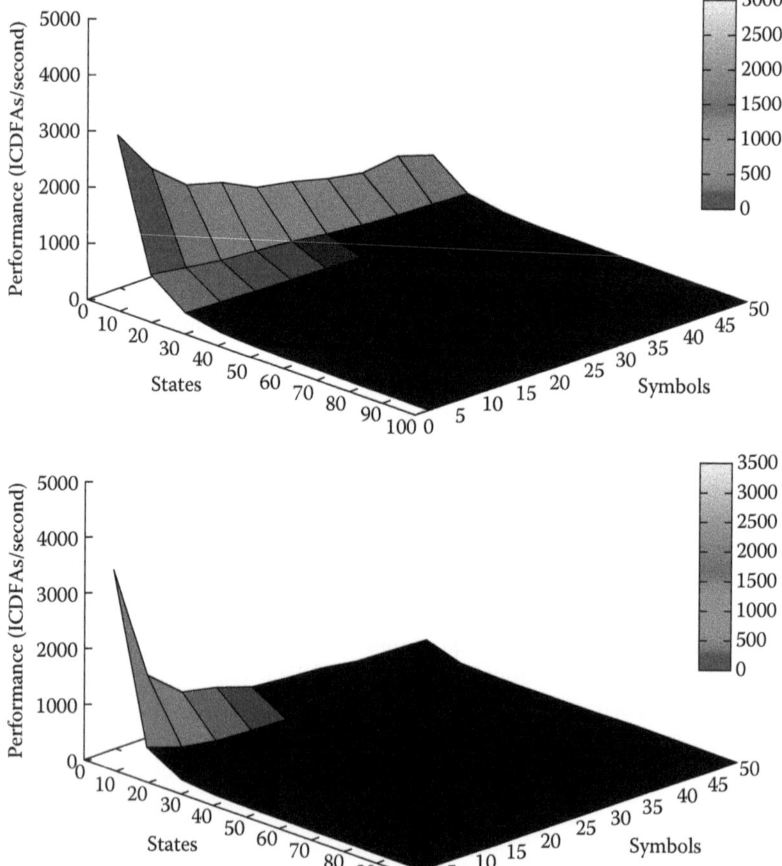

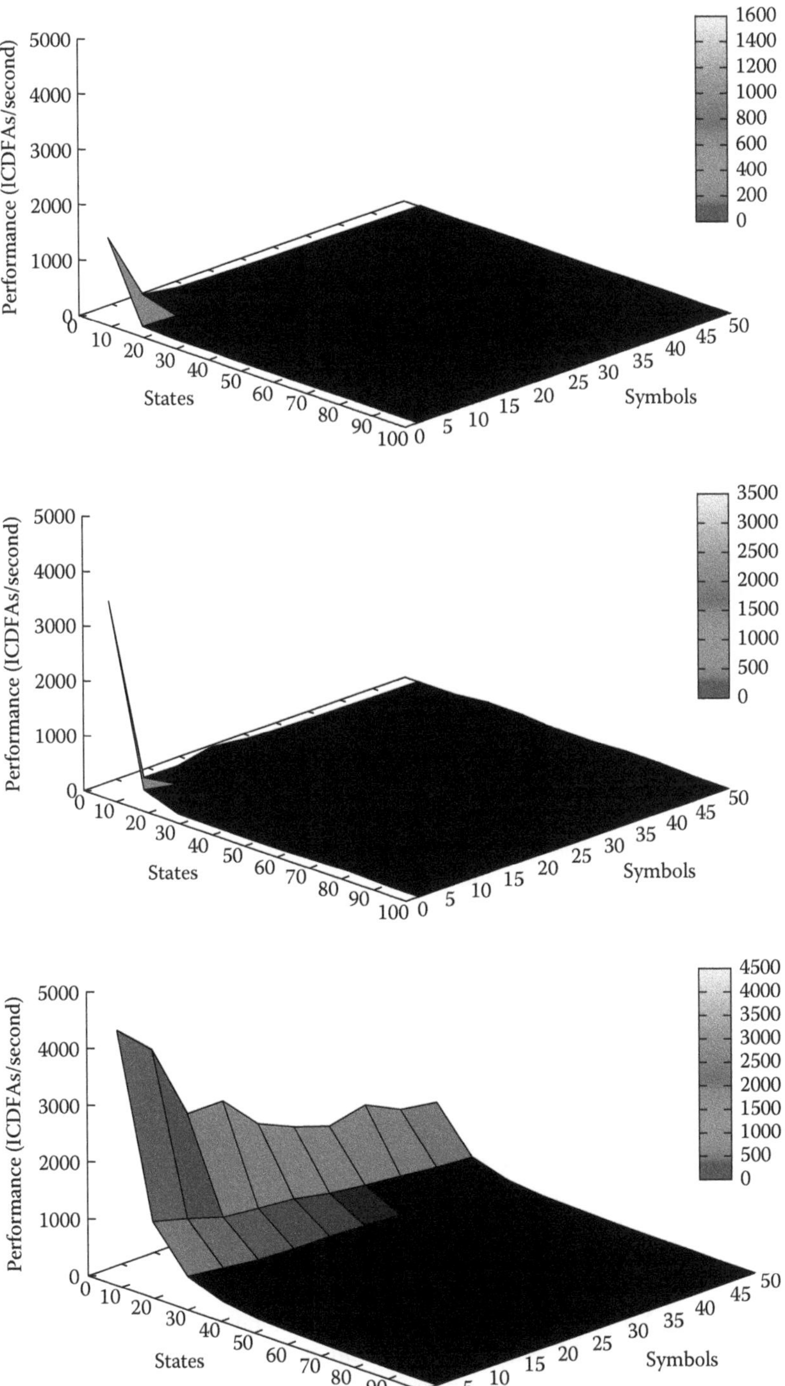

Considering the results of the two classical polynomial algorithms, Dfa-Minimise-Moore and Dfa-Minimise-Hopcroft, we can see that, although the difference is not huge, Dfa-Minimise-Moore performs slightly better.

As for the two exponential algorithms, Dfa-Minimise-Brzozowski and Dfa-Minimise-Watson, while both present very similar results, Dfa-Minimise-Watson displays a slightly superior performance especially when dealing with small alphabets (25 symbols or less). For larger alphabets (30 or more symbols), Dfa-Minimise-Brzozowski outperforms Dfa-Minimise-Watson.

The incremental algorithm Dfa-Minimise-Incremental, while presenting a quadratic worst-case running-time, clearly outperforms even Hopcroft's $O(kn \log(n))$ approach—at least in the average case—on all the tested DFAs.

7.9 Conclusions

Concerning finite automata minimization, several different algorithms have been proposed over the course of the last decades. Nonetheless, apart from the theoretical worst-case run-time analysis, which authors typically present, not much is known about the practical performance of finite automata minimization algorithms. Such information is essential, however, when, for example, one has to choose an appropriate algorithm for a given application.

Using data sets of randomly generated deterministic finite automata it is possible to experimentally compare several algorithms and to extract statistically significant conclusions on its practical performance.

When applied to Moore's, Hopcroft's, Watson's, Brzozowski's, and an incremental quadratic algorithm, these tests show that although having different worst-case running-times, Moore and Hopcroft's algorithms, present similar results in the average case. The algorithms due to Brzozowski and Watson, do reveal their exponential character and also present very similar results. All tested algorithms are clearly outperformed by the quadratic incremental method.

References

Aho, A. V., Hopcroft, J. E., and Ullman, J. D. 1974, *The Design and Analysis of Computer Algorithms*, Addison-Wesley, Boston, MA.

Almeida, M. 2009–2010, *psmon*. http://savannah.nongnu.org/projects/psmon/.

Almeida, M. 2011, Equivalence of regular languages: An algorithmic approach and complexity analysis, PhD thesis, University of Porto.

Almeida, M., Moreira, N., and Reis, R. 2010, Incremental dfa minimisation, in M. Domaratzki, ed., *Proceedings of the 15th International Conference on Implementation and Application of Automata*, Vol. 6482 of *Lecture Notes on Computer Science*, Springer, New York, NY, pp. 39–48.

Almeida, M., Moreira, N., and Reis, R. 2007, Enumeration and generation with a string automata representation, *Theoretical Computer Science* **387**(2), 93–102. Special issue "Selected papers of DCFS 2006".

Almeida, M., Moreira, N., and Reis, R. 2008, On the performance of automata minimization algorithms, in A. Beckmann, C. Dimitracopoulos, and B. Löwe, eds, CiE 2008: Abstracts and extended abst. of unpublished papers.

Baclet, M., and Pagetti, C. 2006, Around hopcroft's algorithm, in O. H. Ibarra, and H.-C. Yen, eds, *Implementation and Application of Automata*, Vol. 4096 of *Lecture Notes on Computer Science*, Springer, New York, NY, pp. 114–125.

Bassino, F., David, J., and Nicaud, C. 2007, REGAL: A library to randomly and exhaustively generate automata, in J. Holub, and J. Žd'árek, eds, *Implementation and Application of Automata*, Vol. 4783 of *Lecture Notes on Computer Science*, Springer, New York, NY, pp. 303–305.

Bassino, F., David, J., and Nicaud, C. 2009, On the average complexity of Moore's state minimization algorithm, in Susanne Albers, and Jean-Yves Marion, eds, *26th International Symposium on Theoretical Aspects of Computer Science STACS 2009 Proceedings of the 26th Annual Symposium on the Theoretical Aspects of Computer Science*, IBFI Schloss Dagstuhl, Freiburg Germany, pp. 123–134.

Brzozowski, J. A. 1962, Canonical regular expressions and minimal state graphs for definite events, in J. Fox, ed., *Proceedings of the Symposium on Mathematical Theory of Automata*, Vol. 12 of *MRI*

Symposia Series, Polytechnic Press of the Polytechnic Institute of Brooklyn, Brooklyn, NY, New York, NY, pp. 529–561.

Champarnaud, J.-M., Khorsi, A., and Paranthoën, T. 2002, Split and join for minimizing: Brzozowski's algorithm, in M. Balík, and M. Simánek, eds, *PSC'02 (Prague Stringology Conference), Proceedings*, Research report DC-2002-03, Czech Technical University of Prague, pp. 96–104.

Cochran, W. G. 1977, *Sampling Techniques*, third edn., John Wiley & Sons, Hoboken, NJ.

Conway, J. H. 1971, *Regular Algebra and Finite Machines*, Mathematical Series, Chapman & Hall, 11 New Fetter Lane, London, EC4.

FAdo, P. 1999–2010, Fado: Tools for formal languages manipulation. Available at: http://www.ncc.up.pt/FAdo.

Free Software Foundation 1999, *GNU time*. Available at: http://www.gnu.org/software/time/.

Gries, D. 1972, Describing an algorithm by Hopcroft, Technical Report TR 72-151, Cornell University, Ithaca, New York.

Hopcroft, J. 1971, An $n \log n$ algorithm for minimizing states in a finite automaton, in Z. Kohavi, ed., *The Theory of Machines and Computations*, Academic Press, New York, pp. 189–196.

Hopcroft, J. E., Motwani, R., and Ullman, J. D. 2000, *Introduction to Automata Theory, Languages and Computation*, second edn., Addison-Wesley. International Edition.

Hopcroft, J. E., and Ullman, J. D. 1979, *Introduction to Automata Theory, Languages, and Computation*, first edn., Addison-Wesley, New York, NY.

Huffman, D. A. 1954, The synthesis of sequential switching circuits, Technical Report 274, Massachusetts Institute of Technology. Reprinted from the Journal of the Franklin Institute Vol. 257, Nos. 3 and 4, March and April, 1954.

Huffman, D. A. 1955, The synthesis of sequential switching circuits, *Journal of Symbolic Logic* **20**(1), 69–70.

Knuutila, T. 2001, Re-describing an algorithm by Hopcroft, *Theoretical Computer Science* **250**(1–2), 333–363.

Moore, E. F. 1956, Gedanken-experiments on sequential machines, *Automata Studies* **34**, 129–153. Princeton University Press, Princeton, NJ.

Nicaud, C. 2000, Étude du comportement en moyenne des automates finis et des langages rationnels, PhD thesis, Université de Paris 7.

Rabin, M. O., and Scott, D. 1959, Finite automata and their decision problems, *IBM Journal of Research and Development* **3**(2), 114–125.

Reis, R., Moreira, N., and Almeida, M. 2005, On the representation of finite automata, in *7th International Workshop on Descriptional Complexity of Formal Systems*, Como, Italy, pp. 269–276.

Shallit, J. 2008, *A Second Course in Formal Languages and Automata Theory*, Cambridge University Press.

Tabakov, D., and Vardi, M. Y. 2005, Experimental evaluation of classical automata constructions, in *In LPAR 2005, LNCS 3835*, Springer, pp. 396–411.

Watson, B. W. 1995, Taxonomies and toolkits of regular language algorithms., PhD thesis, Eindhoven University of Technology, Eindhoven, the Netherlands.

Watson, B. W. 2001, An incremental DFA minimization algorithm, in L. Karttunen, K. Koskenniemi, and G. van Noord, eds, *Proceedings of the Second International Workshop on Finite-State Methods in Natural Language Processing*, Helsinki, Finland.

Watson, B. W., and Daciuk, J. 2003, *An efficient DFA minimization algorithm, Natural Language Engineering* **9**(1), 49–64.

Wood, D. 1987, *Theory of Computation*, John Wiley & Sons, Hoboken, NJ.

8

Incremental Construction of Finite-State Automata

Jan Daciuk
Gdańsk University of Technology

8.1 Introduction

Among finite-state automata, acyclic automata play an important role in many applications. For example, they are often used as *dictionaries* in natural language processing and computational linguistics (including speech processing), as well as compiler construction. A language of an acyclic automaton is a finite set of finite-length strings. To speed-up processing, deterministic automata are used as they do not require backtracking. To save space, the automata are minimal.

We have a set of strings as the input, and we want to obtain a minimal, deterministic, finite-state automaton that recognizes precisely those strings. The traditional way of implementing that task is to build a deterministic automaton, and then to minimize it. The disadvantage of that approach is that the nonminimal automaton can be huge. *Incremental construction* handles the processes of building the automaton and of minimizing it in parallel, keeping the intermediate structure close to minimal, while maintaining high speed. There are two incremental construction algorithms. The first one takes advantage of the fact that its data must be sorted. It is faster than the second algorithm that does not impose restrictions on input data.

The rest of the chapter is organized as follows. Basic definitions are given in Section 8.2. Section 8.3 describes insights from the traditional construction. The algorithm for sorted data is presented in Section 8.4, while the algorithm for data coming in arbitrary order—in Section 8.5.

8.2 Basic Definitions

A *deterministic, finite-state automaton* (DFA) is a 5-tuple $A = (Q, \Sigma, \delta, q_0, F)$, where Q is a finite set of states, Σ is a finite set of symbols called the *alphabet*, $q_0 \in Q$ is the *start state*, also called the *initial state*, and $F \subseteq Q$ is the set of *final states*, also called *accepting states*. A mapping $\delta : Q \times \Sigma \rightarrow Q$ is called the *transition function*. If $\delta(q, \sigma) \notin Q$, we write $\delta(q, \sigma) = \bot$. An *extended transition function* $\delta^* : Q \times \Sigma^* \rightarrow Q$ is defined as

$$\delta^*(q, \varepsilon) = q \tag{8.1}$$

$$\delta^*(q, ax) = \begin{cases} \delta^*(\delta(q, a), x) & \text{if } \delta(q, a) \neq \bot \\ \bot & \text{if } \delta(q, a) = \bot \end{cases} \tag{8.2}$$

Let us denote the set of labels on transitions going out of state q as Σ_q.

$$\Sigma_q = \{\sigma \in \Sigma : \delta(q, \sigma) \neq \bot\} \tag{8.3}$$

A set of transitions going out of state q is denoted as δ_q.

$$\delta_q = \{(\sigma, p) : \sigma \in \Sigma_q \wedge \delta(q, \sigma) = p\} \tag{8.4}$$

The *right language* of a state q is defined as

$$\overrightarrow{\mathcal{L}}(q) = \{w \in \Sigma^* : \delta^*(q, w) \in F\} \tag{8.5}$$

It can also be defined recursively:

$$\overrightarrow{\mathcal{L}}(q) = \bigcup_{\sigma \in \Sigma_q} \sigma \overrightarrow{\mathcal{L}}(\delta(q, \sigma)) \cup \begin{cases} \emptyset & \text{if } q \notin F \\ \{\varepsilon\} & \text{if } q \in F \end{cases} \tag{8.6}$$

The language of an automaton is the right language of its start state:

$$\mathcal{L}(A) = \{w \in \Sigma^* : \delta^*(q_0, w) \in F\} \tag{8.7}$$

The size of an automaton is defined as the number of its states:

$$|A| = |Q| \tag{8.8}$$

Among all deterministic automata that recognize the same language, there is one (up to isomorphisms) that has the minimal number of states. It is called the *minimal automaton*.

$$\forall (A : \mathcal{L}(A) = \mathcal{L}(A_{min})) |A_{min}| < |A| \tag{8.9}$$

Equality of right languages of states is an *equivalence relation*:

$$(p \equiv q) \Leftrightarrow (\overrightarrow{\mathcal{L}}(p) = \overrightarrow{\mathcal{L}}(q)) \tag{8.10}$$

It divides all states of an automaton into *equivalence classes*. In a minimal automaton, each class has exactly one representative:

$$\forall (p, q \in Q_{min})(p \equiv q) \Rightarrow (p = q) \tag{8.11}$$

8.3 Insights from the Traditional Construction

Let us recall that the traditional way of constructing a minimal, deterministic automaton is to first construct a nonminimal deterministic automaton, and then to minimize it. One way to build a deterministic automaton would be to form a regular expression from all the strings that are to constitute the language of the future automaton:

```
string1|string2|string3|...|stringn
```

and to construct a nondeterministic automaton (NFA) (possibly with ε-transitions). Then the NFA would be determinized. However, the NFA would most probably only fit into the operational memory of modern computers when the string collection were small. For natural language dictionaries, such assumption is not realistic. Also, determinization takes exponential time with regard to the size of the NFA being determinized.

A far better approach is to build a deterministic automaton right away, without going through an NFA. Common prefixes are shared, and the result is a tree. When the labels are single letters, such tree is called a *trie*. A trie recognizing endings of a Polish verb *czytać* (in English: *to read*) is depicted in Figure 8.1.

Building a trie automaton means adding strings to the trie one by one:

Algorithm 8.1:

Building a trie from strings. The resulting automaton is returned.

function BUILDTRIE
    Create initial automaton $A = (\{q_0\}, \Sigma, q_0, \emptyset, \emptyset)$
    while there are strings to read **do**
        $w \leftarrow$ next string
        ADDSTRINGTOTRIE(A, w)
    end while
    return A
end function

When adding a string to a trie, the *longest common prefix* (possibly empty) is followed first, and then the remaining *suffix* is added.

Algorithm 8.2:

Adding a string w to an automaton A, which is modified by the procedure.

procedure ADDSTRINGTOTRIE(A, w)
    $(q, v) \leftarrow$ COMMONPREFIX(A, w) ▷ $w = uv$
    ADDSUFFIX(A, q, v)
end procedure

Finding the longest common prefix is implemented by traversing subsequent transitions labeled with subsequent symbols of the string to be added.

Algorithm 8.3:

Traversing the longest common prefix of string w in an automaton A.

function COMMONPREFIX(A, w)
    $q \leftarrow q_0; i \leftarrow 1$
    while $\delta(q, w_i) \neq \bot$ **do**
        $q \leftarrow \delta(q, w_i); i \leftarrow i + 1$
    end while
    return $(q, w_{i+1} \dots w_{|w|})$
end function

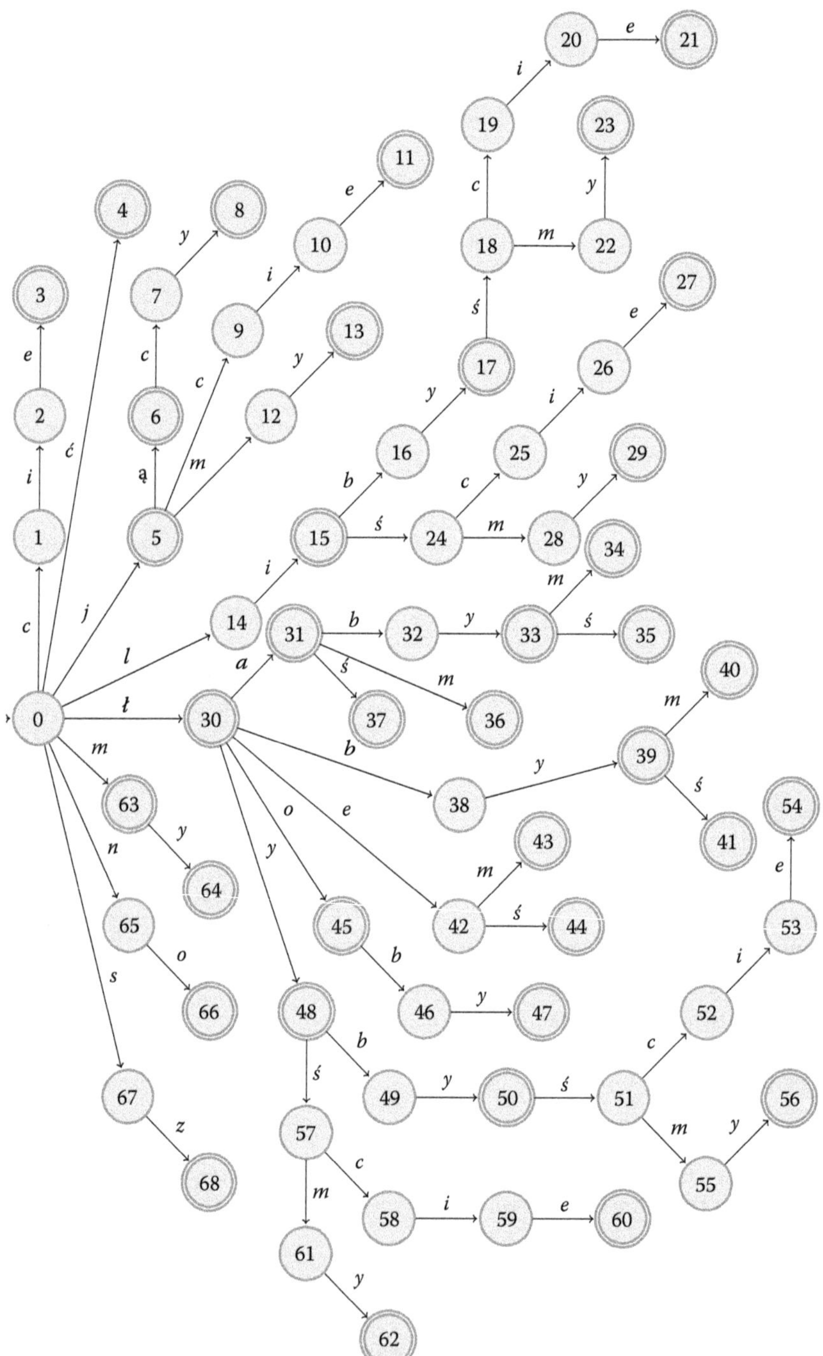

FIGURE 8.1 A trie of endings of a Polish verb *czytać*.

Adding a suffix to a trie is also straightforward. It is done by creating new states and connecting them to the existing automaton by transitions labeled with the subsequent symbols in the suffix.

Algorithm 8.4:

Adding a suffix v to the automaton A starting from state q. The automaton is modified.

procedure ADDSUFFIX(A, q, v)
  while $v \neq \varepsilon$ **do**
    Create new state p
    $\delta(q, v_1) \leftarrow p$
    $v \leftarrow v_{2\ldots|v|}$
    $q \leftarrow p$
  end while
  $F \leftarrow F \cup \{q\}$ ▷ make the last state final
end procedure

Several general minimization algorithms for DFAs exist. Brzozowski algorithm [Brz62] reverses an automaton (it can be an NFA), determinizes it, reverses and determinizes it again. Aho-Sethi-Ullman [ASU86], Hopcroft-Ullman [HU79], and Hopcroft [Hop71] algorithms all divide all states of an automaton into two classes (final and nonfinal states), and then divide the classes farther until all states in every class are equivalent to each other. Then one representative for each class is chosen; the other members are deleted. An incremental algorithm [WD03] compares pairs of states, recursively checking equivalence of other pairs of states if it is needed in order to evaluate equivalence of the original pair.

None of those general algorithms is optimal for an acyclic automaton. Their computational complexity stems from the lack of guidance in the order in which equivalence of states is evaluated. Moreover, except for the case of incremental minimization which would need additional information, they minimize automata globally, while we would like the process to be localized.

Equation 8.6 gives us insight into the proper order of minimization. Combining Equation 8.10 and Equation 8.6 gives us the following conditions for equivalence of states p and q:

1. $(p \in F) \Leftrightarrow (q \in F)$
2. $\delta_p = \delta_q$
3. $\forall(\sigma \in \Sigma_p)(\overrightarrow{\mathcal{L}}(\delta(p,\sigma)) = \overrightarrow{\mathcal{L}}(\delta(q,\sigma)))$

When equivalence of states is evaluated in the proper order, condition 3 becomes:

3′ $\forall(\sigma \in \Sigma_p)(\delta(p,\sigma) = \delta(q,\sigma))$

and eventually conditions 2 and 3′ can be written as:

2′ $\overrightarrow{p} = \overrightarrow{q}$

There are many ways to implement the proper order of evaluation. One solution may be to sort states on the longest path (measured in the number of transitions) from the state to a final state. Final states with no outgoing transitions would be checked first, followed by states with the longest path length being 1, and so on. However, such method would require quite a lot of bookkeeping, which translates both to lower speed and greater memory needs.

A better solution is to visit the nodes of a trie using *postorder*. For a given tree root, we evaluate equivalence of states in all immediate subtrees sorted on labels on branches that lead to them, and then we evaluate equivalence of the root. For example, in the trie in Figure 8.1, we visit state 3 first, followed by 2, 1, 4, 8, 7, 6, 11, 10, 9, 13, 12, 5, 21, and so on.

A phrase "evaluate equivalence of a state" seems cryptic. It means to look for a state equivalent to the given state. If such a state is found, it replaces the current state, and all transitions coming to the current state are redirected toward the equivalent state. It is clear what the current state is, but how is the equivalent state found?

As minimization goes on in the proper order, a part of the automaton is already minimized. Inside that part, all states have a unique right language. It is among them that we search for a state equivalent to the current one. A *hash table* called the *register* and denoted with R is used to speed up the search. The *hash function* uses finality, number of outgoing transitions, as well as labels and identities of target states of transitions as the key. When properly implemented, the register will give us an equivalent state or an information that it does not exist in constant time on average.

Algorithm 8.5:

Minimization of a DFA A starting at state q. The automaton A is modified by the procedure.

procedure MINIMIZE(A, q)
    for all $\sigma \in \Sigma_q$ **do**
        MINIMIZE$(A, \delta(q, \sigma))$
    end for
    if $\exists(r \in R) : q \equiv r$ **then**
        delete q
        redirect a transition coming to q to r ▷ there is one
    else
        $R \leftarrow R \cup \{q\}$
    end if
end procedure

We have two processes that lead to a minimal, deterministic automaton. One builds a trie, another one minimizes it. In the traditional method, minimization is done after building the trie is completed. In incremental construction, those two processes run in parallel, and they have to be synchronized.

8.4 Algorithm for Sorted Data

The crucial issue in synchronization of building a trie and minimizing it is to examine which parts of the automaton will not change as a result of adding new strings. More precisely, we want to know which states in the trie cannot change their right language when new strings are added.

A trie in Figure 8.1 is sorted. All labels on transitions that go out from a state are sorted—lexicographically smaller labels are put higher. Strings recognized along paths in the trie are also sorted—a path for a string that is located later in the lexicographical order is placed lower in the figure. When strings are added in that order (see Algorithm 8.2), the longest prefix of the string already recognized in the trie is traversed, ending in state q. Then procedure ADDSUFFIX creates a chain of states and transitions to recognize the suffix. Two situations are possible. Either the previously added string is a proper prefix of the current string and ADDSUFFIX creates the first transition leaving q, or ADDSUFFIX creates a subsequent transition going out from q, and that transition has a label that follows all labels of other transitions leaving q.

Algorithm 8.6:

Construction of a minimal, deterministic automaton from a sorted set of strings. The constructed automaton is returned.

```
function IncrementalSortedConstruction
    Create initial automaton A = ({q0}, Σ, q0, ∅, ∅)
    R ← ∅                                          ▷ register is empty
    w′ ← ε                                         ▷ previous string
    while there are strings to read do
        w ← next string
        AddStringSorted(A, w, w′)
        w′ ← w
    end while
    q0 ← ReplOrReg(A, q0, w′)
    return A
end function
```

Suppose we have already added strings *cie, ć, j, ją, jący, jcie*. The last added string is *jcie*. The automaton has states 0 to 11. When new strings are added, only states along the path $(0, 5, 9, 10, 11)$ in the already built part of the trie can change their right language. The path recognizes the newly added word. Only they can get more outgoing transitions. States 1, 2, 3, 4, 6, 7, and 8 cannot get new transitions as they cannot lie on a path recognizing any new string. When another string–*jmy*– is added, states 1, 2, 3, 4, 6, 7, and 8 still cannot change their right language. In addition, states 9, 10, and 11 no longer lie in the path of the last added string, so they cannot change their right language too. States 9, 10, and 11 are new states that can undergo minimization. They are those states that lie in the path of the previously added string that are not shared with the path of the last added word. Note that when the previously added string is a prefix of the current string, no new states can undergo minimization, as for example when *ją* or *jący* are added.

When no more strings are to be added, then states in the path recognizing the last string added are still waiting for minimization. As no more strings come, those states should undergo minimization too. The whole algorithm is given as Algorithm 8.6.

The function ReplOrReg, given as Algorithm 8.7, performs local minimization. In contrast to procedure minimize given as Algorithm 8.5, it takes an additional parameter w. It is a string that for all prefixes of w including ε, $\delta^*(q, w)$ is a state is to undergo minimization.

Algorithm 8.7:

Function ReplOrReg performs local minimization along the path of states starting from q and leading to $\delta^*(q, w)$.

```
function ReplOrReg(A, q, w)
    if w ≠ ε then
        δ(q, w1) ← ReplOrReg(A, δ(q, w1), w2...|w|)       ▷ redirect transition
    end if
    if ∃(r ∈ R) : r ≡ q then
```

```
6:        delete q
7:        return r
8:     else
9:        R ← R ∪ {q}
10:       return q
11:    end if
12: end function
```

The first conditional statement in ReplOrReg implements recursive descent toward the end of the argument string. It ends at a final state with no outgoing transitions. Then the second conditional statement checks whether there is a state in the register that is equivalent to the argument state q. If it is so, the current state is deleted, and the function returns the equivalent state r. Note that there is exactly one transition leading to q. That transition is redirected toward r in line 3 of the function. Redirecting a transition does not change the right language of any state present in the register, that is a state in the already minimized part of the automaton. The new target r does not change its right language since the transition is an incoming transition for it, and the automaton is acyclic. The source state of the transition, and any state on a path from q_0 to that state, is not present in the register. If an equivalent state is not found, state q is simply added to the register and returned by the function. Then in line 3 of ReplOrReg, the target does not change, and there is no redirection.

Algorithm 8.8:

Procedure AddStringSorted adds a string w to a deterministic automaton A. The automaton A is minimal except for a path recognizing a previous added string w'. Strings should come in a lexicographical order. The register is modified by the function to include states that lie in the path of the suffix of w'.

```
1: procedure AddStringSorted(A, w, w')
2:    (q, v) ← CommonPrefix(A, w)                          ▷ w = uv
3:    v' ← x ∈ Σ* : w' = xv'
4:    if v' ≠ ε then
5:       δ(q, v'_1) ← ReplOrReg(δ(q, v'_1), v'_{2...|v'|})
6:    end if
7:    AddSuffix(A, q, v)
8: end procedure
```

Function AddStringSorted, given as Algorithm 8.8, does the main job in incremental construction. It synchronizes construction and minimization. In line 2, the longest prefix of the argument string w is traversed. If the prefix is not the entire previously added string w', then the suffix v' of that string undergoes local minimization using ReplOrReg. Finally, the remaining suffix v of the current string w is added using AddSuffix.

Note that ReplOrReg and AddStringSorted use and modify the register R. We assume it to be a global variable. Otherwise it would have to a parameter of those subroutines.

Let us see how the construction works in practice on our example data set consisting of flectional endings of a Polish verb *czytać*. The first statement of IncrementalSortedConstruction creates an empty automaton with the initial state q_0. The previous string w' is set to ε, that is to an empty string, and the while loop begins. The first string *cie* is read. It becomes an argument for AddStringSorted. As the automaton has a single state q_0 and no transitions, CommonPrefix returns (q_0, cie). Then AddSuffix

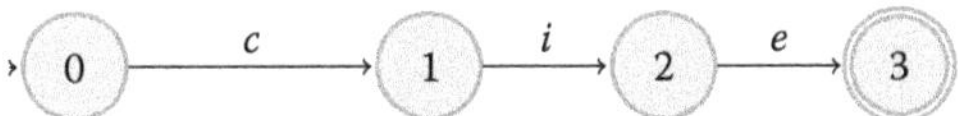

FIGURE 8.2 Example of sorted incremental construction. The string *cie* has just been added. $R = \emptyset$. We rotated the figure to save space.

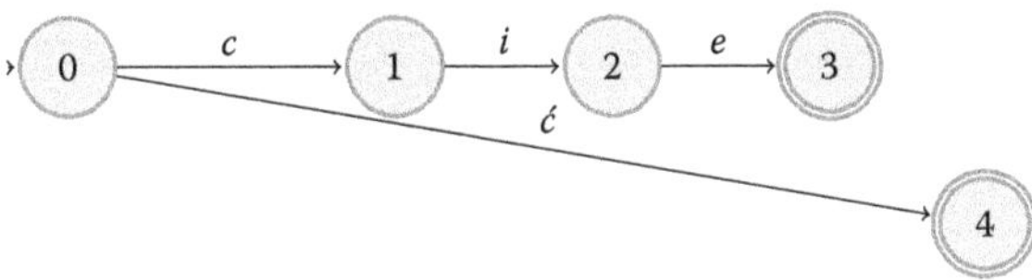

FIGURE 8.3 Example of sorted incremental construction. The string *ć* has just been added. $R = \{1, 2, 3\}$.

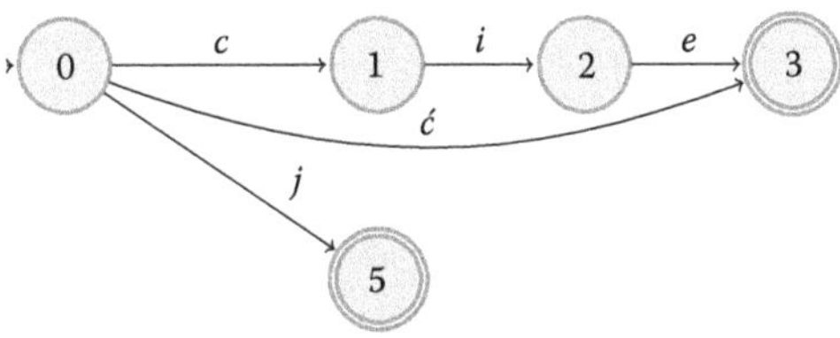

FIGURE 8.4 Example of sorted incremental construction. $\mathcal{L}(A) = \{cie,\ ć,\ j\}$. The string *j* has just been added. $R = \{1, 2, 3\}$.

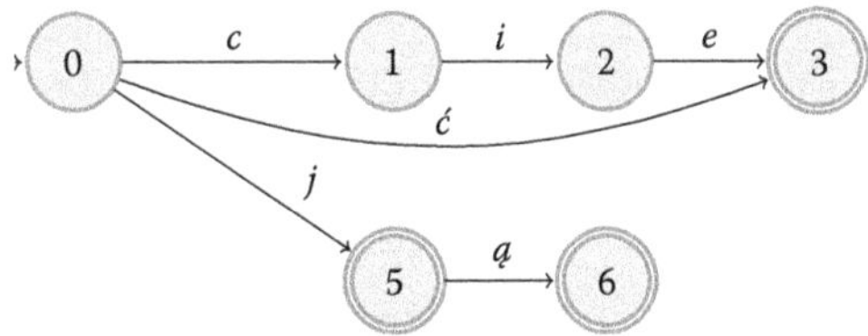

FIGURE 8.5 Example of sorted incremental construction. $\mathcal{L}(A) = \{cie,\ ć,\ j,\ ją\}$. The string *ją* has just been added. $R = \{1, 2, 3\}$.

is called to create a chain of states and transitions that will recognize the string *cie*. The last line of AddSuffix makes the last state of the chain final. The result is shown in Figure 8.2.

The next string to add is *ć*. AddStringSorted is called with A, *cie*, and *ć* as parameters. As there is no transition going out from q_0 labeled with *ć*, CommonPrefix returns $(q_0, ć)$. The previous string w' is *cie*, so v' is *cie* as well. ReplOrReg is called with A, state 1, and *ie* as parameters. As the last parameter is not empty, the function calls itself with A, state 2, and *e*, and then with state 3 and ε. The register is empty, so state 3 is added to the register, and returned. State 2 also has unique right language, so it is added to the register and returned. The same holds for state 1. AddSuffix creates one state depicted in Figure 8.3 as state 4, and a transition from state 0 to state 4 labeled with *ć*.

The next string is *j*. As before, CommonPrefix returns state 0, and the entire string. However, the previous string suffix v' is not empty: $v' = w' = ć$. ReplOrReg is called with A, state 4, and ε as arguments. As the last argument is a null string, no recursive calls take place. There is a state equivalent to state 4 in the register—it is state 3. Both have ε as their right language. Therefore, state 4 is deleted, and the function returns state 3. In line 5 of function AddStringSorted, the transition from state 0 to state 4 is redirected to state 3. Afterward, AddSuffix is called to create a final state 5, and a transition from state 0 to state 5 labeled with *j*. The result is shown in Figure 8.4.

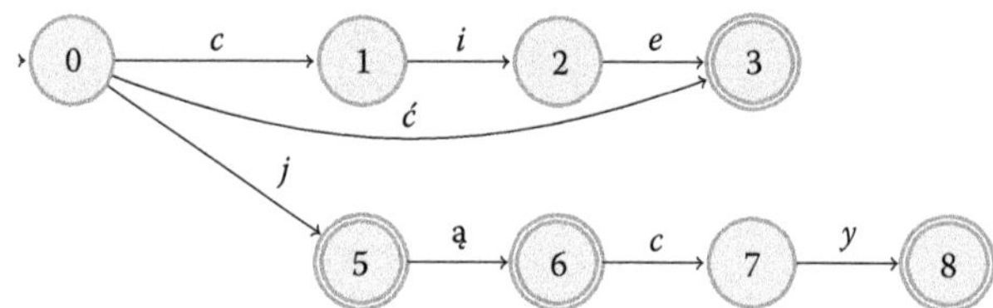

FIGURE 8.6 Example of sorted incremental construction. $\mathcal{L}(A) = \{$*cie, ć, j, ją, jący*$\}$. The string *jący* has just been added. $R = \{1, 2, 3\}$.

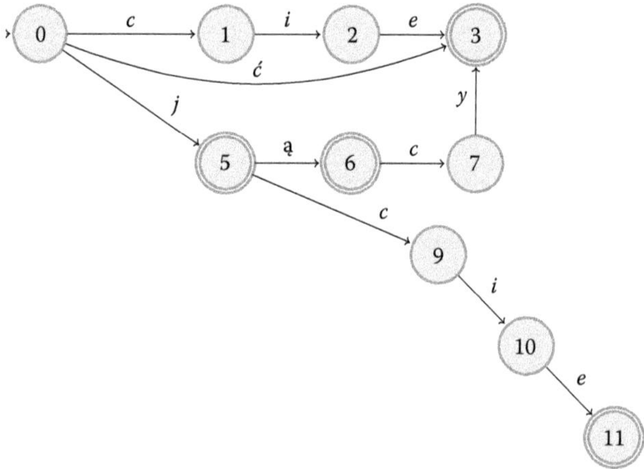

FIGURE 8.7 Example of sorted incremental construction. $\mathcal{L}(A) = \{$*cie, ć, j, ją, jący, jcie*$\}$. The string *jcie* has just been added. $R = \{1, 2, 3, 6, 7\}$.

The next string to be added is *ją*. CommonPrefix returns state 5 and *ą*. As the previous string $w' = j$ is a prefix of the current string, the previous string suffix $v' = \varepsilon$, and ReplOrReg is not called. A call to AddSuffix creates a final state 6, and a transition from state 5 to state 6 labeled with *ą*, as seen in Figure 8.5.

The next string *jący* follows the same pattern. ReplOrReg is not called, and AddSuffix creates states 7 and 8, and two transitions, as demonstrated in Figure 8.6.

The next string is *jcie*. CommonPrefix returns (5, *cie*). As $w' =$ *jący*, v' is now *ący*, and ReplOrReg is called with A, state 6, and *cy* as parameters. It immediately calls itself with A, state 7, and y, and then once again with A, state 8, and ε. State 3 is equivalent to state 8, so state 8 is deleted, and the transition from state 7 to state 8 is redirected to state 3. State 7 is unique, so it is added to the register. The same holds for state 6. Indeed, when any state reachable directly with a single transition from the current state has a unique right language, then the current state has a unique right language as well. The automaton is shown in Figure 8.7.

We finish the example by adding *jmy*. CommonPrefix returns (5, *my*), $v' =$ *cie*, $v =$ *my*. ReplOrReg is called with A, that is, and state 9, then with A, e, and state 10, and finally with A, ε, and state 11. As state 11 is equivalent to state 3, state 11 is deleted and the transition from state 10 to state 11 is redirected to state 3. State 10 is found equivalent to state 2, so state 10 is deleted, and the transition from state 9 to state 10 is redirected to state 2. State 9 is found equivalent to state 1, so state 9 is deleted, and the transition from state 5 to state 9 is redirected to state 1. A call to AddSuffix, creates state 12 and final state 13, as well as transition from state 5 to state 12 labeled with m, and a transition from state 12 to state 13 labeled with y. This result can be seen in Figure 8.8.

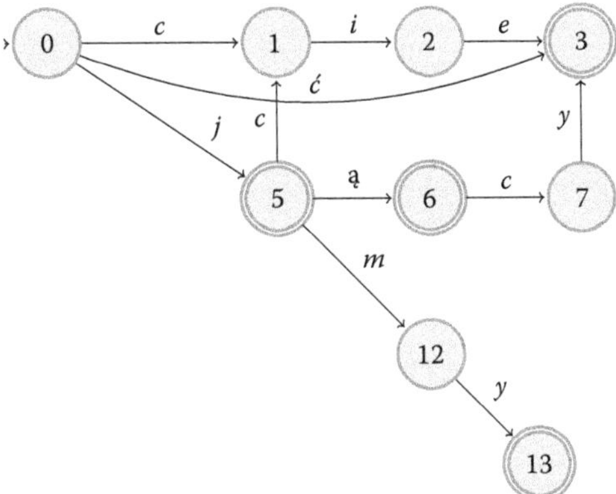

FIGURE 8.8 Example of sorted incremental construction. $\mathcal{L}(A) = \{cie, ć, j, ją, jący, jcie, jmy\}$. The string *jmy* has just been added. $R = \{1, 2, 3, 6, 7\}$.

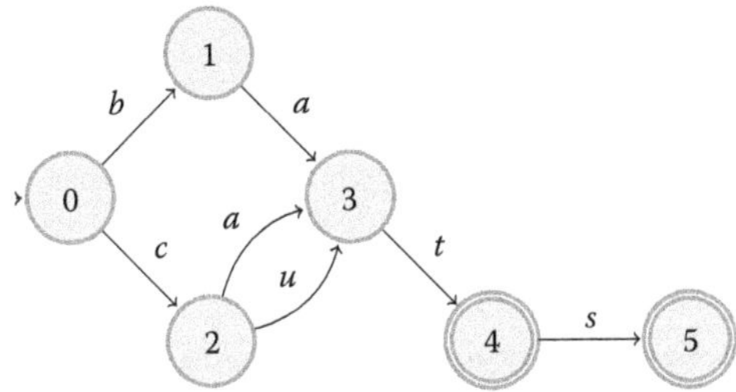

FIGURE 8.9 A minimal DFA recognizing words *bat, bats, cat, cats, cut, cuts.*

8.5 Algorithm for Data in Arbitrary Order

We call a state with more than one incoming transition a *confluence state.* Such states are created by redirecting transitions from states that are deleted to equivalent states, a part of minimization. In the algorithm for sorted data, confluence states cannot lie in the longest common prefix path, as they must already be in the register, which means they cannot change their right language.

When strings come in arbitrary order, the automaton is minimized completely after each string has been added. When a new string is added, confluence states can be on the path in the automaton that recognizes a prefix of the string. If we were to call AddSuffix with the last state of that path as one of the parameters, the language of the automaton would be augmented not only with the string, but possibly with some other strings as well. Suppose we have a minimal, deterministic automaton recognizing words *bat, bats, cat, cats, cut* and *cuts.* The automaton is shown in Figure 8.9. Suppose we want to add *cutting.* If we were to call AddSuffix$(A, 4, ing)$, then the automaton would look like the DFA in Figure 8.10.

The automaton in Figure 8.10 recognizes two additional strings: *batting* and *catting.* This happens because state 3 can be reached not only by following the prefix *cu*, but also *ba* and *ca.* Figure 8.11 shows a trie recognizing the language *bat, bats, cat, cats, cut, cuts,* and *cutting.* State 3 from Figure 8.10 corresponds to states 3, $3'$ and $3''$ in Figure 8.11. Analogically, state 4 from Figure 8.10 corresponds to states 4, $4'$, and $4''$ in Figure 8.11. When we call AddSuffix with an automaton given in Figure 8.10, it is as we were calling AddSuffix$(A', 4, ing)$, AddSuffix$(A', 4', ing)$, and AddSuffix$(A', 4'', ing)$.

A solution to the problem is to extract the proper branch of a trie from the minimal automaton. A process of making a copy of a state with the same finality and exactly the same suite of outgoing transitions is called *cloning.* A copy obtained in that way is a *cloned state.* Extracting a branch means

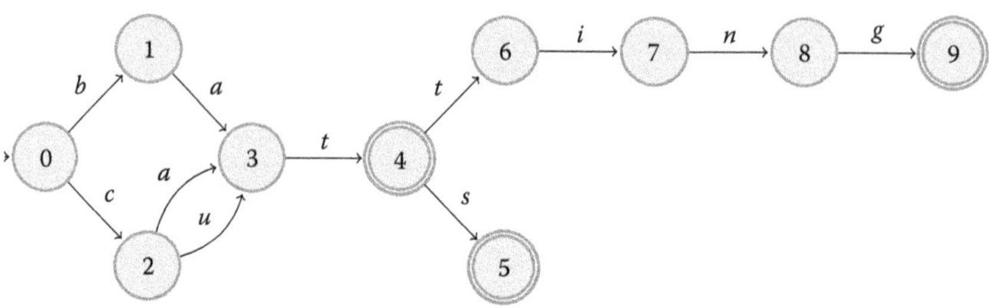

FIGURE 8.10 A DFA from Figure 8.9 after a call to AddSuffix(A, 4, *ing*) in order to make A recognize also *cutting*. Note that the automaton now inadvertently recognizes *batting* and *catting*.

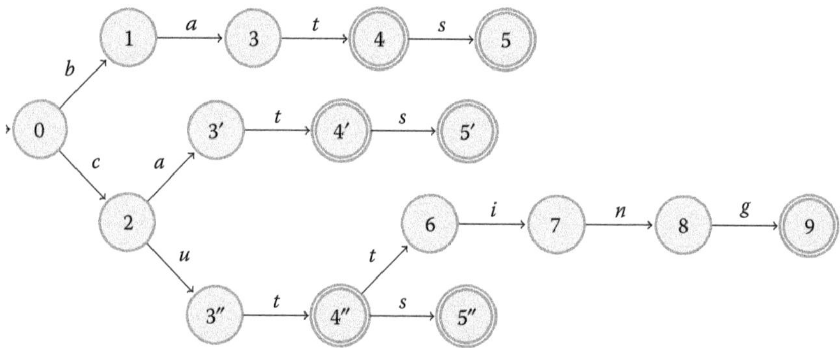

FIGURE 8.11 A DFA A' in form of a trie recognizing all words from the language of a DFA in Figure 8.9, and also *cutting*.

cloning confluence states. Note that all states that are destinations of transitions going out from a cloned state become confluence states.

Algorithm 8.9:

Construction of a minimal, deterministic automaton from an unsorted set of strings. The constructed automaton is returned.

```
function IncrementalUnsortedConstruction
    Create initial automaton
    A = ({q0}, Σ, q0, Ø, Ø)
    R ← Ø                                   ▷ register is empty
    while there are strings to read do
        w ← next string
        AddStringUnsorted(A, w)
    end while
    return A
end function
```

The top level of the algorithm for incremental construction of minimal, deterministic automata from an unsorted set of strings in shown as Algorithm 8.9. It resembles Algorithm 8.6, but there are a few differences. There is no notion of the previous string w' as the data comes in arbitrary order. Since the automaton is minimized completely each time a strings is added, there is no need to final local minimization of the path recognizing the last string added.

Algorithm 8.10:

Procedure AddStringUnsorted adds a string w to the language of an automaton A. A and a global variable R are modified by the procedure.

```
procedure AddStringUnsorted(A, w)
    (q, i) ← TraverseNonconfluence(A, P, w)
    f ← q                                        ▷ first state with modified transitions
    R ← R \ {f}
    j ← i
    (q, i) ← CloneConfluence(A, P, q, w, i)
    AddSuffix(A, q, w_{i...|w|})
    if j ≤ |w| then
        δ(f, w_j) ← ReplOrReg(δ(f, w_j), w_{j+1...|w|})
    end if
    LocMin(A, P, w)
end procedure
```

As with AddStringSorted (p. 178), in AddStringUnsorted, we have three phases. In the first one, we traverse subsequent transitions in the automaton A starting from q_0 labeled with subsequent symbols in the string w to be added. They determine a sequence of states that are stored in P. States in that path can be divided into two possibly empty parts. States in the first part are not confluence states traversed using TraverseNonconfluence. States in the second part are confluence states that are cloned in CloneConfluence. We record in f the first state in the path that changes either its finality or its suite of outgoing transitions. We also record its position on the path in j. State f is thus the first state before a confluence state, or the last state in P in case no confluence states are traversed. State f is withdrawn from the register as it changes its right language.

Algorithm 8.11:

Function TraverseNonconfluence follows transitions from the initial state of an automaton A that have subsequent symbols of the string w as their labels provided that the target states are not confluence states. Visited states are recorded in P. The function returns the state reached and the number of transitions followed.

```
function TraverseNonconfluence(A, P, w)
    P ← ∅                                        ▷ empty path
    q ← q_0
    i ← 1
    Push(P, q)
    while i ≤ |w| ∧ δ(q, w_i) ≠ ⊥ ∧ FanIn(δ(q, w_i)) = 1 do
        q ← δ(q, w_i)
        i ← i + 1
        Push(P, q)
    end while
    return (q, i)
end function
```

Function TRAVERSENONCONFLUENCE simply traverses a path of states in A starting from the start state, so that each subsequent state is reachable from the previous state by a transition labeled with the subsequent symbol from the string to be added w. In contrast to function COMMONPREFIX (presented as Algorithm 8.3 on p. 173), it cannot traverse confluence states. The second difference is that the states it visits are pushed on the stack P. Function FANIN returns the number of incoming transitions of a state.

Algorithm 8.12:

Function CLONECONFLUENCE traverses transitions in the automaton A starting from state q that have labels being subsequent symbols of w starting from ith symbol. State q is either a confluence state or it has no transition labeled with subsequent symbol of w. Confluence states visited in this function are cloned. The first state with no outgoing transition labeled with subsequent (ith) symbol of w along with symbol number in w are returned. Cloned states are recorded in P. A is modified by creating cloned states and their transitions.

```
function CLONECONFLUENCE(A, P, q, w, i)
    while i ≤ |w| ∧ δ(q, w_i) ≠ ⊥ do
        p ← CLONE(δ(q, w_i))
        δ(q, w_i) ← p
        q ← p
        i ← i + 1
        PUSH(P, q)
    end while
    return (q, i)
end function
```

Function CLONECONFLUENCE is called when a confluence state is the target of a transition from the current state q labeled with the next symbol w_i of the input string w. Such target state is cloned. The transition from q to the confluence state is redirected toward the clone. All other transitions coming to the confluence state stay as they were, but as one incoming transition is lost by redirection to the clone, the cloned state may no longer be a confluence state, by contrast, all targets of outgoing transitions of the cloned state and of the clone become confluence states as they have at least two incoming transitions (from the cloned state and from its clone). Cloning continues until either the end of the input string w is reached or until there is no transition from the current state labeled with the subsequent symbol of w. All cloned states are put onto the stack P.

The second phase is adding the remaining part of the string (if any) performed by ADDSUFFIX in line 7 of Algorithm 8.10. Note that when there is no remaining part of the string, ADDSUFFIX makes $\delta^*(q_0, w)$ a final state.

The third phase is local minimization consisting of calls to REPLORREG and LOCMIN. Function REPLORREG performs local minimization of a chain of states created by cloning and those just added by ADDSUFFIX. Those states are not yet in the register. As a result of the minimization, they can either enter into the register or they can be deleted and replaced with equivalent states.

Algorithm 8.13:

Function LOCMIN performs local minimization of the initial segment of a path of states that recognizes the string w in the automaton A. The path is stored on a stack P.

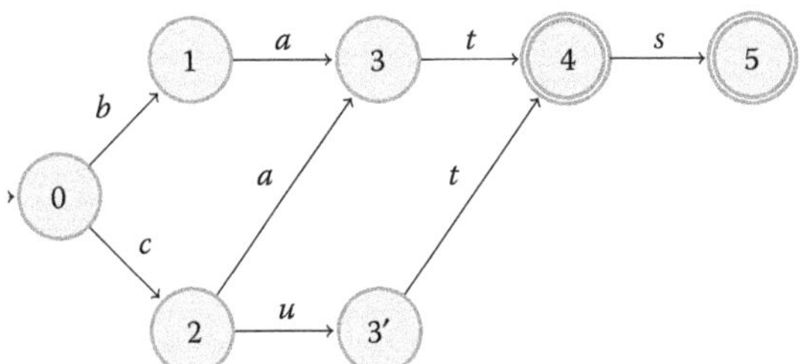

FIGURE 8.12 An automaton from Figure 8.9 during addition of the word cutting. Nonconfluence states in the longest common prefix have been traversed, and the first confluence state has just been cloned.

```
procedure LocMin(A, P, w)
    i = |P|
    q ← Pop(P)
    while i > 0 ∧ ∃r ∈ R : (r ≡ q) do
        if i > 1 then
            delete q
            q ← Pop(P)
            R ← R \ {q}
            δ(q, w_i) ← r
        end if
    end while
    R ← R ∪ {q}
end procedure
```

Recall that when states are processed so that the targets of transitions going out of a state are processed before that state, two states are equivalent when they are both final or both nonfinal, and they have identical suites of outgoing transitions. All states starting from q_0 that lie along the path recognizing the current string change their right language. However, we are not interested in that language itself, and not even in the fact that it changes. We are interested whether the state changes its finality or the suite of outgoing transitions. When they are not changed, then there is no state in the register equivalent to the current state. Function ReplOrReg can lead to replacement of all the states that are arguments of its recursive calls starting with the call in line 9 of AddStringUnsorted. Then, the state q at the top of the path stack P in the top-level call to LocMin is a state that changed its suite of outgoing transitions. If an equivalent state r is found in the register, state q is deleted, and the transition from the previous state to state q is redirected to r. Since the previous state changes its suite of outgoing transitions, it has to be removed from the register. The previous state becomes the current state q under investigation. The changes can percolate toward the initial state q_0. When no equivalent state can be found, which will always be the case with q_0, the state is added to the register, and there is no need to examine any other states located lower on the stack P.

Let us see how the algorithm works in practice. We start with an automaton shown in Figure 8.9. The register contains all states in the Figure. We add the word *cutting*. AddStringUnsorted calls TraverseNonconfluence, which puts states 0 and 2 onto the path stack P, and returns $(2, 2)$. State 2 is removed from the register. Now CloneConfluence is called. It starts by cloning state 3 which gives us state $3'$. The situation is depicted in Figure 8.12. Note that state 4 becomes a confluence state now.

State 4 is cloned in the second run of the loop in CloneConfluence as shown in Figure 8.13.

As there is no transition leaving state $4'$ labeled with t, AddSuffix is called. It produces a chain of states and transitions as shown in Figure 8.14.

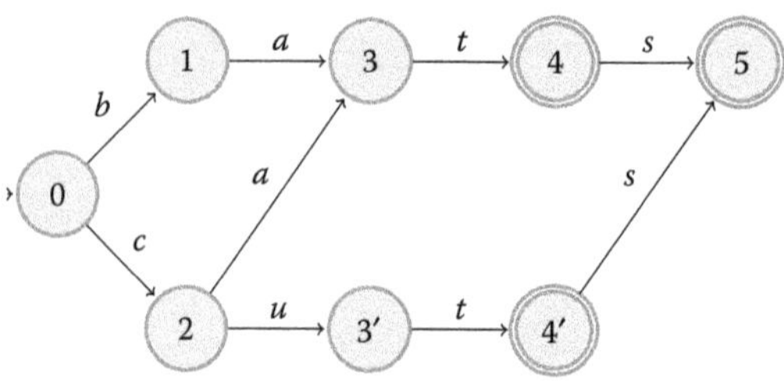

FIGURE 8.13 An automaton from Figure 8.9 during addition of the word cutting. Nonconfluence states in the longest common prefix have been traversed, and states 4 and 5 have just been cloned.

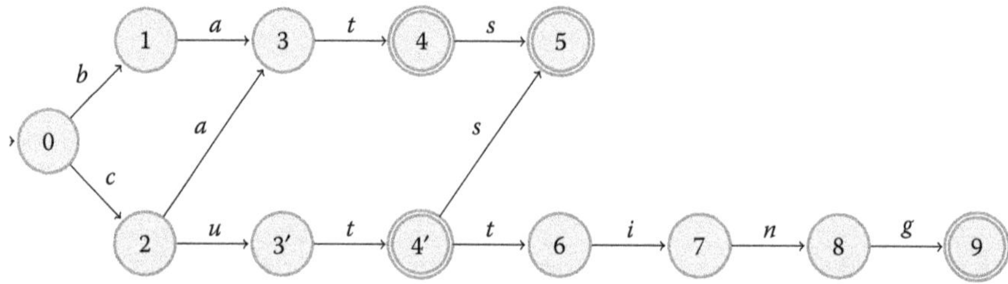

FIGURE 8.14 An automaton from Figure 8.9 during addition of the word cutting. The suffix *ting* has just been added.

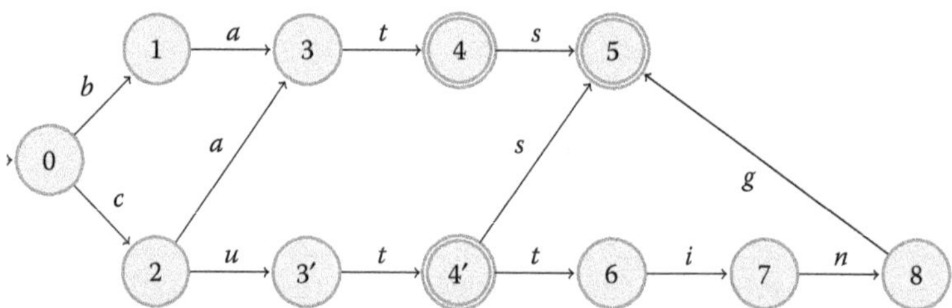

FIGURE 8.15 An automaton from Figure 8.9 during addition of the word cutting. Function REPLORREG has found state 9 to be equivalent to state 5. State 8 proves to be unique, and so must be states 7, 6, $4'$, and $3'$. LOCMIN finds state 2 to be unique as well.

Figure 8.15 shows the minimal automaton after the word *cutting* has been added. The most nested call to REPLORREG deletes state 9 and redirects the transition from state 8 to state 5. All other calls to REPLORREG find their argument states to be unique as they have unique targets of outgoing transitions, so those states—8, 7, 6, $4'$, and 3—are added to the register. In LOCMIN, state 2 is also found unique, which terminates the loop.

Let us consider another example. Figure 8.16 shows an automaton that recognizes words *bat, bats, cat, cats*, and *cuts*. Let us the word *cut* to it. Function ADDSTRINGUNSORTED is called, which immediately calls TRAVERSENONCONFLUENCE. The latter finds no confluence states and consumes the whole word returning $(7,4)$. The path stack P contains the states on the path of *cut*, i.e. 0, 2, 6, and 7. State 7 is removed from the register. Now CLONECONFLUENCE is called, but since there are no more letters in *cut* to be consumed (and no confluence states on the path), it returns immediately. Function ADDSUFFIX is called with an empty suffix. The loop in the function is not executed, but state 7 becomes final. The situation is shown in Figure 8.17.

Since j points behind the end of *cut*, REPLORREG is not called. Function LOCMIN starts with evaluating state 7. It finds that state 7 is equivalent to state 4, so state 6 is removed from the register, state 7 is deleted,

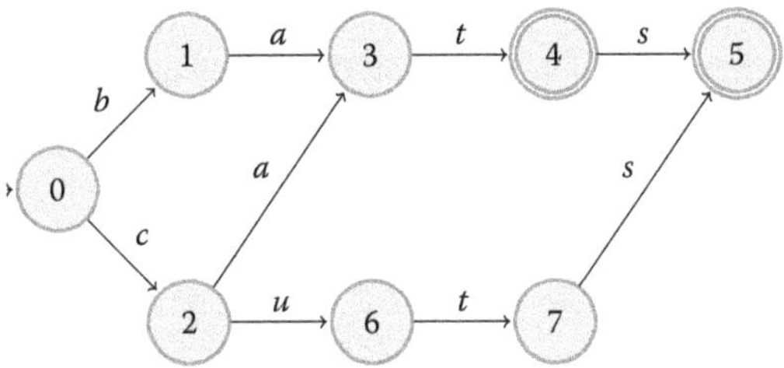

FIGURE 8.16 An automaton recognizing words *bat, bats, cat, cats,* and *cuts.*

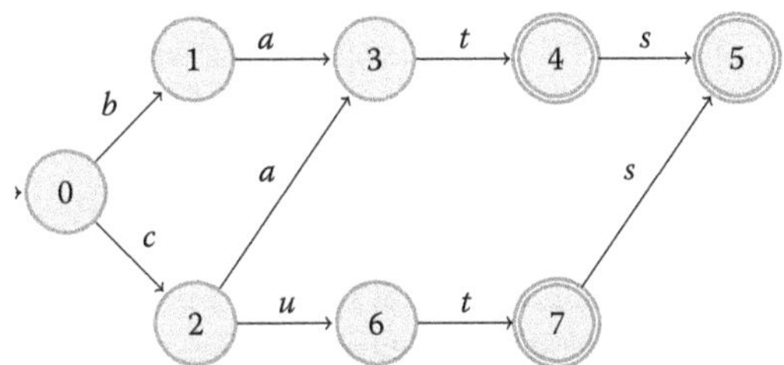

FIGURE 8.17 An automaton recognizing words *bat, bats, cat, cats, cut,* and *cuts.*

and the transition from state 6 to state 7 is redirected to state 4. Now state 6 is under investigation, and it is found equivalent to state 3. State 2 is removed from the register, state 6 is deleted, and the transition from state 2 to state 6 is redirected to state 3. Since state 2 is unique, it is put back into the register, and LocMin terminates. The automaton is now minimal again. It is depicted in Figure 8.9 on p. 181.

8.6 Semi-Incremental Construction

There is a group of algorithms that construct an almost minimal DFA, and then require an additional phase for final minimization. The size of the automaton during construction is close to the size of the almost minimal automaton before the last phase of the algorithm. The final minimization is performed locally so that it takes less time than global minimization. We will not describe semi-incremental algorithms in detail.

The first *semi-incremental* construction algorithm for DFAs was invented by Dominique Revuz [Rev91]. Input words are reversed, sorted, and reversed again. Such sorting facilitates searching for the longest common suffix (LCS). Words are added one by one. First, the longest common prefix (LCP) is found. This is done in a way similar to that used in the algorithm for data in arbitrary order, that is with cloning of confluence states. Then, the LCS between the word being added and the set of words already in the automaton is found in a way similar to finding the LCP. Note that sorting makes finding LCS just as easy as finding LCP, but requires storage of back transitions. Since LCP and LCS can overlap, states of LCP are marked so that they cannot be used simultaneously in LCS. The remaining part of the word is used for creating additional states and transitions necessary to recognize the whole word.

When all words are added, the automaton is *pseudo-minimal.* It has an interesting property: if all words have an end-of-word marker (a special symbol at the end of a word), then all words have their proper transitions in the automaton that are not shared with any other word. This can be used in implementing *dynamic perfect hashing* [DMS05]. Interestingly, a slight modification of the two incremental algorithms can also produce a pseudo-minimal automaton incrementally (the first phase of Revuz'es algorithm also produces a pseudo-minimal automaton fully incrementally, but it requires unusual presorting) [DMS06].

The second phase of Revuz'es algorithm is final minimization. It is done by sorting states on their height in the trie structure. A height $H(q)$ of a state q is defined as the length of the longest word in the right language of a state

$$H(q) = \max_{w \in \overrightarrow{\mathcal{L}}(q)} |w| \tag{8.12}$$

and it can be computed recursively.

$$\begin{aligned} H(q) &= 0 & \text{if } \Sigma_q = \emptyset \\ H(q) &= 1 + \max \sigma \in \Sigma_q H(\delta(q, \sigma)) & \text{if } \Sigma_q \neq \emptyset \end{aligned} \tag{8.13}$$

States with the same height form a layer. When layers are processed in the order of increasing height, $\delta_p = \delta_q$ becomes $\overrightarrow{p} = \overrightarrow{q} \wedge ((p \in F) \equiv (q \in F))$. This is easy to check when states within a layer are sorted on their transitions. The whole Revuz'es algorithm has linear complexity wrt. the total length of input words.

Another semi-incremental construction algorithm was developed by Bruce Watson. It requires input words to be sorted in the order of decreasing lengths. Words are added one by one, the LCP is found, and the remaining suffix (if any) is used to build a chain of states and transitions that recognize it. The last (final) state of the path that recognizes the word is used as a starting point for a stack-building function. It adds in preorder all states reachable from its starting point, including the starting point itself, and excluding any other final states, and states reachable only through those final states. Then the contents of the stack is subject to minimization. If the word just added is not a prefix of any other word (which must have been added before, as it has greater length), then only the last (final) state of the path that recognizes the last word can be replaced with an equivalent state. If the word is a prefix of another word, then the whole part of the path recognizing the longer word after the prefix is minimized. After the last word is added, the remaining states not yet subject to minimization, that is those between the start state and final states reachable from the start state without going through other final states, are put onto the stack and minimized. This is the final minimization. This algorithm has also linear complexity. It seems most suitable for situations when the input is sorted on decreasing length.

8.7 Extensions of the Incremental Algorithms

Several extensions to the incremental and semi-incremental algorithms exist. The first one is to consider adding (finite-length) words to cyclic automata. In such situation, one starts with an automaton that is already cyclic. One can build it using other algorithms, for example, the Thompson construction. Then strings are added to the cyclic automaton. Apart from trivial changes, like accepting an initial cyclic automaton as the starting point, there are two modifications to the construction algorithms that adapt them to the new challenge:

1. Clone the start state if it has any incoming transitions.
2. Clone any confluence states that are visited while recognizing the longest common prefix.

Confluence states appear in the common prefix path even in extensions of algorithms specifically designed to eliminate them (like, e.g., the sorted data algorithm or semi-incremental algorithms). They form a frontier between the old, original states of the automaton, and the new ones, either added or modified when new words are added.

A cover automaton is a cyclic automaton with an associated counter that determines the maximal length of recognized words. It is an attempt to use a smaller, cyclic automaton with an added feature to recognize a language that is normally recognized by a larger acyclic automaton. The idea is appealing, but there seems to be no evidence that at least natural language dictionaries can benefit from this approach. However, it might turn out that in other domains cover automata reduce memory footprint of applications.

Incremental construction is similar to the unsorted data construction for normal automata. The key factor is to find similarity between states so as to "fold" longer words into loops of the cyclic automaton. A difference between two states p and q can be computed by finding *gaps* between them, defined as the length of the shortest word that distinguishes them, that is, a word that leads to a final state from state p and a nonfinal from q or the other way round.

Normal automata process strings in one direction, which is usually visualized as from left to right. Tree automata can process trees either from the root to leaves, or from leaves to the root. The first ones are called *bottom-up* or *frontier-to-root*. The incremental construction algorithm has been developed for the bottom-up variant. The *top-down* or *root-to-frontier* variant is strictly less powerful. Deterministic tree automata (DTAs) are useful as an efficient data structure for storing large collection of trees, for example, data in XML, or collections of parse trees.

There are several substantial differences between (bottom-up) DTAs and DFAs. For example, DTAs have no start state, as processing begins in many leaves at once. Cloning a state is more complicated. For example, if a state q_1 is being cloned as q_2, and it has a sole transition $\delta(a, q_1, q_1) = q_3$, then not only $\delta(a, q_2, q_2) = q_3$ is added to the set of transitions, but also $\delta(a, q_1, q_2) = q_3$ and $\delta(a, q_2, q_1) = q_3$. Similarly, finding an equivalent state is also more complex.

8.8 Bibliographical Notes

The incremental construction algorithm for sorted data was independently invented by Jan Daciuk [Dac98, DWW98], and Stoyan Mihov [Mih98], as well as later on by Ciura and Deorowicz [CD01]. The algorithm for unsorted data was invented independently by many researchers, starting from J. Aoe and K. Morimoto and M. Hase [AMH92], K. Sgarbas and N. Fakotakis and G. Kokkinakis [SFK95], Jan Daciuk, Bruce Watson, and Richard Watson [DWW98], and Dominique Revuz [Rev00].

The semi-incremental algorithm acting on strings sorted in reversed order was developed by Dominique Revuz [Rev91]. Bruce Watson developed the semi-incremental algorithm for strings sorted on decreasing lengths [Wat98].

Rafael Carrasco and Mikel Forcada [CF02] extended the algorithm for unsorted data to the case of adding strings to a cyclic automaton. Jan Daciuk [Dac04a, Dac04b] showed how to do that with the algorithm for sorted data, and with Watson's semi-incremental algorithm. Rafael Carrasco, Jan Daciuk, and Mikel Forcada extended the incremental algorithm for data in arbitrary order to deterministic, bottom-up tree automata [CDF09].

A good but incomplete survey of incremental and semi-incremental algorithms can be found in [Wat10].

References

[AMH92] J. Aoe, K. Morimoto, and M. Hase. An algorithm for compressing common suffixes used in trie structures. *Trans. IEICE*, J75-D-II(4):770–779, 1992.

[ASU86] A. V. Aho, R. Sethi, and J. D. Ullman. *Compilers. Principles, Techniques and Tools*. Addison Wesley, Reading, MA, 1986.

[Brz62] J. A. Brzozowski. Canonical regular expressions and minimal state graphs for definite events. In *Mathematical Theory of Automata*, pp. 529–561. Polytechnic Press, Polytechnic Institute of Brooklyn, NY, 1962. Vol. 12 of MRI Symposia Series.

[CD01] M. Ciura and S. Deorowicz. How to squeeze a lexicon. *Software—Practice and Experience*, 31(11):1077–1090, 2001.

[CDF09] R. C. Carrasco, J. Daciuk, and M. L. Forcada. Incremental construction of minimal tree automata. *Algorithmica*, 55(1):95–110, September 2009.

[CF02] R. C. Carrasco and M. L. Forcada. Incremental construction and maintenance of minimal finite-state automata. *Computational Linguistics*, 28(2):207–216, June 2002.

[Dac98] J. Daciuk. *Incremental Construction of Finite-State Automata and Transducers, and their Use in the Natural Language Processing.* PhD thesis, Technical University of Gdańsk, Poland, 1998.

[Dac04a] J. Daciuk. Comments on incremental construction and maintenance of minimal finite-state automata by R. C. Carrasco and M. Forcada. *Computational Linguistics*, 30(2):227–235, June 2004.

[Dac04b] J. Daciuk. Semi-incremental addition of strings to a cyclic finite automaton. In K. Trojanowski M. A. Kopotek, S. T. Wierzchoń, editors, *Proceedings of the International IIS: IIP WM'04 Conference*, Advances in Soft Computing, pp. 201–207, Zakopane, Poland, May 2004, Springer, Berlin/Heidelberg, Germany.

[DMS05] J. Daciuk, D. Maurel, and A. Savary. Dynamic perfect hashing with pseudo-minimal automata. In M. A. Kopotek, S. Wierzchoń, and K. Trojanowski, editors, *Intelligent Information Processing and Web Mining, Proceedings of the International IIS: IIPWM'05 Conference*, Gdańsk, Poland, June 13–16, 2005, vol. 31 of *Advances in Soft Computing.* Springer, New York, NY, 2005.

[DMS06] J. Daciuk, D. Maurel, and A. Savary. Incremental and semiincremental construction of pseudo-minimal automata. In J. Farre, I. Litovsky, and S. Schmitz, editors, *Implementation and Application of Automata: 10th International Conference*, CIAA 2005, volume 3845 of LNCS, pp. 341–342. Springer, New York, NY, 2006.

[DWW98] J. Daciuk, R. E. Watson, and B. W. Watson. Incremental construction of acyclic finite-state automata and transducers. In *Finite State Methods in Natural Language Processing*, Bilkent University, Ankara, Turkey, June–July 1998.

[Hop71] J. E. Hopcroft. An $n \log n$ algorithm for minimizing the states in a finite automaton. In Z. Kohavi, editor, *The Theory of Machines and Computations*, pp. 189–196. Academic Press, New York, NY, 1971.

[HU79] J. E. Hopcroft and J. D. Ullman. *Introduction to Automata Theory, Languages, and Computation.* Addison-Wesley Publishing Company, Reading, MA, 1979.

[Mih98] S. Mihov. Direct building of minimal automaton for given list. In *Annuaire de l'Université de Sofia* "St. Kl. Ohridski," Vol. 91. Faculté de Mathematique et Informatique, Sofia, Bulgaria, livre 1 edition, February 1998.

[Rev91] D. Revuz. *Dictionnaires et lexiques: méthodes et algorithmes.* PhD thesis, Institut Blaise Pascal, Paris, France, 1991. LITP 91.44.

[Rev00] D. Revuz. Dynamic acyclic minimal automaton. In *CIAA 2000, Fifth International Conference on Implementation and Application of Automata*, pp. 226–232, London, Canada, July 2000.

[SFK95] K. Sgarbas, N. Fakotakis, and G. Kokkinakis. Two algorithms for incremental construction of directed acyclic word graphs. *International Journal on Artificial Intelligence Tools*, World Scientific, 4(3):369–381, 1995.

[Wat98] B. Watson. A fast new (semi-incremental) algorithm for the construction of minimal acyclic DFAs. In *Third Workshop on Implementing Automata*, pp. 91–98, Rouen, France, September 1998. Lecture Notes in Computer Science, Springer, Berlin/Heidelberg, Germany.

[Wat10] B. Watson. *Constructing Minimal Acyclic Deterministic Finite Automata.* PhD thesis, University of Pretoria, November 2010.

[WD03] B. W. Watson and J. Daciuk. An efficient incremental DFA minimization algorithm. *Natural Language Engineering*, 9(1):49–64, March 2003.

9

Esterel and the Semantics of Causality

Mohammad Reza Mousavi
Eindhoven University of Technology

9.1 Introduction

9.1.1 Semantic Models of Embedded Systems

Embedded systems are being used in critical applications, ranging from transportation (e.g., in various aircrafts and train signaling systems) to healthcare (e.g., in pacemakers and medical diagnostic systems). Hence, rigorous development and verification of embedded systems is of utmost importance and can save people's lives. Building rigorous models of such systems, analyzing them for various properties and generating code or refining them into lower-level models is a rigorous engineering practice, often called *model-based* development, which can be employed to develop more reliable embedded systems. The first step model-based development is building models and assigning a well-defined semantics to them. *Finite-state machines* (FSMs) have been traditionally used both as models and for assigning well-defined semantics to other models. Using FSMs as a modeling language is feasible for smaller and lower-level systems. For more complex systems, however, abstract models are usually specified in domain-specific modeling languages with suitable modeling features; then, at the semantic level FSMs can be used to represent the operational semantics (i.e., the behavioral meaning) of such systems, analyze it and turn it into an executable system. In this chapter, we study Esterel, a widely used modeling language for embedded systems, and its operational semantics.

9.1.2 Esterel and the Synchronous Assumption

Esterel [Ber99, PBEB07] is an imperative synchronous language used for the specification and programming of embedded systems. Signals are the first-class citizens in Esterel and their presence denote both the input to the system and the reaction of the system to the inputs. An Esterel program has a reactive style, that is, it defines an ongoing interaction between the system and its environment. Esterel is based on the synchronous hypothesis, which assumes instantaneous reaction to signals and immediate propagation of signals in each time instant. (Examples of other languages following the synchronous hypothesis include Signal [GGBM91, GTL03] and Lustre [HLR92, Halbwachs05].) By abstracting from the execution time of reactions and communication time of signals, this assumption simplifies the design of embedded systems and makes their design much more amenable to rigorous analysis. To better illustrate the synchronous assumption, we consider the following statement in Esterel.

`pres` trigger `?` `emit` reaction ◇ `0` `end`

It checks the "trigger" signal and upon its presence, it emits the "reaction" signal. The synchronous assumption dictates that if the trigger is present, the reaction is emitted immediately and the effect of the reaction is propagated immediately to other program components who may be checking for its presence.

Esterel has been widely used in practice in modeling and programming various embedded systems (see, e.g., [BBFLNS00] for its application to avionic systems) and is backed by tool-sets enabling automated analysis of designs as well as code generation for various software and hardware platforms [PBEB07].

In this chapter, we give a brief introduction to the core language of Esterel. Then, we focus on the issue of causality in Esterel and present a formal semantics for causality using Structured Operational Semantics.

9.1.3 Related Work

The intention of this chapter is not to cover all practical aspects concerning Esterel programs and their compilation. We refer to [PBEB07] for an excellent text on this subject. There are issues concerning loops and timing, which we only touch upon informally in this chapter. We refer to [BG92, Ber99, TdS05, PB02, PBEB07] for a formal account of these aspects of Esterel semantics.

The combination of the imperative programming style and the synchronous hypothesis in Esterel has led to semantic challenges addressed in the literature [BG92, Ber99, Tin00, Tin01, TdS05, PB02]. In this chapter, we address the issue of causality in more depth and illustrate how it is reminiscent of the semantic challenges [Gro93, BG96, Gla04] in Structured Operational Semantics [AFV01] (esp. in the setting with negative premises; the same challenges were encountered before in logic programming [AB94]). We then show that using the known solutions for the latter simplifies the presentation of the semantics of the former substantially and leads to the desired intuitive properties set forth by the language designers.

9.1.4 Structure of the Chapter

The rest of this chapter is organized as follows. In Section 9.2, we present a brief overview of the syntax of Esterel language and its intuitive semantics. In Section 9.3, we introduce the issues concerning causality; by means of a few simple Esterel code snippets, we present the informal intuition behind the semantics of causality in this section. Section 9.4 introduces Structured Operational Semantics and notions of semantics and well-definedness associated with SOS specifications. Section 9.5 connects these two worlds by first presenting an SOS specification for Esterel and then studying the notions of semantics and well-definedness for the given specification. There, we show that certain notions of semantics for SOS formalize the intuitive criteria given by the language designers. Section 9.6 concludes the chapter and presents directions for future research.

9.2 Esterel Language: A Cook's Tour

The abstract syntax of Esterel is given by the grammar in Figure 9.1.

A short introduction to the intuitive semantics of each of these constructs follows.

An Esterel program usually suffixed by a header declaring input and output signals. The syntax of this header is of the form `input` *i*; `output` *o*; and we assume that the set of input and output variables in a

$$
\begin{aligned}
m \;::=\; & \texttt{input}\ s;^* \\
& \texttt{output}\ s;^* \\
& p^* \\
p,q ::=\; & \texttt{0} \mid \texttt{emit}\ s \mid \texttt{pres}\ s\ \texttt{?}\ p \diamond q\ \texttt{end} \mid \\
& p\ \texttt{;}\ q \mid p \ \texttt{||}\ q \mid \texttt{sign}\ s\ \texttt{in}\ p\ \texttt{end} \mid \\
& \texttt{1} \mid \texttt{susp}\ p\ \texttt{when}\ s \mid \texttt{trap}\ t\ \texttt{in}\ p\ \texttt{end} \mid \texttt{exit}\ t \mid \texttt{loop}\ p\ \texttt{end}
\end{aligned}
$$

FIGURE 9.1 The abstract syntax of Esterel.

program is disjoint from the set of its local signals. Moreover, to unclutter the syntax, we assume fixed sets ι, ω and λ, respectively, of input, output and local variables. We pick typical members $i, i', i_0, \ldots \in \iota$, $o, o', o_0, \ldots \in \omega$ and $s, s', s_0 \in \lambda$. This way, we do not consider the input and out declaration throughout the rest of this chapter since input and output (and local) variables are recognized by their names. In some cases output and local variables can be treated uniformly, in which case we denote them by $x, x', x_0 \in \omega \cup \lambda$.

In this grammar, `0` stands for the terminated process. Emitting signal s is denoted by `emit` s, which is instantaneously visible to all parts of the system (and may in turn cause more signals to be emitted). Reacting to present and absent signals is done via the if–then–else construct `pres` s `?` $p \diamond q$ `end`, where if s is currently present (emitted by some other part of the system), p is executed, otherwise if s is absent q is executed. The combination of synchronous assumption, that is, instantaneous propagation of signals, and checking for absence/presence of signals leads to semantic complications, presented shortly. Parallel composition of p and q is denoted by p `||` q. Process `sign` s `in` p `end` encapsulates s in p, that is, declares s local to p. Another way of reacting to signals is by using the suspend construct `susp` p `when` s, which initially acts as p, but after one synchronous round will stop p as soon as signal s is emitted (suspension may happen after a number of rounds). Process `1` stands for a process that passes one unit of time and then terminates. One can define traps (exit points, exception handlers) to which a program can jump to by `trap` t `in` p `end`. The actual jump (raising the exception) is performed by executing `exit` t. A program can engage in a loop by means of `loop` p `end` (it can either keep on executing in the loop or exit the loop using `exit` t). The combination of synchronous assumption with the loop construct can lead to non-trivial puzzels in the semantics of Esterel.

9.3 Causality: An Informal Introduction

In this section, we give an overview of the issue of causality in the semantics of Esterel programs. Most of the examples and definitions in this section are taken from [Ber99].

A causality relation between events s and s' (signals in this case), means that the presence and absence of s directly influences the presence or absence of s'. For example, consider the following Esterel program:

```
P0  pres i ? emit s ◇ 0 end ; pres s ? 0 ◇ emit o end
```

In the above program, there is a causality chain starting from the input variable i to the local variable s and from s to the output variable o, namely the presence of i determines the presence of s and eventually leads to the absence of o, while the absence of s (caused by the absence of i), determines the presence of o. Using the syntax of Esterel, one can easily write programs with cyclic dependencies (e.g., s is present if and only if s is present) or even worse, cyclic dependencies of a paradoxical nature (e.g., s is present if and only if s is absent). To illustrate these issues in Esterel, consider the following simple programs, which are all due to [Ber99]. These programs are canonical examples of different issues concerning causality in Esterel programs.

P1 `pres s ? emit s ◇ 0 end`

Program P1 relies on the presence of *s* in order to emit signal *s*. The logical semantics of Esterel *rejects* this program on the ground that it has two "models." The first one is by assuming that *s* is present, which leads to a justification of this assumption by emitting *s*. The other one is by assuming that *s* is absent, which is supported by that `0` does not emit (denies emitting) signal *s*.

In each synchronous round, the "model" of an Esterel program is defined by a *global status*, which defines the status (presence/absence) of signals in this round. A global status of a program is called *coherent* when the presence/absence of signals are determined consistently by the emit statements in the program [Ber99]:

> The global status of a program is *logically coherent* iff at least one `emit` statement is executed for each signal assumed present and no `emit` statement is executed for each signal assumed absent.

For example, program P1 has two logically coherent global statuses, namely presence of *s* and absence of *s*, as motivated above. The basis for rejecting program P1 is called "logical determinism" and is defined as follows [Ber99].

> A program is logically deterministic if it has at most one logically coherent global status.

P2 `pres s ? 0 ◇ emit s end`

Program P2 relies on the absence of *s* in order to emit signal *s*. According to the logical semantics of Esterel, the above-given program has no logically coherent global status. Assuming that *s* is absent leads to *s* being emitted and hence, incoherency. Likewise, assuming that *s* is present requires emission of *s*, which is only justified when *s* is absent.

The basis for rejecting program P2 is called "logical reactivity" and is defined as follows [Ber99].

> A program is logically reactive if it has at least one logically coherent global status.

The conjunction of logical determinism and logical reactivity is called logical coherency and is the main well-definedness criterion for the *logical semantics* of Esterel.

P3 `pres s ? emit s ◇ emit s end`

The program above has only one logically coherent global status, namely that *s* is present. This global status is also coherent since assuming the presence of *s* leads to emitting it and moreover, it is not logically coherent to assume the absence of *s*, because it leads to its emission. Hence, as far as logical coherency is concerned this program is accepted and the logical semantics defines the semantics sketched above for this program.

However, the semantics of Esterel used for its compiler, called the *constructive semantics* [Ber99, PBEB07], has further constraints which lead to the rejection of the above program. In this chapter, we consider the issue of causality in both variants of the semantics and hence, also study the issue of constructiveness defined below.

> A program is *constructive*, if for each signal, it either proves its presence (must emit the signal) or proves its absence (cannot emit it).

Program P3 is rejected by the above criterion since it can neither prove the emission of *s* (its only possible proof is cyclic since relies on the assumption that *s* is emitted), nor can it coherently prove its absence, since to prove the absence of *s* it should prove that neither of the two emit statements can be executed, thus it should prove that *s* can neither be present nor absent.

P4 `pres` s_0 `? emit` $s_0 \diamond$ `0 end ||`
`pres` s_0 `? pres` s_1 `? 0` $\diamond$ `emit` s_1 `end` $\diamond$ `0 end`

Note that P4 is logically coherent, since its only logically coherent global status is that both s_0 and s_1 are absent. To check its constructiveness, let us focus on the emission of s_0. It definitely does not have to emit s_0, since the only reason for emitting s_0 is the `emit` s_0 statement in the left-hand side of the parallel composition, which is guarded by the check on the presence of s_0. Hence, the only proof for emitting s_0 is cyclic. But it can potentially emit s_0 (because it contains an `emit` s_0 statement) and the only way to make sure that s_0 cannot be emitted is to prove that the guard for `emit` s_0 never becomes true, that is, we need again to show that s_0 cannot be emitted, which is also a cyclic reasoning. Hence, we conclude that P4 is not constructive because it neither must emit s_0 nor it can deny its emission.

P5 `pres` s_0 `? emit` $s_1 \diamond$ `0 end ; emit` s_0

The above program is logically coherent and its unique logically coherent global status is that both s_0 and s_1 are present. However, it is again rejected by the constructive semantics of Esterel. The reason is that in order to reach the emit statement for s_0, we should first make sure that the first statement has a well-defined semantics in this context, that is, it either takes the if branch or the else branch and then terminates. However, giving a constructive proof for the transition of the conditional requires a constructive proof for the emission of s_0. This is another instance of the cyclic proof phenomenon rejected by the constructive semantics.

9.4 Structured Operational Semantics

Structural Operational Semantics (SOS) was originally proposed by Plotkin [Plo04] as a syntax-directed and compositional way of defining semantics. Gradually, SOS has gained popularity and by now has become a de facto standard in defining operational semantics. This popularity has called for a richer syntax for SOS deduction rules and thus, in some applications, SOS deduction rules lost their structural, that is, inductive, nature. Some authors then decided to use the same acronym for Structured Operational Semantics [AFV01, GV92]. With the richer syntax of SOS rules, one can write deduction rules whose meaning is not clear any more.

Example 9.1:

Examples of cyclic rules are the deduction rules (**r1**) and (**r2**) given below.

$$(\mathbf{r1})\frac{p \xrightarrow{s} p}{p \xrightarrow{s} p} \quad (\mathbf{r2})\frac{p \stackrel{s}{\nrightarrow}}{p \xrightarrow{s} p}$$

The reader may already note the curious similarity between program P1 and deduction rule (**r1**) on one hand and program P2 and deduction rule (**r2**) on the other hand. Moreover, program P3 resembles the combination of (**r1**) and (**r2**). These similarities materialize as formal definitions in the remainder of this chapter.

To formalize the syntax and semantics of SOS, we first formalize the concepts of formulae and (transition) formulae.

Definition 9.1: **(Signature and (Sub)terms)**

We let V represent an infinite set of variables. A *signature* Σ is a set of function symbols (operators), each with a fixed arity. An operator with arity zero is called a *constant*. We define the set $\mathbb{T}(\Sigma)$ of *terms* over Σ as the smallest set satisfying the following constraints.

- A variable $x \in V$ is a term
- If $f \in \Sigma$ has arity n and $t_1, \ldots, t_n$ are terms, then $f(t_1, \ldots, t_n)$ is a term

We write $t_1 \equiv t_2$ if t_1 and t_2 are syntactically equal. The function $vars : \mathbb{T}(\Sigma) \rightarrow 2^V$ gives the set of variables appearing in a term. The set $\mathbb{C}(\Sigma) \subseteq \mathbb{T}(\Sigma)$ is the set of *closed terms*, that is, terms that contain no variables. A *substitution* σ is a function of type $V \rightarrow \mathbb{T}(\Sigma)$. We extend the domain of substitutions to terms homomorphically. If the range of a substitution lies in $\mathbb{C}(\Sigma)$, we say that it is a *closing substitution*.

A term s is considered a *subterm* of itself; if s is a subterm of t_i, then s is also a subterm of $f(t_0, \ldots, t_i, \ldots, t_{n-1})$, for each $s, t_i \in \mathbb{T}(\Sigma)$, $0 \leq i < n$, and n-ary $f \in \Sigma$. The set of subterms of a term t are denoted by *subterms*(t).

Next, we formalize the syntax of SOS in terms of Transition System Specifications.

Definition 9.2: **(Transition System Specifications (TSS))**

A *transition system specification* is a triplet (Σ, L, D) where

- Σ is a signature.
- L is a set of labels. If $l \in L$, and $t, t' \in \mathbb{T}(\Sigma)$ we say that $t \xrightarrow{l} t'$ is a *positive formula* and $t \stackrel{l}{\nrightarrow}$ (also denoted by $\neg t \xrightarrow{l}$) and $t \stackrel{l}{\nrightarrow} t'$ are *negative formulae*. A formula, typically denoted by ϕ, ψ, ϕ', ϕ_i, $\ldots$ is either a negative formula or a positive one.
- D is a set of *deduction rules*, that is, tuples of the form (Φ, ϕ) where Φ is a set of formulae and ϕ is a positive formula. We call the formulae contained in Φ the *premises* of the rule and ϕ the *conclusion*.

We write *vars*(r) to denote the set of variables appearing in a deduction rule (**r**). We say a formula is *closed* if all of its terms are closed. Substitutions are also extended to formulae and sets of formulae in the natural way.

A deduction rule (Φ, ϕ) is typically written as $\frac{\Phi}{\phi}$. For a deduction rule r, we write *conc*(r) to denote its conclusion and *prem*(r) to denote its premises. A set of positive closed formulae is called a *transition relation*. Given a transition relation T, L'-labeled transitions of closed term p, denoted by $T \downarrow (p, L')$ is the subset of T containing all formulae in T that have p as their source and some $l \in L'$ as their label. A TSS is supposed to define a transition relation but for the TSSs such as those given by deduction rules (**r0**) and (**r1**), it is not clear what the associated transition relation is. Several proposals are given in the literature, of which [Gla04] gives a comprehensive overview and comparison. In this chapter, we shall use some of these proposals to define the semantics of Esterel. In order to facilitate the presentation of these proposals, we need two auxiliary definitions, namely contradiction and contingency, which are given below.

Definition 9.3: **(Contradiction and Consistency)**

Formula $t \xrightarrow{l} t'$ is said to *contradict* both $t \stackrel{l}{\nrightarrow}$ and $t \stackrel{l}{\nrightarrow} t'$, and vice versa. Φ is *consistent* w.r.t. Ψ, denoted by $\Phi \vDash \Psi$, when for each positive formula $\psi \in \Psi$, it holds that $\psi \in \Phi$ and for each negative formula $\psi \in \Psi$, there is no $\phi \in \Phi$ such that ϕ contradicts ψ.

In the remainder, we only use negative formulae of the form $t \stackrel{l}{\nrightarrow}$ in our specifications. We now have all the necessary ingredients to present different proposals for the semantics of TSSs. The first proposal is the following notion of supported model, which is a slight modification of the definition in [Gla04] (restricting it to particular sets of terms and labels).

Definition 9.4: **(Supported Model)**

Given A TSS, a transition relation T is a supported model for a set $P \subseteq \mathbb{C}(\Sigma)$ of closed terms and a set $L' \subseteq L$ of labels, when

1. For each $q, q' \in \mathbb{T}(\Sigma)$ and $l \in L$ if $q \xrightarrow{l} q' \in T$, then there exists a deduction rule $\frac{\Phi}{\phi}$ and a substitution σ such that $\sigma(\phi) = q \xrightarrow{l} q'$ and $T \vDash \Phi$, and
2. For each $p \in P$ and $l \in L'$, $p' \in \mathbb{T}(\Sigma)$, if there exists a deduction rule $\frac{\Phi}{\phi}$ and a substitution σ such that $\sigma(\phi) = p \xrightarrow{l} p'$ and $T \vDash \Phi$, then $p \xrightarrow{l} p' \in T$.

A transition relation T is a supported model for a TSS when it is a supported model for $\mathbb{C}(\Sigma)$ and L.

Note that in the above definition and throughout the rest of the chapter, we only consider the "immediate transitions" of p as its semantics. One can adapt the above definitions (and the subsequent ones) to consider the "transition system" (i.e., the finite state machine) associated with p as its semantics. For the subset of Esterel considered in this chapter, these two notions lead to the same conclusion concerning the well-definedness and the semantics of a program.

Semantics 9.1: **(Unique Supported Model Semantics)**

Given a set $P \in \mathbb{C}(\Sigma)$ of closed terms and a set $L' \subseteq L$ of labels a TSS is *meaningful* w.r.t. P and L' when it has a unique supported model for P and L'; the *transition system* associated with P and L' is the unique supported model for P and L'. A TSS is meaningful when it has a unique supported model; the transition relation associated with a TSS is its unique supported model.

To illustrate these concepts, we give a few simple TSSs and study their supported models.

Example 9.2:

Consider the deduction rules given in Example 9.1.

Consider the TSS comprising only deduction rule (**r1**). This TSS is not meaningful (w.r.t. $\{p\}$ and $\{s\}$) according to Semantics 9.1 because it has two supported models, namely $\emptyset$ and $\{p \xrightarrow{s} p\}$.

Also according to Semantics 9.1, the TSS comprising only deduction rule (**r2**) is not meaningful (w.r.t. $\{p\}$ and $\{s\}$) either, because it has no supported model. Particularly, $T = \emptyset$ is not a supported model because it follows from the right-to-left implication of Definition 9.4 that $p \xrightarrow{s} p \in T$. $T = \{p \xrightarrow{s} p\}$ is not a supported model either since the only deduction rule providing a reason for $p \xrightarrow{s} p \in T$ is (**r2**) but it does not hold that $T \vDash p0 \stackrel{s}{\nrightarrow}$.

The TSS comprising both (**r1**) and (**r2**) is indeed meaningful and its associated transition relation is $T = \{p \xrightarrow{s} p\}$. Transition relation T is indeed a supported model since (**r1**) now provides a reason for $p \xrightarrow{s} p \in T$. Moreover, $T' = \emptyset$ is not a supported model for this TSS because it then follows from (**r2**) and the right-to-left implication of Semantics 9.1 that $p \xrightarrow{s} p \in T'$.

If one takes the transition system of a program as a formalization of its global state, then the TSS comprising of deduction rule (**r1**) is rejected because it has no coherent global state and the TSS with only (**r2**) is rejected because it does equivocally define a coherent global state.

This suggests that Semantics 9.1 provides a suitable formalization for logical coherency. Next, we give a formalization of constructiveness in terms of supported proofs and denials.

Definition 9.5: (Supported Proofs)

A TSS $\mathcal{T}$ provides a supported proof for a formula ϕ, denoted by $\mathcal{T} \vdash_s \phi$, when there is a well-founded upwardly branching tree with formulae as nodes and of which

- The root is labeled by ϕ
- If a node is labelled by a positive formula ψ and the nodes above it form the set K then $\frac{K}{\psi}$ is an instance of a deduction rule in $\mathcal{T}$
- If a node is labelled by a negative formula ψ, and the nodes above it form the set K, then for each instance of a deduction rule $\frac{K_i}{\psi_i}$ in $\mathcal{T}$ such that ψ_i contradicts ψ, there exists a formula $\psi_i' \in K$ contradicting a formula in K_i

Semantics 9.2: (S-Complete Semantics)

A TSS is s-complete for a set of closed terms P when for each formula ϕ with a $p \in P$ as its source, either ϕ or a formula contradicting it has a supported proof. A TSS is s-complete when it is s-complete for the set $\mathbb{C}(\Sigma)$ of all closed terms and its transition relation is the set of positive formulae, for which it provides supported proofs.

The following theorem is taken from [Gla04], which shows that constructiveness is indeed stronger than logical coherency.

Theorem 9.1:

A program (TSS) is s-complete only if it is meaningful according to Semantics 9.1 (has a unique supported model) and its associated transition system (unique supported model) coincides with the set of all positive formulae with a supported proof.

Another useful property of supported proofs is their consistency [Gla04], stated below.

Theorem 9.2:

The notion of supported proof is consistent, that is, for each formula ϕ with a supported proof, its negation does not have a supported proof.

Next, we reexamine the TSS of Example 9.2 using our new notion of semantics.

Example 9.3:

The two TSSs comprising only (**r1**) and only (**r2**) are both rejected by Semantics 9.2, as well, since neither $p \xrightarrow{s} p$, nor $p \stackrel{s}{\nrightarrow}$ can be proven from either of them. (This is also an immediate consequence of Theorem 9.1.)

In the case of the TSS comprising only (**r1**), any attempt to build a supported proof for $p \xrightarrow{s} p$ has the same formula as its premise. Moreover, $p \stackrel{s}{\nrightarrow}$ cannot be proven because its proof tree should prove a negation of a premise of (**r1**), that is, again $p \stackrel{s}{\nrightarrow}$. In other words, both $p \xrightarrow{s} p$ and $p \stackrel{s}{\nrightarrow}$ only have cyclic, and thus unsupported, proofs.

Similarly, in the case of the TSS comprising only (**r2**), neither $p \xrightarrow{s} p$, nor $p \stackrel{s}{\nrightarrow}$ have a supported proof.

Consider the TSS comprising both (**r1**) and (**r2**); it does have a unique supported model $T = \{p \xrightarrow{s} p\}$ but it is *not* s-complete and is thus rejected by Semantics 9.2. Any proof for $p \xrightarrow{s} p$ or its negation leads to a cycle, that is, repeating the node below in the node above, and are thus not supported.

Again drawing an analogy with Esterel programs, Semantics 9.2 requires the existence of a "constructive" (supported) proof for presence/absence of signals and thus rejects a program which uses both *possibilities* for a signal in order to establish its own presence.

9.5 Structured Operational Semantics for Esterel

Our semantic specification of Esterel is presented in Figures 9.2 and 9.3. The state of the SOS comprises the syntax of the program currently being executed (defined by the grammar in Figure 9.1). The semantics is supposed to define two predicate, $p\checkmark_{I,c}$, $p \uparrow^{I,c,s}$, respectively, where the former means that p terminates with input evaluation I and under context c (if p is part of program c), and the latter means that p emits signal s (in the present time-instant) under the same assumptions. (A predicate formula can be formally interpreted as a transition formula with a dummy right-hand-side; in our case one can take 0 to be the dummy target of all predicate formulae, that is, read $p \uparrow^{I,c,s}$ and $p\checkmark_{I,c}$ as $p \uparrow^{I,c,s} 0$ and $p\checkmark_{I,c} 0$, respectively.) In addition to the two predicates, the semantics is supposed to define a transition relation of the form $p \xrightarrow{I,c,s} p'$, which denotes that program p emits signal s under input evaluation I and context c. Next, we briefly describe the deduction rules in Figures 9.2 and 9.3 and then show how they formalize the intuitive properties of Esterel programs discussed before. In all labels (of predicates and transitions) of Figures 9.2 and 9.3, $I \subseteq \{i^+, i^- \mid i \in \iota\}$ such that for each $i \in \iota$, either $i^- \in I$ or $i^+ \in \iota$ (but not both), $c \in \mathbb{C}(\Sigma)$, $i \in \iota$, $x \in \omega \cup \lambda$ and $s, s', s'' \in \lambda$.

In Figure 9.2, (**e0**) states that `emit` s can emit signal s under any arbitrary input evaluation and context. Deduction rules (**s0**), (**s1**), and (**s2**) describe when a sequential composition emits a signal, namely, when either the first component of the composition emits it, or when the first component terminates (possibly after a transition) and the second component emits the signal.

The notions of termination and transition are defined in Figure 9.3. A parallel composition emits a signal if one of its components emits the signal, which is captured by deduction rules (**p0**) and (**p1**). An if-then-else constructs emits a signal, if either, according to deduction rule (**f0**), the local

$$\text{(e0)}\frac{}{\texttt{emit}\ x \uparrow^{I,c,x}}$$

$$\text{(s0)}\frac{p \uparrow^{I,c,x}}{p\ ;\ q \uparrow^{I,c,x}} \quad \text{(s1)}\frac{p\checkmark_{I,c} \quad q \uparrow^{I,c,x}}{p\ ;\ q \uparrow^{I,c,x}} \quad \text{(s2)}\frac{p \xrightarrow{I,c,x'} p' \quad p'\checkmark_{I,c} \quad q \uparrow^{I,c,x}}{p\ ;\ q \uparrow^{I,c,x}}$$

$$\text{(p0)}\frac{p \uparrow^{I,c,x}}{p \mid\mid q \uparrow^{I,c,x}} \quad \text{(p1)}\frac{q \uparrow^{I,c,x}}{p \mid\mid q \uparrow^{I,c,x}}$$

$$\text{(f0)}\frac{c \uparrow^{I,c,s} \quad p \uparrow^{I,c,x}}{\texttt{pres}\ s\ ?\ p \diamond q\ \texttt{end} \uparrow^{I,c,x}} \quad \text{(f1)}\frac{\neg c \uparrow^{I,c,s} \quad q \uparrow^{I,c,x}}{\texttt{pres}\ s\ ?\ p \diamond q\ \texttt{end} \uparrow^{I,c,x}}$$

$$\text{(f2)}\frac{i^+ \in I \quad p \uparrow^{I,c,x}}{\texttt{pres}\ i\ ?\ p \diamond q\ \texttt{end} \uparrow^{I,c,x}} \quad \text{(f3)}\frac{i^- \in I \quad q \uparrow^{I,c,x}}{\texttt{pres}\ i\ ?\ p \diamond q\ \texttt{end} \uparrow^{I,c,x}}$$

$$\text{(en0)}\frac{p[s''/s] \uparrow^{I,c,s'}}{\texttt{sign}\ s\ \texttt{in}\ p\ \texttt{end} \uparrow^{I,c,s'[s/s'']}} \quad s'' \text{ fresh in } p \text{ and } r$$

FIGURE 9.2 Structured operational semantics for Esterel (Part I: signal emission).

signal in its condition is emitted and the if-branch emits the signal or, according to deduction rule **(f1)**, the local signal in the condition cannot be emitted and the else-branch is taken. Deduction rules **(f2)** and **(f3)** take care of the case where the condition is an input signal. In such cases, the condition is checked against the given input evaluation. A program p with a local signal s can emit a signal s', if p with a fresh signal s'' substituted for s can emit s' (but if s' is s, then p should be able to emit s'').

In Figure 9.3, the concept of termination is defined through the predicate $\checkmark_{I,c}$ in a straightforward manner. Exceptions are deduction rules **(if4)** and **(if5)**, which rely on (the impossibility of) the emission of the condition signal for proving termination. In Figure 9.3, the deduction rules specifying a transition relation are almost identical to their counterparts in Figure 9.2. The most notable exceptions are deduction rules **(seq0)** to **(seq4)** and **(par0)** to **(par3)**, which should consider all possible combinations of simultaneous transitions and individual transitions with (non-)termination in order to record the right target for the transition.

One advantage of our approach to the semantics of Esterel presented in [Ber99, Tin00, Tin01] is that we can capture both the logical semantics and constructive semantics of Esterel using the same TSS (by using two generic notions of semantics for TSS already known in the literature). Another advantage is that it establishes a clear link between, respectively, the logical and the constructive approaches to Esterel semantics, on the one hand and the model- and proof-theoretic semantics of TSSs on the other.

Definition 9.6: (Logical Semantics of Esterel)

An Esterel program p is logically coherent if the above given TSS is meaningful according to Semantics 9.1 for *subterms*(p) and (predicates and) transitions labeled $\{\checkmark_{I,p}, \uparrow^{I,p,x}, \xrightarrow{I,p,x}\}$. The semantics of p is the set of above-mentioned predicates and transitions associated with *subterms*(p).

$$\textbf{(nil)}\ \frac{}{0\checkmark_{I,c}} \qquad \textbf{(em)}\ \frac{}{\texttt{emit}\ x \xrightarrow{I,c,x} 0}$$

$$\textbf{(seq0)}\ \frac{p \xrightarrow{I,c,x} p' \quad p'\checkmark_{I,c} \quad q \xrightarrow{I,c,x'} q'}{p\ \texttt{;}\ q \xrightarrow{I,c,x} q'} \qquad \textbf{(seq1)}\ \frac{p \xrightarrow{I,c,x} p' \quad p'\checkmark_{I,c} \quad q \xrightarrow{I,c,x'} q'}{p\ \texttt{;}\ q \xrightarrow{I,c,x'} q'}$$

$$\textbf{(seq2)}\ \frac{p\checkmark_{I,c} \quad q \xrightarrow{I,c,x} q'}{p\ \texttt{;}\ q \xrightarrow{I,c,x} q'} \qquad \textbf{(seq3)}\ \frac{p \xrightarrow{I,c,x} p' \quad p'\checkmark_{I,c} \quad q\checkmark_{I,c}}{p\ \texttt{;}\ q \xrightarrow{I,c,x} p'} \qquad \textbf{(seq4)}\ \frac{p\checkmark_{I,c} \quad q\checkmark_{I,c}}{p\ \texttt{;}\ q\checkmark_{I,c}}$$

$$\textbf{(par0)}\ \frac{p \xrightarrow{I,c,x} p' \quad q \xrightarrow{I,c,x'} q'}{p\ ||\ q \xrightarrow{I,c,x} p'\ ||\ q'} \qquad \textbf{(par1)}\ \frac{p \xrightarrow{I,c,x} p' \quad q \xrightarrow{I,c,x'} q'}{p\ ||\ q \xrightarrow{I,c,x'} p'\ ||\ q'}$$

$$\textbf{(par2)}\ \frac{p\checkmark_{I,c} \quad q \xrightarrow{I,c,x} q'}{p\ ||\ q \xrightarrow{I,c,x} q'} \qquad \textbf{(par3)}\ \frac{p \xrightarrow{I,c,x} p' \quad q\checkmark_{I,c}}{p\ ||\ q \xrightarrow{I,c,x} p'} \qquad \textbf{(par4)}\ \frac{p\checkmark_{I,c} \quad q\checkmark_{I,c}}{p\ ||\ q\checkmark_{I,c}}$$

$$\textbf{(if0)}\ \frac{c \uparrow^{I,c,s} \quad p \xrightarrow{I,c,x} p'}{\texttt{pres}\ s\ \texttt{?}\ p \diamond q\ \texttt{end} \xrightarrow{I,c,x} p'} \qquad \textbf{(if1)}\ \frac{\neg c \uparrow^{I,c,s} \quad q \xrightarrow{I,c,x} q'}{\texttt{pres}\ s\ \texttt{?}\ p \diamond q\ \texttt{end} \xrightarrow{I,c,x} q'}$$

$$\textbf{(if2)}\ \frac{i^+ \in I \quad p \xrightarrow{I,c,x} p'}{\texttt{pres}\ i\ \texttt{?}\ p \diamond q\ \texttt{end} \xrightarrow{I,c,x} p'} \qquad \textbf{(if3)}\ \frac{i^- \in I \quad q \xrightarrow{I,c,x} q'}{\texttt{pres}\ i\ \texttt{?}\ p \diamond q\ \texttt{end} \xrightarrow{I,c,x} q'}$$

$$\textbf{(if4)}\ \frac{c \uparrow^{I,c,s} \quad p\checkmark_{I,c}}{\texttt{pres}\ s\ \texttt{?}\ p \diamond q\ \texttt{end}\checkmark_{I,c}} \qquad \textbf{(if5)}\ \frac{\neg c \uparrow^{I,c,s} \quad q\checkmark_{I,c}}{\texttt{pres}\ s\ \texttt{?}\ p \diamond q\ \texttt{end}\checkmark_{I,c}}$$

$$\textbf{(if6)}\ \frac{i^- \in I \quad p\checkmark_{I,c}}{\texttt{pres}\ i\ \texttt{?}\ p \diamond q\ \texttt{end}\checkmark_{I,c}} \qquad \textbf{(if7)}\ \frac{i^+ \in I \quad q\checkmark_{I,c}}{\texttt{pres}\ i\ \texttt{?}\ p \diamond q\ \texttt{end}\checkmark_{I,c}}$$

$$\textbf{(enc0)}\ \frac{p[s''/s] \xrightarrow{c,s'} p'}{\texttt{sign}\ s\ \texttt{in}\ p\ \texttt{end} \xrightarrow{c,s'[s/s'']} \texttt{sign}\ s\ \texttt{in}\ p'[s/s'']\ \texttt{end}} \qquad \textbf{(enc1)}\ \frac{p[s''/s]\checkmark_{c}}{\texttt{sign}\ s\ \texttt{in}\ p\ \texttt{end}\checkmark_{I,c}}$$

s'' fresh in p and r

FIGURE 9.3 Structured operational semantics for Esterel (Part II: transition and termination).

Next, we show that Definition 9.6 indeed satisfies the intuition behind logical coherency by reexamining the examples introduced in Section 9.2.

Example 9.4:

Consider program P0, recalled below.

```
P0  pres i ? emit s ◇ 0 end ; pres s ? 0 ◇ emit o end
```

It is straightforward to check that the following is the semantics of $P0$:
$\{P0 \uparrow^{\{i^+\},P0,s},\ P0 \uparrow^{\{i^-\},P0,o},\ P0 \xrightarrow{\{i^+\},P0,i} 0,\ P0 \xrightarrow{\{i^-\},P0,o} 0,\ \texttt{emit}\ s \xrightarrow{\{i^+\},P0,s} 0,\ \texttt{emit}\ s \xrightarrow{\{i^-\},P0,s} 0,\ \texttt{emit}\ o \xrightarrow{\{i^+\},P0,s} 0, \texttt{emit}\ o \xrightarrow{\{i^-\},P0,s} 0, 0\checkmark_{\{i+\},P0}, 0\checkmark_{\{i-\},P0}\}$.*

Consider program P1 quoted below.

```
P1 pres s ? emit s ◇ 0 end
```

It has two supported models, namely $\{P1 \uparrow^{\emptyset,P1,s},\ P1 \xrightarrow{\emptyset,P1,s} 0,\ \texttt{emit}\ s \uparrow^{\emptyset,P1,s},\ \texttt{emit}\ s \xrightarrow{\emptyset,P1,s} 0,\ 0\checkmark_{\emptyset,P1}\}$ and $\{\texttt{emit}\ s \xrightarrow{\emptyset,P1,s} 0, \texttt{emit}\ s \uparrow^{\emptyset,P1,s}, 0\checkmark_{\emptyset,P1}\}$. Hence, $P1$ is not meaningful according to Semantics 9.1.

Consider program P2 recalled below.

```
P2 pres s ? 0 ◇ emit s end
```

Program P2 does not have any supported model: Assume, toward a contradiction, that $P2 \uparrow^{\emptyset,P2,s}$ is in the purported supported model of T. It then follows from item 1 in Definition 9.4 that there exists a deduction rule whose conclusion can match $P2 \uparrow^{\emptyset,P2,s}$ and whose premises are consistent with T. The only candidates are (**f0**) and (**f1**); we analyze both cases below and show that they both lead to a contradiction.

- (**f0**) The premises of the instance of (**f0**) are $P2 \uparrow^{\emptyset,P2,s}$ and $0 \uparrow^{\emptyset,P2,s}$. It follows from item 1 of Definition 9.4 that both predicates should be in T and hence, item 1 again applies to both predicates and in particular to $0 \uparrow^{\emptyset,P2,s}$. Hence, there should exist a deduction rule whose conclusions matches with the above predicate. A simple syntactic check on the deduction rules of Figures 9.2 and 9.3 reveals that none of the conclusions can be unified with the above predicate and hence a contradiction follows.
- (**f1**) The premises of the instance of (**f0**) are $\neg P2 \uparrow^{\emptyset,P2,s}$ and $\texttt{emit}\ s \uparrow^{\emptyset,P2,s}$, both of which should be in T. Again item 1 of Definition 9.4 applies and thus, $\neg P2 \uparrow^{\emptyset,P2,s}$ should be consistent with T, or in other words, $P2 \uparrow^{\emptyset,P2,s} \notin T$, which contradicts our initial assumption.

The next program to consider is P3, quoted below.

```
P3 pres s ? emit s ◇ emit s end
```

Program P3 is indeed meaningful and has the following unique supported model.
$\{P3 \uparrow^{\emptyset,P3,s}, P3 \xrightarrow{\emptyset,P3,s} 0, \texttt{emit}\ s \uparrow^{\emptyset,P3,s}, \texttt{emit}\ s \xrightarrow{\emptyset,P3,s} 0, 0\checkmark_{\emptyset,P3}\}$.

Note that $P3 \uparrow^{\emptyset,P3,s}$ (and/or the transition of $P3$) cannot be removed from the supported model; to see this, it follows from item 2 of Definition 9.4 and deduction rule (**e0**) that $\texttt{emit}\ s \uparrow^{\emptyset,P3,s} \in T$, and following the same reasoning and deduction rule (**f1**), we have that $P3 \uparrow^{\emptyset,P3,s} \in T$.

Program P4 is considered logically coherent but not constructive by the language designers. Next, we show that this intuition is indeed supported by our formal definitions.

```
P4    pres s0 ? emit s0 ◇ 0 end ||
    pres s0 ? pres s1 ? 0 ◇ emit s1 end ◇ 0 end
```

Program P4 has a unique supported model, given below.
$\{\texttt{emit}\ s_0 \uparrow^{\emptyset,P3,s_0}, \texttt{emit}\ s_0 \xrightarrow{\emptyset,P3,s_0} 0, \texttt{emit}\ s_1 \uparrow^{\emptyset,P3,s_1}, \texttt{emit}\ s_1 \xrightarrow{\emptyset,P3,s_1} 0, 0\checkmark_{\emptyset,P3}\}$.

Note that neither emission of s_0 nor s_1 cannot be present in a supported model. First, concerning s_1, suppose that s_1 can be emitted, then it follows from item 1 of Definition 9.4 that there should be a deduction rule supporting this emission. This can only be due to (**p1**) and thus, the right-hand-side component of the parallel composition. This component, in turn can only emit s_1 (due to deduction rules (**f0**) and then (**f1**)) if s_0 is present and s_1 is absent under the same context. The latter contradicts our

* For each program, we choose ι, ω and λ, respectively, to comprise only the input, output and local variables mentioned in the program at hand. This allows us to focus only on the possibly relevant part of I when considering supported models.

assumption. Similarly, suppose that the supported model contains a predicate (or transition) to the effect that s_0 can be emitted. We already know that no predicate for emitting s_1 can be in the supported model. Hence, it follows from successive application of item 2 of Definition 9.4 using deduction rules (**f0**), (**f1**) and (**e0**) that s_1 can be emitted under the same context, which is already shown to lead to contradiction.

P5 `pres` s_0 `? emit` $s_1 \diamond$ `0 end ; emit` s_0

Program P5 is also meaningful and has a unique supported model, given below.

$\{P5 \uparrow^{\emptyset,P5,s_0}, P5 \uparrow^{\emptyset,P5,s_1}, P5 \xrightarrow{\emptyset,P5,s_0} 0, P5 \xrightarrow{\emptyset,P5,s_1} 0,$
`pres` s_0 `? emit` $s_1 \diamond$ `0 end` $\uparrow^{\emptyset,P5,s_1}$, `pres` s_0 `? emit` $s_1 \diamond$ `0 end` $\xrightarrow{\emptyset,P5,s_1} 0,$
`emit` $s_0 \uparrow^{\emptyset,P5,s_0}$, `emit` $s_0 \xrightarrow{\emptyset,P3,s_0} 0$, `emit` $s_1 \uparrow^{\emptyset,P5,s_1}$, `emit` $s_1 \xrightarrow{\emptyset,P3,s_1} 0, 0\checkmark_{\emptyset,P5}\}$.

Note that none of the predicates or transitions concerning the emission of s_0 and s_1 can be omitted from the supported model. If the predicate (transition) concerning the emission of s_0 is omitted then the first component of sequential composition terminates and hence s_0 should be emitted due to the second component. Since s_0 should always be emitted, the emission of s_1 is guaranteed by the first component of sequential composition.

Definition 9.7: (Constructive Semantics of Esterel)

An Esterel program p is constructive if for each signal s and each input evaluation I either $p \uparrow^{I,p,s}$ and $p \xrightarrow{I,p,s} p'$ (for some p') or $\neg p \uparrow^{I,p,s}$ and $p \overset{I,c,s}{\not\rightarrow}$ has a supported proof and moreover, either $p\checkmark_{I,p}$ or $\neg p\checkmark_{I,p}$ has a supported proof.

To illustrate this semantics and identify its differences with the logical semantics, we reconsider those programs whom are considered nonconstructive but logically coherent in Section 9.2.

Example 9.5:

Consider program P3. This program is both intuitively and formally shown to be logically coherent. Moreover, in Section 9.2, we introduced this program as a canonical example of a nonconstructive program. Next, we show that it is also formally nonconstructive since neither $P3 \uparrow^{\emptyset,P3,s}$ nor $\neg P3 \uparrow^{\emptyset,P3,s}$ have a supported proof (a similar reasoning shows that neither $P3 \xrightarrow{\emptyset,P3,s} p'$ for any p' nor $P3 \overset{\emptyset,P3,s}{\not\rightarrow}$ have a supported proof). Suppose $P3 \uparrow^{\emptyset,P3,s}$ has a supported proof, then its proof is either due to (**f0**) or (**f1**). In the former case, the nodes placed above our proof obligation are $P3 \uparrow^{I,P3,s}$ and `emit` $s \uparrow^{\emptyset,P3,s}$. While the latter has a supported proof (due to (**e0**)), the former was our original proof obligation, thus, it only remains to check the alternative option due to (**f1**). The premises of (**f1**) are then $\neg P3 \uparrow^{I,P3,s}$ and `emit` $s \uparrow^{\emptyset,P3,s}$. Again the latter formula has a supported proof but the former is the negation of our proof obligation and thanks to Theorem 9.2, we know that if $\neg P3 \uparrow^{I,P3,s}$ has a supported proof then $P3 \uparrow^{I,P3,s}$ cannot have a supported proof. Similarly, if $\neg P3 \uparrow^{I,P3,s}$ has a supported proof, then a negation of a premise of all deduction rules that can match $P3 \uparrow^{I,P3,s}$ must have a supported proof. These two rules are again (**f0**) and (**f1**). The negation of the common premise of these two rules, that is, `emit` $s \uparrow^{\emptyset,P3,s}$ cannot have a supported proof (following Theorem 9.2, because the premise itself has a supported proof). Hence a negation of both $P3 \uparrow^{I,P3,s}$ and $\neg P3 \uparrow^{I,P3,s}$ should have supported proofs, which is again impossible due to Theorem 9.2.

Program P4 is not constructive since neither $P4 \uparrow^{\emptyset,P4,s_0}$ nor its negation has a supported proof. The only possible proof for the emission predicate can be due to (**p0**) or (**p1**). The case for (**p1**) does not lead to a supported proof since the right-hand side does not contain any emit statement for s_0. If the supported proof is due to (**p0**), then it should hold that $P4 \uparrow^{\emptyset,P4,s_0}$ which was to be proven. The negation of the predicate, that is, $P4 \uparrow^{\emptyset,P4,s_0}$ does not have a supported proof, either. Since then a negation of a premise of (**p0**) and (**p1**) should have a supported proof. The negation of the only premise of (**p0**) is $\texttt{pres}\ s_0\ \texttt{?}\ \texttt{emit}\ s_0 \diamond \texttt{0 end} \uparrow^{\emptyset,P4,s_0}$, which in turn means that a negation of a premise of (**f0**) or (**f1**) must have a supported proof. Consider (**f0**), its two premises are $P4 \uparrow^{\emptyset,P4,s_0}$, but we were seeking a proof of its negation and $\texttt{emit}\ s_0 \uparrow^{\emptyset,P4,s_0}$, whose negation cannot be proven.

Program P5 is not constructive, either. We next show that neither $P5 \uparrow^{\emptyset,P5,s_0}$ nor its negation are provable. The purported supported proof for predicate $P5 \uparrow^{\emptyset,P5,s_0}$ is due to one of the rules (**s0**) to (**s2**). Next, we analyze each case and show that it leads to a contradiction.

- (**s0**) Then, it should hold that $\texttt{pres}\ s_0\ \texttt{?}\ \texttt{emit}\ s_1 \diamond \texttt{0 end} \uparrow^{\emptyset,P5,s_0}$. This, in turn, can be either due to (**f0**) or (**f1**). If the predicate is due to (**f0**), then we should have a supported proof for $P5 \uparrow^{\emptyset,P5,s_0}$, which was to be proven. If the proof is due to (**f1**), then $\neg P5 \uparrow^{\emptyset,P5,s_0}$ should have a supported proof, which is impossible due to Theorem 9.2.
- (**s1**) Then, it should hold that $\texttt{pres}\ s_0\ \texttt{?}\ \texttt{emit}\ s_1 \diamond \texttt{0 end}\checkmark_{\emptyset,P5}$. This termination can be due to either (**if4**) or (**if5**). None of these two are possible since otherwise, respectively, $P5 \uparrow^{\emptyset,P5,s_0}$ or $\neg P5 \uparrow^{\emptyset,P5,s_0}$ should have a supported proof.
- (**s2**) Then, it should hold that $\texttt{pres}\ s_0\ \texttt{?}\ \texttt{emit}\ s_1 \diamond \texttt{0 end}\checkmark_{\emptyset,P5,s'}p'$ for some s' and p'. This transition is due to either (**if0**) or (**if1**). Again, both cases lead to a contradiction due to a similar reasoning as in item (**s0**).

As a side note, the common intuition and the similarities between deduction rules of Figures 9.2 and 9.3 may suggest that we can replace deduction rules of Figure 9.2 with the following rule (or even do without the emission predicates and make the same changes in the deduction rule for if-then-else statements in Figure 9.3):

$$(\textbf{emit})\dfrac{p \xrightarrow{I,c,x} p'}{p \uparrow^{I,c,x}}$$

This change leads to a much more restrictive semantics, which is unable to provide supported proofs for transitions of perfectly acceptable programs such as the following:

P6 $\texttt{pres s ? emit o} \diamond \texttt{0 end || emit s}$

To see this, the reader may try to prove that P6 can emit signal o using deduction rule (**par0**). The proof of the premise of (**par0**) then should rely on (**if0**) and hence due to deduction rule (**emit**), we need to prove that s can be emitted (for the if–then–else to be able to take a transition). In turn, this can only be due to (**par1**). But to apply (**par1**), we need to know that the left-hand-side component can take a transition (in order to record its target), which is what we wanted to prove initially. This cycle is broken in our semantics, by deduction rule (**p1**) which only considers one of the two components to infer the emission of s_0 (without trying to record the target of the transition). The following proof illustrates why this program is indeed constructive.

$$\dfrac{\dfrac{\dfrac{\dfrac{}{\texttt{emit s} \uparrow^{\emptyset,P6,s}}}{P6 \uparrow^{\emptyset,P6,s}} \qquad \dfrac{}{\texttt{emit o} \xrightarrow{\emptyset,P6,o} 0}}{\texttt{pres s ? emit o} \diamond \texttt{0 end} \xrightarrow{\emptyset,P6,o} 0} \qquad \dfrac{}{\texttt{emit s} \xrightarrow{\emptyset,P6,s} 0}}{P6 \xrightarrow{\emptyset,P6,o} 0 \mid\mid 0}$$

9.6 Conclusions and Future Work

In this chapter, we presented a brief overview of the Esterel core language. Subsequently, we presented a link between the intuitive notions of logical coherency and constructiveness in the semantics of Esterel on the one hand, and the formal notions of supported models and supported proofs in the semantics of Structured Operational Semantics, on the other hand. By means of several canonical examples from the literature, we showed that our formal definitions indeed capture the intuitive criteria put forward by the language designers.

Several formalizations of these two intuitive criteria exist in the literature. For example [Ber99, PBEB07] present three formalizations of constructive semantics of Esterel. In [Tin00, Tin01] another formalization of constructive semantics of Esterel is presented and is proven to coincide with one of the notions in [Ber99]. A rigorous comparison between all these notions and the ones presented in this chapter remains as a topic for future research.

In the formal semantics presented in this chapter, we abstracted from the issues of exceptions (traps), loops and time. We expect that one can include these aspects without any substantial change in the semantics presented in this chapter using the modular semantics approach of [Mos04, MN08]. This remains as another interesting exercise for the future.

Acknowledgments

Inspiring discussions with Jean-Pierre Talpin and Paul Le Guernic are gratefully acknowledged. The author would like to thank the anonymous reviewers of SOS 2009 for their insightful reviews.

References

[AB94] K. R. Apt and R. N. Bol. Logic programming and negation: A survey. *Journal of Logic Programming (JLAP)*, 19/20:9–71, 1994.

[AFV01] L. Aceto, W. Jan (Wan) Fokkink, and C. Verhoef. Structural operational semantics. In J. A. Bergstra, A. Ponse, and S. A. Smolka, editors, *Handbook of Process Algebra, Chapter 3*, pp. 197–292. Elsevier Science, Dordrecht, The Netherlands, 2001.

[Ber99] G. Berry. *The Constructive Semantics of Pure Esterel*. 1999. Draft version, available from: ftp://ftp-sop.inria.fr/meije/esterel/papers/constructiveness3.ps.gz.

[BG92] G. Berry and G. Gonthier. The Esterel synchronous programming language: Design, semantics, implementation. *Science of Computer Programming (SCP)*, 19(2):87–152, 1992.

[BBFLNS00] G. Berry, A. Bouali, X. Fornari, E. Ledinot, E. Nassor, and R. de Simone. ESTEREL: A formal method applied to avionic software development. *Sci. Comput. Program.* 36(1): 5–25, 2000.

[BG96] R. N. Bol and J. F. Groote. The meaning of negative premises in transition system specifications. *Journal of the ACM (JACM)*, 43(5):863–914, September 1996.

[Gla04] R. J. (Rob) van Glabbeek. The meaning of negative premises in transition system specifications II. *Journal of Logic and Algebraic Programming (JLAP)*, 60–61:229–258, 2004.

[Gro93] J. F. Groote. Transition system specifications with negative premises. *Theoretical Computer Science (TCS)*, 118(2):263–299, 1993.

[GV92] J. F. Groote and F. W. Vaandrager. Structured operational semantics and bisimulation as a congruence. *Information and Computation (I&C)*, 100(2):202–260, October 1992.

[GGBM91] P. Le Guernic, T. Gautier, M. Le Borgne, and C. Le Maire. Programming real-time applications with SIGNAL. *Proceedings of the IEEE*, 79(9):1321–1336, 1991.

[GTL03] P. Le Guernic, J.-P. Talpin, and J.-C. Le Lann. Polychrony for system design. *Journal for Circuits, Systems and Computers*, 12(3): 261–304, 2003.

[Halbwachs05] N. Halbwachs. A synchronous language at work: The story of Lustre. In *Proceedings of the 3rd ACM & IEEE International Conference on Formal Methods and Models for Co-Design (MEMOCODE 2005)*, pp. 3–11, Verona, Italy, IEEE, 2005.

[HLR92] N. Halbwachs, F. Lagnier, and C. Ratel. Programming and verifying real-time systems by means of the synchronous data-flow language LUSTRE. *IEEE Trans. Software Eng.*, 18(9):785–793, 1992.

[MN08] P. D. Mosses and M. J. New. Implicit propagation in structural operational semantics. In *Proceedings of the 5th Workshop on Structural Operational Semantics (SOS'08)*, Reykjavik, Iceland, pp. 78–92, 2008.

[Mos04] P. D. Mosses. Modular structural operational semantics. *Journal of Logic and Algebraic Programming (JLAP)*, 60–61:195–228, 2004.

[PB02] D. Potop-Butucaru. *Optimizations for Faster Simulation of Esterel Programs*. PhD thesis, École des Mines de Paris, CMA, Paris, France, 2002.

[PBEB07] D. Potop-Butucaru, S. A. Edwards, and G. Berry. *Compiling Esterel*. Springer-Verlag, Berlin-Heidelberg, Germany, 2007.

[Plo04] G. D. Plotkin. The origins of structural operational semantics. *Journal of Logic and Algebraic Programming (JLAP)*, 60:3–15, 2004.

[TdS05] O. Tardieu and R. de Simone. Loops in Esterel. *ACM Transactions on Embedded Computing Systems (ACM TECS)*, 4:708–750, 2005.

[Tin00] S. Tini. *Structural Operational Semantics for Synchronous Languages*. PhD thesis, Dipartimento di Informatica, Università degli Studi di Pisa, Pisa, Italy, 2000.

[Tin01] S. Tini. An axiomatic semantics for Esterel. *Theoretical Computer Science (TCS)*, 269(1–2):231–282, 2001.

10

Regular Path Queries on Graph-Structured Data

Alex Thomo and S. Venkatesh
University of Victoria

10.1 Introduction

Graph-Structured Data (GSD) are abundant today in areas such as communication and traffic networks, web information systems, digital libraries, social networks, biological data management, and so on. GSD is easy to deal with as there is no need to worry about making the data fit rigid relational tables. They can be constructed on the fly using underlying data sources that we already have in hand. They can also be easily merged to create bigger databases, once they agree in a common node referencing scheme. Also, there are ongoing, enormous efforts for capturing the immense knowledge available on the Web and representing it in the form of "subject-predicate-object" triples, which are in essence "node-label-node" graph edges. The graphs they form are the well-known RDF graphs (often called triplestores) and they are immensely popular today (cf. [8,16]). Semistructured data (SSD) is another example of graph-structured data. Querying and reasoning on SSD has been the theme of a very extensive and multifaceted research for the last decade.

10.1.1 Querying GSD

GSD allows for flexible, user-friendly, query mechanisms. As pointed out by seminal works in the field (cf. [3,5,7,9,14]), regular path queries (RPQs) are the "winner" when it comes to expressing navigational recursion over graph data. Notably, RPQs are the ubiquitous building blocks of virtually all the query languages for GSD. RPQs are in essence regular expressions (REs) over the alphabet of the database edges. As REs have been around from the early days of computers, RPQs provide an easy to learn mechanism to query GSD.

Let us illustrate these concepts with an example from social networks. Consider the user network of Facebook. This network is already a graph, but let us suppose that we are interested in studying the influence of one user on other users. In order to detect influence, let us consider the sharing of videos posted online. Once a user posts a video, his friends can click the "share" link underneath the posted video and share the video with their own friends in turn. In this way, a video can become viral and be known to a large set of users, far beyond the set of friends of the the user who originally posted the video. We can say that "a user *A* probably influences a user *B*," if *B* has clicked "share" on the videos posted by *A* a considerable amount of times, say 10 or more times. In the same vein, if this number is, say from 8 to 10, we can say "*A* perhaps influences *B*." What is interesting is that we can create an overlay graph-structured database over the Facebook social network. The nodes of this database are the users. The edges are directed and express user influence. For instance, in our example we introduce an edge between *A* and *B*. We also label this edge depending on the number of times *B* has clicked "share" on the videos posted by *A*, specifically, *red* for 10 or more times, *orange* for 8 or 9 times, *yellow* for 6 or 7 times, and *green* for 4 or 6 times. A regular path query now can be

$$Q = (red + orange)^* \parallel (yellow + \epsilon)^3 \parallel (green + \epsilon)^2,$$

where $+$, $*$, and $||$ are the "or," "Klene star," and "shuffle" operators in regular expressions, ϵ is the empty word, and exponents 2 and 3 are shorthand for expressing concatenation, that many times, of the subexpressions they apply to with themselves. This query will bring as a result the set of all the pairs (A, C) of users such that there exists a path from *A* to *C* whose edges are *red* or *orange* and tolerating up to three *yellow* and up to two *green* edges. Plausibly, for each such (A, C) pair, we can say that *A* transitively influences *C*. Also, observe that the graph-structured database in this example does not need to be explicitly computed and stored. The existence of an edge, and its color, can be determined on the fly.

10.1.2 Database Services

We focus on three important database services for RPQs. These are: (i) Query Answering (QA), (ii) Query Containment (QC), and (iii) Query answering using Views (QV). The first two are easy. QA is casted to the rechability problem in a special Cartesian product graph constructed from the database graph and a query automaton. QC is shown to coincide with the containment of the underlying regular languages for the given RPQs. On the other hand, QV is much more challenging and, therefore, the thrust of our chapter will be mostly on this problem.

Let us discuss QV in some more detail. Suppose that we do not have a database available. Rather, what we have is a set of views on the possible data. These views represent partial information about the database and are expressed by regular expressions as well. For example, we could be given two views with definitions $V_1 = (red + orange)^*$ and $V_2 = (red + orange + yellow + green)^2$. The view definitions are nothing else but regular path queries. Additionally, for each view, we are given a set of pairs that represent the answer to these views (considering them as RPQs).

This is the classical scenario in view-based query answering (cf. [2,3,5,7,10,12,13]). The basic problem in this setting is to be able to answer a given query using only the available view information. Answering queries using views is typically achieved by reformulating the query in terms of the view definitions and then evaluating it on the provided view data. For example, the above query *Q* can be reformulated (or rewritten) as $Q' = V_1 \cdot V_2 \cdot V_1$. Then, if there are pairs (a, b), (b, c), and (c, d) associated with V_1, V_2, and (again) V_1, respectively, we produce (a, d) as an answer to *Q*.

There are two lines of research for answering RPQs using views. Works in the first line of research study the computation of "rewritings," which are RPQs over the view names (cf. [5,11]). Having such rewritings is desirable because they enable query answering in polynomial time with respect to the size of the data. On the other hand, works in the second line of research study the "perfect" (or "certain") answering of RPQs, which asks for all the answer tuples obtainable on every possible database consistent with the views (cf. [3,4,12]). However, obtaining all the certain answer is much more computationally expensive. As it has been shown in Ref. [3], to decide whether a tuple belongs in the certain answer is coNP-complete with respect to the size of data.

The answer obtained by using a (query) rewriting is a subset (sometimes strict) of the certain answer, and thus, by using a rewriting we get a lower approximation, which might be an acceptable compromise given the data complexity of computing the full certain answer (see [7] for a discussion).

The most important cornerstone in the rewriting of RPQ's using views is the work by Calvanese et al. [5], which shows that the rewriting is indeed possible by giving an algorithm for computing it. The complexity of computing the (maximal) view-based rewriting of a regular path query Q is shown to be in 2EXPTIME (in the size of the query) and this bound is also shown to be tight ([5]). Also, in Ref. [5] it is shown that the size of the automaton for the rewriting can be doubly exponential in the size of the query Q as measured by the size of a simple NFA for Q. This is a high complexity, and therefore, one has to seek practical ways to be able to get the benefits of the rewritings without explicitly constructing it in full entirety. Such an approach is given by Ref. [17], and explained in this chapter as well.

The rest of the chapter is organized as follows. In Section 10.2, we formally define graph-structured databases, regular path queries, and the QA and QC problems. In Section 10.3, we discuss the QV problem. Next, in Section 10.4, we examine the algorithm of [5] for obtaining maximal view-based rewritings. Then, in Section 10.5 we present the optimization techniques of Ref. [17].

10.2 Graph-Structured Databases and Regular Path Queries

We consider a database to be an edge labeled graph. This graph model is typical in GSD, where the nodes of the database graph represent the objects and the edges represent the attributes of the objects, or relationships between the objects.

Example 10.1:

Let us consider here another example, this time about an online store, which sells books and software products. We show in Figure 10.1 a database with information about books and software products. The edges can be stored as a "node-label-node" triples, also called "subject–predicate–object" triple in the RDF world (cf. [8,16]).

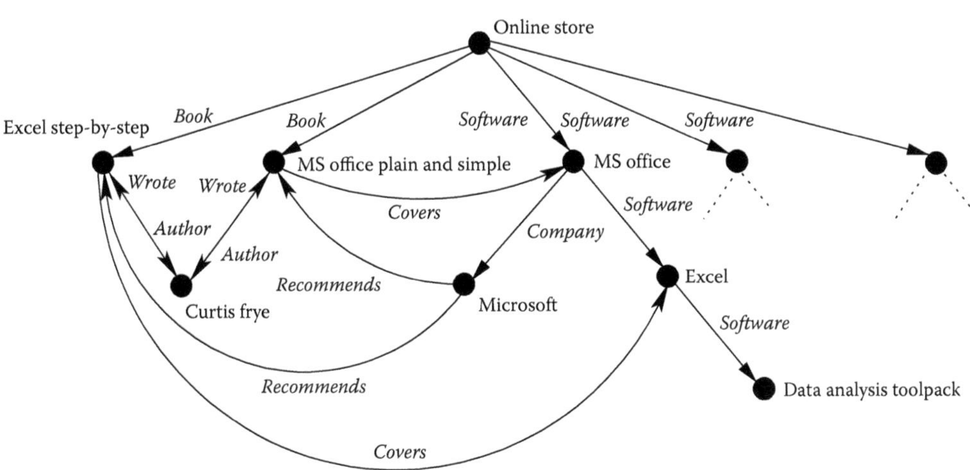

FIGURE 10.1 A graph database.

Now, a book has an author and covers some software product(s). A software product has a company and possibly other software subproducts. A company might recommend some books for its products. The database is semistructured because the schemas of its objects are not rigid. For example, a company can only optionally recommend books, or we might be missing information about what products a book might cover.

Formally, let Δ be a finite alphabet. We shall call Δ the *database alphabet*. Elements of Δ will be denoted $R, S, \ldots$. As usual, Δ^* denotes the set of all finite words over Δ. Words will be denoted by $u, w, \ldots$. We also assume that we have a universe of objects, and objects will be denoted $a, b, c, \ldots$. A *database DB* over Δ is a subset of $N \times \Delta \times N$, where N is a set of objects, that we usually will call nodes. We view a database as a directed labeled graph, and interpret a triple (a, R, b) as a directed edge from object a to object b, labeled with R.

If there is a path labeled $R_1, R_2, \ldots, R_k$ from a to b, we write $a \stackrel{R_1R_2\ldots R_k}{\longrightarrow} b$.

A *(user) query* Q is a regular language over Δ. For the ease of notation, we will blur the distinction between regular languages and regular expressions that represent them. Let Q be a query and DB a database. Then, the *answer* to Q on DB is defined as

$$\text{ans}(Q, DB) = \{(a, b) : a \stackrel{w}{\longrightarrow} b \text{ in } DB \text{ for some } w \in Q\}.$$

Example 10.2:

Consider again the database in Figure 10.1. Suppose that the user would like to know for each software product, all the books that might have some useful information about the product. For this, the user can give the regular path query $Q = covers \cdot software^*$. This query, on the database DB in Figure 10.1, will have as answer

$$\begin{aligned}\text{ans}(Q, DB) = \{&(\text{MS Office Plain \& Simple, MS Office}),\\ &(\text{MS Office Plain \& Simple, Excel}),\\ &(\text{MS Office Plain \& Simple, Data Analysis Toolpack}),\\ &(\text{Excel Step-by-Step, Excel}),\\ &(\text{Excel Step-by-Step, Data Analysis Toolpack})\}\end{aligned}$$

10.2.1 Query Answering

The well-known method for answering RPQ's on a given database (cf. [1]) is as follows. In essence, we create state-object pairs from the query automaton and the database. For this, let $\mathcal{A}$ be an NFA that accepts an RPQ Q. Starting from an object a of a database DB, we first create the pair (p_0, a), where p_0 is the initial state in $\mathcal{A}$. Then, we create all the pairs (p, b) such that there exist a transition from p_0 to p in $\mathcal{A}$, and an edge from a to b in DB, and furthermore the labels of the transition and the edge match. In the same way, we continue to create new pairs from existing ones, until we are not anymore able to do so. In essence, what is happening is a lazy construction of a Cartesian product graph of the query automaton with the database graph. Of course, only a small (hopefully) part of the Cartesian product is really constructed depending on the selectivity of the query.

After obtaining the above Cartesian product graph, producing query answers becomes a question of computing reachability of nodes (p, b), where p is a final state, from (p_0, a), where p_0 is the initial state. Namely, if (p, b) is reachable from (p_0, a), then (a, b) is a tuple in the query answer.

10.2.2 Query Containment and Equivalence

We say that a query Q_1 is contained in (equivalent to) another query Q_2 if the answer to the former is always contained in (equal to) the answer to the latter on any database *DB*. Formally, we have

---

Definition 10.1:

1. $Q_1 \sqsubseteq Q_2$ if $\text{ans}(Q_1, DB) \subseteq \text{ans}(Q_2, DB)$ for each *DB* over Δ.
2. $Q_1 \equiv Q_2$ if $Q_1 \sqsubseteq Q_2$ and $Q_2 \sqsubseteq Q_1$.

In the following, we give the language theoretic lemma of Ref. [6], which shows that for RPQs, query containment (equivalence) coincides with language containment (equality).

Theorem 10.1: ([6])

$Q_1 \sqsubseteq Q_2$, if and only if, $Q_1 \subseteq Q_2$.

Proof. The "if" direction is straightforward. The "only if" direction is also easy. Let $w \in Q_1$. Consider the "chain" database *DB*, with a single path spelling w and connecting an object a with an object b. We have $(a, b) \in \text{ans}(Q_1, DB)$. From $Q_1 \sqsubseteq Q_2$, we get $(a, b) \in \text{ans}(Q_2, DB)$. From the definition of query answering, we get in turn that $w \in Q_2$. □

Therefore, we can use $\subseteq$ ($=$) to denote both query containment (equivalence) and language containment (equality).

10.3 Query Answering Using Views

Let $V_1, \ldots, V_n$ be languages (queries) on alphabet Δ. We will call them *views* and associate with each V_i a view name v_i.

We call the set $\Omega = \{v_1, \ldots, v_n\}$ the *outer alphabet*, or *view alphabet*. For each $v_i \in \Omega$, we set $def(v_i) = V_i$. The substitution *def* associates with each view name v_i in the Ω alphabet the language V_i. The substitution *def* is applied to words, languages, and regular expressions in the usual way (see, e.g., Ref. [18]).

A *view graph* is database $\mathcal{V}$ over Ω. In other words, a view graph is a database where the edges are labeled with symbols from Ω. View graphs can also be queried by regular path queries over Ω. However, as explained below these are not queries given by the user, but rather rewritings computed by the system.

The user queries are expressed on alphabet Δ, and the system has to answer based solely on the information provided by the views. In order to do this, the system has to reason with respect to the set of *possible databases* over Δ that $\mathcal{V}$ could represent. Under the *sound view* assumption, a view graph $\mathcal{V}$ defines a set $poss(\mathcal{V})$ of databases as follows:

$$poss(\mathcal{V}) = \{DB : \mathcal{V} \subseteq \bigcup_{i \in \{1,\ldots,n\}} \{(a, v_i, b) : (a, b) \in \text{ans}(V_i, DB)\}\}.$$

(Recall that $V_i = \text{def}(v_i)$.) The above definition reflects the intuition about the connection between an edge (a, v_i, b) in $\mathcal{V}$ with some path from a to b in the possible *DB*'s, labeled by some word in V_i.

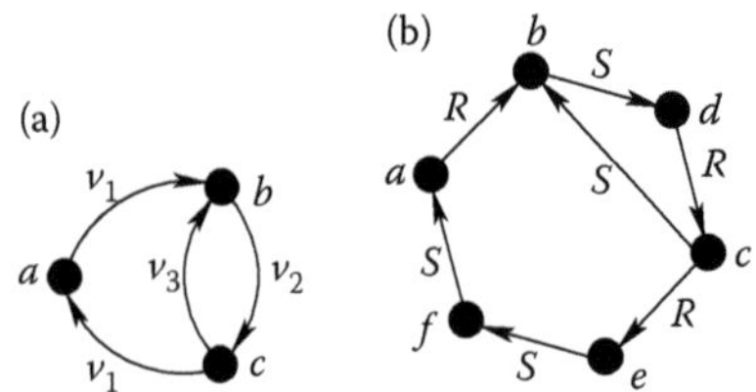

FIGURE 10.2 A view graph (a) and a possible database (b).

Example 10.3:

Consider the view graph in Figure 10.2a, and view definitions $V_1 = \text{def}(v_1) = RS^*$, $V_2 = \text{def}(v_2) = S^*R$, and $V_3 = \text{def}(v_3) = S^+$. Then, a possible database is shown in Figure 10b. Observe that the views are sound only. They are not required to be complete. For example, we do not have a v_2-edge from f to b in the view graph. In fact, we do not even have an f object in the view graph. We remark that view soundness is usually the only "luxury" that we have in information integration systems, where the information is often incomplete. □

The meaning of querying a view graph through the global schema Δ is defined as follows. Let Q be a query over Δ. Then

$$\text{ANS}(Q, \mathcal{V}) = \bigcap_{DB \in poss(\mathcal{V})} \text{ans}(Q, DB).$$

There are two approaches for computing $\text{ANS}(Q, \mathcal{V})$. The first one is to use an *exponential* procedure in the size of the data (i.e. $\mathcal{V}$) in order to completely compute $\text{ANS}(Q, \mathcal{V})$ (see Ref. [3]). There is little that one can better hope for, since in the same paper it has been proven that to decide whether a tuple belongs to $\text{ANS}(Q, \mathcal{V})$ is co-NP complete with respect to the size of data.

The second approach is to compute first a view-based rewriting Q' for Q, as in Ref. [5]. Such rewritings are regular path queries on Ω. Then, we can approximate $\text{ANS}(Q, \mathcal{V})$ by $\text{ans}(Q', \mathcal{V})$, which can be computed in polynomial time with respect to the size of data ($\mathcal{V}$). In general, for a view-based rewriting Q' computed by the algorithm of [5], we have that

$$\text{ans}(Q', \mathcal{V}) \subseteq \text{ANS}(Q, \mathcal{V}),$$

with equality when the rewriting is exact ([3]). In the rest of the chapter, we will focus on the rewriting approach.

10.4 Maximal View-Based Rewritings

We first examine "maximal view-based rewritings" and the method of Ref. [5] for their computation.

Formally, for a given query Q, the maximal view-based rewriting Q', is the set of *all* words on Ω such that their substitution through def is contained in the query language Q, that is,

$$Q' = \{w : w \in \Omega^* \text{ and } def(w) \subseteq Q\}.$$

Interestingly, as shown in Ref. [5], the above set is a regular language on Ω and the algorithm of Ref. [5] for computing an automaton for this language is as follows.

Algorithm 10.1:

1. Construct a DFA $\mathcal{A} = (\Delta, S, s_0, \tau_{\mathcal{A}}, F)$ such that $Q = L(\mathcal{A})$.
2. Construct automaton $\mathcal{B} = (\Omega, S, s_0, \tau_{\mathcal{B}}, S - F)$, where $(s_i, v_x, s_j) \in \tau_{\mathcal{B}}$ iff there exists $w \in V_x$ such that $(s_i, w, s_j) \in \tau^*_{\mathcal{A}}$.
3. The rewriting Q' is the Ω language accepted by an automaton $\mathcal{C}$ obtained by complementing automaton $\mathcal{B}$.

Step 2 can also be expressed equivalently as: Consider each pair of states (s_i, s_j). If in $\mathcal{A}$ there is a path from s_i to s_j, which spells a word in some view language V_x, then insert a corresponding v_x-transition from s_i to s_j in $\mathcal{B}$.

Example 10.4:

Let query be $Q = (RS^*)^2 + S^+$ and the view definitions be as in Example 10.3, that is, $V_1 = \text{def}(v_1) = RS^*$, $V_2 = \text{def}(v_2) = S^*R$, and $V_3 = \text{def}(v_3) = S^+$. The DFA $\mathcal{A}$ for the query Q is shown in Figure 10.3a and the corresponding automaton $\mathcal{B}$ is shown in Figure 10.3b. The resulting complement automaton $\mathcal{C}$ is shown in Figure 10.3c. Note that the "trap" and unreachable states have been removed for clarity. □

Observe that, if $\mathcal{B}$ accepts an Ω-word $v_1 \ldots v_m$, then there exist m Δ-words $w_1, \ldots, w_m$ such that $w_i \in V_i$ for $i = 1, \ldots, m$ and such that the Δ-word $w_1 \ldots w_m$ is rejected by $\mathcal{A}$. On the other hand, if there exists a Δ-word $w_1 \ldots w_m$ that is rejected by $\mathcal{A}$ such that $w_i \in V_i$ for $i = 1, \ldots, m$, then the Ω-word

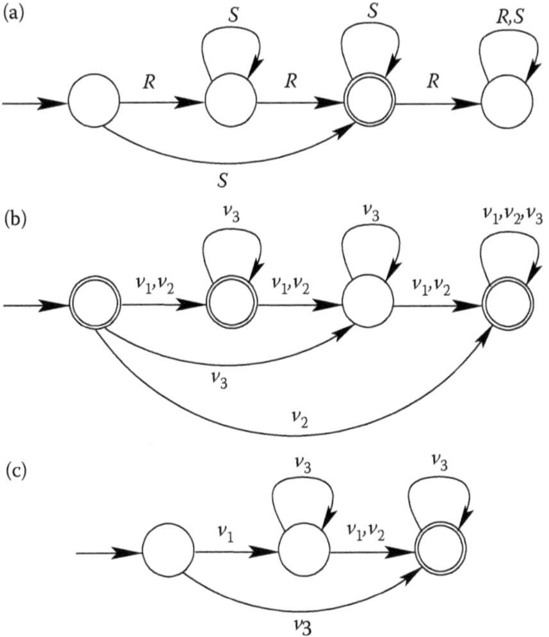

FIGURE 10.3 Automata $\mathcal{A}$ (a), $\mathcal{B}$ (b), and $\mathcal{C}$ (c).

$v_1 \dots v_m$ is accepted by $\mathcal{B}$. That is, $\mathcal{B}$ accepts an Ω-word $v_1 \dots v_m$ if and only if there is a Δ-word in $\text{def}(v_1 \dots v_m)$ that is rejected by $\mathcal{A}$. Hence, $\mathcal{C}$ is being the complement of $\mathcal{B}$ accepts an Ω-word if and only if all Δ-words $w = w_1 \dots w_m$ such that $w_i \in V_i$ for $i = 1, \dots, m$, are accepted by $\mathcal{A}$.

As mentioned in the previous section, the rewriting Q' represented by automaton $\mathcal{C}$ is evaluated on a view graph $\mathcal{V}$ obtaining $\text{ans}(Q', \mathcal{V})$ which is an approximation of $\text{ANS}(Q, \mathcal{V})$.

Example 10.5:

Consider the rewriting Q' represented by the automaton $\mathcal{C}$ in Figure 10.3c, and the view graph $\mathcal{V}$ in Figure 10.2a. It is easy to see that $\text{ans}(Q', \mathcal{V}) = \{(a, b), (a, c), (c, b)\}$.

Assuming that the user query is given by means of a regular expression, [5] showed, using the algorithm above, that the complexity of computing the maximal view-based rewriting is in 2EXPTIME. Moreover, this bound was shown to be tight by constructing a query instance Q, whose rewriting has a doubly exponential size compared to the size of a simple NFA for Q.

10.5 Optimization Techniques

The above 2EXPTIME bound tells us that to obtain a view-based rewriting is computationally hard except for very small query instances. While the first determinization [for obtaining automaton $\mathcal{A}$] is in practice quite tolerable for typical user queries, the second determinization [for obtaining automaton $\mathcal{C}$ by complementing $\mathcal{B}$] is often prohibitively expensive. However, it is possible to argue that the analysis in Ref. [5] is worst case and hence Algorithm 10.1 might take only reasonable amount of time on "typical" instances (or on the average). Our experimental results indicate that this is not the case (see Section 10.6). Experimentally, we were unable to compute automaton $\mathcal{C}$, in reasonable time and space, for about one third of the time while working on "randomly generated" instances.

In this section, we first describe how to optimize the construction of automaton $\mathcal{B}$ in step 2, which is a significant bottleneck when the number of views is considerable. Then, we deal with step 3 of the algorithm, and propose a technique which essentially eliminates this step.

10.5.1 Computing Automaton $\mathcal{B}$ Efficiently

We present an optimization technique for the step 2 of the above algorithm for computing automaton $\mathcal{B}$. In our experiments we observed that this step, although a polynomial one, is very time consuming if implemented in the straightforward manner.

Taking a closer look at step 2, let s_i and s_j be two arbitrary states in automaton $\mathcal{A}$. Now consider automaton $\mathcal{A}_{ij}$, which is obtained by keeping all the states and transitions in $\mathcal{A}$, but making state s_i and s_j initial and final, respectively. All the other states in $\mathcal{A}_{ij}$ are neither initial nor final.

In step 2 of the algorithm, we want to determine whether there should be transition v_x between states s_i and s_j in $\mathcal{B}$. It is easy to see that this is in fact achieved by testing for the emptiness of the intersection $L(\mathcal{A}_{ij}) \cap V_x$. Namely, we insert a transition (s_i, v_x, s_j) in $\mathcal{B}$ iff $L(\mathcal{A}_{ij}) \cap V_x \neq \emptyset$. Recall that the intersection $L(\mathcal{A}_{ij}) \cap V_x$ is obtained by constructing the Cartesian product $\mathcal{A}_{ij} \times \mathcal{A}_{V_x}$, where $\mathcal{A}_{V_x}$ is an automaton for V_x.

However, the automata $\mathcal{A}_{ij}$ for different i's and j's have the same states and transitions [namely those of automaton $\mathcal{A}$]. Only their initial and final states are different. Thus, we construct only one Cartesian product $\mathcal{A} \times \mathcal{A}_{V_x}$ for a given view V_x. Then, we test emptiness on this Cartesian product automaton for $|\mathcal{A}|^2$ different combinations of [one] initial and [one] final states. Although asymptotically there is no gain in doing this, experimentally, we found that for typical queries and views, the speedup achieved by this optimization is often more than sixfold. This is explained by a better utilization of the CPU cache because there is only one Cartesian product automaton to be constructed and examined.

10.5.2 Answer Computation Using Automaton $\mathcal{B}$

In this subsection, we describe how to essentially eliminate step 3 of the algorithm of Ref. [5].

Recall that the "riskier penalty" in the algorithm of Ref. [5] is the computation of automaton $\mathcal{C}$ in step 3 by complementing the automaton $\mathcal{B}$ obtained in step 2. $\mathcal{C}$ might be doubly exponential in the size of the query. Once $\mathcal{C}$ is computed, the final step is to compute ans$(Q', \mathcal{V})$ by constructing the Cartesian product of the automaton $\mathcal{C}$ and a viewgraph $\mathcal{V}$. We ask if it still possible to compute ans$(Q', \mathcal{V})$ directly without first computing the DFA for $L(\overline{\mathcal{B}})$? We achieve this by merging the underlying determinization procedure of step 3 and the subsequent computation of the Cartesian product graph into a single step. We illustrate this using an example.

Example 10.6:

Consider the NFA $\mathcal{B}$ and the viewgraph $\mathcal{V}$ shown in Figure 10.4 (top–left) and (top–right), respectively. To compute the set of all objects reachable, say from a in $\mathcal{V}$ and following paths spelling words rejected by $\mathcal{B}$, we will build a lazy Cartesian product graph, whose nodes are object–bitvector pairs and edges are labeled with Ω symbols.

We start with the pair $(a, 100)$, where 100 is an abbreviation for $(1, 0, 0)$. This bitvector says that automaton $\mathcal{B}$ is now in state s_0. Next, we construct the pair $(b, 110)$ and put a v_1–edge from $(a, 100)$ to $(b, 110)$. This is because when reading symbol v_1, we hope to object b in $\mathcal{V}$, and in states s_0 and s_1 in $\mathcal{B}$. Continuing in this way, we obtain the Cartesian product graph shown in Figure 10.4 (bottom).

Building of the above bitvectors is reminiscent of the classical subset construction for converting an NFA into a DFA. In fact, each bitvector corresponds to a state in a DFA for $\mathcal{B}$. However, we only build those bitvectors, which are asked for by the input viewgraph. Thus, observe that in this example only 4 bitvectors are needed, namely 100, 110, 101, and 111. On the other hand, the minimum size DFA corresponding to $\mathcal{B}$ has $2^3 = 8$ states.

Now, once the Cartesian product graph is constructed, it is easily seen that b is reachable from a using a string not in $L(\mathcal{B})$ but c is not.

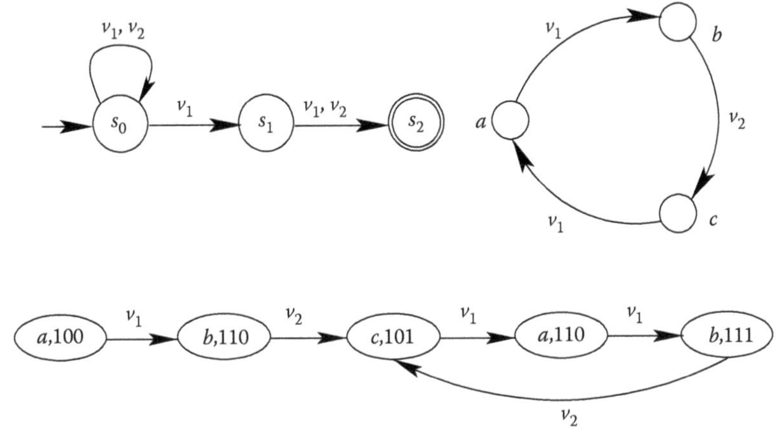

FIGURE 10.4 Automaton $\mathcal{B}$ (top-left), viewgraph $\mathcal{V}$ (top-right), and Cartesian product graph (bottom).

In general, for a $\mathcal{B}$ automaton with set S of states, we use bitvectors of size $|S|$ to keep track of the states that $\mathcal{B}$ can be when reaching some object of the viewgraph. As illustrated by the above example, the nodes of the (lazy) Cartesian product graph are of the form (a, u), where a is an object in the viewgraph and u is a bitvector of size $|S|$. Since the input is a graph as opposed to a string, there can be different bitvectors associated with the same given object (for instance with objects a and b in the example).

We want to stress that we build the Cartesian product graph starting from all the viewgraph objects. In the above example, for clarity, we showed the Cartesian product constructed starting from one object only. However, these Cartesian products overlap, and thus, in order to not generate the same object–bitvector pair twice, we maintain a hashtable of the pairs generated so far. In fact, even for a single Cartesian product, the same pair might be needed more than once, and the hashtable is necessary for this case as well in order for the method to terminate.

The edge labels in the Cartesian product graph are of no importance when it comes to generating the query answers. The only thing that matters in this graph is pure reachability. Namely, we produce a pair (a, b) as an answer, if there exists a path [in the Cartesian product graph] from (a, u_0) to (b, w), where u_0 is the initial bitvector $10\ldots0$, and w is a bitvector having no bit set to 1 for any final state in $\mathcal{B}$.

Formally, our algorithm is as follows.

Algorithm 10.2:

Input: Automaton $\mathcal{B}$ and a viewgraph $\mathcal{V}$.
Output: $\text{ans}(Q', \mathcal{V})$, *where* $Q' = L(\overline{\mathcal{B}})$.
Method:

1. Denote by u_0 the bitvector $10\ldots0$ corresponding to the initial state s_0 in $\mathcal{B}$.
2. Initialize
   a. A processing queue $P = \{(a, u_0) : a \text{ object in } \mathcal{V}\}$.
   b. A hashtable $H = \emptyset$.
   c. A Cartesian product graph $\mathcal{G} = \emptyset$.
3. Repeat (a), (b), and (c) until queue P becomes empty.
   a. Dequeue a pair (a, u) from P.
      Lookup (a, u) in H.
      If (a, u) is not yet in H, then insert it in H and $\mathcal{G}$.
      Otherwise, discard (a, u) as we have already dealt with it.
   b. For each outgoing edge from a to b in $\mathcal{V}$, labeled by some symbol, say v_{ab}, compute the "next" bitvector w by procedure

$$w = Next(u, v_{ab}).$$

      [We discuss this procedure soon.]
   c. If w is different from the all zero's vector, then insert (b, w) in P.
      Also, insert edge $((a, u), v_{ab}, (b, w))$ in $\mathcal{G}$
4. Finally, set

$$\text{ans}(Q', \mathcal{V}) = \{(a, b) : \text{there exists a path in } \mathcal{G} \text{ from } (a, u_0) \text{ to } (b, w) \text{ such that } w \text{ has no bit set to 1 for any final state in } \mathcal{B}\}.$$

Implementation of $Next(u, v)$.

We optimize the amount of time taken to compute adjacent bitvectors as follows.

Normally, each entry in the transition table of $\mathcal{B}$ is just a list of next possible states of the NFA given the current state and input symbol. Instead of storing this list, we store a bitvector α of $|S|$ bits, that is, the characteristic vector of this list of states. Using the various values of α in the transition table, given an object-vector pair of the form (a, u) and an input symbol v, we can compute $Next(u, v)$ in only $O(n)$ time using a sequence of bitwise-OR operations [compared to the naive method of updating vectors that takes $O(n^2)$ in the worst case]. In particular, without loss of generality, suppose the set of indices in u which have a 1 is exactly $\{i_1, i_2 \ldots, i_k\}$. Then it is easy to see that

$$Next(u, v) = \alpha_{i_1,v} \vee \alpha_{i_2,v} \vee \ldots \vee \alpha_{i_k,v}$$

where $\alpha_{i_j,v}$ is the bitvector α in the transition table corresponding to the state q_{i_j} and the input symbol v.

In the next section, we show that our ideas give substantial improvement in running time making it possible to solve the problem of view-based answering on much larger instances compared to the naive implementation.

10.6 Experimental Results

We conducted experiments in order to assess the improvements offered by answering the query using automaton $\mathcal{B}$ over answering using automaton $\mathcal{C}$ ([17]).

First, we give some details on how we generated queries, views, and viewgraphs. For this we used a simple DataGuide (cf. [1]). DataGuides are essentially finite state automata capturing all the words spelled out by the database paths. In general, DataGuides are compact representations of graph databases. They are small automata presented to the user in order to guide his/her in writing queries. Each word in a DataGuide could possibly represent many paths that spell that word in a database. For example, a DataGuide, capturing databases such as the one shown in Figure 10.1, contains a word *software·company·recommends*. Certainly, there are many such paths in databases about online stores.

In our experiments, we used the DataGuide given in Figure 10.5a, where all the states are both initial and final.

For generating view language definitions, we obtained a right linear grammar from the DataGuide of Figure 10.5a. The rules of this grammar are given in the same figure (b), where A, B, C, D, and E are the non terminal symbols corresponding to the DataGuide states.

Then, we randomly generated partial derivations using the above grammar. Such a partial derivation is for example $B \rightarrow company \cdot recommends \cdot D$. By randomly selecting such partial derivations, we created new right linear grammars. We kept only those grammars generating nonempty languages. Clearly, the grammars generated in this way capture sublanguages of the DataGuide. By this random procedure, we created 50 test sets of 40 view definitions each.

For each set of views, we created random queries as follows. Let $\mathbf{V} = \{V_1, \ldots, V_{40}\}$ be a view set. Then, the outer-alphabet is $\Omega = \{v_1, \ldots, v_{40}\}$. First, we randomly created a regular expression on Ω of length not more than 10. For instance, such a regular expression could be $re = v_1 \cdot v_{13}^* + v_{40}$. Next, we set $Q = \text{def}(re)$, which is a language on Δ, and computed its view-based rewriting using set $\mathbf{V}$ of views.

We could certainly generate queries in a similar fashion as for generating view languages, that is, directly from the DataGuide. However, doing so generates many cases when the rewriting is empty, and the experiments would be uninteresting. On the other hand, generating queries as above guarantees that the rewritings will not be empty.

Regarding the generation of view graphs, we first randomly generated databases from the Data-Guide, and then evaluated on these databases each of the generated views. In this way, we obtained an "answer" for each view. For instance, we could have $\{(a, b), (b, c), \ldots\}$ as the answer for V_1 in some randomly generated database. Then, we inserted edges (a, v_1, b), (b, v_1, c), ...in the the corresponding viewgraph. For each of the 50 sets of views, we randomly generated as above a viewgraph of more than 10,000 nodes.

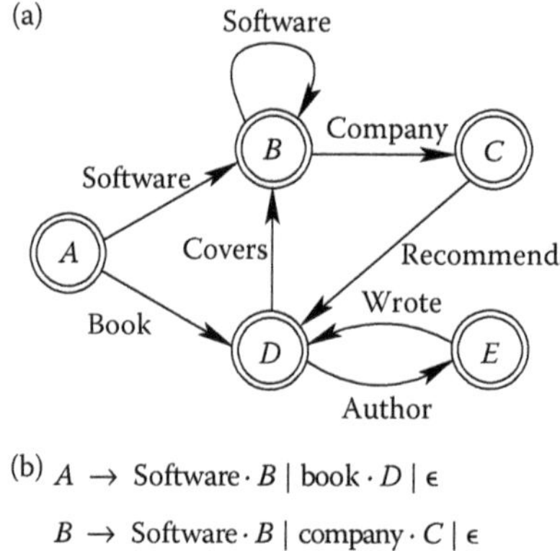

(b) $A \rightarrow \text{Software} \cdot B \mid \text{book} \cdot D \mid \epsilon$

$B \rightarrow \text{Software} \cdot B \mid \text{company} \cdot C \mid \epsilon$

$C \rightarrow \text{Recommends} \cdot D \mid \epsilon$

$D \rightarrow \text{Covers} \cdot B \mid \text{author} . E \mid \epsilon$

$E \rightarrow \text{Wrote} \cdot D \mid \epsilon$

FIGURE 10.5 (a) DataGuide corresponding to the database in Figure 10.1. (b) Grammar for the given DataGuide.

Then, we computed automaton $\mathcal{B}$ for each set of views, and evaluated it [as described in Section 10.5] on the corresponding viewgraph. Also, we tried to compute automaton $\mathcal{C}$ accepting $L(\overline{\mathcal{B}})$. We used GRAIL+ (see Ref. [15]), which is a well-engineered automata package written in C++. As already mentioned, computing $\mathcal{C}$ was not always possible. Out of our 50 cases, computing $\mathcal{C}$ timed out in 15 of them. We used a big timeout of 4 h. Whenever we were able to obtain a DFA $\mathcal{C}$, we evaluated it on the corresponding viewgraph. For these cases, we compared the times of evaluating $\mathcal{B}$ versus evaluating $\mathcal{C}$ on the viewgraphs.

In all the test cases, we computed automaton $\mathcal{B}$ using the technique described in Section 10.5.1. It was this technique that made possible the computation of $\mathcal{B}$ in a reasonable amount of time for each test case (of 40 views each). As mentioned in Section 10.5.1, using our technique we were able to achieve a speedup of more that sixfold in computing $\mathcal{B}$. Our times for computing the $\mathcal{B}$ automata range between 10 and 15 min.

We have tabulated our time and size results in Figure 10.6. The results were obtained using a modern Sun-Blade-1000 machine with 1 GB of RAM. In the following, we describe the column headers of our result table.

ID: ID of test set.

B-NFA-size: Size of automaton (NFA) $\mathcal{B}$.

C-DFA-size: Size of automaton (DFA) $\mathcal{C}$.

C-DFA-time: Time (in secs) to compute automaton (DFA) $\mathcal{C}$.

C-DFA-V-time: Time (in secs) to evaluate automaton (DFA) $\mathcal{C}$ on the corresponding viewgraph.

C-DFA-V-TTime: Total time (in secs) to compute and then evaluate automaton (DFA) $\mathcal{C}$ on the corresponding viewgraph. [This is the sum of the above two times.]

B-BitNFA-V-time: Time (in secs) to bitwise evaluate automaton (NFA) $\mathcal{B}$ on the corresponding viewgraph.

B-BitNFA-V-size: Size of the input-aware Cartesian product of automaton (NFA) $\mathcal{B}$ with the corresponding viewgraph.

Ratio: Ratio of the time to obtain the answers using bitwise evaluation of automaton (NFA) $\mathcal{B}$ to the time to obtain the answers using automaton (DFA) $\mathcal{C}$ whenever possible. The last number of 1.3 in this column is the average of the column.

ID	B-NFA Size	C-DFA-size	C-DFA-time	C-DFA-V-time	C-DFA-VT Time	B-BitNFA-V-time	B-BitNFA-V-size	Ratio
1	11	35	2	317	319	348	27887	1.1
2	9	16	1	396	397	390	23967	1
3	12	67	7	330	337	396	31875	1.2
4	11	34	3	410	413	417	29003	1
5	9	40	1	407	409	419	26172	1
6	12	74	5	489	494	530	31766	1.1
7	13	57	7	573	580	651	32501	1.1
8	15	83	11	630	641	678	35332	1.1
9	23	462	393	454	847	773	40899	0.9
10	12	69	8	805	813	887	37850	1.1
11	14	114	8	703	711	901	39252	1.3
12	17	166	29	540	569	911	44261	1.6
13	13	72	9	905	914	1037	38095	1.1
14	13	221	19	642	661	1159	50413	1.8
15	16	513	87	609	696	1180	48698	1.7
16	20	319	153	743	896	1247	47582	1.4
17	12	82	8	1067	1074	1457	47051	1.4
18	35	1442	2148	824	2972	1505	52686	0.5
19	33	3316	4126	592	4718	1593	61860	0.3
20	16	266	61	1058	1119	1867	56361	1.7
21	21	552	296	859	1154	1879	58879	1.6
22	35	723	710	1106	1816	2074	55939	1.1
23	31	831	461	913	1374	2113	61329	1.5
24	21	1316	526	867	1392	2121	66651	1.5
25	20	1098	379	1046	1425	2206	65004	1.5
26	18	238	60	1372	1432	2846	63561	2
27	18	523	104	1056	1160	3061	74273	2.6
28	20	550	177	1083	1260	3403	83515	2.7
29	26	2001	855	1245	2099	3512	80085	1.7
30	33	3578	2197	1106	3303	3599	83338	1.1
31	38	3492	2937	1628	4565	3666	76546	0.8
32	35	1720	1210	959	2169	3674	84318	1.7
33	28	2894	2515	1330	3845	4477	102625	1.2
								1.3
34	49	N/P	N/P	N/A	N/A	697	103251	
35	42	N/P	N/P	N/A	N/A	820	101291	
36	44	N/P	N/P	N/A	N/A	892	92852	
37	53	N/P	N/P	N/A	N/A	1224	50903	
38	41	N/P	N/P	N/A	N/A	1554	56048	
39	52	N/P	N/P	N/A	N/A	1754	44805	
40	48	N/P	N/P	N/A	N/A	2033	66406	
41	53	N/P	N/P	N/A	N/A	2239	66052	
42	43	N/P	N/P	N/A	N/A	2270	76941	
43	42	N/P	N/P	N/A	N/A	2549	85026	
44	48	N/P	N/P	N/A	N/A	3358	80330	
45	44	N/P	N/P	N/A	N/A	3468	83515	
46	30	N/P	N/P	N/A	N/A	3542	86632	
47	42	N/P	N/P	N/A	N/A	3816	81133	
48	40	N/P	N/P	N/A	N/A	3890	84563	
49	45	N/P	N/P	N/A	N/A	4872	103183	
50	47	N/P	N/P	N/A	N/A	5985	123831	

FIGURE 10.6 Table of results.

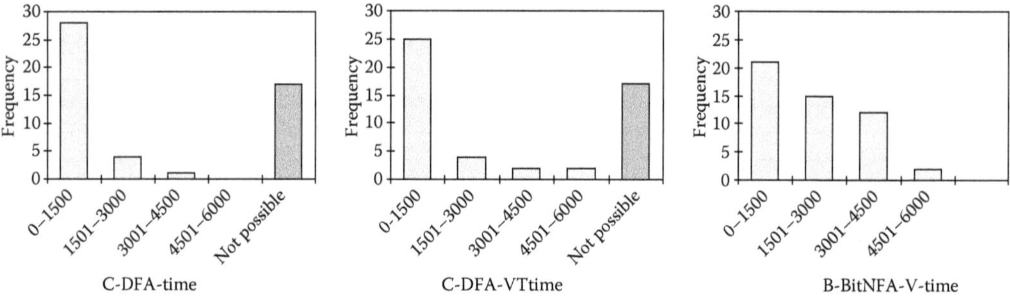

FIGURE 10.7 Number of instances for different time ranges of C-DFA-V-time, C-DFA-V-TTime, and B-BitNFA-V-time. The darker bars for C-DFA-V-time and C-DFA-V-TTime show the fraction of instances for which the computation of automaton $\mathcal{C}$ was not possible.

We have sorted the results in ascending order of the B-BitNFA-V-time. The first part of the table contains the results for the cases when the computation of automaton $\mathcal{C}$ succeeded. The second part of the table contains the results for the cases when the computation of automaton $\mathcal{C}$ failed. As such, the second part of the table has results which relate to the use of automaton $\mathcal{B}$ only. The shaded area of this part of the table is marked by N/P (Not Possible) or (N/A) (Not Applicable) as appropriate.

Also, we have graphed columns C-DFA-V-time, C-DFA-V-TTime, and B-BitNFA-V-time in Figure 10.7 in order to more clearly show the fractions of instances corresponding to different time ranges, as well as the fraction of instances for which the computation of rewritings is impossible in reasonable time and space.

Based on the table of results, we are able to draw the following natural conclusions.

1. Computing in full the view-based rewriting represented by automaton $\mathcal{C}$ is hard and fails in a considerable number of cases (30% of them). Hence, one should not pursue this route for producing view-based query answers.
2. Even when constructing $\mathcal{C}$ is possible, the performance advantage offered by the determinism of $\mathcal{C}$ over automaton $\mathcal{B}$ is small. In average, bitwise evaluation of $\mathcal{B}$ is only 1.3 times slower on the average, while in some cases it can be even faster. This is due to the smaller footprint of $\mathcal{B}$ having so a better hardware cache utilization.
3. For all the test cases, the size of the input-aware bitwise Cartesian product of automaton $\mathcal{B}$ with the corresponding viewgraph $\mathcal{V}$ is very far from the worst case of $2^{|\mathcal{B}|} \cdot |\mathcal{V}|$.

From all the above, one can see that by employing the proposed techniques, the view-based answering of RPQ's becomes (fairly) feasible in practice.

References

1. S. Abiteboul, P. Buneman, and D. Suciu. *Data on the Web: From Relations to Semistructured Data and XML*. Morgan Kaufmann, San Francisco, 1999.
2. L. Bravo and L. E. Bertossi. Deductive databases for computing certain and consistent answers from mediated data integration systems. *J. Applied Logic*, 3(1):329–367, 2005.
3. D. Calvanese, G. D. Giacomo, M. Lenzerini, and M. Y. Vardi. Answering regular path queries using views. In *Proceedings of the 16th International Conference on Data Engineering*, San Diego, California, February 28–March 3, pp. 389–398, 2000.
4. D. Calvanese, G. D. Giacomo, M. Lenzerini, and M. Y. Vardi. View-based query processing and constraint satisfaction. In *Proceedings of 15th Annual IEEE Symposium on Logic in Computer Science*, Santa Barbara, California, June 26–29, pp. 361–371, 2000.

5. D. Calvanese, G. D. Giacomo, M. Lenzerini, and M. Y. Vardi. Rewriting of regular expressions and regular path queries. *J. Comput. Syst. Sci.*, 64(3):443–465, 2002.
6. D. Calvanese, G. D. Giacomo, M. Lenzerini, and M. Y. Vardi. Reasoning on regular path queries. *SIGMOD Record*, 32(4):83–92, 2003.
7. D. Calvanese, G. D. Giacomo, M. Lenzerini, and M. Y. Vardi. View-based query processing: On the relationship between rewriting, answering and losslessness. *Theor. Comput. Sci.*, 371(3):169–182, 2007.
8. M. P. Consens. Managing linked data on the web: The linkedmdb showcase. In *Proceedings of the Latin American Web Conference*, LA-WEB 2008, Vila Velha, Espírito Santo, Brasil, October 28–30, pp. 1–2, 2008.
9. M. P. Consens and A. O. Mendelzon. Graphlog: A visual formalism for real life recursion. In *Proceedings of the Ninth ACM SIGACT-SIGMOD-SIGART Symposium on Principles of Database Systems*, Nashville, Tennessee, April 2–4, pp. 404–416, 1990.
10. G. Grahne and A. O. Mendelzon. Tableau techniques for querying information sources through global schemas. In *Proceedings of 7th International Conference on Database Theory*, Jerusalem, Israel, January 10–12, pp. 332–347, 1999.
11. G. Grahne and A. Thomo. An optimization technique for answering regular path queries. In *Proceedings of the Third International Workshop on the Web and Databases*, WebDB 2000, Adam's Mark Hotel, Dallas, Texas, May 18–19, pp. 215–225, 2000.
12. G. Grahne, A. Thomo, and W. W. Wadge. Preferentially annotated regular path queries. In *Proceedings of 11th International Conference on Database Theory*, Barcelona, Spain, January 10–12, pp. 314–328, 2007.
13. M. Lenzerini. Data integration: A theoretical perspective. In *Proceedings of the Twenty-first ACM SIGACT-SIGMOD-SIGART Symposium on Principles of Database Systems*, Madison, Wisconsin, June 3–5, pp. 233–246, 2002.
14. A. O. Mendelzon and P. T. Wood. Finding regular simple paths in graph databases. *SIAM J. Comput.*, 24(6):1235–1258, 1995.
15. D. R. Raymond and D. Wood. Grail: A c++ library for automata and expressions. *J. Symb. Comput.*, 17(4):341–350, 1994.
16. T. Segaran, J. Taylor, and C. Evans. *Programming the Semantic Web*. O'Reilly, Cambridge, MA, 2009.
17. M. Tamashiro, A. Thomo, and S. Venkatesh. Towards practically feasible answering of regular path queries in lav data integration. In *Proceedings of the Sixteenth ACM Conference on Information and Knowledge Management*, Lisbon, Portugal, November 6–10, pp. 381–390, 2007.
18. S. Yu. Regular languages. In: *Handbook of Formal Languages*, G. Rozenberg and A. Salomaa (Eds.), vol. I, pp. 41–110. Springer, Berlin, Germany, 1997.

11

Applying Timed Automata to Model Checking of Security Protocols

Mirosław Kurkowski
Czestochowa University of Technology

Wojciech Penczek
Polish Academy of Sciences
University of Natural Sciences and Humanities

11.1 Introduction to Security Protocols

Security protocols are an important and very often crucial element of security in computer systems. In open computer networks, especially in systems realizing goals of the public key infrastructure, these concurrent algorithms are often used during an exchange of information for, generally speaking, ensuring an adequate level of security of data transmitted over networks and electronic transactions. Their principal tasks include, among others, mutual authentication of communicating parties (users, servers) or a new session key distribution. Basically, they are most often applied as essential components of large systems, such as widely used communication protocols. Important examples of such systems are Kerberos, SSL/TLS, and Zfone. However in literature and applications, one can find errors in a protocol design. This justifies a vital need to develop methods to specify and to verify their properties in order to find faults allowing unwanted behavior, which can lower the security level or even discredit it completely.

The main objectives that should be achieved by security protocols are as follows:

1. Mutual authentication of communicating entities
2. Confidentiality of information transmitted
3. Integrity of this information
4. New session key distribution

Security protocols that are applied in practice can realize one or more of the mentioned objectives. They may also use both types of cryptography: symmetric and asymmetric. This is sometimes described in names of protocols: symmetric, authentication or session key distribution protocols. For the sake of simplicity, in further sections of this chapter we will call all security protocols simply protocols.

The actions performed by parties during the protocol execution can be divided into internal and external ones. By the external actions, we mean actions of mutual transmission of information by the parties. Specification of these actions must contain a source of each information (a sender), a receiver of the information (a recipient) and of course, its content, indicating, respectively, which parts of the information have to be encrypted and how. The internal actions are all the other actions that each party performs on its own during the protocol execution. They include, for example, generating new confidential information, encryption and decryption of cryptograms, comparing data or performing mathematical operations on locally stored data.

Applying protocols in security systems need to fulfil several requirements:

1. Each user must be aware of its use and has to agree for applying the protocol
2. Each user has to know the protocol and execute consecutively all the steps of the protocol
3. All steps of the protocol must be precisely defined so that the users could not be confused about the way or the order of their performance

11.2 Examples of Protocols

In order to better understand problems connected to design and practical use of protocols in this section, we introduce three examples of them: (i) the Needham Schroeder Public Key Authentication Protocol (NSPK), (ii) the Wide Mouth Frog Protocol (WMF), and (iii) the Kerberos Protocol. The examples presented below help to understand formalisms that are introduced later and which aim at specification and automatic verification of protocol correctness. The examples of the protocols are accompanied by a discussion of their history and correctness.

11.2.1 The Needham Schroeder Public Key Protocol

A pioneering role in the area of the protocols and their application in open computer networks was played by the paper [36] by Needham and Schroeder published in 1978. This took place at the time of the early development of computer networks and, therefore, communication protocols, as well. In the cited article, its authors presented an idea of using cryptographic techniques for solving problems which concern authentication of the parties communicating in networks. They also suggested schemes of authentication protocols that use symmetric and asymmetric cryptography.

The protocols described below are written in the so-called Common Language—a protocol specification language widely used in the literature. Unfortunately, as it can be noticed, this language does not fully describe a protocol. It presents only a scheme of sending messages during the protocol execution and constructing the messages, that is, the external actions of the protocol. The description of the internal actions and the other conditions of the protocol execution is usually written in a natural language.

Now, we show one of the protocols proposed in Ref. [36]. In the literature, it is simply called the Needham Schroeder Public Key protocol. In the notation presented below, the symbols I_A and I_B denote the identifiers of the users A and B, which want to communicate safely with each other. By $\langle X\rangle_{K_A}$ we mean a ciphertext containing a message X encrypted with the public key of the user A. Analogously, by $\langle X\rangle_{K_B}$ we mean a ciphertext encrypted by B's public key.

In the described protocol, as in many others, there are used so-called nonces, that is, large (consisting of hundreds of digits) disposable pseudo-random numbers generated and used for only one execution of a protocol or one communication session. The notion *nonce* is an abbreviation of *number used once*. Randomness of these numbers aims at ensuring that no one has ever known the information before, not to mention transmitting it, and in practice (i.e., with a sufficiently very high probability) it is unfeasible for anybody to reconstruct the number used. In this sense, it is said that the information is fresh. Such numbers are denoted by N_A (the number generated by A) and N_B (the number generated by B). These numbers, as well as appropriate cryptographic keys, have to remain secret during the protocol execution. In next parts of this chapter, we call these nonces and secret cipher keys simply secrets.

The user A initiates the protocol. The objective, which should be achieved after the protocol execution, is a mutual authentication of the parties A and B, that is, the mutual confirmation of their identities. The notation $A \rightarrow B : X$ means that the message X is transmitted from A to B. In our approach, it is often assumed that sending information implies its receipt by the receiver.

The paper [36] proposed a protocol with public keys distribution in server environment. Nowadays, when the Public Key Infrastructure is widely used, NSPK without key servers is particularly interesting. The scheme of this version of the protocol is the following:

1. $A \rightarrow B : \langle N_A, I_A\rangle_{K_B}$
2. $B \rightarrow A : \langle N_A, N_B\rangle_{K_A}$
3. $A \rightarrow B : \langle N_B\rangle_{K_B}$

In the first step of the protocol, the user A generates its random number N_A and sends it together with its identifier to B, encrypting everything with the public key belonging to B. Let us recall that according to asymmetric cryptography rules only B, as the owner of the appropriate private key, is able to decrypt the message and obtain the unique number N_A. In the second step, B generates its own number N_B and encrypts it together with the obtained number N_A with A's public key. Then, B sends the so-prepared ciphertext to A, which can decrypt it (A is the only one which is able to do it). A compares the number received from B with its nonce N_A and at this moment A acknowledges B to be authenticated, since B sent to A the nonce N_A. Let us observe that at this stage A gets the number N_B. A encrypts this number with B's public key and sends it to B. B decrypts the ciphertext (only it is able to do that) and compares the number sent by A with it's own N_B. At this moment A is authenticated by B.

Due to the structure of transferring data used in the protocol and properties of the asymmetric cryptography, after executing the protocol the users A and B should be sure of their identity. Let us also pay attention to the fact that after the protocol execution the users should also be sure that they are the only holders of the numbers N_A and N_B and now they may use them in further communication session as identifiers. These numbers may also be the basis for determining a new common symmetric key for further communication.

The history of this protocol is interesting and instructive, as well. It was used in practice in its original version for 17 years. However, in 1995 it turned out that the protocol can be broken. Gavin Lowe, at that time a professor at the Oxford University, presented an attack upon the protocol [31], that is, such its execution that does not realize the aim of the protocol, namely the mutual authentication. The attack is performed by the Intruder denoted by I. The Intruder is a user of the computer network which has, therefore, its own identifier and asymmetric keys. However, I does not need to execute the protocol according to its scheme. In many ways, it may cheat other parties communicating with them. Below we present the attack given by Lowe. After its execution, one of its parties is cheated as concerns the identity of the communicating party.

α 1. $A \rightarrow I : \langle N_A, I_A \rangle_{K_I}$
 β 1. $I(A) \rightarrow B : \langle N_A, I_A \rangle_{K_B}$
 β 2. $B \rightarrow I(A) : \langle N_A, N_B \rangle_{K_A}$
α 2. $I \rightarrow A : \langle N_A, N_B \rangle_{K_A}$
α 3. $A \rightarrow I : \langle N_B \rangle_{K_I}$
 β 3. $I(A) \rightarrow B : \langle N_B \rangle_{K_B}$

The above scheme shows two simultaneous runs of the protocol. The execution α corresponds to communication of the user A with the Intruder I, who impersonates A (we denote it by $I(A)$) during the execution β. In the step α 1, A starts communicating with I. The latter, however, abuses A's confidence and improperly uses the nonce N_A, given to I by the user A, in order to start a session with B (step β 1). According to the protocol, B sends a ciphertext $\langle N_A, N_B \rangle_{K_A}$ to I (believing it is A). Of course, I is not able to decrypt this ciphertext, since it is encrypted with A's public key. Therefore, the Intruder forwards the message to A (step α 2). Unaware of the Intruder's presence, A decrypts the ciphertext, obtains the number N_B, encrypts it and sends to I (step α 3). Now, I is able to decrypt the message and knowing the value of N_B sends it to B (step β 3). After that, B believes it is communicating with A, while in fact B exchanges information with the Intruder I. In further potential communication, the Intruder may use the known numbers N_A and N_B, for example, to deceive B.

The fixed version of the protocol, also developed by Lowe, is as follows:

1. $A \rightarrow B : \langle N_A, I_A \rangle_{K_B}$
2. $B \rightarrow A : \langle N_A, N_B, I_B \rangle_{K_A}$
3. $A \rightarrow B : \langle N_B \rangle_{K_B}$

The modification affects the second step. Simply adding the responder's identifier I_B to the ciphertext $\langle N_A, N_B, I_B \rangle_{K_A}$ excludes the possibility of its deceptive use by the potential Intruder. This easy correction prevents the protocol from an attack presented above. The Needham Schroeder protocol revised by Lowe has passed all correctness tests known so far.

11.2.2 The Wide Mouth Frog Protocol

One of the aims of using disposable numbers in protocols is the possibility of identifying subsequent communication sessions between users. This technique, however, does not prevent from repetition or continuation of the protocol, for example, after a long break, since the cryptographic key may have already been broken. It turns out that it is necessary to introduce to protocol schemes the so-called timestamps confirming the time of key generation or transmitting a message. This allows to reject a prolonged or continued after a long break communication session in which keys are already dangerously old. Timestamps protect the system also against the so-called replay attacks during which the Intruder can use old ciphertexts in the way that enables breaking them.

A simple example of a protocol that uses timestamps is the Wide Mouth Frog Protocol. This protocol is designed for establishing a new secret cryptographic key, which later enables the secure (encrypted) exchange of information between communicating parties. This key is called a session key, because it is used during one session only. Such keys are useful, as they are disposable, that is, exist during the communication period only.

WMF was designed by Michael Burrows [12]. It is probably the simplest symmetric protocol for key management and a very good example for scientific investigations. In this protocol, two users and a third trusted entity take part; this third party is usually represented by the main server which task is to facilitate authentication and to exchange a session key, that will be later used by parties to encrypt messages transmitted. Independently of each other, two parties A and B use S (the trusted server) to set up a session key between them for secure communication. This key is used for distributing keys, but it is not applied to encryption of any messages transmitted between the parties. By transmitting two messages, the session key is transferred from A to B.

The protocol consists of two steps, and it aims at authentication of the participant A by B and at the exchange of a new session key K_{AB}. Each party A and B uses a common secret key shared with S denoted by K_{AS} and K_{BS}, respectively, T_A is a timestamp generated by A, T_S is a timestamp generated by S, I_A is the identifier of the participant A, and I_B is the identifier of B.

The scheme of the protocol is the following:

1. $A \to S : I_A, \langle T_A, I_B, K_{AB} \rangle_{K_{AS}}$
2. $S \to B : \langle T_S, I_A, K_{AB} \rangle_{K_{BS}}$

The protocol analysis is as follows. In the first step, the participant A concatenates its timestamp T_A with the identifier I_B, with which it wants to communicate and a freshly generated random session key K_{AB} and it encrypts the whole message using the secret key shared with S. The timestamp T_A contains information about the exact time of generating the key K_{AB}. This key can be used only within a fixed period of time. Next, A sends an encrypted package to the trusted server S together with it's identifier I_A. At this point, it should be noted that the user S will continue this protocol session if only the key is valid. Thus, in order to execute the second step of the protocol, there must be preserved an appropriate time dependence between the timestamp T_A and the lifetime L. In the second step, S decrypts the message received from A, and then concatenates a new timestamp T_S, the identifier I_A and the new session key K_{AB}, as well. When such a message is constructed, it is encrypted with the secret key that S shares with B and sent to the latter. The timestamp T_S and its independently fixed lifetime inform the user B about the expiration date of the key K_{AB}. The user B may continue communication with A after comparing the values of T_S and L and being certain that the proper time dependence is preserved. At the end, let us emphasize that timestamping aims at protecting the protocol against replay attacks.

11.2.3 The Kerberos Protocol

As the next example, we present one of the currently most popular and widely used protocols which also applies timestamps, namely the Kerberos protocol [34]. One of its versions has the following scheme:

1. $A \to S : I_A, I_B$
2. $S \to A : \langle T_S, L, K_{AB}, I_B \rangle_{K_{AS}}, \langle T_S, L, K_{AB}, I_A \rangle_{K_{BS}}$
3. $A \to B : \langle T_S, L, K_{AB}, I_A \rangle_{K_{BS}}, \langle I_A, T_A \rangle_{K_{AB}}$
4. $B \to A : \langle T_A \rangle_{K_{AB}}$

This protocol has two objectives. The first one is to distribute a new session key for the users A and B, while the second is mutual authentication. The protocol is initiated by A which by sending of the identifiers I_A and I_B in the first step requests the server S to establish a new symmetric session key for communication between A and B. As it can be noticed, in the second step the server sends two ciphertexts to A. The first one contains a freshly generated session key K_{AB}, a timestamp determining when this key was generated, a lifetime value L (i.e., an expiration date of the key), and the identifier of B (with which the key will be shared). This ciphertext is encrypted with a key which is known only to A and S. The second ciphertext, which is designed for B, also contains the key K_{AB}, its lifetime L and a timestamp T_S. This ciphertext is encrypted with a key shared by the server and B, therefore A cannot decrypt it. Upon receiving this message, A checks whether an appropriate time dependence between the timestamp T_S and the lifetime L is maintained. Only in case of positive verification of this dependence, A sends to B both the ciphertext $\langle T_S, K_{AB}, L, I_A \rangle_{K_{BS}}$ and a new timestamp T_A, which determines the time of setting up the session and which is encrypted with a new key K_{AB}. Due to this timestamp, B knows since when A has been using the key K_{AB} and B will use it until its expiration date which is determined by T_A and L. As it can be seen, using timestamps protects the protocol against replay attacks, which might be executed by the Intruder. Indeed, unless the ciphertext $\langle I_A, T_A \rangle_{K_{AB}}$ contained a timestamp, the Intruder could have intercepted this message, waited a certain period of time needed to break the key and sent it again knowing already the key K_{AB}. Additionally, Kerberos not only prevents from replaying a part of the protocol after a long time, but also enables replaying the end of the protocol within the validity period of the key.

11.3 Specification and Verification of Protocol Properties

The above examples show that constructing a correct protocol is not a trivial task. The seemingly safe NSPK turned out to be vulnerable to a simple attack. The crucial role in this attack, as well as in executing protocols, is played by the knowledge the user has acquired about cryptographic keys, nonces, and so on. Systems for automatic and semi-automatic protocol verification require, therefore, not only modeling protocol executions themselves, but also modeling the user's knowledge gained during the system performance. Verification of timed protocols needs to take into account also satisfying appropriate time conditions between times of generating keys or timestamps and times of executing actions comprising the protocol. A formal verification has to be based on the full formal description of the protocol. As it can be seen in the examples mentioned above, the Common Language does not fulfill this property. Hence, also a formal language is needed in order to specify internal and external protocol actions and required time dependencies fully. The methodology proposed in the following parts of this chapter meets these requirements.

Observe that a protocol specification should contain:

1. A number of entities participating in the protocol
2. Roles they play in the protocol
3. An aim of the protocol
4. A full description of the external and internal actions of which the protocol execution comprises

Usually while testing the correctness of protocols it is assumed that the ciphers used in them are safe, that is, we do not consider attacks on the encryption algorithms themselves. We leave this task to cryptanalysts. In our approach, it is assumed that no one is able to read the content of an encrypted document, unless it has the proper key. When verifying the protocol we also assume that no one can guess just generated pseudorandom numbers.

Another specification method is based on the timestamp technique. The use of timestamps is related to numerous technical problems, such as the necessity of reliable synchronization of clocks. In case of protocols using timestamps, it is more difficult to verify their correctness. Until recently, all known verification methods have treated a timestamp value, that is, the time of a transaction, as yet another disposable number. The order of their generation simulated the passage of time.

Automated verification of protocols is a very active and important area of computer science, which has been an object of an intensive research for several years in both academic and commercial institutions. There are numerous approaches to verification of untimed protocols [4,5,22,29] as well as of time-dependent ones [13,15,18,20,21,33]. Algorithmic approaches include mainly methods based on model checking. Intuitively, model checking of a protocol consists in checking whether a model of the protocol contains an execution or a reachable state that is representing an attack on the protocol. Comparing to standard model checking methods for communicating protocols or for distributed systems, the main difficulty is caused by the need to model both the Intruder which is responsible for generating attacks as well as changes of knowledge (about keys, nonces, etc.) of the participants.

In order to verify the correctness of a protocol, one can apply software verification tools designed for a broader class of programs, which do not need to use cryptographic techniques and do not necessarily run concurrently. Examples of such tools can be found, for example, in Ref. [6]. However, the best verification results are achieved by means of specialised tools designed to verify an appropriate class of protocols.

Basically, there are two main methods of verification:

1. Testing real and/or virtual systems (simulations)
2. Modeling and formal verification

In the first case, the method is based on simple testing of already implemented systems, or simulating their executions, for example, in virtual machines. After many such tests or simulations, however, we can only be sure that the system has worked properly so far. The second method, namely modeling and formal verification, bases on constructing special mathematical structures representing the actions that occur during executions of protocols.

Another important issue is the way of modeling the behavior of the potential Intruder attacking the system. Mostly, for this purpose there is used a model proposed quite a long time ago in [16] called, after the names of their authors, the Dolev–Yao model. The Intruder has an access to all the information transmitted, can collect and use it with a precision according to its skills and resources and is able to send it somehow through the network. Of course, the Intruder does need not to comply with the requirements of using the protocols and, in particular, does not have to use fresh information and to meet time requirements.

Among formal methods, depending on the applied mathematical structures, one can distinguish the following types:

1. Inductive
2. Deductive (axiomatic)
3. Model checking

The inductive method was proposed in Ref. [38] in the 1990s by Larry Paulson. This method considers all the strings of the possible execution steps of a protocol of the form: A sends a message M to B. Briefly, this method uses the principle of mathematical induction in order to prove that the protocol satisfies a given property. Properties of protocols are confirmed here by positive statements, that is, if one can prove some property, then it means that the system satisfies it. At the same time, when an attempt to prove the property fails, this may indicate an error in the protocol design. The inductive method was used, among others, because of relatively good existing tools for automated theorem proving by means of the mathematical induction. This method has been successfully applied to the analysis of real and complex protocols such as TLS (Transport Layer Security/Secure Sockets Layer) [38] or SET (Secure Electronic Transactions) [7].

The deductive method consists in constructing a specially chosen formal system of deduction, that is, a logic. Usually, however, these logics are significantly different from the classical one. The first system of this type was formulated in 1990 in Ref. [12]. Other logics, interesting from the application or theoretical point of view, can be found, for example, in Refs. [1,24,30].

Application of these logics to protocol verification allows to discover and correct several attacks upon protocols. It is important to note that these systems cannot discover new kinds of attacks. Another problem is that the use of a deductive method often requires referring to the semantics of a given logic, that is, in some sense, a mathematical model of protocol executions. Thus, nothing prevents us from considering only these models without building deductive systems. Perhaps then, in order to verify a protocol, a good model is enough?

Model checking is the last of the discussed methods. It consists in developing a formal mathematical model of the executions of an examined protocol and checking in a formal and strictly mathematical way that no program executions and/or their interleaves lead to unwanted situations. Due to the rather large size of such models, the construction can be, and usually is, virtual, while checking is done automatically or at least semi-automatically. One way of searching the executions space is checking whether the model can reach a state satisfying an unwanted property.

The main problem with this verification method is the state explosion, that is, an exponential explosion of the state space in the number of sessions, participants, and messages. Most often, during executions of protocols we have so many possible snapshots and model states corresponding to them that it is impossible to examine them all. An important reason of such a situation can be a huge amount of possible information that may be generated and sent through the network by the potential Intruder. Computational complexity of model checking algorithms is typically exponential in parameters of the verified protocol, particularly in the number of participants and execution steps.

11.4 Automata-Based Methods of Protocols Verification

One way of building formal models of protocol executions is by means of automata-based methods. It is well known that automata appear to be a natural formal way for modeling executions of various types of computer systems. Similarly, they can be applied to modeling executions and supporting verification of protocols.

In 1999 and 2000 two similar approaches exploiting automata were put forward. Monniaux [35] and Genet together with Klay [19] used tree automata for modeling executions of protocols. The main idea of these approaches is as follows. Tree automata that accept languages corresponding to the behaviors of the users running a protocol are constructed. These are languages of infinite words corresponding to infinite communication sessions. The method starts with constructing a basic tree automaton A_0, which models an initial configuration of the network considered and different types of Intruder behaviors during a protocol execution. The automaton A_0 is then gradually extended according to a specially built term rewriting system *TRS* to get an automaton modeling all the protocol executions to be verified. The system *TRS* determines all the protocol steps, the behavior of the protocol users including the Intruder, as well as the way how the Intruder can encrypt/decrypt and compose messages using these which have been intercepted.

The above ideas have been implemented in several tools for automatic or semi-automatic verification. Implementations can be found in Ref. [35], Timbuk tool [19], ACTAS [37], and TA4SP [11]. Several protocols have been modeled and tested in these works, for example Otway Rees Protocol in Ref. [35] and NSPK, NSPKL, and EKE in Ref. [11]. These are still ongoing works, see, for example, [10].

A slightly different approach was proposed by Corin et al. in Refs. [13,14]. There, timed automata [2] are used for modeling protocol executions. The main difference between this approach and the one mentioned before is that the behaviors of the protocol participants (Sender, Receiver, Server, and the Intruder) are modelled as separate synchronizing timed automata. Synchronization of two or more automata is via sharing a synchronization channel, that is, the automata execute transitions labeled with corresponding channel symbols (numbers). These transitions can be executed together at the same time moment only. The protocol executions are modeled by the runs of the product automaton of a network of the synchronizing timed automata. Such an approach allows for verification with UppAal [3]—a well-known real-time model checker. An authentication property and a resistance to a reflexion replay attack of a simplified version of NSPK and Yahalom Protocol were verified and discussed in Refs. [13,14].

The paper [29] offers another method for verifying untimed protocols, where the notion of a computational structure and an interpretation was based on Ref. [24]. The main idea consists in using networks of automata for modeling separately the participants and their knowledge about secrets. Thanks to that it was developed a very distributed representation of the protocol executions, which is crucial for an efficient symbolic encoding and model checking. Then, the above approach was extended in Ref. [28] to timed protocols. This approach is described in detail in this chapter.

Another method [21] also uses timed automata. Timed protocols are modeled in the higher-level language *IL*[*] [17], and only then translated to timed automata. The paper [21] offers a new method of verifying protocols, but the translation from *IL* can suffer from the state explosion in the resulting timed automata.

[*] *IL* is the acronym for the Intermediate Language (ver. 1.0).

Following our papers [26,28], in this chapter we give a method for representing the executions of a timed security protocol (within a computational structure for a bounded number of sessions) by the runs of the product timed automaton of a network of the timed automata for the participants and their knowledge. Then, we show how to look for attacks on authentication. To this aim we use Bounded Model Checking (*BMC*), which consists in translating the problem of reachability in the product timed automaton to satisfiability of some propositional formula.

The rest of the chapter is organized as follows. In Section 11.5, we introduce syntax for dealing with timed protocols. A computational structure generating all the runs of the protocols considered is defined in Section 11.6. A method for finding attacks by analysing computations of the protocol is shown in Section 11.6.1. Section 11.7 defines a network of automata for representing the participants of a protocol and their knowledge about secrets. Then, experimental results are given in Section 11.8 and some concluding remarks in Section 11.9.

11.5 Syntax of Timed Cryptographic Protocols

As we have noticed before, a formal verification of protocols requires a proper specification. In the approach proposed here, we consider protocols strictly as algorithms and define them as abstract concepts. In this way we can distinguish a protocol from its particular execution or executions. In order to be able to define a protocol as an abstract object, first we have to define a formal language. Such a language should be appropriately designed so that it could express all the properties which characterize the protocol, as well as its internal and external actions.

In this section, we introduce the syntax for dealing with timed protocols. We start with the following notations that are used in the whole chapter:

- $\mathcal{R}_+$ is the set of all the positive real numbers
- $\mathcal{R}_* = \mathcal{R}_+ \cup \{0\}$
- for any set Z by 2^Z_{fin} we denote a set of all the finite subsets of Z
- $n_P, n_N, n_\tau, n_L, n_R, k_N, n_T$ are some fixed natural numbers

The above notations are used for defining the following sets of symbols that are basic syntactic notions of our model.

- $\mathcal{T}_P = \{\mathcal{P}_1, \mathcal{P}_2, \ldots, \mathcal{P}_{n_P}\}$—the symbols for denoting the *users* of the computer network
- $\mathcal{T}_I = \{\mathcal{I}_{\mathcal{P}_1}, \mathcal{I}_{\mathcal{P}_2}, \ldots, \mathcal{I}_{\mathcal{P}_{n_P}}\}$—the symbols for denoting the *identifiers* of the users
- $\mathcal{T}_K = \bigcup_{i=1}^{n_P}\{\mathcal{K}_{\mathcal{P}_i}, \mathcal{K}_{\mathcal{P}_i}^{-1}\} \cup (\bigcup_{i=1}^{n_P}\bigcup_{j=1}^{n_P}\{\mathcal{K}_{\mathcal{P}_i\mathcal{P}_j}\} \setminus \bigcup_{i=1}^{n_P}\{\mathcal{K}_{\mathcal{P}_i\mathcal{P}_i}\})$—the symbols for denoting the *cryptographic keys* (public, private, and symmetric ones) of the users
- $\mathcal{T}_N = \bigcup_{i=1}^{n_P}\{\mathcal{N}_{\mathcal{P}_i}^1, \ldots, \mathcal{N}_{\mathcal{P}_i}^{n_N}\}$—the symbols for denoting the *nonces* of the users
- $\mathcal{T}_T = \bigcup_{i=1}^{n_P}\{\tau_{\mathcal{P}_i}^1, \ldots, \tau_{\mathcal{P}_i}^{n_\tau}\}$—the symbols for denoting the *timestamps* of the users
- $\mathcal{T}_L = \{\mathcal{L}_{\mathcal{F}1}, \mathcal{L}_{\mathcal{F}2}, \ldots, \mathcal{L}_{\mathcal{F}n_L}\}$—the symbols for denoting the *lifetimes*
- $\mathcal{T}_R = \{\tau_1, \tau_2, \ldots, \tau_{n_R}\}$—the real (time) variables representing the *times* of transmitting the messages
- {"(", ")", "{", "}", "," "⟨", "⟩"}—the *auxiliary* symbols

Below, we construct a special algebra of terms, which is used for representing messages transmitted during the protocol execution.

Definition 11.1:

The set $\mathcal{T}$ of *letter terms* is defined as the smallest set satisfying the following conditions:

1. $\mathcal{T}_P \cup \mathcal{T}_I \cup \mathcal{T}_K \cup \mathcal{T}_N \cup \mathcal{T}_T \cup \mathcal{T}_L \subseteq \mathcal{T}$
2. If $X \in \mathcal{T}$ and $Y \in \mathcal{T}$, then the concatenation $X \cdot Y \in \mathcal{T}$

3. If $X \in \mathcal{T}$ and $\mathcal{K} \in \mathcal{T}_{\mathcal{K}}$, then $\langle X \rangle_{\mathcal{K}} \in \mathcal{T}$*
4. If $X \in \mathcal{T}$, then $h(X) \in \mathcal{T}$, where $h(X)$ is the hash value of the letter term X

Since the behavior of the users during the protocol execution is related to the construction of messages transmitted and the possibility of using them, we introduce now two relations that allow to describe certain dependencies between message terms.

Definition 11.2:

The *immediate subterm relation* $\prec_{\mathcal{T}} \subseteq \mathcal{T} \times \mathcal{T}$ is the smallest relation, which satisfies the following conditions:

1. If $X, Y \in \mathcal{T}$, then $X \prec_{\mathcal{T}} X \cdot Y$ and $Y \prec_{\mathcal{T}} X \cdot Y$
2. If $X \in \mathcal{T}$ and $\mathcal{K} \in \mathcal{T}_{\mathcal{K}}$, then $X \prec_{\mathcal{T}} \langle X \rangle_{\mathcal{K}}$ and $\mathcal{K} \prec_{\mathcal{T}} \langle X \rangle_{\mathcal{K}}$
3. If $X \in \mathcal{T}$, then $X \prec_{\mathcal{T}} h(X)$

By $\preceq_{\mathcal{T}}$ we denote the transitive and reflexive closure of $\prec_{\mathcal{T}}$.
The relation $\preceq_{\mathcal{T}}$ is simply the relation of being a subterm in the algebra constructed above.

Example 11.1:

Below we show several examples of dependencies between message terms expressed by means of relations $\prec_{\mathcal{T}}$ and $\preceq_{\mathcal{T}}$:

- $\mathcal{N}_{\mathcal{B}} \prec_{\mathcal{T}} \langle \mathcal{N}_{\mathcal{B}} \rangle_{\mathcal{K}_{\mathcal{B}}}$—the term $\mathcal{N}_{\mathcal{B}}$ is an immediate subterm of the term $\langle \mathcal{N}_{\mathcal{B}} \rangle_{\mathcal{K}_{\mathcal{B}}}$
- $\mathcal{I}_{\mathcal{A}} \prec_{\mathcal{T}} \mathcal{I}_{\mathcal{A}} \cdot \langle \tau_{\mathcal{A}} \cdot \mathcal{I}_{\mathcal{B}} \cdot \mathcal{K}_{\mathcal{AB}} \rangle_{\mathcal{K}_{\mathcal{AS}}}$—the term $\mathcal{I}_{\mathcal{A}}$ is an immediate subterm of the term $\mathcal{I}_{\mathcal{A}} \cdot \langle \tau_{\mathcal{A}} \cdot \mathcal{I}_{\mathcal{B}} \cdot \mathcal{K}_{\mathcal{AB}} \rangle_{\mathcal{K}_{\mathcal{AS}}}$
- $\mathcal{N}_{\mathcal{A}} \preceq_{\mathcal{T}} \langle \mathcal{N}_{\mathcal{A}} \cdot \mathcal{N}_{\mathcal{B}} \rangle_{\mathcal{K}_{\mathcal{A}}}$—the term $\mathcal{N}_{\mathcal{A}}$ is a subterm of the term $\langle \mathcal{N}_{\mathcal{A}} \cdot \mathcal{N}_{\mathcal{B}} \rangle_{\mathcal{K}_{\mathcal{A}}}$
- $\mathcal{K}_{\mathcal{AB}} \preceq_{\mathcal{T}} \langle \tau_{\mathcal{S}} \cdot \mathcal{L}_{\mathcal{F}} \cdot \mathcal{K}_{\mathcal{AB}} \cdot \mathcal{I}_{\mathcal{A}} \rangle_{\mathcal{K}_{\mathcal{BS}}}$—the term $\mathcal{K}_{\mathcal{AB}}$ is a subterm of the term $\langle \tau_{\mathcal{S}} \cdot \mathcal{L}_{\mathcal{F}} \cdot \mathcal{K}_{\mathcal{AB}} \cdot \mathcal{I}_{\mathcal{A}} \rangle_{\mathcal{K}_{\mathcal{BS}}}$

The behavior of the protocol users depends on their knowledge, and particularly on which messages they are able to generate on the basis of available information. This is important especially in case of the Intruder which aim is to break the protocol, for example, by composing a message against the protocol scheme. The construction presented below allows to represent knowledge resources (terms) which can be built out by means of a given set $\mathcal{X}$ of terms.

Next, for any $\mathcal{X} \subseteq \mathcal{T}$ we define inductively a sequence of the sets $(\mathcal{X}^n)_{n \in \mathbb{N}}$ that are subsets of $\mathcal{T}$ as follows:

- $\mathcal{X}^0 \stackrel{def}{=} \mathcal{X}$
- $\mathcal{X}^{n+1} \stackrel{def}{=} \mathcal{X}^n \cup \{Z \in \mathcal{T} \mid (\exists X, Y \in \mathcal{X}^n, \mathcal{K} \in \mathcal{X} \cap \mathcal{T}_{\mathcal{K}})\, Z = X \cdot Y \vee Z = \langle X \rangle_{\mathcal{K}} \vee Z = h(X)\}$

* $\langle X \rangle_{\mathcal{K}}$ is a term that is interpreted as a ciphertext containing the letter X encrypted with the key $\mathcal{K}$.

Intuitively, the set $\mathcal{X}^{n+1}$ contains the, gradually built, letter terms from $\mathcal{X}^n$ using the operations of composition, encryption, and hashing. For $\mathcal{X} \in 2^{\mathcal{T}}_{fin}$ the set $Comp(\mathcal{X}) \stackrel{def}{=} \bigcup_{n \in \mathrm{N}} \mathcal{X}^n$ is composed of all the letter terms that can be constructed out of elements of $\mathcal{X}$ only.*

Example 11.2:

Below we present examples of applying the *Comp* operator. Let $\mathcal{X} = \{\mathcal{P}, \mathcal{Q}, \mathcal{K}, \mathcal{T}, \mathcal{L}\}$. The respective letter terms are then constructable from $\mathcal{X}$:

- $\mathcal{P} \cdot \mathcal{Q} \in Comp(\mathcal{X})$
- $\langle \mathcal{P} \rangle_{\mathcal{K}} \in Comp(\mathcal{X})$
- $\langle \mathcal{P} \cdot \mathcal{T} \rangle_{\mathcal{K}} \in Comp(\mathcal{X})$
- $\langle \mathcal{P} \cdot \mathcal{T} \rangle_{\mathcal{K}} \cdot \langle \mathcal{Q} \cdot \mathcal{L} \rangle_{\mathcal{K}} \in Comp(\mathcal{X})$

In our approach we deal with specifications of time-dependent protocols, therefore we have to be able to express time dependencies associated with the correct execution of each protocol step. To this end we define the set of time conditions to be used in specifications of time dependences of protocol instructions.

Definition 11.3:

The set of time conditions C is defined by the following grammar:

$$\mathrm{tc} ::= true \mid \tau_i - \tau_j \leq \mathcal{L}_{\mathcal{F}} \mid \mathrm{tc} \wedge \mathrm{tc}, \;\; where\ \tau_i \in \mathcal{T}_R, \tau_j \in \mathcal{T}_T, and\ \mathcal{L}_{\mathcal{F}} \in \mathcal{T}_L.$$

Notice that we have introduced the three basic constructions of time conditions. The simplest condition is *true*, which characterises executions of steps independent from the time conditions. The condition $\tau_i - \tau_j \leq \mathcal{L}_{\mathcal{F}}$ expresses that the time τ_i of sending a message minus the time τ_j of generating the timestamp is less or equal than the lifetime $\mathcal{L}_{\mathcal{F}}$. The above conditions can be combined through logical conjunction operator.

Now, we are in a position to define formally the syntax of a protocol step and then the syntax of a protocol itself. Our notion of a step is clearly more complicated than in Common Language as it provides the information not only about the sender $\mathcal{P}$, the receiver $\mathcal{Q}$, and the letter $\mathcal{L}$ transmitted from $\mathcal{P}$ to $\mathcal{Q}$, but also about the letters and the generated secrets that are necessary to compose $\mathcal{L}$. The intended aim of this extra information is to point out to additional actions of the sender like generating new secrets or composing the letter $\mathcal{L}$.

Definition 11.4:

By a *(protocol) step* α we mean an ordered pair (α^1, α^2), where

1. α^1 is a triple $(\mathcal{P}, \mathcal{Q}, \mathcal{L}) \in \mathcal{T}_{\mathcal{P}} \times \mathcal{T}_{\mathcal{P}} \times \mathcal{T}$
2. α^2 is a 4-tuple $(\tau, \mathcal{X}, \mathcal{G}, \mathrm{tc}) \in \mathcal{T}_R \times 2^{\mathcal{T}}_{fin} \times 2^{\mathcal{T}_K \cup \mathcal{T}_N \cup \mathcal{T}_T}_{fin} \times C$

having the following intuitive meaning:

- $\mathcal{P}$—the sender
- $\mathcal{Q}$—the receiver
- $\mathcal{L}$—the letter transmitted from $\mathcal{P}$ to $\mathcal{Q}$

* Decryption is not allowed here.

- τ—the time of the step execution*
- $\mathcal{X}$—the set of letters necessary to compose $\mathcal{L}$
- $\mathcal{G}$—the set of freshly generated secrets necessary to compose $\mathcal{L}$
- tc—the step time condition

which satisfies the following conditions:

1. $\mathcal{P} \neq \mathcal{Q}$ (no sending to itself)
2. $\mathcal{L} \in Comp(\mathcal{X})$ ($\mathcal{L}$ is composable from $Comp(\mathcal{X})$)
3. $(\forall \mathcal{Y} \subseteq \mathcal{X})(\mathcal{L} \in Comp(\mathcal{Y}) \Rightarrow \mathcal{Y} = \mathcal{X})$ ($\mathcal{X}$ is a minimal set from which $\mathcal{L}$ can be constructed)
4. $\mathcal{G} \subseteq \mathcal{X}$ (the secrets of $\mathcal{G}$ are elements of $\mathcal{X}$)

Notice that $\alpha^1 = (\mathcal{P}, \mathcal{Q}, \mathcal{L})$ describes a protocol step having the following intuitive meaning: *Sender $\mathcal{P}$ sends the message $\mathcal{L}$ to Receiver $\mathcal{Q}$*. This is denoted in Common Language by $\mathcal{P} \rightarrow \mathcal{Q} : \mathcal{L}$.

Definition 11.5:

By a *protocol* Σ we mean a finite sequence of steps $(\alpha_1, \ldots, \alpha_n)$.

Example 11.3:

NSPK is used as our first working example. Let us recall that the original version of this protocol does not include timestamps. But, since we are able to investigate timed protocols, we consider below a time version of NSPK, in which instead of the nonces N_A and N_B there occur timestamps T_A and T_B. The use of timestamps does not restrict in any way properties of the protocol. Pseudo-randomly generated nonces may be, after all, integral components of the timestamps T_A and T_B. We denote the time version of NSPK by NSPK$_{Time}$. Additionally, we assume that the lifetime $\mathcal{L}_\mathcal{F}$ is fixed in the system as some constant. The syntax of NSPK$_{Time}$ in our language is given as follows.

- $\mathcal{T}_P = \{\mathcal{A}, \mathcal{B}\}$
- $\mathcal{T}_I = \{\mathcal{I}_\mathcal{A}, \mathcal{I}_\mathcal{B}\}$
- $\mathcal{T}_K = \{\mathcal{K}_\mathcal{A}, \mathcal{K}_\mathcal{B}\}$
- $\mathcal{T}_T = \{\tau_\mathcal{A}, \tau_\mathcal{B}\}$
- $\mathcal{T}_L = \{\mathcal{L}_\mathcal{F}\}$

NSPK$_{Time}$ is defined by the sequence of the steps $(\alpha_1, \alpha_2, \alpha_3)$, where:

1. $\alpha_1 = (\alpha_1^1, \alpha_1^2)$
   - $\alpha_1^1 = (\mathcal{A};\ \mathcal{B};\ \langle \mathcal{I}_\mathcal{A} \cdot \tau_\mathcal{A} \rangle_{\mathcal{K}_\mathcal{B}})$
   - $\alpha_1^2 = (\tau_1; \{\mathcal{I}_\mathcal{A}, \tau_\mathcal{A}, \mathcal{K}_\mathcal{B}\}; \{\tau_\mathcal{A}\}; \tau_1 - \tau_\mathcal{A} \leq \mathcal{L}_\mathcal{F})$
2. $\alpha_2 = (\alpha_2^1, \alpha_2^2)$
   - $\alpha_2^1 = (\mathcal{B};\ \mathcal{A};\ \langle \tau_\mathcal{A} \cdot \tau_\mathcal{B} \rangle_{\mathcal{K}_\mathcal{A}})$
   - $\alpha_2^2 = (\tau_2; \{\tau_\mathcal{A}, \tau_\mathcal{B}, \mathcal{K}_\mathcal{A}\}; \{\tau_\mathcal{B}\}; \tau_2 - \tau_\mathcal{A} \leq \mathcal{L}_\mathcal{F} \wedge \tau_2 - \tau_\mathcal{B} \leq \mathcal{L}_\mathcal{F})$
3. $\alpha_3 = (\alpha_3^1, \alpha_3^2)$
   - $\alpha_3^1 = (\mathcal{A};\ \mathcal{B};\ \langle \tau_\mathcal{B} \rangle_{\mathcal{K}_\mathcal{B}})$
   - $\alpha_3^2 = (\tau_1; \{\tau_\mathcal{B}, \mathcal{K}_\mathcal{B}\}; \emptyset; \tau_3 - \tau_\mathcal{A} \leq \mathcal{L}_\mathcal{F} \wedge \tau_3 - \tau_\mathcal{B} \leq \mathcal{L}_\mathcal{F})$

The time condition $\tau_1 - \tau_\mathcal{A} \leq \mathcal{L}_\mathcal{F}$ says that B can receive the message $\langle \mathcal{I}_\mathcal{A}, \tau_\mathcal{A} \rangle_{\mathcal{K}_\mathcal{B}}$ only under the condition that the difference between the time τ_1 of sending/receiving the message and the time $\tau_\mathcal{A}$ of

* For simplicity reasons we assume that the time of sending is equal to the time of receiving the message, so there is no delay on delivery of the message. However, delays are allowed between two consecutive steps of the protocol.

generating the timestamp is smaller than $\mathcal{L_F}$. Note that this condition is clearly satisfied when A is an honest user. Similarly, the time condition $\tau_2 - \tau_\mathcal{A} \leq \mathcal{L_F}$ expresses that $\mathcal{B}$ can send the message only under the condition that the difference between the time τ_2 of transmitting the message and the time $\tau_\mathcal{A}$ of generating the timestamp is smaller than $\mathcal{L_F}$. The remaining time conditions concern similar properties. □

Example 11.4:

WMF is used as our second working example. The syntax of WMF is given as follows.

- $\mathcal{T}_P = \{\mathcal{A}, \mathcal{B}, \mathcal{S}\}$
- $\mathcal{T}_I = \{\mathcal{I_A}, \mathcal{I_B}\}$
- $\mathcal{T}_K = \{\mathcal{K_{AS}}, \mathcal{K_{BS}}, \mathcal{K_{AB}}\}$
- $\mathcal{T}_T = \{\tau_\mathcal{A}, \tau_\mathcal{S}\}$
- $\mathcal{T}_L = \{\mathcal{L_F}\}$

WMF is defined by the sequence of the two steps (α_1, α_2), where:

1. $\alpha_1 = (\alpha_1^1, \alpha_1^2)$
   - $\alpha_1^1 = (\mathcal{A};\ \mathcal{S};\ \mathcal{I_A} \cdot \langle \tau_\mathcal{A} \cdot \mathcal{I_B} \cdot \mathcal{K_{AB}} \rangle_{\mathcal{K_{AS}}})$
   - $\alpha_1^2 = (\tau_1; \{\mathcal{I_A}, \mathcal{I_B}, \tau_\mathcal{A}, \mathcal{K_{AB}}, \mathcal{K_{AS}}\}; \{\tau_\mathcal{A}, \mathcal{K_{AB}}\}; \tau_1 - \tau_\mathcal{A} \leq \mathcal{L_F})$
2. $\alpha_2 = (\alpha_2^1, \alpha_2^2)$
   - $\alpha_2^1 = (\mathcal{S};\ \mathcal{B};\ \langle \tau_\mathcal{S} \cdot \mathcal{I_A} \cdot \mathcal{K_{AB}} \rangle_{\mathcal{K_{BS}}})$
   - $\alpha_2^2 = (\tau_2; \{\mathcal{I_A}, \tau_\mathcal{S}, \mathcal{K_{AB}}, \mathcal{K_{BS}}\}; \{\tau_\mathcal{S}\}; \tau_2 - \tau_\mathcal{A} \leq \mathcal{L_F} \wedge \tau_2 - \tau_\mathcal{S} \leq \mathcal{L_F})$

The time condition $\tau_1 - \tau_\mathcal{A} \leq \mathcal{L_F}$ says that S can receive the message $\langle \tau_\mathcal{A}, \mathcal{I_B}, \mathcal{K_{AB}} \rangle_{\mathcal{K_{AS}}}$ only under condition that the difference between the time τ_1 of the message transmission and the time $\tau_\mathcal{A}$ of generating the timestamp is smaller than $\mathcal{L_F}$. Note that this condition is clearly satisfied when A is an honest user. Similarly, the time condition $\tau_2 - \tau_\mathcal{A} \leq \mathcal{L_F}$ expresses that S can send (and B—receive) the message only under the condition that the difference between the time τ_2 of transmitting the message and the time $\tau_\mathcal{A}$ of generating the timestamp is smaller than $\mathcal{L_F}$. □

Example 11.5:

Kerberos is used as our next working example. The syntax of Kerberos is as follows:

- $\mathcal{T}_P = \{\mathcal{A}, \mathcal{B}, \mathcal{S}\}$
- $\mathcal{T}_I = \{\mathcal{I_A}, \mathcal{I_B}\}$
- $\mathcal{T}_K = \{\mathcal{K_{AS}}, \mathcal{K_{BS}}, \mathcal{K_{AB}}\}$
- $\mathcal{T}_T = \{\tau_\mathcal{B}, \tau_\mathcal{S}\}$
- $\mathcal{T}_L = \{\mathcal{L_F}\}$

Kerberos is defined by the following sequence of the four steps: $(\alpha_1, \alpha_2, \alpha_3, \alpha_4)$, where:

1. $\alpha_1 = (\alpha_1^1, \alpha_1^2)$
   - $\alpha_1^1 = (\ \mathcal{A};\ \mathcal{S};\ \mathcal{I_A} \cdot \mathcal{I_B}\ )$
   - $\alpha_1^2 = (\tau_1; \{\mathcal{I_A}, \mathcal{I_B}\}; \emptyset; \mathit{true})$
2. $\alpha_2 = (\alpha_2^1, \alpha_2^2)$
   - $\alpha_2^1 = (\ \mathcal{S};\ \mathcal{A};\ \langle \tau_\mathcal{S} \cdot \mathcal{L_F} \cdot \mathcal{K_{AB}} \cdot \mathcal{I_B} \rangle_{\mathcal{K_{AS}}} \cdot \langle \tau_\mathcal{S} \cdot \mathcal{L_F} \cdot \mathcal{K_{AB}} \cdot \mathcal{I_A} \rangle_{\mathcal{K_{BS}}})$
   - $\alpha_2^2 = (\tau_2; \{\tau_\mathcal{S}, \mathcal{L_F}, \mathcal{K_{AB}}, \mathcal{I_A}, \mathcal{I_B}, \mathcal{K_{AS}}, \mathcal{K_{BS}}\}; \{\tau_\mathcal{S}, \mathcal{K_{AB}}\}; \tau_2 - \tau_\mathcal{S} \leq \mathcal{L_F})$
3. $\alpha_3 = (\alpha_3^1, \alpha_3^2)$
   - $\alpha_3^1 = (\ \mathcal{A};\ \mathcal{B};\ \langle \tau_\mathcal{S} \cdot \mathcal{L_F} \cdot \mathcal{K_{AB}} \cdot \mathcal{I_A} \rangle_{\mathcal{K_{BS}}} \cdot \langle \tau_\mathcal{S} \cdot \mathcal{I_A} \rangle_{\mathcal{K_{AB}}})$
   - $\alpha_3^2 = (\tau_3; \{\tau_\mathcal{S}, \mathcal{I_A}, \mathcal{K_{AB}}, \langle \tau_\mathcal{S} \cdot \mathcal{L_F} \cdot \mathcal{K_{AB}} \cdot \mathcal{I_A} \rangle_{\mathcal{K_{BS}}}\}; \emptyset; \tau_3 - \tau_\mathcal{S} \leq \mathcal{L_F})$

4. $\alpha_4 = (\alpha_4^1, \alpha_4^2)$
   - $\alpha_4^1 = (\ \mathcal{B};\ \mathcal{A};\ \langle\tau_{\mathcal{A}}\rangle_{\mathcal{K}_{\mathcal{AB}}}\ )$
   - $\alpha_4^2 = (\tau_4; \{\tau_{\mathcal{A}}, \mathcal{K}_{\mathcal{AB}}\}; \emptyset; \tau_4 - \tau_{\mathcal{S}} \leq \mathcal{L}_{\mathcal{F}})$ □

The above examples show how our formalism can be used for specification of all the properties of protocols which determine its correct execution. The internal and external actions of the protocol, as well as the appropriate time conditions, are described in a fully formal way. Introducing the above formal structures allows now to interpret them adequately in a mathematical model of protocol executions. Next, we add the Intruder to our approach.

11.6 Computational Structure

In this section, we define a computational structure, which generates all the computations under our interpretation of a protocol. Next, we represent these computations by the runs of the product automaton generated by a network of timed automata.

Many constructions presented below are similar to syntactic ones. Moreover, several new structures and notions, associated with introducing the Intruder and its behavior, are defined.

Now, we begin to construct a structure that reflects different protocol executions and, what is more important, their interleavings. A special attention should be paid to the introduction of the Intruder, which can execute the protocol on its behalf, as well as impersonate other users.

Similarly to Section 11.5, we start with defining the following basic sets:

- $\mathbf{P} = \{p_1, p_2, \ldots, p_{n_p}\}$—the honest *participants* in the network
- $\mathbf{P_\iota} = \{\iota, \iota(p_1), \iota(p_2), \ldots, \iota(p_{n_p})\}$—the *Intruder* ι and the *Intruder impersonating* the participant p_i for $1 \leq i \leq n_p$
- $\mathbf{I} = \{i_{p_1}, \ldots, i_{p_{n_p}}, i_\iota\}$—the *identifiers* of the participants in the network
- $\mathbf{K} = \bigcup_{i=1}^{n_p}\{k_{p_i}, k_{p_i}^{-1}\} \cup (\bigcup_{i=1}^{n_p}\bigcup_{j=1}^{n_p}\{k_{p_ip_j}\} \setminus \bigcup_{i=1}^{n_p}\{k_{p_ip_i}\})$—the cryptographic *keys* of the participants
- $\mathbf{N} = \bigcup_{i=1}^{n_p}\{n_{p_i}^1, \ldots, n_{p_i}^{k_N}\} \cup \{n_\iota^1, \ldots, n_\iota^{k_N}\}$—the *nonces**
- $\mathbf{T} = \bigcup_{i=1}^{n_p}\{t_{p_i}^1, \ldots, t_{p_i}^{n_T}\} \cup \{t_\iota^1, \ldots, t_\iota^{n_T}\}$—the *timestamps* of the participants
- $\Gamma = \{l_1, l_2, \ldots, l_{n_L}\} \subseteq \mathcal{R}_+$—the *lifetimes*

Now, we introduce a special algebra of terms that enables to define the messages and the relations among them.

Definition 11.6:

A set of *letters* $\mathbf{L}$ is defined as the smallest set satisfying the following conditions:

- $\mathbf{P} \cup \mathbf{P_\iota} \cup \mathbf{I} \cup \mathbf{K} \cup \mathbf{N} \cup \mathbf{T} \cup \Gamma \subseteq \mathbf{L}$
- If $x, y \in \mathbf{L}$, then the concatenation $x \cdot y \in \mathbf{L}$
- If $x \in \mathbf{L}$ and $k \in \mathbf{K}$, then $\langle x\rangle_k \in \mathbf{L}$, where $\langle x\rangle_k$ is a ciphertext consisting of the letter x encrypted with the key k
- If $x \in \mathbf{L}$, then $h(x) \in \mathbf{L}$, where $h(x)$ is the hash value of the letter x

Next, we define some auxiliary relations over the set $\mathbf{L}$.

---

* As before, we assume that n_p and k_N are some fixed natural numbers. For simplicity, we take the same number of nonces for each participant.

Definition 11.7:

The *(immediate) subletter relation* $\prec\!\cdot \subseteq \mathbf{L} \times \mathbf{L}$ is the smallest relation satisfying the following conditions:

1. If $x, y \in \mathbf{L}$, then $x \prec\!\cdot\, x \cdot y$ and $y \prec\!\cdot\, x \cdot y$
2. If $x \in \mathbf{L}$ and $k \in \mathbf{K}$, then $x \prec\!\cdot\, \langle x \rangle_k$ and $k \prec\!\cdot\, \langle x \rangle_k$
3. If $x \in \mathbf{L}$, then $x \prec\!\cdot\, h(x)$

Similarly to the previous section by $\preceq$ we denote the transitive and reflexive closure of $\prec\!\cdot$. Next, for any $X \subseteq \mathbf{L}$ we define a sequence of the sets $(X^n)_{n\in\mathbf{N}}$ that are also subsets of $\mathbf{L}$, where

- $X^0 \stackrel{def}{=} X$
- $X^{n+1} \stackrel{def}{=} X^n \cup \{z \in \mathbf{L} \mid (\exists x, y \in X^n, k \in X \cap \mathbf{K})\; z = x \cdot y \vee z = \langle x \rangle_k \vee z = h(x)\}$

The intuition behind this definition is the same as for the corresponding one in Section 11.5, that is, the set X^{n+1} contains the, gradually built, letters from X^n using the operations of composition, encryption and hashing.

Next, we define the set $Comp(X) \stackrel{def}{=} \bigcup_{n\in\mathbf{N}} X^n$, which consists of all the letters that can be composed out of elements of X only,* and the new construction: the set $Sublet(X) \stackrel{def}{=} \{l \in \mathbf{L} \mid (\exists x \in X)\; l \preceq x\}$, which contains all the subletters of X.

As concerns detecting an incorrect behaviour of the Intruder, which might lead to an attack on the protocol, it is essential to examine exactly the knowledge that the Intruder is able to gain during the protocol execution. This knowledge can be derived directly from the messages transmitted to the Intruder, eavesdropping, or using information obtained with respect to possessed cryptographic keys. The construction below allows to specify precisely the knowledge acquired by the users, and especially by the Intruder.

Definition 11.8:

Let $X \subseteq \mathbf{L}$ and $K \subseteq \mathbf{K}$. The set $\xi_K(X) \subseteq \mathbf{L}$ is defined as the smallest set of letters satisfying the following conditions:

1. $X \subseteq \xi_K(X)$
2. if $l \cdot m \in \xi_K(X)$, then $l \in \xi_K(X)$ and $m \in \xi_K(X)$
3. if $\langle l \rangle_k \in \xi_K(X)$ and $k \in \xi_K(X) \cup K$, then $l \in \xi_K(X)$

The set $\xi_K(X)$ contains all the letters which can be retrieved from X by decomposing a concatenation or decrypting a letter using a key, which is either in $\xi_K(X)$ or in K. By $\xi(X)$ we mean the set $\xi_\emptyset(X)$.

As we have mentioned before, in our approach we treat a protocol as an algorithm, that is, as an abstract concept. Since our interest lies in different executions of the same protocol and its possible interleavings, it is necessary to introduce functions that map abstract/syntactic notions into the semantic model built. The two definitions given below are crucial for the approach used here aimed at building models of protocol executions and their verification. The notion of a partial interpretation determines the roles that the users play in each protocol execution, but neglects the execution time. The notion of interpretation specifies the protocol execution not only by defining the roles played in it by the users and what messages are transmitted, but also by determining strictly the time of the protocol execution.

Now, we define partial interpretations and interpretations of the letter terms of $\mathcal{T}$ which are used for defining the runs of protocols.

* Decryption is not allowed here.

Definition 11.9:

A *partial interpretation* of the set of the letter terms $\mathcal{T}$ is defined as any injection $\bar{f} : \mathcal{T} \rightarrow \mathbf{L}$ satisfying the following conditions:

1. $\bar{f}(\mathcal{T}_P) \subseteq \mathbf{P} \cup \mathbf{P_i}, \bar{f}(\mathcal{T}_I) \subseteq \mathbf{I}, \bar{f}(\mathcal{T}_K) \subseteq \mathbf{K}, \bar{f}(\mathcal{T}_N) \subseteq \mathbf{N}, \bar{f}(\mathcal{T}_L) \subseteq \mathbf{\Gamma}, \bar{f}(\mathcal{T}_T) \subseteq \mathbf{T}$
2. $(\forall X, Y \in \mathcal{T})\, \bar{f}(X \cdot Y) = \bar{f}(X) \cdot \bar{f}(Y)$ (homomorphism)
3. $(\forall X \in \mathcal{T})(\forall \mathcal{K} \in \mathcal{T}_K)\, \bar{f}(\langle X \rangle_{\mathcal{K}}) = \langle \bar{f}(X) \rangle_{\bar{f}(\mathcal{K})}$ (homomorphism)
4. $(\forall X \in \mathcal{T})\, \bar{f}(h(X)) = h(\bar{f}(X))$ (homomorphism)
5. If $\bar{f}(\mathcal{P}) = p$ for $p \in \mathbf{P}$, then $\bar{f}(\mathcal{I}_{\mathcal{P}}) = i_p, \bar{f}(\mathcal{N}_{\mathcal{P}}) \in \{n_p^1, \ldots, n_p^{k_N}\}, \bar{f}(\mathcal{K}_{\mathcal{P}}) = k_p, \bar{f}(\mathcal{K}_{\mathcal{P}}^{-1}) = k_p^{-1}$, and $\bar{f}(\tau_{\mathcal{P}}) \in \{t_p^1, \ldots, t_p^{n_T}\}$
6. If $\bar{f}(\mathcal{P}) = \iota$, then $\bar{f}(\mathcal{I}_{\mathcal{P}}) = i_\iota, \bar{f}(\mathcal{K}_{\mathcal{P}}) = k_\iota$ and $\bar{f}(\mathcal{K}_{\mathcal{P}}^{-1}) = k_\iota^{-1}$
7. If $\bar{f}(\mathcal{P}) = \iota(p)$, then $\bar{f}(\mathcal{I}_{\mathcal{P}}) = i_p, \bar{f}(\mathcal{K}_{\mathcal{P}}) = k_p$ and $\bar{f}(\mathcal{K}_{\mathcal{P}}^{-1}) = k_p^{-1}$
8. $\bar{f}(\mathcal{T}_P) \setminus \mathbf{P_i} \neq \emptyset$

Condition 1 states that the atomic terms are mapped into the corresponding objects of the computational structure, that is, the symbols representing the participants are mapped into the participants, and so on.

The next three conditions 2, 3, and 4 guarantee the homomorfical separation between the symbols mapped.

The condition 5 says that the symbols related to a given participant are mapped into the corresponding objects (the identifiers, the keys, the nonces, the timestamps) in the structure.

The condition 6 determines that if the Intruder ι behaves honestly in an execution of the protocol, then it uses its own identifier and keys. Since the Intruder can use an arbitrary nonce, there is no condition regarding its nonces.

Condition 7 states that if the Intruder ι impersonates another participant p in some interpretation, then in any execution under this interpretation the keys of p and its identifier need to be used by ι. Then, due to Condition 1, no participant symbol is mapped to p in this interpretation.

The last condition says that at least one honest participant takes part in each interpretation.

Definition 11.10:

For a given partial interpretation $\bar{f}$ by an *interpretation* (associated with $\bar{f}$) of the set of the letter terms $\mathcal{T}$ and the set of time variables $\mathcal{T}_R$ we mean any injection $f : \mathcal{T} \cup \mathcal{T}_R \rightarrow \mathbf{L} \cup \mathcal{R}_*$ such that $f\,|_{\mathcal{T} \setminus \mathcal{T}_T} = \bar{f}$ and $f(\mathcal{T}_T \cup \mathcal{T}_R) \subseteq \mathcal{R}_*$.

As we can see from the introduced definitions, the timestamps and the times of the messages transmitted are finally mapped into time instances represented by positive real numbers.

In a correctly constructed protocol, the honest users execute protocol steps according to its scheme using their knowledge. The Intruder can try to use different data and apply them in the protocol execution. What is more, the Intruder may also try to compose messages, which are needed for deceits, in different manners. The following definition describes what can be generated from a given set of messages.

In order to define later an interpretation of a protocol step in which the Intruder is the sender, we need the notion of a set of generators for a letter.

Definition 11.11:

Let $l \in \mathbf{L}$ be a letter and $X \subseteq \mathbf{L}$. The set X is said to be a set of *generators* of l (denoted by $X \vdash l$) if the following conditions are met:

1. $X \subseteq Sublet(\{l\})$
2. $l \in Comp(X)$
3. $(\forall m \in X)(m \notin Comp(X \setminus \{m\})$
4. $(\forall m \in X)(l \notin Comp(X \setminus \{m\})$

Intuitively, we have $X \vdash l$ if all the elements of X are subletters of l, l can be composed out of the elements of X, and X is a minimal such a set.

Example 11.6:

Consider the letter $l = \langle t_s \rangle_{k_{bs}} \cdot \langle k_{ab} \rangle_{k_{as}}$. Consider the following sets $X_1 = \{t_s, k_{ab}, k_{bs}, k_{as}\}$, $X_2 = \{\langle t_s \rangle_{k_{bs}}, k_{ab}, k_{as}\}$, $X_3 = \{t_s, k_{bs}, \langle k_{ab} \rangle_{k_{as}}\}$, $X_4 = \{\langle t_s \rangle_{k_{bs}}, \langle k_{ab} \rangle_{k_{as}}\}$. Observe that all these sets are generators of l, that is, we have $X_i \vdash l$, for $i = 1, 2, 3, 4$. □

Next, we extend an interpretation f to deal with the time conditions of $\mathcal{C}$. This is defined inductively as follows:

- $f(true) := true$
- $f(\tau_i - \tau_j \leq \mathcal{L}_{\mathcal{F}}) := f(\tau_i) - f(\tau_j) \leq f(\mathcal{L}_{\mathcal{F}})$
- $f(tc_1 \wedge tc_2) := f(tc_1) \wedge f(tc_2)$

Having defined a set of letter generators and an interpretation of $\mathcal{T}$, we are now in a position to apply these notions to a protocol step and then to the whole protocol.

Definition 11.12:

Consider a step $\alpha = (\alpha^1, \alpha^2) = ((\mathcal{P}, \mathcal{Q}, \mathcal{L}), (\tau_\alpha, \mathcal{X}, \mathcal{G}, tc))$ of a given protocol Σ and an interpretation f of $\mathcal{T} \cup \mathcal{T}_R$, which satisfies the condition $\bigwedge_{\tau \in \mathcal{G} \cap \mathcal{T}_T} (f(\tau_\alpha) = f(\tau))$, that is, the time assigned to each timestamp generated is equal to the time of the protocol step. By the *f-interpretation* of the step α (denoted by $f(\alpha)$) we mean the following tuple:

- $((f(\mathcal{P}), f(\mathcal{Q}), f(\mathcal{L})), (f(\tau_\alpha), f(\mathcal{X}), f(\mathcal{G}), f(tc)))$, if $f(\mathcal{P}) \in \mathbf{P}$
- $((f(\mathcal{P}), f(\mathcal{Q}), f(\mathcal{L})), (f(\tau_\alpha), \{X \mid X \vdash f(\mathcal{L})\}, \emptyset, f(tc)))$, if $f(\mathcal{P}) \in \mathbf{P_i}$

In the case of the Intruder being the sender, we assume that it can compose a letter $f(\mathcal{L})$ from any set which generates $f(\mathcal{L})$. We also assume that the Intruder has got a set of nonces, keys and timestamps at its disposal and it does not need to generate them. The reason is that the Intruder can use the same nonce, key, or timestamp many times and in many sessions.

In order to define the protocol executions and knowledge of the participants as well as of the Intruder we introduce the following auxiliary notions.

Let $f(\alpha_i) = ((p, q, l), (t, X, G, tc))$, for some $p, q \in \mathbf{P} \cup \mathbf{P_i}$, $X \in 2^{\mathbf{L}}_{fin}$, $G \in 2^{\mathbf{K} \cup \mathbf{N}}_{fin}$, and $l \in \mathbf{L}$.

Then, we define the following notations:

- $Send^{f(\alpha_i)} = p$ (the sender of $f(\alpha_i)$)
- $Lett^{f(\alpha_i)} = l$ (the letter of $f(\alpha_i)$)
- $Gen^{f(\alpha_i)} = G$ (the set of generated new secrets in $f(\alpha_i)$)
- $Resp^{f(\alpha_i)} = q$ (the responder of $f(\alpha_i)$)

- $Part^{f(\alpha_i)} = \{Send^{f(\alpha_i)}, Resp^{f(\alpha_i)}\}$
- $Time^{f(\alpha_i)} = t$ and $TConstr^{f(\alpha_i)} = tc$

In addition, if $Send^{f(\alpha_i)} \in \mathbf{P}$, then let $Comp^{f(\alpha_i)} = X$ (the set of letters that are sufficient to compose $Lett^{f(\alpha_i)}$) and if $Send^{f(\alpha_i)} \in \mathbf{P_i}$, then let $Comp^{f(\alpha_i)} = \bigcup\{X \subseteq L \mid X \vdash Lett^{f(\alpha_i)}\}$ (the union of sets which generate $Lett^{f(\alpha_i)}$). Similarly, for a partial interpretation $\overline{f}$ the interpretation f is associated with, we use notations: $Send^{\overline{f}(\alpha_i)}, Lett^{\overline{f}(\alpha_i)}, Gen^{\overline{f}(\alpha_i)}, Resp^{f(\alpha_i)}$ for $\overline{f}(\mathcal{P}), \overline{f}(\mathcal{L})), \overline{f}(\mathcal{G}), \overline{f}(\mathcal{Q})$, respectively.

The constructions presented in this section allow now to define formally the concrete steps of the protocol as well as its full execution.

Definition 11.13:

Let f be an interpretation satisfying the following conditions:

- $f(\alpha_1), f(\alpha_2), \ldots, f(\alpha_n)$ are f-interpretations of the steps of the protocol $\Sigma = (\alpha_1, \alpha_2, \ldots, \alpha_n)$
- $\bigwedge_{i=1}^{n} TConstr^{f(\alpha_i)} \equiv true$

By *the f-execution of a protocol* Σ we mean the sequence $f(\Sigma) = (f(\alpha_1), f(\alpha_2), \ldots, f(\alpha_n))$.

Example 11.7:

Consider NSPK$_{Time}$ given in Example 11.3 and the interpretation f, where $f(\mathcal{A}) = a, f(\mathcal{B}) = b, f(\mathcal{L_F}) = l$, and $f(\tau_i) = t_i$ for $i = 1, 2, 3$. Observe that from Definition 11.9, we have for this interpretation $f(\mathcal{I_A}) = i_a, f(\tau_\mathcal{A}) = t_a, f(\tau_\mathcal{B}) = t_b, f(\mathcal{K_A}) = k_a$, and $f(\mathcal{K_B}) = k_b$. $f-$interpretations of the NSPK$_{Time}$ steps are the following:

1. $f(\alpha_1) = ((a, b, \langle i_a, t_a \rangle_{k_b}), (t_1, \{i_a, t_a, k_b\}, \{t_a\}, t_1 - t_a \leq l))$
2. $f(\alpha_2) = ((b, a, \langle t_a, t_b \rangle_{k_a}), (t_2, \{t_a, t_b, k_a\}, \{t_b\}, t_2 - t_a \leq l \wedge t_2 - t_b \leq l))$
3. $f(\alpha_3) = ((a, b, \langle t_b \rangle_{k_a}), (t_3, \{t_b, k_a\}, \emptyset, t_3 - t_a \leq l \wedge t_3 - t_b \leq l))$ □

Example 11.8:

Consider another execution of NSPK$_{Time}$ involving the Intruder given by the following interpretation g, where $g(\mathcal{A}) = b, g(\mathcal{B}) = \iota(a), g(\tau_\mathcal{B}) = t_\iota, g(\mathcal{L_F}) = l$, and $g(\tau_i) = t_i$ for $i = 1, 2, 3$. Observe that from Definition 11.9, we have for this interpretation $g(\mathcal{I_A}) = i_b, g(\tau_\mathcal{A}) = t_b, g(\mathcal{K_A}) = k_b, g(\mathcal{K_B}) = k_a$. $g-$interpretations of the NSPK$_{Time}$ steps are as follows:

1. $g(\alpha_1) = ((b, \iota(a), \langle i_b, t_b \rangle_{k_a}), (t_1, \{i_a, t_b, k_a\}, \{t_b\}, t_1 - t_b \leq l))$
2. $g(\alpha_2) = ((\iota(a), b, \langle t_b, t_\iota \rangle_{k_b}), (t_2, \{\{t_b, t_\iota, k_b\}, \langle t_b, t_\iota \rangle_{k_b}\}, \emptyset, t_2 - t_b \leq l \wedge t_2 - t_\iota \leq l))$
3. $g(\alpha_3) = ((b, \iota(a), \langle t_\iota \rangle_{k_a}), (t_3, \{t_\iota, k_b\}, \emptyset, t_3 - t_b \leq l \wedge t_3 - t_\iota \leq l))$ □

As it can be seen on the basis of the above examples, each interpretation corresponds to a concrete protocol execution. Since we are interested in examining interleavings of different protocol executions, it is necessary to introduce some auxiliary constructions enabling to analyze sets of interpretations. The following notions allow, among others, to express the knowledge of the users participating in an interleaving of multiple protocol executions.

For a set of interpretations $\mathcal{F}$, we define the set $Comp^{p}_{\mathcal{F}}$ ($Comp^{\iota}_{\mathcal{F}}$) of the letters that are needed by the participant $p \in \bigcup_{f \in \mathcal{F}} f(\mathcal{T_P}) \setminus \mathbf{P}_\iota$ (the Intruder ι, resp.) in order to compose all the letters transmitted in an execution under any interpretation $f \in \mathcal{F}$.

Definition 11.14:

For an honest user p, the set $Comp^p_{\mathcal{F}} = \bigcup_{i=1}^n \bigcup\{Comp^{f(\alpha_i)} \mid f \in \mathcal{F} \wedge Send^{f(\alpha_i)} = p\}$.
For the Intruder ι, the set $Comp^{\iota}_{\mathcal{F}} = \bigcup_{i=1}^n \bigcup\{Comp^{f(\alpha_i)} \mid f \in \mathcal{F} \wedge Send^{f(\alpha_i)} \in \mathbf{P}_{\iota}\}$.
Consider any finite sequence of interpretations of k protocol steps $\tau = (f^1(\alpha_{i_1}), f^2(\alpha_{i_2}), \ldots, f^k(\alpha_{i_k}))$. For every $p \in \bigcup_{i=1}^k f^i(\mathcal{T}_P)$ we define a sequence of the participant's knowledge $(\kappa^j_p)_{j=1,\ldots,k}$ at the steps of the protocol.

Definition 11.15:

For an honest user $p \in \bigcup_{f \in \mathcal{F}} f(\mathcal{T}_P) \setminus \mathbf{P_i}$ its knowledge at the step j is given inductively as follows:
$\kappa^0_p = \mathbf{I} \cup \{k_p^{-1}\} \cup \{k_q \mid q \in \mathbf{P}\} \cup \{k_{\iota}\}$,

$$\kappa_p^{j+1} = \begin{cases} \kappa_p^j & \text{if} \quad p \notin Part^{f^{j+1}(\alpha_{i_{j+1}})} \\ \kappa_p^j \cup Gen^{f^{j+1}(\alpha_{i_{j+1}})} & \text{if} \quad p = Send^{f^{j+1}(\alpha_{i_{j+1}})} \\ Comp^p_{\mathcal{F}} \cap \xi_{\{k_p^{-1}\}}(\kappa_p^j \cup \{Lett^{f^{j+1}(\alpha_{i_{j+1}})}\}) & \text{if} \quad p = Resp^{f^{j+1}(\alpha_{i_{j+1}})} \end{cases}$$

The intuition behind the above definition is as follows. The knowledge of a user not participating in a protocol step is not changing. If a user is the initiator of a step, then his knowledge is extended with the set of the generated nonces. If a user is the responder of a step, then his knowledge is extended by all the letters, which can be retrieved from the former knowledge and the letter actually received. But, for efficiency reasons it is restricted to a subset of $Comp^p_{\mathcal{F}}$, that is, to the letters that the user needs in order to compose any letter in any execution determined by $\mathcal{F}$.

Example 11.9:

Consider f-execution of NSPK$_{Time}$ given in Example 11.7 and a sequence of interpretations of the protocol steps $\tau = (f(\alpha_1), f(\alpha_2), f(\alpha_3))$. The sequences of knowledge $(\kappa^i_a)_{i=0,1,2,3}$ and $(\kappa^i_b)_{i=0,1,2,3}$ are as follows:

- $\kappa^0_a = \{i_a, i_b, k_a, k_a^{-1}, k_b\}$ (the initial knowledge of the user a)
- $\kappa^1_a = \kappa^0_a \cup \{t_a\}$ (after executing the first step the knowledge of the user a increases by the timestamp t_a generated by it)
- $\kappa^2_a = \kappa^1_a \cup \{t_b\}$ (after executing the second step a receives the ciphertext $\langle t_a, t_b \rangle_{k_a}$, it can decrypt it and its knowledge increases by the timestamp t_b contained in the ciphertext)
- $\kappa^3_a = \kappa^2_a$
- $\kappa^0_b = \{i_b, i_a, k_b, k_b^{-1}, k_a\}$
- $\kappa^1_b = \kappa^0_b \cup \{t_a\}$
- $\kappa^2_b = \kappa^1_b \cup \{t_b\}$
- $\kappa^3_b = \kappa^2_b$ □

The knowledge of the Intruder is defined in a slightly different way, which is explained after the definition.

Definition 11.16:

For the Intruder, the knowledge at each step j of the protocol is common for all $p \in \bigcup_{f \in \mathcal{F}} f(\mathcal{T}_P) \cap \mathbf{P_i}$ and it is given inductively as follows:

$$\kappa_\iota^0 = \mathbf{I} \cup \{k_\iota^{-1}, k_\iota\} \cup \{k_q \mid q \in \mathbf{P}\} \cup \{n_\iota^1, \ldots, n_\iota^{k_N}\} \cup \{t_\iota^1, \ldots, t_\iota^{n_T}\}$$

$$\kappa_\iota^{j+1} = \begin{cases} \kappa_\iota^j & \text{if} \quad Resp^{f^{j+1}(\alpha_{i_{j+1}})} \notin P_\iota \\ Comp_{\mathcal{F}}^\iota \cap \xi_{\{k_\iota^{-1}\}}(\kappa_\iota^j \cup \{Lett^{f^{j+1}(\alpha_{i_{j+1}})}\}) & \text{if} \quad Resp^{f^{j+1}(\alpha_{i_{j+1}})} \in P_\iota \end{cases}$$

Some explanation about the above definition is in place. If the Intruder is not the responder of a letter, then his knowledge does not change. Otherwise, the Intruder is retrieving all the possible letters from his knowledge and the letter he has received (restricted to a subset of $Comp_{\mathcal{F}}^\iota$ for efficiency reasons).

For simplicity we assume that the Intruder does not generate its nonces as it can use them many times in all executions. Moreover, we assume that the Intruder is in a possession of timestamps that are ready to use. When used they need to be assigned a current time. These assumptions do not introduce any limitations.

Example 11.10:

Consider g-execution of NSPK$_{Time}$ given in Example 11.8 and a sequence of interpretations of the protocol steps $\tau = (g(\alpha_1), g(\alpha_2), g(\alpha_3))$. The sequences of knowledge $(\kappa_b^i)_{i=0,1,2,3}$ and $(\kappa_\iota^i)_{i=0,1,2,3}$ are as follows:

- $\kappa_b^0 = \{i_b, i_a, k_b, k_b^{-1}, k_a\}$
- $\kappa_b^1 = \kappa_b^0 \cup \{t_b\}$
- $\kappa_b^2 = \kappa_b^1 \cup \{t_\iota\}$
- $\kappa_b^3 = \kappa_b^2$
- $\kappa_\iota^0 = \{i_b, i_a, k_b, k_b^{-1}, k_a, t_\iota\}$
- $\kappa_\iota^1 = \kappa_\iota^0 \cup \{\langle i_b, t_b \rangle_{k_a}\}$
- $\kappa_\iota^2 = \kappa_\iota^1$
- $\kappa_\iota^3 = \kappa_\iota^2 \cup \{\langle t_\iota \rangle_{k_a}\}$

In the following definition we formulate the conditions, which guarantee that a sequence of protocol step interpretations is a computation of the protocol.

Definition 11.17:

By *a computation* of the protocol Σ we mean any finite sequence of protocol step interpretations $\tau = (f^1(\alpha_{i_1}), f^2(\alpha_{i_2}), \ldots, f^k(\alpha_{i_k}))$, which meets the following conditions:

1. $(\forall k \in \mathcal{N}_+)[i_k > 1 \Rightarrow (\exists j < k)(f^j = f^k \wedge i_j = i_k - 1)]$
2. $(\forall k, j \in \mathcal{N}_+)[k \neq j \Rightarrow Gen^{f^k(\alpha_{i_k})} \cap Gen^{f^j(\alpha_{i_j})} = \emptyset]$
3. $(\forall j \in \mathbf{N}_+)[Lett^{f^j(\alpha_{i_j})} \in Comp(\kappa^{j-1}_{Send^{f^j(\alpha_{i_j})}} \cup Gen^{f^j(\alpha_{i_j})})]$
4. $\forall_{n=1,\ldots,k-1}(Time^{f^n(\alpha_{i_n})} < Time^{f^{n+1}(\alpha_{i_{n+1}})})$
5. $\forall_{n=1,\ldots,k} TConstr^{f^n(\alpha_{i_n})} = true$

The first condition states that for each protocol step (except for the first one) in interpretation f, there is a preceding step in the same interpretation. The second one says that the sets of the generated nonces are disjoint, whereas the third one guarantees that the letter $Lett^{(f^j(\alpha_{i_j}))}$ can be transmitted by $Send^{(f^j(\alpha_{i_j}))}$ only if it can be composed from the set of currently generated nonces and the knowledge of the participant $Send^{(f^j(\alpha_{i_j}))}$ at the step $j-1$. The next condition guarantees that the time is progressing between two successive steps of the execution. The last condition says that all time conditions for all interpretations hold true.

Example 11.11:

Observe that the sequence $\tau = (g(\alpha_1), g(\alpha_2), g(\alpha_3))$ given in Example 11.10 is not a computation of NSPK_{Time}. This sequence does not meet the third condition put on a computation. □

11.6.1 Attacks upon Protocols

The protocols are used in order to establish a secure communication channel between two parties involved in the communication. This is obtained by ensuring that each party is confident about several security properties: the other party is who they say they are (*authentication*), a confidential information is not visible to nonauthorized parties (*secrecy*), the information exchanged by two parties cannot be altered by the Intruder (*integrity*), and finally the parties taking part in the transaction cannot deny it later (*nonrepudiation*). In our approach, we focus on checking authentication and secrecy only. We say that a given protocol is *correct* if it cannot be executed in such a way that identifiers or keys of one participant are used by someone else. Depending on the protocol and its goal, we verify authentication, secrecy, or a combination of both of them. Having this in mind, we give the following definition of an attack.

Definition 11.18:

By *an attacking execution* we mean any execution under an interpretation f, where $f(\mathcal{P}) = \iota(p)$, for some $\mathcal{P} \in \mathcal{T_P}$ and $p \in \mathbf{P}$. We define two types of attacks:

- An *authentication attack* upon a protocol is any of its computations such that an attacking execution is its subsequence.
- A *secrecy attack* upon a protocol is any of its computations in which a secret information is a part of the knowledge of the Intruder.

Example 11.12:

Consider the following two interpretations of NSPK_{Time}:

- $h_1(\mathcal{A}) = a,\ h_1(\mathcal{B}) = \iota,\ h_1(\tau_\mathcal{A}) = t_a,\ h_1(\tau_\mathcal{B}) = t_b,\ h_1(\mathcal{K_A}) = k_a,\ h_1(\mathcal{K_B}) = k_\iota, h_1(\mathcal{L_F}) = l$ and $h_1(\tau_i) = t_i^1$ for $i = 1, 2, 3$, and
- $h_2(\mathcal{A}) = \iota(a),\ h_2(\mathcal{B}) = b,\ h_2(\tau_\mathcal{A}) = t_a,\ h_2(\tau_\mathcal{B}) = t_b,\ h_2(\mathcal{K_A}) = k_a,\ h_2(\mathcal{K_B}) = k_b, h_2(\mathcal{L_F}) = l$ and $h_2(\tau_i) = t_i^2$ for $i = 1, 2, 3$.

Consider the following sequence of steps interpretations:

- $h_1(\alpha_1) = ((a, \iota, \langle i_a, t_a\rangle_{k_\iota}), (t_1^1, \{i_a, t_a, k_\iota\}, \{t_a\}, t_1^1 - t_a \leq l))$
- $h_2(\alpha_1) = ((\iota(a), b, \langle i_a, t_a\rangle_{k_b}), (t_1^2, \{\{i_a, t_a, k_b\}, \{\langle i_a, t_a\rangle_{k_b}\}\}, \emptyset, t_1^2 - t_a \leq l))$
- $h_2(\alpha_2) = ((b, \iota(a), \langle t_a, t_b\rangle_{k_a}), (t_2^2, \{t_a, t_b, k_a\}, \{t_b\}, t_2^2 - t_a \leq l \wedge t_2^2 - t_b \leq l))$
- $h_1(\alpha_2) = ((\iota, a, \langle t_a, t_b\rangle_{k_a}), (t_2^1, \{\{t_a, t_b, k_a\}, \{\langle t_a, t_b\rangle_{k_a}\}\}, t_2^1 - t_a \leq l \wedge t_2^1 - t_b \leq l))$
- $h_1(\alpha_3) = ((a, \iota, \langle t_b\rangle_{k_\iota}), (t_3^1, \{t_b, k_\iota\}, \emptyset, t_3^1 - t_a \leq l \wedge t_3^1 - t_b \leq l))$
- $h_2(\alpha_3) = ((\iota(a), b, \langle t_b\rangle_{k_b}), (t_3^2, \{\{t_b, k_b\}, \{\langle t_b\rangle_{k_b}\}\}, \emptyset, t_3^2 - t_a \leq l \wedge t_3^2 - t_b \leq l))$

Observe that the sequences of knowledge $(\kappa_a^i)_{i=0,\ldots,6}$, $(\kappa_b^i)_{i=0,\ldots,6}$, $(\kappa_\iota^i)_{i=0,\ldots,6}$ are as follows:

- $\kappa_a^0 = \{i_a, i_b, i_\iota, k_a, k_a^{-1}, k_b, k_\iota\}$
- $\kappa_a^1 = \kappa_a^0 \cup \{t_a\} = \kappa_a^2 = \kappa_a^3$
- $\kappa_a^4 = \kappa_a^3 \cup \{t_b\} = \kappa_a^5 = \kappa_a^6$

- $\kappa_b^0 = \{i_b, i_a, i_\iota, k_b, k_b^{-1}, k_a, k_\iota\} = \kappa_b^1$
- $\kappa_b^2 = \kappa_b^1 \cup \{t_a\}$
- $\kappa_b^3 = \kappa_b^2 \cup \{t_b\} = \kappa_b^4 = \kappa_b^5 = \kappa_b^6$
- $\kappa_\iota^0 = \{i_\iota, i_a, i_b, k_a, k_b, k_\iota, k_\iota^{-1}\}$
- $\kappa_\iota^1 = \kappa_\iota^0 \cup \{t_a\} = \kappa_\iota^2$
- $\kappa_\iota^3 = \kappa_\iota^2 \cup \{\langle t_a, t_b \rangle_{k_a}\} = \kappa_\iota^4$
- $\kappa_\iota^5 = \kappa_\iota^4 \cup \{t_b\} = \kappa_\iota^6$

Observe that for some time moments $t_1^1, t_1^2, t_2^1, t_2^2, t_3^1, t_3^2$, which satisfy $t_1^1 \leq t_1^2 \leq t_2^2 \leq t_2^1 \leq t_3^1 \leq t_3^2$, the sequence $(h_1(\alpha_1), h_2(\alpha_1), h_2(\alpha_2), h_1(\alpha_2), h_1(\alpha_1), h_1(\alpha_1))$ is a computation of the protocol $NSPK_{Time}$. This computation represents the Lowe's attack upon the original (untimed) version of NSPK given in Ref. [31]. □

11.7 Networks of Synchronizing Timed Automata

In this section, we give definitions of a timed automaton and a network of synchronizing timed automata. Next, we show how to model the executions of a protocol by the runs of a network of synchronizing timed automata, where each timed automaton represents one component of the protocol. We also introduce automata representing messages transmitted during a protocol execution and automata representing users' knowledge. An appropriate synchronization between the automata of the introduced network enables to model the executions of the protocol according to the knowledge acquired by its users.

Let C be a set of time conditions defined in a similar way to Definition 11.3, but using the clocks $\mathcal{X}$ and integers.

Definition 11.19:

A *timed automaton* (TA, for short) is a five-tuple $\mathcal{A} = (A, L, l^0, E, \mathcal{X})$, where

- A is a finite set of *actions*, where $A \cap \mathcal{R}_* = \emptyset$
- L is a finite set of *locations*
- $l^0 \in L$ is the *initial location*
- $\mathcal{X}$ is a finite set of clocks
- $E \subseteq L \times A \times C \times 2^{\mathcal{X}} \times L$ is a *transition relation*

Each element e of E is denoted by $l \xrightarrow{a,cc,X} l'$, which represents a transition from the location l to the location l', executing the action a, with the set $X \subseteq \mathcal{X}$ of clocks to be reset, and with the clock condition $cc \in C$ defining the enabling condition (guard) for e.

Given a transition $e = l \xrightarrow{a,cc,X} l'$, we write *source*$(e)$, *target*$(e)$, *action*$(e)$, *guard*$(e)$ and *reset*(e) for l, l', a, cc and X, respectively.

The clocks of a timed automaton allow to express the timing properties. An enabling condition constrains the execution of a transition without forcing it to be taken.

Several examples of timed automata used for modeling of protocols executions are given at the end of this section.

11.7.1 Semantics of Timed Automata

Let $\mathcal{A} = (A, L, l^0, E, \mathcal{X})$ be a timed automaton. A *concrete state* of $\mathcal{A}$ is defined as an ordered pair (l, v), where $l \in L$ and $v \in \mathcal{R}_*^{\mathcal{X}}$ is a clock valuation. In what follows, by $[\![\mathfrak{cc}]\!]$ we mean a set of all the valuations v that satisfy a time condition $\mathfrak{cc}$. By $v[X := 0]$ we mean the valuation v' such that it assigns 0 to all clocks from the set X and it agrees with v on all the remaining clocks, that is, $v'(X) = \{0\}$ and $v'(x) = v(x)$ for each $x \in \mathcal{X} \setminus X$. The formal definitions of these notations can be found in Ref. [39]. The *state space* of $\mathcal{A}$ is a transition system $C(\mathcal{A}) = (Q, \mathfrak{s}^0, \rightarrow_c)$, where

- $Q = L \times \mathcal{R}_*^{\mathcal{X}}$ is the set of all the *concrete states*
- $\mathfrak{s}^0 = (l^0, v^0)$ with $v^0(x) = 0$ for all $x \in \mathcal{X}$ is the *initial state*
- $\rightarrow_c \subseteq Q \times (E \cup \mathcal{R}_*) \times Q$ is the *transition relation*, defined by action- and time successors as follows:
  - for any $\delta \in \mathcal{R}_*$, $(l, v) \xrightarrow{\delta}_c (l, v + \delta)$ (*time successor*)
  - for $a \in A$, $(l, v) \xrightarrow{a}_c (l', v')$ iff $(\exists \mathfrak{cc} \in C)(\exists X \subseteq \mathcal{X})$ such that $l \xrightarrow{a, \mathfrak{cc}, X} l' \in E$, $v \in [\![\mathfrak{cc}]\!]$ and $v' = v[X := 0]$ (*action successor*)

Intuitively, a time successor does not change the location l of a concrete state, but it increases the clocks. An action successor corresponding to an action a is executed when the guard $\mathfrak{cc}$ holds for v and the valuation v' obtained by resetting the clocks in X.

For $(l, v) \in Q$ and $\delta \in \mathcal{R}_*$, let $(l, v) + \delta$ denote $(l, v + \delta)$. A $\mathfrak{s}_0$-*run* ρ of $\mathcal{A}$ is a maximal sequence $\rho = \mathfrak{s}_0 \xrightarrow{\delta_0}_c \mathfrak{s}_0 + \delta_0 \xrightarrow{a_0}_c \mathfrak{s}_1 \xrightarrow{\delta_1}_c \mathfrak{s}_1 + \delta_1 \xrightarrow{a_1}_c \mathfrak{s}_2 \xrightarrow{\delta_2}_c \ldots$, where $a_i \in A$ and $\delta_i \in \mathcal{R}_*$, for each $i \geq \mathcal{N}$.

Now, we are going to use networks of timed automata for modeling executions of the protocol as well as for modeling the knowledge of the participants.

11.7.2 Product of a Network of Timed Automata

A network of timed automata can be composed into a global (*product*) timed automaton [39] in the following way. The transitions of the timed automata that do not correspond to a shared action are interleaved, whereas the transitions labeled with a shared action are synchronized.

Let $\mathfrak{I} = \{\mathfrak{i}_1, \ldots, \mathfrak{i}_{n_\mathfrak{I}}\}$ be a finite ordered set of indices, and $\mathfrak{A} = \{\mathcal{A}_\mathfrak{i} \mid \mathfrak{i} \in \mathfrak{I}\}$, where $\mathcal{A}_\mathfrak{i} = (A_\mathfrak{i}, L_\mathfrak{i}, l_\mathfrak{i}^0, E_\mathfrak{i}, \mathcal{X}_\mathfrak{i})$, be a network of timed automata indexed with $\mathfrak{I}$. The automata in $\mathfrak{A}$ are called *components*. Let $A(a) = \{\mathfrak{i} \in \mathfrak{I} \mid a \in A_\mathfrak{i}\}$ be a set of the indices of the components containing the action $a \in \bigcup_{\mathfrak{i} \in \mathfrak{I}} A_\mathfrak{i}$.

Definition 11.20:

A composition (*product*) of the timed automata $\mathcal{A}_{\mathfrak{i}_1}, \ldots, \mathcal{A}_{\mathfrak{i}_{n_\mathfrak{I}}}$ is the timed automaton $\mathcal{A} = (A, L, l^0, E, \mathcal{X})$, where

- $A = \bigcup_{\mathfrak{i} \in \mathfrak{I}} A_\mathfrak{i}$
- $L = \prod_{\mathfrak{i} \in \mathfrak{I}} L_\mathfrak{i}$
- $l^0 = (l_{\mathfrak{i}_1}^0, \ldots, l_{\mathfrak{i}_{n_\mathfrak{I}}}^0)$
- $\mathcal{X} = \bigcup_{\mathfrak{i} \in \mathfrak{I}} \mathcal{X}_\mathfrak{i}$

and the transition relation is given by

$$((l_{\mathfrak{i}_1}, .., l_{\mathfrak{i}_{n_\mathfrak{I}}}), a, \bigwedge_{\mathfrak{i} \in A(a)} \mathfrak{cc}_\mathfrak{i}, \bigcup_{\mathfrak{i} \in A(a)} X_\mathfrak{i}, (l'_{\mathfrak{i}_1}, .., l'_{\mathfrak{i}_{n_\mathfrak{I}}})) \in E \Leftrightarrow$$

$$\Leftrightarrow (\forall \mathfrak{i} \in A(a))(l_\mathfrak{i}, a, \mathfrak{cc}_\mathfrak{i}, X_\mathfrak{i}, l'_\mathfrak{i}) \in E_\mathfrak{i} \wedge (\forall \mathfrak{i} \in \mathfrak{I} \setminus A(a))\ l'_\mathfrak{i} = l_\mathfrak{i}.$$

11.7.3 Automata for Modeling Executions of the Protocol

Now we show a key construction of an automaton that models a transmission of messages during executions of a given protocol. Such automata will be synchronized with automata that reflect users' knowledge. An adequate synchronisation will provide a proper way of modeling a full protocol execution. The time conditions related to a protocol execution under an interpretation f will be given for this type of automata.

Consider a protocol $\Sigma = (\alpha_1, \ldots, \alpha_n)$ and its partial interpretation $\overline{f}$. All the executions $f(\Sigma)$, where f is associated with $\overline{f}$, are modeled by the timed automaton $A_{\overline{f}} = (\Sigma_{\overline{f}}, Q_{\overline{f}}, s_0^{\overline{f}}, \delta_{\overline{f}}, \mathcal{X}_{\overline{f}})$, where:

- $\Sigma_{\overline{f}} = \{k_{\overline{f},i} \mid 1 \leq i \leq n \wedge Send^{f(\alpha_i)} \in \mathbf{P}\} \cup \bigcup_{i=1}^{n} \bigcup \{\{k^X_{\overline{f},i}\} \mid X \subseteq \mathcal{L} \wedge Send^{f(\alpha_i)} \in \mathbf{P_I} \wedge X \vdash Lett^{f(\alpha_i)}\}$
- $Q_{\overline{f}} = \{s_0^{\overline{f}}, s_1^{\overline{f}}, s_2^{\overline{f}}, \ldots, s_n^{\overline{f}}\}$ is the set of states, where $s_0^{\overline{f}}$ is the initial state
- $\mathcal{X}_{\overline{f}} = \bigcup_{i=1}^{n} \{z_t \mid t \in (\mathbf{T} \cap Gen^{f(\alpha_i)})\}$
- $\delta_{\overline{f}} = \{(s_{i-1}^{\overline{f}}, k_{\overline{f},i}, Z(TConstr^{\overline{f}(\alpha_i)}), \{z_t \mid t \in (\mathbf{T} \cap Gen^{f(\alpha_{i-1})})\}, s_i^{\overline{f}}) \mid 1 \leq i \leq n \wedge k_{\overline{f},i} \in \Sigma_{\overline{f}}\} \cup$
  $\cup \{(s_{i-1}^{\overline{f}}, k^X_{\overline{f},i}, Z(TConstr^{\overline{f}(\alpha_i)}), \{z_t \mid t \in (\mathbf{T} \cap Gen^{f(\alpha_{i-1})})\}, s_i^{\overline{f}}) \mid 1 \leq i \leq n \wedge k^X_{\overline{f},i} \in \Sigma_{\overline{f}}\}$

The time conditions $Z(TConstr^{\overline{f}(\alpha_i)})$ are defined inductively as follows:

1. $Z(true) = true$
2. $Z(\tau_i - \tau_p \leq l) = z_{t_p} \leq l$
3. $Z(TConstr_1 \wedge TConstr_2) = Z(TConstr_1) \wedge Z(TConstr_2)$

The intuition behind the above definition is as follows. Each state $s_i^{\overline{f}}$ of the automaton is reached after executing one of the steps $f(\alpha_i)$ of the execution $f(\Sigma)$ for some f associated with $\overline{f}$. Additionally the state $s_i^{\overline{f}}$ can be reached iff $TConstr^{f(\alpha_i)} = true$ and all the clocks $\{z_t \mid t \in (\mathbf{T} \cap Gen^{\overline{f}(\alpha_{i-1})})\}$ are reset.

If the sender of this step is honest, then there is only one possibility to execute this step as the sender needs to have the required knowledge for composing the letter transmitted in this step. However, if the Intruder is the sender of this step, then there are many possibilities to execute this step determined by the sets of generators of the letter to be sent. Each of these cases is labeled with a different label $k^X_{\overline{f},i}$.

Below we present examples of execution automata for some selected interpretations.

Example 11.13:

The execution automaton for NSPK$_{Time}$ given in Example 11.3 for the partial interpretation $\overline{f}$, where $\overline{f}(\mathcal{A}) = a, \overline{f}(\mathcal{B}) = b, \overline{f}(\mathcal{L_F}) = l$, is as follows:

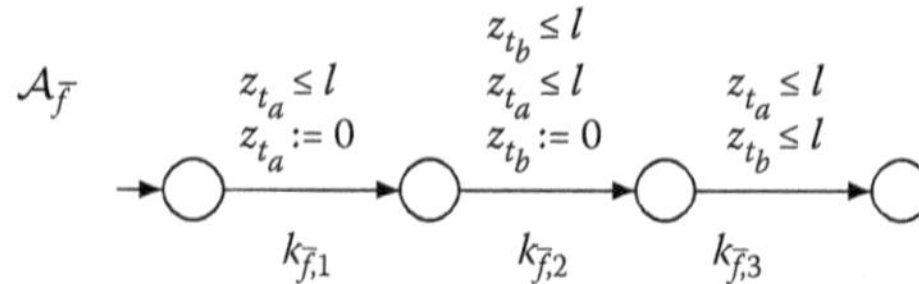

□

As we have mentioned before, the above automaton models messages transmitted during executions of NSPK$_{Time}$ according to the interpretation f. However, this automaton cannot be treated as a full model of the execution, as it does not model the users' knowledge needed for proper execution of the protocol. Below we show two more examples of similar automata for WMF and Kerberos.

Example 11.14:

The execution automaton for WMF given in Example 11.4 with the partial interpretation $\overline{f}$, where $\overline{f}(\mathcal{A}) = a, \overline{f}(\mathcal{B}) = b, \overline{f}(\mathcal{S}) = s, \overline{f}(\tau_A) = t_a, \overline{f}(\tau_S) = t_s, \overline{f}(\mathcal{L_F}) = l, \overline{f}(\mathcal{K_{AS}}) = k_{as}, \overline{f}(\mathcal{K_{BS}}) = k_{bs}, \overline{f}(\mathcal{K_{AB}}) = k_{ab}$, is as follows:

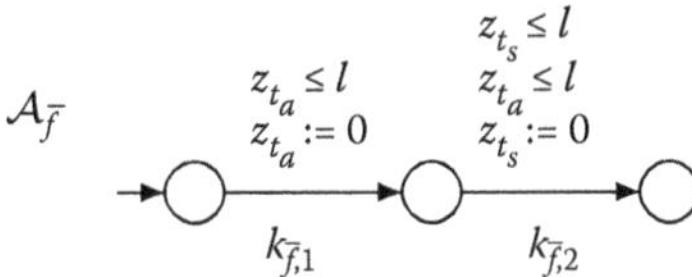

Example 11.15:

The execution automaton for Kerberos given in Example 11.5 for the partial interpretation $\overline{f}$, where $\overline{f}(\mathcal{A}) = a, \overline{f}(\mathcal{B}) = b, \overline{f}(\mathcal{S}) = s, \overline{f}(\mathcal{L_F}) = l, \overline{f}(\tau_{\mathcal{A}}) = t_a, \overline{f}(\tau_S) = t_s$, is as follows:

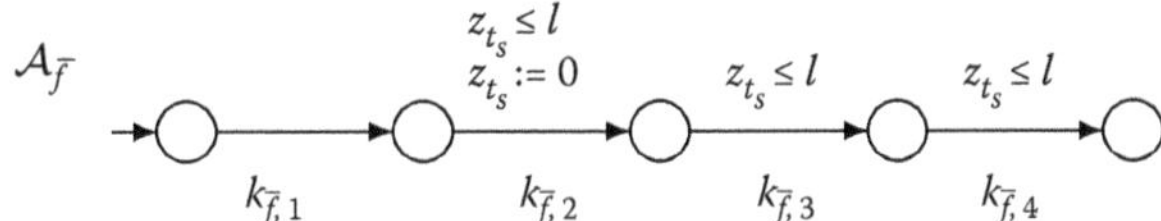

□

11.7.4 Automata for Modeling Knowledge of the Participants

Now, we give a construction of automata that model the users' knowledge acquired during a protocol execution. This knowledge is essential for correct execution of the protocol. Each automaton is related to a user and its concrete information. The automata act as characteristic functions. They indicate whether given information belongs to the knowledge set of a user. Suitable synchronization with execution automata allows to associate execution automata with knowledge automata. The network, constructed this way, fully models executions of a given protocol: the messages transmitted and the users' knowledge sets.

Consider a finite set of protocol partial interpretations $\mathcal{F}$. For each honest participant $p \in (\bigcup_{\overline{f} \in \mathcal{F}} \overline{f}(\mathcal{T_P}) \setminus \mathbf{P_i})$ and each element $l \in Comp^p_{\mathcal{F}} \setminus \kappa^0_p$, we define the following (knowledge) automaton $\mathcal{A}^p_l = (\Sigma^p_l, Q^p_l, q^p_l, \delta^p_l, \mathcal{X}^p_l)$, where

- $\Sigma^p_l \stackrel{def}{=} \{k \in \bigcup_{\overline{f} \in \mathcal{F}} \Sigma_{\overline{f}} \mid Cond_1(k) \vee Cond_2(k)\}$ with
  $Cond_1(k) := (\exists \overline{f} \in \mathcal{F})(\exists \sigma \in \delta_{\overline{f}})$
  $\sigma = (s^{\overline{f}}_{i-1}, k, TConstr^{\overline{f}(\alpha_i)}, \{z_t \mid t \in (\mathbf{T} \cap Gen^{f(\alpha_{i-1})})\}, s^{\overline{f}}_i) \wedge$
  i. $(p = Send^{f(\alpha_i)} \wedge l \in Gen^{f(\alpha_i)}) \vee$
  ii. $(p = Resp^{f(\alpha_i)} \wedge l \in \xi_{\{k_p^{-1}\}}(\{Lett^{f(\alpha_i)}\}) \wedge$
  $\wedge\ (\forall j \in \{1, \ldots, i-1\})((p = Resp^{f(\alpha_j)} \Rightarrow l \notin \xi_{\{k_p^{-1}\}}(Lett^{f(\alpha_j)})) \wedge$
  $\wedge\ (p = Send^{f(\alpha_j)} \Rightarrow l \notin Gen^{f(\alpha_j)}))$
  $Cond_2(k) := (\exists \overline{f} \in \mathcal{F})(\exists \sigma \in \delta_{\overline{f}})$
  $\sigma = (s^{\overline{f}}_{i-1}, k, TConstr^{\overline{f}(\alpha_i)}, \{z_t \mid t \in (\mathbf{T} \cap Gen^{f(\alpha_{i-1})})\}, s^{\overline{f}}_i) \wedge$
  iii. $(p = Send^{f(\alpha_i)} \wedge l \in Comp^{f(\alpha_i)} \setminus Gen^{f(\alpha_i)}) \vee$

iv. $(p = Resp^{f(\alpha_i)} \wedge l \in \xi_{\{k_p^{-1}\}}(\{Lett^{f(\alpha_i)}\}) \wedge$
$\wedge (\forall j \in \{1, \dots, i-1\})((p = Resp^{f(\alpha_j)} \Rightarrow l \notin \xi_{\{k_p^{-1}\}}(Lett^{f(\alpha_j)})) \wedge$
$\wedge (p = Send^{f(\alpha_j)} \Rightarrow l \notin Gen^{f(\alpha_j)}))$

- $Q_l^p = \{q_l^p, s_l^p\}$ is the set of states (q_l^p is the initial state)
- δ_l^p is the transition relation given as follows
  $(q_l^p, k, True, \emptyset, s_l^p) \in \delta_l^p$ iff $Cond_1(k)$, $(s_l^p, k, True, \emptyset, s_l^p) \in \delta_l^p$ iff $Cond_2(k)$

If the automaton A_l^p is in the state q_l^p, then this means that the participant p does not know l. If the automaton A_l^p moves to the state s_l^p, then this corresponds to the fact that p learns about l and can use it. The condition (i) specifies that l is generated by p at the step $\overline{f}(\alpha_i)$. The condition (ii) says that p learns about l at the step $\overline{f}(\alpha_i)$. This is modeled only once in order to reduce the number of the transitions. The condition (iii), which defines the loop, enables p to use l while composing new letters. The condition (iv) enables to receive l in a different execution that the one, which was used to define the condition (ii).

For the Intruder ι and each letter $l \in Comp^{\iota}_{\mathcal{F}} \setminus \kappa_\iota^0$ we define the knowledge automaton $A_l^\iota = (\Sigma_l^\iota, Q_l^\iota, q_l^\iota, \delta_l^\iota, \mathcal{X}_l^\iota)$ in a similar way to the definition of A_l^p. So, we have:

1. $Q_l^\iota = \{q_l^\iota, s_l^\iota\}$ is a set of the states
2. q_l^ι is the initial state
3. δ_l^ι is the transition relation given as follows:
   - $(q_l^\iota, k, True, \emptyset, s_l^\iota) \in \delta_l^\iota$ iff $Cond_1^\iota(k)$
   - $(s_l^\iota, k, True, \emptyset, s_l^\iota) \in \delta_l^\iota$ iff $Cond_2^\iota(k)$

The differences w.r.t A_l^p are as follows:

- The condition (i) is omitted
- In (ii, iv) we replace $p = Send^{f(\alpha_i)}$ with $Send^{f(\alpha_i)} \in \mathbf{P_i}$ and $p = Resp^{f(\alpha_i)}$ with $Resp^{f(\alpha_i)} \in \mathbf{P_i}$
- We replace (iii) with $(Send^{f(\alpha_i)} \in \mathbf{P_i} \wedge (\exists X \subseteq \mathbf{L})(X \vdash Lett^{f(\alpha_i)} \wedge l \in X \wedge k = k_{\overline{f},i}^X))$

The new condition (iii) enables ι to use l while composing new letters.

Below we show examples of networks of knowledge automata for the protocols NSPK$_{Time}$, WMF, and Kerberos. These automata are synchronized with the corresponding execution automata, previously defined.

Example 11.16:

The network of knowledge automata for NSPK$_{Time}$ given in Example 11.3 for the partial interpretation $\overline{f}$, where $\overline{f}(\mathcal{A}) = a$, $\overline{f}(\mathcal{B}) = b$, and $\overline{f}(\mathcal{L}_{\mathcal{F}}) = l$, is as follows:

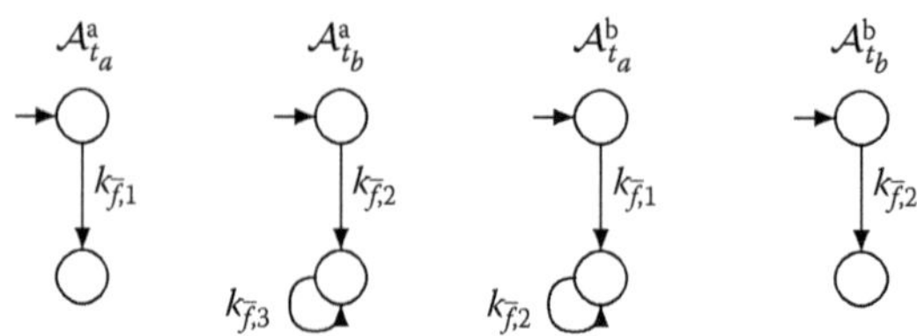

Example 11.17:

The network of knowledge automata for WMF given in Example 11.4 with the partial interpretation $\overline{f}$, where $\overline{f}(\mathcal{A}) = a, \overline{f}(\mathcal{B}) = b, \overline{f}(\mathcal{S}) = s$, and $\overline{f}(\mathcal{L}_{\mathcal{F}}) = l$, is as follows:

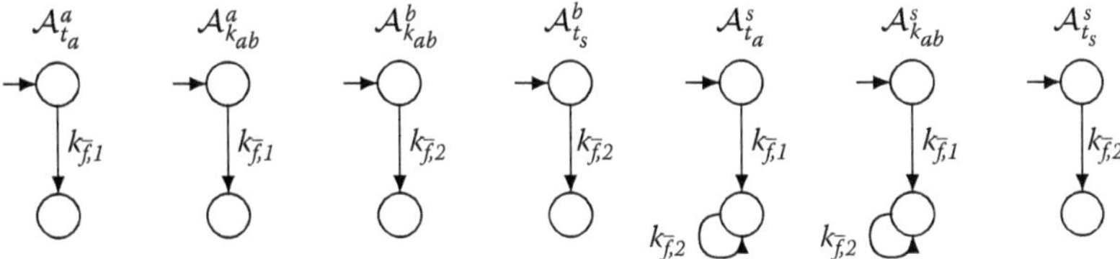

Example 11.18:

The network of automata for Kerberos given in Example 11.5 for the partial interpretation $\overline{f}$, where $\overline{f}(\mathcal{A}) = a, \overline{f}(\mathcal{B}) = b, \overline{f}(\mathcal{S}) = s$, and $\overline{f}(\mathcal{L}_{\mathcal{F}}) = l$, is as follows:

Now we proceed to formulate the main theorem of this chapter. This theorem guarantees the equivalence between the mathematical structure defined in Section 11.6 and the network of the execution and knowledge timed automata developed for the set of the interpretations of all the protocol executions.

Recall that we are dealing with the protocol Σ and a finite set $\mathcal{F}$ of its partial interpretations. Let $\mathcal{A}_{\mathcal{P}} = (N, Q, s^0, \delta, X)$ be the product automaton of the following set of the automata $\{\mathcal{A}_{\overline{f}} \mid \overline{f} \in \mathcal{F}\} \cup \{\mathcal{A}^p_l \mid p \in \bigcup_{\overline{f} \in \mathcal{F}} \overline{f}(\mathcal{T}_{\mathcal{P}}) \cap (\mathcal{P} \cup \{\iota\}) \wedge l \in Comp^p_{\mathcal{F}}\}$.

Consider the *state space* $C(\mathcal{A}_{\mathcal{P}}) = (Q, \mathfrak{s}^0, \rightarrow_c)$ of the product automaton $\mathcal{A}_{\mathcal{P}}$. Let $(k_{\overline{f^1},i_1}, k_{\overline{f^2},i_2}, \ldots, k_{\overline{f^k},i_k})$ be any sequence of actions from the set $\bigcup_{\overline{f} \in \mathcal{F}} \Sigma_f$. By a run on the word $(k_{\overline{f^1},i_1}, k_{\overline{f^2},i_2}, \ldots, k_{\overline{f^k},i_k})$ in the state space $C(\mathcal{A}_{\mathcal{P}})$ we mean the following sequence:

$$\rho = \mathfrak{s}_0 \stackrel{\delta_1}{\rightarrow}_c \mathfrak{s}_0 + \delta_1 \stackrel{a_1}{\rightarrow}_c \ldots \mathfrak{s}_{k-1} \stackrel{\delta_k}{\rightarrow}_c \mathfrak{s}_{k-1} + \delta_k \stackrel{a_k}{\rightarrow}_c \mathfrak{s}_k,$$

where $a_j = k_{\overline{f^j},i_j}$ for all $j = 1, \ldots, k$, $\delta_1 = Time^{f^1(\alpha_{i_1})}$ and $\delta_j = Time^{f^j(\alpha_{i_j})} - Time^{f^{j-1}(\alpha_{i_{j-1}})}$ for all $j = 2, \ldots, k$ and for some interpretations f^j associated with partial interpretations $\overline{f^j}$ (for $j = 1, \ldots, k$). Additionally if $\mathfrak{s}_j = (s_j, v_j)$, then $s_j \xrightarrow{a_j, cc_j, \mathcal{X}_j} s_{j+1} \in \delta$, where $cc_j = Z(TConstr^{f^j(\alpha_{i_j})})$ and $v_j \in [\![cc_j]\!]$ and $v_{j+1} = v_j[\{z_t \mid t \in (\mathbf{T} \cap Gen^{\overline{f}(\alpha_{i_j})})\} := 0]$.

The following theorem says that for each computation in the computation structure there is a corresponding run in the the state space $C(\mathcal{A}_{\mathcal{P}})$ built for this structure and moreover each run of $C(\mathcal{A}_{\mathcal{P}})$ corresponds to some computation.

Theorem 11.1:

Let $f^i \in \mathcal{F}$ for $1 \leq i \leq k$. A sequence of protocol steps $\tau = (f^1(\alpha_{i_1}), f^2(\alpha_{i_2}), \ldots, f^k(\alpha_{i_k}))$ is a computation iff there exists a run

$$\rho = \mathfrak{s}_0 \xrightarrow{\delta_1}_c \mathfrak{s}_0 + \delta_1 \xrightarrow{a_1}_c \ldots \mathfrak{s}_{k-1} \xrightarrow{\delta_k}_c \mathfrak{s}_{k-1} + \delta_k \xrightarrow{a_k}_c \mathfrak{s}_k,$$

in the space $C(\mathcal{A}_{\mathcal{P}}) = (Q, \mathfrak{s}^0, \rightarrow_c)$ on the word $(a_1, a_2, \ldots, a_k)$, where:

$$a_j \in \begin{cases} \{k_{\overline{f^j}, i_j}\} & \text{if} \quad Send^{f^j(\alpha_{i_j})} \in \mathbf{P}, \\ \{k^X_{\overline{f^j}, i_j} \mid X \vdash Lett^{f^j(\alpha_{i_j})}\} & \text{if} \quad Send^{f^j(\alpha_{i_j})} \in \mathbf{P_i}, \end{cases}$$

$\delta_1 = Time^{f^1(\alpha_{i_1})}$ and for all $j = 2, \ldots, k$ $\delta_j = Time^{f^j(\alpha_{i_j})} - Time^{f^{j-1}(\alpha_{i_{j-1}})}$.

The proof can be found in Ref. [27]. □

Thanks to above theorem, we can reduce an analysis of a protocol for the interpretations assumed to verification of the corresponding product automaton. Specifically, there is an attack on the protocol iff there is a run in the product automaton corresponding to some attacking execution.

Next, consider the following example.

Example 11.19:

The network of the execution and knowledge automata for NSPK_{Time} for the partial interpretations (see Example 11.12) $\overline{h_1}$ and $\overline{h_2}$, where $\overline{h_1}(\mathcal{A}) = a$, $\overline{h_1}(\mathcal{B}) = \iota$, $\overline{h_2}(\mathcal{A}) = \iota(a)$, $\overline{h_2}(\mathcal{B}) = b$, $\overline{h_1}(\tau_{\mathcal{A}}) = \overline{h_2}(\tau_{\mathcal{A}}) = t_a$, $\overline{h_1}(\tau_{\mathcal{B}}) = \overline{h_2}(\tau_{\mathcal{B}}) = t_b$, $\overline{h_1}(\mathcal{K}_{\mathcal{A}}) = k_a$, $\overline{h_1}(\mathcal{K}_{\mathcal{B}}) = k_\iota$, $\overline{h_2}(\mathcal{K}_{\mathcal{A}}) = k_a$, $\overline{h_2}(\mathcal{K}_{\mathcal{B}}) = k_b$ is as follows:

$\mathcal{A}_{\overline{h_1}}$: $k_{\overline{h_1},1}$ ($z_{t_a} \leq l$, $z_{t_a} := 0$), $k_{\overline{h_1},2}$ ($z_{t_a} \leq l$, $z_{t_a} \leq l$), $k_{\overline{h_1},3}$ ($z_{t_a} \leq l$, $z_{t_a} \leq l$)

$\mathcal{A}_{\overline{h_2}}$: $k_{\overline{h_2},1}$ ($z_{t_a} \leq l$), $k_{\overline{h_2},2}$ ($z_{t_a} \leq l$, $z_{t_a} \leq l$, $z_{t_a} := 0$), $k_{\overline{h_2},3}$ ($z_{t_a} \leq l$, $z_{t_a} \leq l$)

$\mathcal{A}^a_{t_a}$: $k_{\overline{h_1},1}$; $\mathcal{A}^a_{t_b}$: $k_{\overline{h_1},2}$, $k_{\overline{h_1},3}$; $\mathcal{A}^b_{t_a}$: $k_{\overline{h_2},1}$, $k_{\overline{h_2},2}$; $\mathcal{A}^b_{t_b}$: $k_{\overline{h_2},2}$; $\mathcal{A}^\iota_{t_a}$: $k_{\overline{h_1},1}$, $k_{\overline{h_2},1}$; $\mathcal{A}^\iota_{t_b}$: $k_{\overline{h_1},3}$, $k_{\overline{h_2},3}$; $\mathcal{A}^\iota_{\langle t_a \cdot t_b \rangle k_a}$: $k_{\overline{h_2},2}$, $k_{\overline{h_1},2}$

The above partial interpretations $\overline{h_1}$ and $\overline{h_1}$, correspond to the computation given in Example 11.12. The automaton $\mathcal{A}_{\overline{h_1}}$ models the executions over the partial interpretation $\overline{h_1}$, whereas $\mathcal{A}_{\overline{h_2}}$ models $\overline{h_2}$. The executions over $\overline{h_2}$ are dishonest as the Intruder is impersonating a. The partial interpretations describe the roles played by the users in the above executions. The knowledge automata model the knowledge of the users, which is gathered while executing the protocol steps. Automated verification is aimed at checking whether the local automata of the network can reach the states marked with the dots in the figure. These states correspond to completing the dishonest run by the Intruder and to its possession of some secret information. The reader can easily notice that the above network can exhibit the following run: $k_{\overline{h_1},1}, k_{\overline{h_2},1}, k_{\overline{h_2},2}, k_{\overline{h_1},2}, k_{\overline{h_1},3}, k_{\overline{h_2},3}$, which represents the known attack of Lowe [31]. □

11.8 Experimental Results

The algorithms discussed in this chapter have been implemented and used for verification with the tool Verics [23]. First, networks of timed automata corresponding to the protocols have been automatically generated, optimized, and then translated to propositional formulas in CNF. Satisfiability of the propositional formulas have been tested using the SAT-solver MiniSat.

The translation was defined in such a way that satisfiability of a formula corresponds to the existence of an attack, whereas nonsatisfiability means that there is no attack. We have tested the susceptibility of the protocols on reflection, authentication, and Man-in-the-middle attacks. The computer used to perform the experiments was equipped with the processor Intel Pentium D (3000 MHz), 2 GB main memory, the operating system Linux, and the SAT-solver MiniSat.

The following protocols have been verified: CCITT(1), WMF—the original and the corrected version [32], Denning Sacco, Kerberos, Timed Yahalom, Zhou-Golmann, NSPK—the original and the corrected version as well as their time extensions.

Typically, the state space considered consists of the two honest participants A and B, the honest server S (if needed), and the Intruder I. The experiments show that our method captures all the mentioned above types of attacks upon the protocols as well as reports correctness for flawless protocols.

In the following tables, we compare our experimental results with these obtained with TPMC [8,9] and SATMC [5]—a SAT-based model checker module of the state-of-the-art tool AVISPA [4].

Since SATMC supports verification of untimed protocols only, for comparing SATMC with Verics we consider untimed protocols and untimed versions of the timed protocols verified with our method. The results are displayed in Table 11.1, which gives also the results received for these untimed protocols, verified with TPMC. The columns (from the left) give the size of a propositional formula encoding the network (the number of variables and clauses), the memory used and the time of verification.

Table 11.2 displays the results of verification of two versions of WMF in which the number of instances of the users is scaled. In the first column, WFM is followed by, respectively, the number of instances of Alice, Bob, and Server.

TABLE 11.1 Experimental Results for NSPK, $NSPK_{Fix}$ and Kerberos

	Verics				SATMC	TPMC
Protocol	Variables	Clauses	Memory (MB)	Time (s)	Time (s)	Time (s)
NSPK	5226	15,124	2.56	0.036	0.85	0.33
$NSPK_{Fix}$	9723	28,171	2.72	1.06	0.90	0.04
Kerberos	7560	20,426	2.80	0.132	6.24	NA

TABLE 11.2 Experimental Results for WFM with Multiple Instances of the Users

	Verics				TPMC
Protocol	Variables	Clauses	Memory (MB)	Time (s)	Time (s)
TZhouG	9769	25,957	3.04	0.188	0.04
WMF 1-1-3	4331	11,432	2.33	0.056	0.34
WMF 5-5-15	22,835	62,533	4.85	0.364	628.8
WMFL 1-1-3	19,408	52,526	4.25	0.232	0.01
WMFL 4-4-12	230,493	661,217	35.49	42.990	25.83

TABLE 11.3 Experimental Results for Timed Protocols

	VerICS			
Protocol	Variables	Clauses	Memory (MB)	Time (s)
TNSPK	11,242	30,633	3.30	0.588
TNSPKL	11,638	31,753	3.16	0.644
Denning Sacco	1983	5102	2.05	0.020
CCITT	1440	3702	1.91	0.008
TYahalom	28,616	79,518	5.60	2.590
WMF	3641	9581	2.19	0.032
WMFL	20,501	56,257	4.52	0.456
UnTWMF	4965	12,677	1.97	0.016

Next, Table 11.3 gives the results of verification of timed protocols for one instance of the users. Unfortunately, we cannot compare our results with these of other tools, as they are unavailable.

The experimental results prove efficiency of our verification method. In case of the protocols susceptible to attacks our results are better than these obtained with other tools. In the other case, our results are sometimes worse. The reason for this is efficiency of SAT-solvers mainly in case of checking satisfiable propositional formulas, which takes places when an attack is found. It is worth discussing the results of verifying WFM with multiple users. In case of finding an attack, even for the increased number of the users, the times of our verification are superior to TPMC. But, the corrected version is much more difficult for verifying and requires more time than in case of TPMC.

11.9 Conclusions and Perspectives

In this chapter we described a novel methodology for verifying secrecy and authentication of timed security protocols. The method consists in modeling a finite number of participants (including the Intruder) and their knowledge about secrets, by a network of synchronizing timed automata. Due to a distributed representation our approach seems to be quite efficient. The experimental results also look very promising in comparison with SATMC (linear encoding) and TPMC.

Future research directions will include several extensions of our formalism. We would like to cover more types of attacks as well as a larger class of security protocols. Independently, we plan to investigate the computational limits of our method in terms of the number of sessions and the number of participants covered. Another direction of research consists in applying the backward induction method [24,25] for searching the structures constructed. It seems also interesting to encode our structures directly into propositional formulas avoiding an intermediate step of using timed automata. This should make our method even more efficient. We plan also to apply parametric model checking in order to synthesise the values of time parameters like lifetimes and timeouts which guarantee correctness of protocols. Finally we would like to integrate all the modules developed using our methodology into one tool for a SAT-based automated specification and verification of protocols.

Acknowledgments

The authors wish to thank Paweł Dudek and Andrzej Zbrzezny for their help in implementation.

References

1. M. Abadi and M. R. Tuttle. A semantics for a logic of authentication (extended abstract). In *PODC*, New York, NY, pp. 201–216, 1991.
2. R. Alur and D. L. Dill. A theory of timed automata. *Theoretical Computer Science*, 126: 183–235, 1994.
3. T. Amnell, G. Behrmann, J. Bengtsson, P. D'Argenio, A. David, A. Fehnker, T. Hune et al. UPPAAL—Now, next, and future. In *Proc. of the 4th Summer School 'Modelling and Verification of Parallel Processes' (MOVEP'00)*, Vol. 2067 of LNCS, pp. 99–124. Springer-Verlag, London, 2001.
4. A. Armando, D. Basin, Y. Boichut, Y. Chevalier, L. Compagna, J. Cuellar, P. Hankes Drielsma et al. The AVISPA tool for the automated validation of internet security protocols and applications. In *Proc. of 17th International Conference on Computer Aided Verification (CAV'05)*, Vol. 3576 of *LNCS*, Edinburgh, Scotland, pp. 281–285. Springer-Verlag, 2005.
5. A. Armando and L. Compagna. Sat-based model-checking for security protocols analysis. *International Journal of Information Security*, 7(1):3–32, 2008.
6. D. A. Basin and B. Wolff, editors. *Theorem Proving in Higher Order Logics, 16th International Conference, TPHOLs 2003, Roma, Italy, September 8–12, 2003, Proceedings*, Vol. 2758 of *Lecture Notes in Computer Science*. Springer, 2003.
7. G. Bella, F. Massacci, and L. C. Paulson. Verifying the set registration protocols. *IEEE Journal on Selected Areas in Communications*, 20(1):77–87, 2003.
8. M. Benerecetti, N. Cuomo, and A. Peron. TPMC: A model checker for time-sensitive security protocols. In: *Proceedings of the 2007 High Performance Computing and Simulation Conference (HPCS 2007)*, Prague, Czech Republic, pp. 742–749, 2007.
9. M. Benerecetti, N. Cuomo, and A. Peron. TPMC: A model checker for time-sensitive security protocols. *JCP*, 4(5):366–377, 2009.
10. Y. Boichut, P. C. Hoam, and O. Kouchnarenko. Tree automata for detecting attacks on protocols with algebraic cryptographic primitives. *Electronic Notes in Theoretical Computer Science*, 239:57–72. Elsevier Science, 2009.
11. Y. Boichut, P.C. Hoam, and O. Kouchnarenko. Automatic verification of security protocols using approximations. INRIA Research Report, 2005.
12. M. Burrows, M. Abadi, and R. M. Needham. A logic of authentication. *ACM Trans. Comput. Syst.*, 8(1):18–36, 1990.
13. R. Corin, S. Etalle, P. H. Hartel, and A. Mader. Timed model checking of security protocols. In *Proc. of the 2004 ACM Workshop on Formal Methods in Security Engineering (FMSE'04)*, Washington, DC, pp. 23–32. ACM, 2004.
14. R. Corin, S. Etalle, P. H. Hartel, and A. Mader. Timed analysis of security protocols. *Journal of Computer Security*, 6(1–2):53–84, 2005.
15. G. Delzanno and P. Ganty. Automatic verification of time sensitive cryptographic protocols. In *Proc. of the 10th International Conference on Tools and Algorithms for the Construction and Analysis of Systems (TACAS'04)*, Vol. 2988 of *LNCS*, Barcelona, Spain, pp. 342–356. Springer, 2004.
16. D. Dolev and A. Yao. On the security of public key protocols. *IEEE Transactions on Information Theory*, 29(2):198–207, 1983.
17. A. Doroś, A. Janowska, and P. Janowski. From specification languages to timed automata. In *Proc. of the Int. Workshop on Concurrency, Specification and Programming (CS&P'02)*, volume 161(1) of *Informatik-Berichte*, Berlin, Germany, pp. 117–128. Humboldt University, 2002.
18. N. Evans and S. Schneider. Analysing time dependent security properties in CSP using PVS. In *Proc. of the 6th European Symposium on Research in Computer Security (ESORICS'00)*, Vol. 1895 of *LNCS*, Toulouse, France, pp. 222–237. Springer, 2000.
19. T. Genet and F. Klay. Rewriting for cryptographic protocol verification. In: *Proc. of CADE 2000*, Vol. 1831 of *LNCS*, Pittsburgh, PA, pp. 271–290. Springer-Verlag, 2000.

20. R. Gorrieri, E. Locatelli, and F. Martinelli. A simple language for real-time cryptographic protocol analysis. In *Proc. of the 12th European Symposium on Programming (ESOP'03)*, Vol. 2618 of *LNCS*, Warsaw, Poland, pp. 114–128. Springer, 2003.
21. G. Jakubowska and W. Penczek. Is your security protocol on time? In *Proc. of FSEN'07*, Vol. 4767 of *LNCS*, Tehran, Iran, pp. 65–80. Springer-Verlag, 2007.
22. G. Jakubowska, W. Penczek, and M. Srebrny. Verifying security protocols with timestamps via translation to timed automata. In *Proc. of the International Workshop on Concurrency, Specification and Programming (CS&P'05)*, pp. 100–115. Warsaw University, Warsaw, Poland, 2005.
23. M. Kacprzak, W. Nabialek, A. Niewiadomski, W. Penczek, A. Półrola, M. Szreter, B. Woźna, and A. Zbrzezny. Verics 2007—A model checker for knowledge and real-time. *Fundamenta Informaticae*, 85(1–4):313–328, 2008.
24. M. Kurkowski. *Deductive methods for verification of correctnes of the authentication protocols (in polish)*. PhD thesis, Institute of Computer Science, Polish Academy of Sciences, Warsaw, Poland, 2003.
25. M. Kurkowski and W. Mackow. Using backward strategy to the Needham–Schroeder Public Key Protocol verification. In *Proc. of the ACS'03*, pp. 249–260. Kluwer Academic Publishers, Boston, MA, 2003.
26. M. Kurkowski and W. Penczek. Verifying security protocols modelled by networks of automata. *Fundamenta Informaticae*, 79(3–4):453–471, 2007.
27. M. Kurkowski and W. Penczek. Timed automata based model checking of timed security protocols. ICS PAS Report, No. 1010, ICS PAS, Warsaw, 2008.
28. M. Kurkowski and W. Penczek. Verifying timed security protocols via translation to timed automata. *Fundamenta Informaticae*, 93(1–3):245–259, 2009.
29. M. Kurkowski, W. Penczek, and A. Zbrzezny. SAT-based verification of security protocols via translation to networks of automata. In *MoChart IV*, Vol. 4428 of *LNAI*, Riva del Garda, Italy, pp. 146–165. Springer-Verlag, 2007.
30. M. Kurkowski and M. Srebrny. A quantifier-free first-order knowledge logic of authentication. *Fundamenta Informaticae*, 72(1–3):263–282, 2006.
31. G. Lowe. Breaking and fixing the Needham–Schroeder public-key protocol using FDR. In *TACAS*, *Lecture Notes in Computer Science*, Passau, Germany, pp. 147–166. Springer, 1996.
32. G. Lowe. A family of attacks upon authentication protocols. Technical report 1997/5, Department of Mathematics and Computer Science, University of Leicester, 1997.
33. G. Lowe. Casper: A compiler for the analysis of security protocols. *Journal of Computer Security*, 6(1–2):53–84, 1998.
34. A. Menezes, P. C. van Oorschot, and S. A. Vanstone. *Handbook of Applied Cryptography*. CRC Press, 1996.
35. D. Monniaux. Abstracting cryptographic protocols with tree automata. In: *Proc. of 6th International Static Analysis Symposium (SAS'99)*, Vol. 1694 of *LNCS*, Venice, Italy, pp. 149–163. Springer-Verlag, 1999.
36. R. M. Needham and M. D. Schroeder. Using encryption for authentication in large networks of computers. *Commun. ACM*, 21(12):993–999, 1978.
37. H. Ohsaki and T. Takai. ACTAS: A system design for associative and commutative tree automata theory. In: *Proc. RULE'04*, Vol. 124 of *Electronic Notes in Theoretical Computer Science*, Aachen, Germany, pp. 97–111. Elsevier Science, 2004.
38. L. C. Paulson. Inductive analysis of the internet protocol tls. *ACM Trans. Inf. Syst. Secur.*, 2(3):332–351, 1999.
39. W. Penczek and A. Półrola. *Advances in Verification of Time Petri Nets and Timed Automata: A Temporal Logic Approach*, Vol. 20 of *Studies in Computational Intelligence*. Springer-Verlag, 2006.

12

Optimal Adaptive Pattern-Matching Using Finite State Automata

Nadia Nedjah and
Luiza de Macedo Mourelle
State University of Rio de Janeiro

12.1 Introduction

Pattern matching of terms using the left-to-right tree automaton identifies a match (or lack thereof) after a single scan of the target term. Unnecessary positions may be inspected to ensure that no backtracking is needed when pattern matching fails. Using the left-to-right automaton with the adaptive strategy, the evaluation of the subterms rooted at inspected positions is forced in the left-to-right order. If such evaluations do not terminate, the evaluation of the subject term does not terminate either. However, the order of pattern matching terms need not be left-to-right. Sometimes, the evaluation of the same subject term would terminate if subterms were evaluated in a different order so that non-terminating evaluations might be avoided. We illustrate this through Example 12.1:

Example 12.1:

Consider the equational program with the following set of rewrite rules:

$$\begin{array}{lll} f(a,\omega,a) & \to a & (r_1) \\ f(a,b,b) & \to b & (r_2) \\ c & \to c & (r_3) \end{array}$$

where $\#f = 3$ and as usual a, b, and c are constants. Consider the matching automata of Figure 12.1, wherein nonfinal states are labeled inside with the positions inspected and final states are labeled with the name of the matched rule.

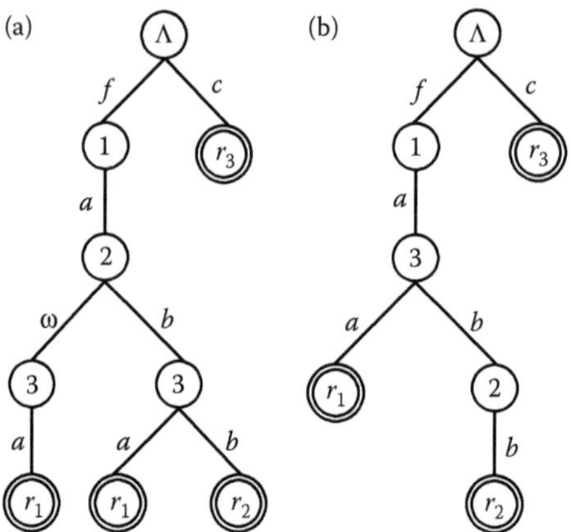

FIGURE 12.1 Matching automata for $\{r_1\colon fa\omega a, r_2\colon fabb, r_3\colon c\}$. (a) Left-to-right automaton, (b) adaptive automaton.

With appropriate data structures, pattern-matching can be performed in any given order. So, one way to improve efficiency, that is, matching times, which may improve termination too, consists of adapting the traversal order to suit the input patterns. We identify such an adaptive traversal order for a given pattern set and construct the corresponding automaton.

In general, pattern-matching of terms can be performed according to a prescribed traversal order. The pattern-matching order in the lazy reduction strategy may affect the size of the matching automaton, the matching time and in the worst case, the termination properties of term evaluations. Recall that the lazy strategy is the top-down left-to-right lazy strategy used in most lazy functional languages (Augustsson 1984; Dershowitz and Jouannaud 1990; Field and Harrison 1988; O'Donnell 1985; Paul et al. 1992; Sekar et al. 1995; Turner 1985). It selects the leftmost–outermost redex but may force the reduction of a subterm if the root symbol of that subterm fails to match a function symbol in the patterns. For more details, the reader can refer to Nedjah (1997), and Gräf (1991). So, the order of such reductions coincides with that of pattern-matching. Using left-to-right pattern matching, a subject term evaluation may fail to terminate only because of forcing reductions of subterms when it is unnecessary before declaring a match. For the left-to-right traversal order, such unnecessary reductions are required to ensure that no backtracking is needed when matching fails (Christian 1993; Gräf 1991; Hoffman and O'Donnell 1982; Laville 1991; Wadler 1987).

We first introduce from Nedjah and Mourelle (2001) a method that for any given traversal order constructs the corresponding adaptive matching automaton. *Indexes* are positions whose inspection is necessary to declare a match. Inspecting them first for patterns which are not essentially strongly sequential (Nedjah 1997) allows us to engineer adaptive traversal orders that should improve space usage and matching times as shown in Nedjah and Mourelle (2001) and Sekar et al. (1995). When no index can be found for a pattern set, we showed (Nedjah 1997) that a position, which is an index for a maximal number of high priority patterns can always be identified. Selecting such a position attempts to improve matching times for terms that match patterns of high priority. A *good* traversal order, one which improves space, time and termination properties (Nedjah and Mourelle 2001), inspects positions that are indexes/partial indexes for a pattern set.

We propose a practical technique to compile pattern-matching for prioritized overlapping patterns, generally used in equational programming languages, into a minimal, deterministic, adaptive, matching automaton. First, we present a method for constructing an adaptive tree matching automaton for such patterns. Compared with left-to-right matching automata, adaptive automata have a smaller size and shorter matching time. They may improve termination properties as well. Here, space requirements are reduced further by using a directed acyclic graph (*dag*) automaton that shares all the isomorphic subautomata, which are duplicated in the tree automaton. We design an efficient method to identify such subautomata and avoid duplicating their construction while generating the dag automaton.

In this chapter, space requirements of matching automata are further reduced by using a directed acyclic graph (*dag*) automaton that shares all the isomorphic subautomata, which are duplicated in the tree automaton. We design an efficient method to identify such subautomata and avoid duplicating their construction while generating the dag automaton. We generalize some results of Nedjah et al. (1999) to be able to construct adaptive dag automata directly. Then, we compare left-to-right matching automata to obtained adaptive automata.

12.2 Preliminaries

In the rest of the chapter, we will use some notation and concepts defined as follows: symbols in a *term* are either function or variable symbols; the nonempty set of function symbols that appear in the patterns $F = \{a, b, f, g, h, \ldots\}$ is *ranked*, that is, every function symbol f in F has an *arity* which is the number of its arguments and is denoted $\#f$; a term is either a constant, a variable or has the form $ft_1t_2 \ldots t_{\#f}$ where each t_i, $1 \leq i \leq \#f$, is itself a term. Terms are represented by their corresponding abstract tree. We abbreviate terms by removing the usual parentheses and commas. This is unambiguous in our examples since the function arities will be kept unchanged throughout, namely $\#f = 3, \#g = 1, \#h = 1, \#a = \#b = 0$. Variable occurrences are replaced by ω, a meta-symbol that is used since the actual symbols are irrelevant here. A term containing no variables is said to be a *ground* term. We generally assume that patterns are linear terms, that is, each variable symbol can occur at most once in them. Pattern sets will be denoted by L and patterns by $\pi_1, \pi_2, \ldots$, or simply by π. A term t is said to be an *instance* of a (linear) pattern π if t can be obtained from π by replacing the variables of π by corresponding subterms of t. If term t is an instance of pattern π then we denote this by $t \lhd \pi$.

Here, we assume that we are free to choose any order for pattern-matching terms and their evaluation proceeds using the adaptive strategy. In particular, if $f(t_1, \ldots, t_n)$ is the term being matched at the root, the argument t_n may be reduced before the argument $t_j, j \leq i$ if the pattern-matcher visits the position at which t_n is rooted before that at which t_j is rooted.

Definition 12.1:

A term t *matches* a pattern $\pi \in L$ if, and only if, t is an instance of π, that is, $t \lhd \pi$ and t is not an instance of any other pattern in$L$, that is, higher priority than π.

Definition 12.2:

A *position* in a term is a path specification, which identifies a node in the parse tree of the term. Position is specified here using a list of positive integers. The empty list Λ denotes the position of the root of the parse tree and the position $p.k$ $(k \geq 1)$ denotes the root of the kth argument of the function symbol at position p.

Definition 12.3:

A *matching item* is a pattern in which all the symbols already matched are now *ticked*, that is, they have the *check-mark* ✓. Moreover, it contains the hollow *matching dot* "•" that only designates the *matching symbol*, that is, the symbol to be accepted next. The position of the matching symbol is called the *matching position*. A *final* matching item, namely one of the form π•, has the *final* matching position, which we write ∞. Final matching item may contain unchecked positions. These positions are irrelevant for announcing a match and so must be labeled with the symbol ω. Matching items are associated with a rule name.

The term obtained from a given item by replacing all the terms with an *unticked* root symbol by the placeholder "_" is called the *context* of the items. For instance, the context of the item $f\checkmark a \bullet g\omega aa\checkmark$ is the term $f(\_,\_,a)$ where the arities of f, g, and a are as usual 3, 2, and 0. In fact, no symbol will be checked until all its parents are all checked. So, the positions of the placeholders in the context of an item are the positions of the subterms that have not been checked yet. The set of such positions for an item i is denoted by $up(i)$ (short for unchecked positions).

Definition 12.4:

A *matching set* is a set of matching items that have the same context and a common matching position. The *initial* matching set contains items of the form $\bullet\pi$ because we recognize the root symbol (which occurs first) first whereas, *final* matching sets contain items of the form $\pi\bullet$, where π is a pattern. For initial matching sets, no symbol is ticked. A final matching set must contain a final matching item, that is, in which all the unticked symbols are ωs. Furthermore, the rule associated with that item must be of the highest priority amongst the items in the matching set.

Definition 12.5:

Suppose $L \cup \{\pi\}$ is a prioritized pattern set. Then π is said to be *relevant for L* if there is a term that matches π in $L \cup \{\pi\}$. Otherwise, π is *irrelevant for L*. Similarly, an item π is *relevant for (the matching set) M* if there is a term that *deterministically* matches π in $M \cup \{\pi\}$.

Since the items in a matching set M have a common context, they all share a common list of unchecked positions and we can safely write $up(M)$. The only unchecked position for an initial matching set, is clearly the empty position Λ.

12.3 Adaptive Tree Matching Automata

States of adaptive tree automata are labeled by matching sets. Since here the traversal order is not fixed *a priori* (i.e., it will be computed during the automaton construction procedure), the symbols in the patterns may be accepted in any order. When the adaptive order coincides with the left-to-right order, matching items and matching sets coincide with matching items and matching set, respectively (Nedjah 1997).

We describe adaptive automata by a 4-tuple $\langle S_0, S, Q, \delta\rangle$. S is the state set, $S_0 \in S$ is the initial state, $Q \subseteq S$ is the set of final states and δ is the state transition function. The states are labeled by matching sets which consist of original patterns together with extra instances of the patterns which are added to avoid backtracking in reading the input. In particular, the matching set for S_0 contains the initial matching items formed from the original patterns and labeled by the rules associated with them. The transition function δ of an adaptive automaton is defined using three functions namely, *accept*, *choose*, and *close*: $\delta(M,s) = close(accept(M,s,choose(M,s)))$. The function *accept* and *close* are similar to those of the same name in Nedjah and Mourelle (2001), and *choose* picks the next matching position. For each of these functions, we give an informal description followed by a formal definition, except for *choose* for which we present an informal description. Its formal definition will be discussed in detail in the next section.

accept: For a matching set and an unchecked position, this function accepts and ticks the symbol immediately after the matching dot in the items of the matching set and inserts the matching dot immediately before the symbol at the given unchecked position. Let t be a term in which some symbols are ticked. We denote by $t_{\bullet p}$ the matching item which is t with the matching dot inserted immediately before the symbol $t[p]$. The definition is $accept(M, s, p) = \{(\alpha s\checkmark\beta)_{\checkmark p} | \alpha \bullet s\beta \in M\}$.

choose: This function selects a position among those that are unchecked for the matching set obtained after accepting the symbol given. For a matching set M and a symbol s, the *choose* function selects the position which should be inspected next. The function may also return the final position ∞. In general, the set of such positions is denoted by $\overline{up}(M, s)$. It represents the unchecked positions of $\delta(M, s)$, and consists of $up(M)$ with the position of the symbol s removed. Moreover, if the arity of s is positive then there are $\#s$ additional unchecked terms in the items of the set $\delta(M, s)$, assuming that ωs are of arity 0. Therefore, the positions of these terms are added (see Nedjah and Mourelle (2001) for details).

close: Given a matching set, this function computes its closure in the same way as for the left-to-right matching automaton. As it is shown in the following, the function adds an item $\alpha \bullet f\omega^{\#f}\beta$ to the given matching set M whenever an item $\alpha \bullet \omega\beta$ is in M together with at least one item of the form $\alpha \bullet f\beta'$. The definition of function *close* allows us to avoid introducing irrelevant items (Nedjah and Mourelle 2000).

For a matching set M and a symbol $s \in F \cup \{\omega\}$, the transition function for an adaptive automaton can now be formally defined by the composition of the three functions accept, choose, and close is $\delta(M, s) = close\ (accept\ (M, s, choose\ (M, s)))$.

Example 12.2:

Consider the set $L = \{1 : hggа\omega\omega a, 2 : hgfa\omega\omega aa, 3 : h\omega b\}$. An adaptive automaton for L using matching sets is shown in Figure 12.2. The choice of the positions to inspect will be explained later. Transitions corresponding to failures are omitted, and the ω-transition is only taken when there is no other available transition, which accepts the current symbol. Notice that for each of the matching set in the automaton, the items have a common set of symbols checked. Accepting the function symbol h from the initial matching set (i.e., that which labels state 0) yields the set $\{1{:}h\checkmark \bullet gga\omega\omega a, 2{:}h\checkmark \bullet gfa\omega\omega aa, 3{:}h\checkmark \bullet \omega b, 3{:}h\checkmark \bullet g\omega\omega b\}$ if position 1 is to be chosen next. Then, the closure function adds the item $h\checkmark \bullet g\omega\omega b$. Provided with a matching set M, the *choose* function selects positions that are labeled with a function symbol in at least one item of M. If more than one position is available then the leftmost is selected. In short, the traversal order used avoids positions labeled with ω in every item of M.

12.4 Optimal Adaptive Matching Automata

The tree automaton described above is time efficient during operation because it avoids symbol reexamination. However, it achieves this at the cost of increased space requirements. The unexpanded automaton corresponding to the pattern set of Figure 12.2, and to which no patterns are added, is given in Figure 12.3. In that automaton, states are also labeled with the matching position. For instance, state 0 investigates position Λ and states 4 and 5 both scans position 1.1.1. Furthermore, final states are also labeled with the number of matched rule. The nondeterministic automaton is much smaller. For instance, $hg\omega\omega b$ is only recognized by backtracking from state 2 to state 1 and then taking the branch through state 3 instead. But in Figure 12.2, a branch recognizing $hg\omega\omega b$ has been added to avoid backtracking, thereby duplicating the existing sub-branch which recognizes the b in $h\omega b$. We can see similar duplication in several other branches of Figure 12.2; those identified by sharing the same main state numbers.

By sharing duplicated branches, tree automata can be converted into an equivalent but smaller directed acyclic graph (*dag*) automata. States, which recognize the same inputs and assign the same rule numbers

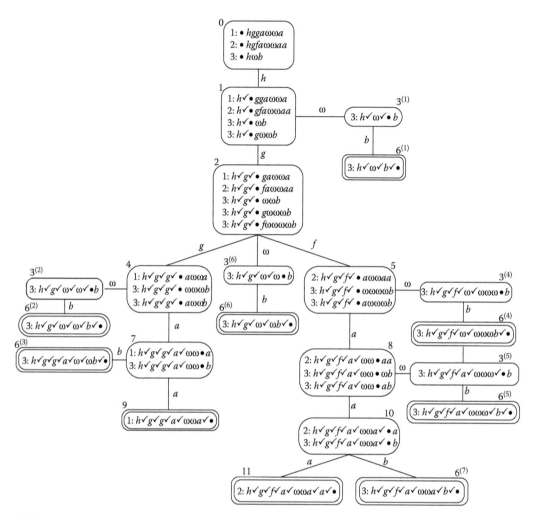

FIGURE 12.2 An adaptive tree automaton for {1:*hggaωωa*, 2:*hgfaωωaa*, 3:*hωb*}.

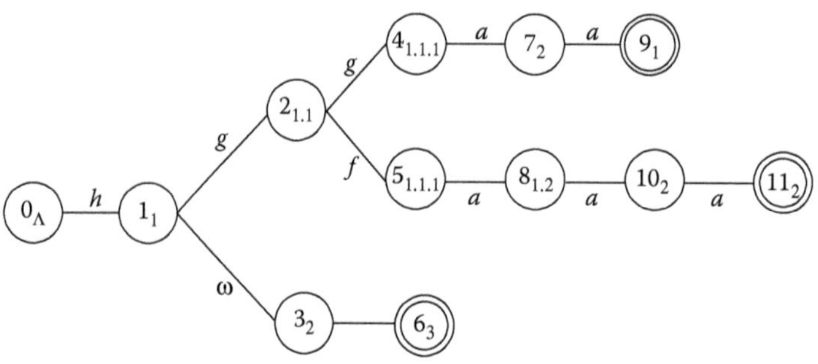

FIGURE 12.3 Unexpanded adaptive tree automaton for {1:*hggaωωa*, 2:*hgfaωωaa*, 3:*hωb*}.

to them are functionally equivalent, and can be identified. For instance, the dag automaton corresponding to Figure 12.2 is given in Figure 12.4. The state number is thereby reduced from 22 to 12.

In tree-based adaptive matching automata, functionally identical (or isomorphic) subautomata may be duplicated. Using directed acyclic graphs instead of trees can then improve the size of adaptive automata. In this section, we define the notion of matching item equivalence so that a dag-based adaptive automaton can be generated.

The illustrative example hides the complexity of recognizing duplication where a number of suffixes are being recognized, not just one. The required dag automaton can be generated using finite state automaton minimization techniques but this may require a lot of memory and time. The obvious alternative approach consists of using the matching sets to check new states for equality with existing ones while generating the automaton. In the case of equality, the new state is discarded and the existing one is shared. However, comparison of matching sets may be prohibitively expensive and it may well require bookkeeping for all previously generated matching sets. A major aim of this chapter is to show how to avoid much of this work. First, we must characterize states that would generate isomorphic subautomata.

Definition 12.6:

Let $i_1 = r_1 : \alpha_1 \bullet \beta_1$ and $i_2 = r_2 : \alpha_2 \bullet \beta_2$ be two matching items and p a position in $up(i_1) \cup up(i_2)$. i_1 and i_2 are *equivalent* if, and only if, $r_1 = r_2$ and the symbols labeling unchecked positions in i_1 and i_2 are as follows, otherwise, i_1 and i_2 are *inequivalent.*

$$\begin{cases} \alpha_1\beta_1[p] = \alpha_2\beta_{21}[p] & \text{if } p \in up(i_1) \cap up(i_2) \\ \alpha_1\beta_1[p] = \omega & \text{if } p \in up(i_1) \backslash up(i_2) \\ \alpha_2\beta_2[p] = \omega & \text{if } p \in up(i_2) \backslash up(i_1) \end{cases}$$

Definition 12.7:

Two matching sets M_1 and M_2 are *equivalent* if, and only if, to every item i in $M_1 \cup M_2$ there correspond items $i_1 \in M_1$ and $i_2 \in M_2$ which are equivalent to i. Otherwise, the sets are *inequivalent.*

For instance, in the adaptive matching automaton of Figure 12.2, the matching sets labeling the states $3^{(1)}$ and $3^{(2)}$ are equivalent. So, if two matching sets are equivalent, then their items differ only in those unchecked positions that occur in either of the matching sets. Moreover, for every item in which such positions occur, they are labeled with the symbol ω and so irrelevant for declaring a match. Therefore, equivalence as formalized in Definitions 12.5 and 12.6 is the right criterion for coalescing nodes of the adaptive tree automaton to obtain the equivalent adaptive dag automaton.

Lemma 12.1:

Two matching sets generate identical automata if they are equivalent.

For equivalent sets, using the same traversal order strategy, function *choose* will certainly select the same positions. They must occur in the intersection of their unchecked position sets. So, equivalent matching sets generate equivalent adaptive automata. We believe this equivalence is actually necessary as well as sufficient to combine corresponding states in the automaton. Equivalent matching sets may have different contexts, as can be seen in Figure 12.2.

12.5 Adaptive Dag Automata Construction

In this section, we describe how to build the minimized dag automaton efficiently without constructing the tree automaton first. This requires the construction of a list of matching sets in a suitable order to ensure that every possible state is obtained, and a means of identifying potentially equivalent states.

The items in matching sets all share a common context. Hence, the matching position of any item is an invariant of the whole matching set. The states of the tree automaton can, therefore, be ordered using the left-to-right total ordering on the common matching positions of their matching sets. The dag automaton is constructed as in Algorithm 12.1. We iteratively construct the machine using a list l of matching sets in which the sets are ordered according to the matching position of the set. So, the initial matching set is first and final matching sets come last. Each set in l is paired with its state in the automaton and a pointer is kept to the current position in l.

Algorithm 12.1: Constructing Adaptive Dag Automata

Algorithm *DagAutomaton*(pattern set L);
  A $\leftarrow \emptyset$; $l \leftarrow \{\langle M_0 \leftarrow (\forall \pi \in L, \bullet\pi), S_0\rangle\}$; current $\leftarrow$ 0;
  do for each $s \in FU\{\omega\}$ **do**
    compute $\delta(M,\ s)$;
    if $\nexists\langle M', S'\rangle \in l|\delta(M,s) \equiv M'$ **then**
      create state S'labelled with $\delta(M,s)$;
      add transition $S \xrightarrow{s} S'$to A;
      add pair $\langle\delta(M,s),\ S'\rangle$ to l;
    else add transition $S \xrightarrow{s} S'$ to A;
    end if;
    current $\leftarrow$ current + 1;
  while current $\neq$ null;
end algorithm;

The list l represents the equivalence classes of matching states where the assigned matching position is that of the representative which is generated first. In each new set which is generated, the position of the current matching set is incremented at least one place to the right. So, new members of l are always inserted to the right of the current position. This ensures that all necessary transitions will eventually be generated without moving the pointer backwards in l. It is easy to see from the definition of the *close* function that added patterns cannot contain positions that were not in one of the original patterns. So, l only contains sets with matching positions from a finite collection and, as each set can only generate a finite number of next states all of which are to the right, the total length of l is bounded and the algorithm must terminate. New members of l are always inserted to the right of the current position. This ensures that all necessary transitions will eventually be generated without moving the pointer backwards in l. It is easy to see from the definition of the *close* function that added patterns cannot contain positions that were not in one of the original patterns. So, l only contains sets with matching positions from a finite collection and, as each set can only generate a finite number of next states, the total length of l is bounded and the algorithm must terminate. Furthermore, the tree and dag automata clearly accept the same language and the automaton is minimal, in the sense that, by construction, none of the matching sets labeling the states in the automaton are equivalent.

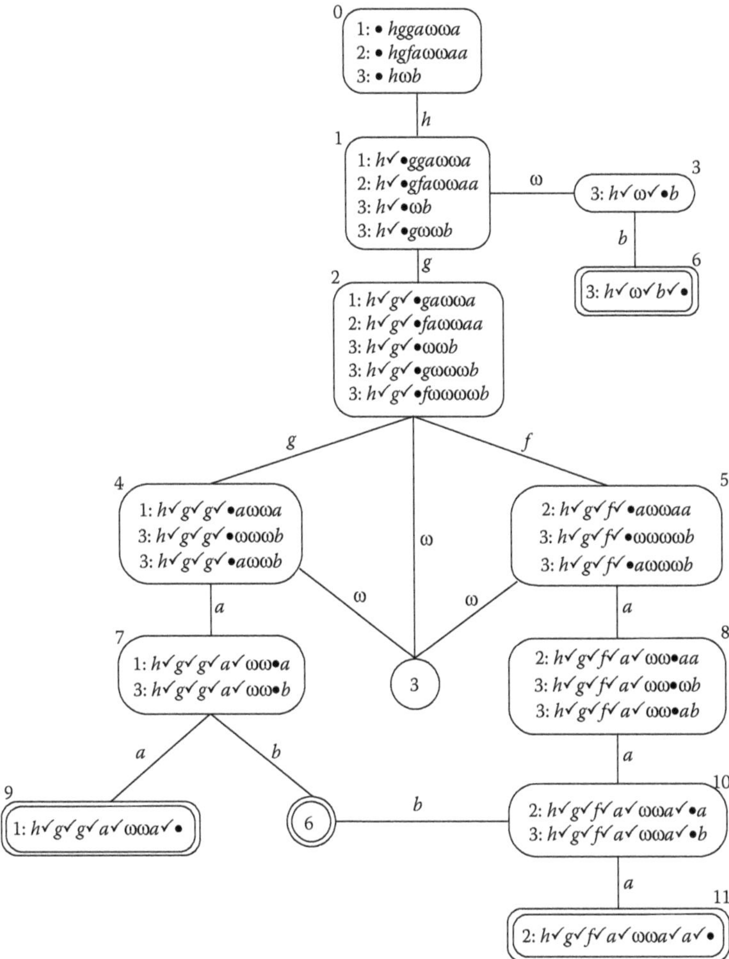

FIGURE 12.4 Adaptive dag automaton for {1:*hgga*ωω*a*, 2:*hgfa*ωω*aa*, 3:*h*ω*b*}.

12.6 Automaton State Equivalence

In this section, we show how matching sets can frequently be discriminated easily so that the cost of checking for equivalence is minimized. Comparison of suffixes in the matching sets is completely avoided. We look at two properties to help achieve this. One is the set of rules represented by patterns in the matching set and the other is the matching position.

In Figure 12.4, all inequivalent matching sets are distinguished by the use of the rule set. Thus, the criterion is useful in practice. However, it will clearly not be sufficient in general. In the opposite direction, it is also sometimes easy to establish equivalence. Combining it with the matching position, we have the following very useful result, which enables the direct checking of equivalence to be avoided entirely in Example 12.2.

Theorem 12.1:

Matching sets that share a common matching position and rule set are equivalent.

Proof. It suffices to show the kernels of the matching sets are equivalent, since then the function *close* will add equivalent items to both sets. Let M_1 and M_2 be two matching sets that share the same rule set and common matching position p. Let $i_1 = r : \alpha_1 \bullet \beta_1$ and $i_2 = r : \alpha_2 \bullet \beta_2$ be any two items associated with the same rule in their, respectively, kernels. The definitions of *accept* and *close* guarantee that the suffixes consist of a suffix of the original pattern π_r of the rule preceded by a number of copies of ω. To identify this suffix, let p' be the maximal prefix of the position p corresponding to a symbol in π_r. This is either the whole of p or is the position of a variable symbol ω, either ways, substitutions made by *close* for variables before p' in either i_1 or i_2 have already been fully passed in the prefix, and no substitution has yet been made for any variable further on in π_r. So, if β is the suffix of π_r that starts at p' then the items i_1 and i_2 must have the form $\alpha_1 \bullet \omega^n 1\beta$ and $\alpha_2 \bullet \omega^n 2\beta$ for some $n_1, n_2 \geq 0$. It is clear that those items are equivalent in the sense of Definition 12.5. We conclude that M_1 and M_2 are equivalent. □

Although matching sets that share these two properties are equivalent, the matching positions of equivalent matching sets are not necessarily identical. Many examples can be found in Nedjah (1997) and Nedjah et al. (2003).

12.7 Automaton Complexity

In this last main section, we evaluate the space complexity of the adaptive dag automaton by giving an upper bound for its size in terms of the number of patterns and symbols in the original pattern set. The bound established considerably improves left-to-right dag automata bound (Gräf 1991; Nedjah et al. 1997). The size of the left-to-right dag automaton for a pattern set L is bounded above by: $1 + |L| + (2^{|\pi|} - 1)(\Sigma_{\pi \in L}(|\pi| - 1))$ (see Nedjah et al. (1997) for details).

Theorem 12.2:

The size of the adaptive dag automaton for a pattern set L is bounded above by $2^{|L|} * |F|$.

Proof. Consider the left-to-right dag automaton for pattern set L. Let M_1 and M_2 be two matching sets labeling two sets S_1 and S_2 in the dag automaton. Assume that M_1 and M_2 have the same rule set. Now, consider the contexts of their relevant items. Since symbols are scanned in the left-to-right order, one of the contexts c_1 must be an instance of the other context c_2. Hence, there exists a total order among theses prefixes. This argument can be generalized to all the states of the dag automaton that share the same rule set with M_1 and M_2. Consequently, the number of such prefixes, and hence the number of states, is bounded above by the size of the largest context, which is in turn bounded above $|F|$. Furthermore, as there are $|L|$ distinct patterns, there are at most $2^{|F|}$ different rule sets. This yields the upper bound $2^{|L|} * |F|$ for the size of the adaptive dag automaton. □

Although the adaptive tree automata that inspect indexes first are smaller (or the same) than the tree left-to-right automaton for any given pattern set, it is not necessarily true that the equivalent adaptive dag automata are smaller than the left-to-right dag automaton. For instance, consider the pattern set $L = \{1 : fa\omega a\omega, 2 : fa\omega b\omega, 3{:}fa\omega c\omega, 4{:}fa\omega d\omega, 5{:}fa\omega e\omega, 6{:}f\omega bbb, 7{:}f\omega a\omega\omega\}$ where f has an arity of 4 and a, b, c, d, and e are constants. Assuming a textual priority rule and using the traversal order, which is described in Section 12.3, the adaptive dag automaton obtained for L (where most of inspected positions are indexes) is shown in Figure 12.5. It has 17 states whereas the left-to-right dag automaton for L, given in Figure 12.6, includes 15 states only.

The matching times, however, using the adaptive dag automaton remain better than that using the left-to-right dag automaton as sharing equivalent subautomata does not affect the matching times.

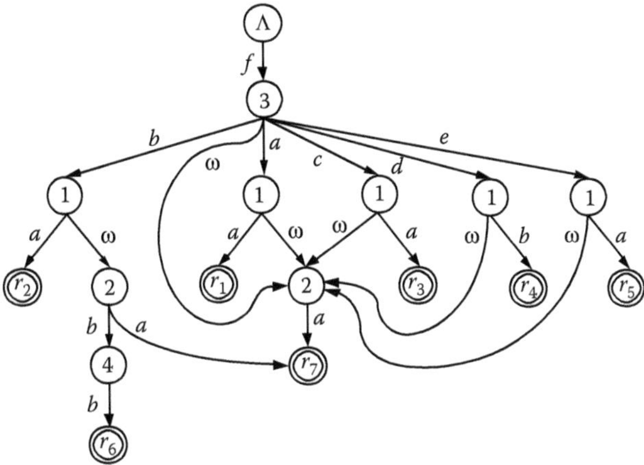

FIGURE 12.5 Adaptive dag automaton for L.

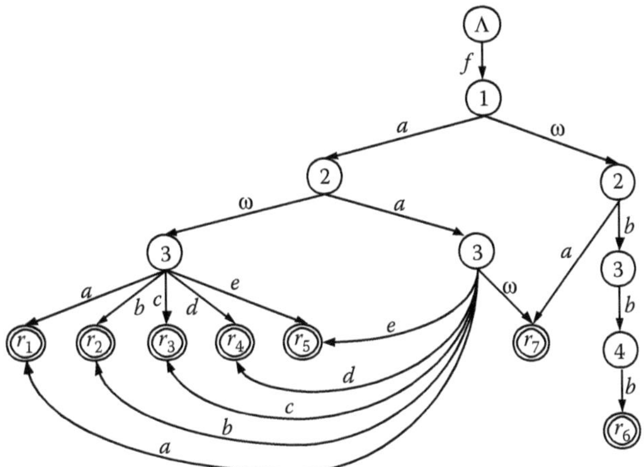

FIGURE 12.6 Left-to-right dag automaton for L.

In particular, with the adaptive dag automaton of Figure 12.5, the matching times for terms that match either of the patterns of all the rules except 6 and 7 are optimal. However, using the left-to-right dag automaton of Figure 12.6, every term needs at least three position inspections. Also, from the contrived nature of the example, this situation of the left-to-right dag automata having lesser states than adaptive ones seems to occur only for rare examples.

12.8 Conclusion

First, we described a practical method that compiles a set of prioritized overlapping patterns into an equivalent deterministic adaptive automaton, which does not need backtracking to announce a match. With ambiguous patterns a subject term may be an instance of more than one pattern. To select the pattern to use, a priority rule is usually engaged. The closure operation of the program patterns adds new patterns which are instances of original ones. Reexamination of symbols while matching terms is completely avoided. The matching automaton can be used to drive the pattern-matching process with any rewriting strategy (Nedjah et al. 1999; Nedjah and Mourelle 2003).

In the main body of the chapter, we described a method to generate an equivalent minimized dag adaptive matching automaton very efficiently without constructing the tree automaton first. We directly built the dag-based automaton by identifying the states of the tree-based automaton that would generate identical subautomata. By using the dag-based automata, we can adaptive pattern-matchers that avoid symbol reexamination without much increase in the space requirements.

Some useful matching set properties were then described for distinguishing inequivalent states when building the adaptive dag automaton. A theorem which guarantees equivalence in terms of several simple criteria was then applied to establish improved upper bounds on the size of the dag automaton in terms of just the number of patterns and symbols in the original pattern set.

References

Augustsson, A. 1984. A compiler for lazy ML. In *Proceedings ACM Conference on Lisp and Functional Programming*, ACM Press, Austin, TX, pp. 218–227.

Christian, J. 1993. Flatterms, discrimination nets and fast term rewriting. *Journal of Automated Reasoning*, 10: 95–113.

Dershowitz, N. and Jouannaud, J. P. 1990. Rewrite systems. In *Handbook of Theoretical Computer Science*. Elsevier Science Publishers, Amsterdam, the Netherlands.

Field, A. J. and Harrison, P. G. 1988. *Functional Programming*, International Computer Science Series. Addison-Wesley, Reading, MA.

Gräf, A. 1991. Left-to-right tree pattern-matching. In *Proceedings Conference on Rewriting Techniques and Applications*, Lecture Notes in Computer Science, Ronald, V. B. (Ed.), Springer-Verlag. Vol. 488, pp. 323–334.

Hoffman, C. M. and O'Donnell, M. J. 1982. Pattern-matching in trees. *Journal of ACM*, 29(1): 68–95.

Laville, A. 1991. Comparison of priority rules in pattern matching and term rewriting, *Journal of Symbolic Computation*, 11: 321–347.

Nedjah, N. 1997. *Pattern-Matching Automata for Efficient Evaluation in Equational Programming*, PhD thesis, University of Manchester-Institute of Science and Technology, Manchester, UK.

Nedjah, N., Walter, C. D., and Eldridge, S. E. 1997. Optimal left-to-right pattern-matching automata. In *Proceedings of the Sixth International Conference on Algebraic and Logic Programming*. Lecture Notes in Computer Science, M. Hanus, J. Heering and K. Meinke Editors, Springer-Verlag, Southampton, UK, Vol. 1298, pp. 273–285.

Nedjah, N., Walter, C. D., and Eldridge, S. E. 1999. Efficient automata-driven pattern-matching for equational programs. *Software-Practice and Experience*, 29(9): 793–813.

Nedjah, N. and Mourelle, L. M. 2001. Improving time, space and termination in term rewriting-based programming. In *Proc. International Conference on Industrial & Engineering Applications of Artificial Intelligence & Expert Systems*. Lecture Notes in Computer Science, Laszlo, M., Józef, V., and Mounis, A. (Eds), Springer-Verlag, Vol. 2070, pp. 880–890.

Nedjah, N. and Mourelle, L. M. 2003. Implementation of term rewriting-based programming languages. In *Advances in Computation: Theory and Practice*, Vol. 13. Nova Science, NY/Hauppauge.

O'Donnell, M. J. 1985. *Equational Logic as Programming Language*, MIT Press, Cambridge, MA.

Paul, H., Simon, P. J., Philip, W., Brian, B., Jon, F., Joseph, F., María, M. G. et al. 1992. Report on the programming language Haskell: A non-strict, purely functional language. *ACM SIGPLAN Notices*, 27(5), May 1992.

Sekar, R. C., Ramesh, R., and Ramakrishnan, I. V. 1995. Adaptive pattern-matching. *SIAM Journal*, 24(6): 1207–1234.

Turner, D. A. 1985. Miranda: A non strict functional language with polymorphic types. In *Proceedings of Conference on Lisp and Functional Languages*, ACM Press, New York, NY, pp. 1–16.

Wadler, P. 1987. Efficient compilation of pattern-matching. In *The Implementation of Functional Programming Languages*, S. L. Peyton-Jones, Editor, Prentice-Hall International, Hempstead, UK, pp. 78–103.

13

Finite State Automata in Compilers

Yang Zhao
Nanjing University of Science and Technology

13.1 Introduction

A compiler is a program that reads a program in one language—the source language—and translates it into an equivalent program in another language—the target language [1]. It may generate an error or warning report if the source program has problems, lexical, syntactical, or semantic. Source languages are usually high-level languages, such as Java or C/C++, while their target languages could be machine or assembly languages depending on different requirements. The traditional structure of compilers consists of some or all of the following phases: lexical analysis, syntax analysis, semantics analysis, intermediate code generation, optimization, and target code generation, where the input of each phase comes from the output of its previous phase except for the first one whose input is the source program.

A finite state automaton [2] is a mathematical model that includes a fixed number of states with some transition edges among them. For all phases in a compiler, the lexical and syntax analysis are the two that directly utilize finite state automata as one of their core technologies. Roughly speaking, the lexical analysis aims at recognizing tokens from a stream of input characters, while the syntax analysis makes use of a sequence of tokens and composes them into a structured representation, a syntax tree usually. A common thing between the lexical and syntax analysis is both of them attempt to dig out a structured output from some unstructured input. Therein lies the capability of finite state automata, which are good recognizers. Although an automaton as a recognizer simply says "YES" or "NO," it is still feasible to achieve more information during the whole process of recognition. In the following sections, we introduce how finite state automata could be applied to the lexical and syntax analysis.

13.2 Lexical Analysis

The lexical analysis (also called *scanning*) is normally the first phase performed by a compiler. It reads input source code (as a stream of characters), groups neighboring related characters, and then generates a sequence of tokens, which could be keywords, operators, identifiers, constants, and so on. For instance, the lexical analyzer may translate a code segment "a0 = a0 + b0 ∗ 10" into tokens: identifier "a0," operator "=," identifier "a0," operator "+," identifier "b0," operator "*," and constant "10." It should not come up with some weird results, such as "= a" or "b0 *." On the whole, the structure of tokens should be key knowledge for the lexical analyzer to recognize tokens correctly. Here, in compilers, *regular expressions* [3] have been widely used to explicitly characterize the structure of tokens in programming languages.

A regular expression is defined recursively from smaller ones. All symbol strings over some alphabet Σ that could be denoted by a regular expression r is called the language of r, written as $L(r)$. The recursive definitions for regular expressions and their corresponding languages over some alphabet Σ are given in the following:

- ε is a regular expression and $L(\varepsilon)$ is {ε}, a set with a sole member ε.
- ϕ is a regular expression and $L(\phi)$ is {}, an empty set.
- For any symbol a in Σ, a is a regular expression and $L(a)$ is $\{a\}$.
- Assuming r_1 and r_2 are regular expressions, and their languages are $L(r_1)$ and $L(r_2)$, respectively, then (r_1), $r_1|r_2$, r_1r_2, $r_1{}^*$ are also regular expressions and their languages are $L(r_1)$, $L(r_1) \cup L(r_2)$, $L(r_1)L(r_2)$, and $(L(r_1))^*$, respectively. It is then obvious that we may add arbitrary pairs of parentheses around expression without changing its language.

For instance, a regular expression (a|b)* in an alphabet Σ={a, b} denotes a language {ε, a, b, aa, bb, ab, ...} which includes each string constitute from any number of a or b. With the help of regular expressions, identifiers in the C language could be given as "$L(L|D)^*$" where L stands for any letter or an underscore and D stands for any digit. Based on this structure definition, the "a0" in the previous example is then considered as a valid identifier, while "= a" is not.

Nevertheless, a regular expression only provides a way to characterize the structure of tokens in a language. It does not tell how to decide whether some neighboring characters could form a valid token. Hence, what we really need is a recognizer to actually read these characters successively and give a "YES" or "NO" answer once in a while. Fortunately, as an abstract machine, a finite state automaton satisfies this requirement, where the fundamental idea is that it is able to say "YES" for some input characters once they follow the structure defined in a regular expression. We will come back to this in more detail later on.

There are two categories of finite state automata: deterministic finite automata and nondeterministic finite automata, distinguished by whether their beneath transitions among states are deterministic or not.

13.2.1 Deterministic Finite Automata

A deterministic finite automaton (DFA) is represented as a five-tuple $M = (S, \Sigma, f, s_0, F)$, where

1. S is a finite set of states
2. Σ is a set of input symbols
3. f is a transition function that gives, for each state in S, and for each symbol in Σ one next state, such that $f{:}S \times \Sigma \rightarrow S$
4. s_0, a state in S, is a start state for M
5. F, a subset of S, is a set of accepting states

A DFA can be represented as a transition graph, where its nodes are states and labeled edges indicate the transition function. Furthermore, the start state has an edge coming in from nowhere, while any double circle node is an accepting state.

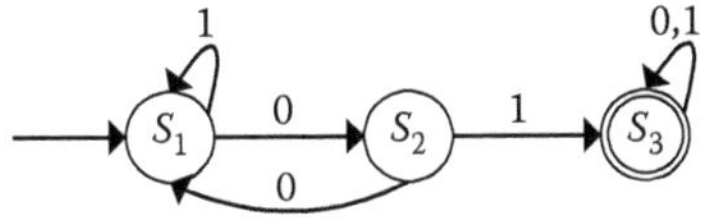

FIGURE 13.1 A deterministic finite automaton.

```
currentState = s0;    // initially set the current state as the start state
inputChar = nextChar();    // read the next character
while (inputChar != eof) {
    currentState = f (currentState, inputChar);    // execute a transition
    inputChar = nextChar();
}
if (currentState ∈ F)
then return "YES"
else return "NO";
```

FIGURE 13.2 The pseudocode of simulating a DFA.

Example 13.1:

A DFA is defined as $M = (\{s_1, s_2, s_3\}, \{0, 1\}, f, s_1, \{s_3\})$, where

$$f(s_1, 1) = s_1 \quad f(s_1, 0) = s_2 \quad f(s_2, 0) = s_1$$
$$f(s_2, 1) = s_3 \quad f(s_3, 0) = s_3 \quad f(s_3, 1) = s_3$$

Its transition graph is demonstrated in Figure 13.1. Given an input string 1100011, this DFA starts from s_1 and goes through the following states in turn: $s_1, s_1, s_2, s_1, s_2, s_3, s_3$. Finally, it says "YES," since the last state s_3 is an accepting state (double cycled).

This DFA can easily be simulated using the pseudocode in Figure 13.2.

13.2.2 Nondeterministic Finite Automata

A nondeterministic finite automaton (NFA) [4] is also a five-tuple model $M = (S, \Sigma, f, s_0, F)$, such that S, Σ, s_0, and F are the same as defined in DFA except for the transition function f which gives, for each state and for each symbol in $\Sigma \cup \{\varepsilon\}$, a set of next states. Here, we assume the symbol ε, standing for an empty string, is never a member of Σ.

According to the definition given above, an NFA state transition graph allows any edge to be labeled ε or multiple same-labeled edges from one state to end at different states. In other words, an NFA could move one step forward without consuming a symbol through an ε-labeled edge or make a choice among multiple same-labeled edges to carry out a transition. Therefore, if there is neither an ε-labeled edge nor multiple same-labeled edges coming out of the current state, the transition behavior of the NFA is certainly deterministic. It is then obvious that the DFA could be considered as a special case of an NFA.

Despite the nondeterminism and flexibility in NFA, it has been demonstrated that both NFA and DFA have the same capabilities in recognizing languages, which could roughly be explained by two statements: (1) any NFA can be converted to an equivalent DFA and (2) the DFA is a special case of an NFA (as mentioned previously).

Generally speaking, people prefer DFAs more than NFAs because the former always carry out deterministic transition which is easy for implementation. But their drawback is also obvious: the flexibility of a language definition is limited due to deterministic state transitions. Consequently, we intend to borrow the flexibility of an NFA to recognize a language, but implement it in an easy DFA way. Besides the conversion from NFAs to DFAs, there is also an algorithm to convert any regular expression to an NFA that defines the same language, which gives us a clue to connect language definition with language recognizer. Actually, the fundamental idea behind the lexical analysis is to define a language based on regular expressions, then to directly construct an NFA which could be converted into a DFA as a lexical recognizer indeed. Given any symbol string, the DFA could answer whether it belongs to the language. In the following subsections, we first show how to convert an NFA to an equivalent DFA, and then illustrate the method of translating any regular expression to an NFA.

13.2.3 From NFAs to DFAs

The key to the conversion from an NFA to a DFA is to convert any DFA state of the latter to be a set of NFA states. After reading the input symbol string $a_1, a_2, \ldots, a_n$, the DFA state will be the one that corresponds to the set of states at which NFA could be, from the very beginning. Then, edges in this path are labeled $a_1, a_2, \ldots, a_n$ sequentially.

There are two core operations (functions) in the conversion:

- *ε-closure*(I) returns a set of NFA states reachable from any state in I through zero or more occurrences of continuous ε transitions. Obviously, $I \subseteq \varepsilon\text{-}closure(I)$.
- *move*(I,a) returns a set of NFA states reachable from any state in I through one transition edge labeled a.

The algorithm that converts an NFA to a DFA is called *subset construction*. Its pseudocode is given in Figure 13.3.

This algorithm gives us a way to compute all transition edges corresponding to the transition function f in the newly created DFA. Besides that, these two automata (NFA and DFA) will share the same symbol set Σ. After this algorithm is executed, all items in C become DFA states, and the $\varepsilon\text{-}closure(\{s_0\})$ is the start state. Any accepting states of the newly created DFA should contain at least one accepting state of the original NFA.

Let us take the NFA in Figure 13.4 as an example. Initially, set C has only one unmarked item which is $\varepsilon\text{-}closure(\{s_0\})$. In this case, {0, 1, 2} (named S_0). In the first **while** loop, we mark it and continue with two **for** loops for two characters in Σ:

1. $U = \varepsilon\text{-}closure(move(\{0, 1, 2\}, a)) = \varepsilon\text{-}closure(\{1, 3\}) = \{1, 2, 3\}$
   Set {1, 2, 3} is not included in C, so we insert the unmarked {1, 2, 3} (named A) into C and there is an edge labeled a coming from S_0 to A

```
Initially, ε-closure({s0}) is the only element in C and it is unmarked
while (there exists an unmarked T in C ) {
    mark T;
    for ( every character a in ∑ ) {
        U = ε-closure( move(T, a) );
        if U is not in C, then insert U to C and U is unmarked;
        create an edge labeled a coming from state T to U in the DFA;
    }
}
;
```

FIGURE 13.3 The *subset construction* algorithm.

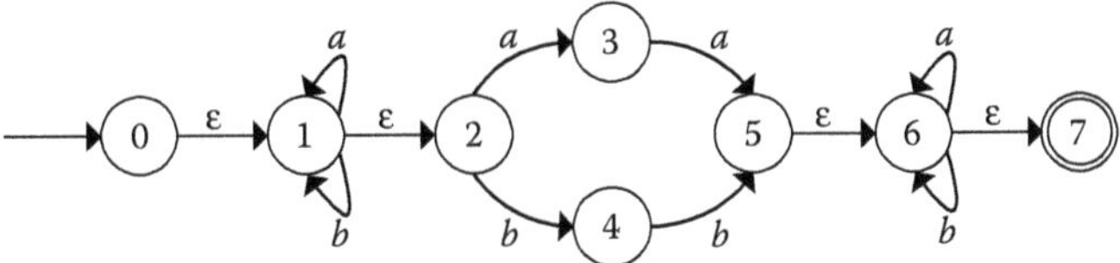

FIGURE 13.4 An NFA.

2. U = ε-*closure*(*move*({0, 1, 2}, b)) = ε-*closure*({1, 4}) = {1, 2, 4}
   Set {1, 2, 4} is not included in *C* either, so we insert the unmarked {1, 2, 4} (named B) into set *C* and there is an edge labeled b coming from S_0 to B.

We skip the remaining steps and directly give intermediate results in the *subset construction* algorithm in Table 13.1. The uppercase letters are used to be state names in the DFA. According to the previous description, S_0 is the start state, while C, D, E, and F are accepting states since all of them contain state 7, the accepting state of the original NFA. Finally, the newly constructed DFA is then given in Figure 13.5.

There are many different DFAs that recognize the same language. For instance, the DFA in Figure 13.6 recognizes the same language as the one in Figure 13.5, but with fewer states. Hence, it is possible to

TABLE 13.1 Intermediate Computing Steps

Current State	Input Character "a"	Input Character "b"	Transition Function in DFA
S_0:{0,1,2}	A:{1,2,3}	B:{1,2,4}	$f(S_0, a)$ = A; $f(S_0, b)$ = B
A:{1,2,3}	C:{1,2,3,5,6,7}	B:{1,2,4}	f(A, a) = C; f(A, b) = B
B:{1,2,4}	A:{1,2,3}	D:{1,2,4,5,6,7}	f(B, a) = A; f(B, b) = D
C:{1,2,3,5,6,7}	C:{1,2,3,5,6,7}	E:{1,2,4,6,7}	f(C, a) = C; f(C, b) = E
D:{1,2,4,5,6,7}	F:{1,2,3,6,7}	D:{1,2,4,5,6,7}	f(D, a) = F; f(D, b) = D
E:{1,2,4,6,7}	F:{1,2,3,6,7}	D:{1,2,4,5,6,7}	f(E, a) = F; f(E, b) = D
F:{1,2,3,6,7}	C:{1,2,3,5,6,7}	E:{1,2,4,6,7}	f(F, a) = C; f(F, b) = E

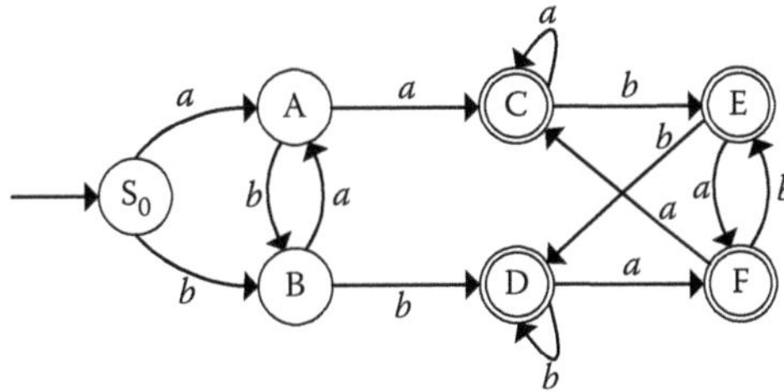

FIGURE 13.5 The newly constructed DFA.

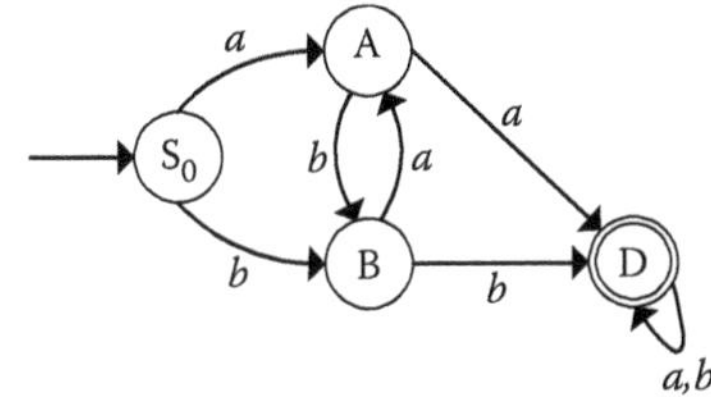

FIGURE 13.6 A DFA.

minimize the number of states in a DFA without changing the language it recognizes. There are quite a lot of methods to achieve this goal, and the basic idea is to get rid of redundant states as well as merging remaining equivalent states. Interested readers could refer to Refs. [1–3] for more information.

When converting an NFA to a DFA, we may face a state explosion issue because every DFA state is a subgroup of the whole NFA states. However, it turns out that the usage of NFA-to-DFA conversion in the lexical analysis for real languages will be reluctant to involve exponential operations [1].

13.2.4 From Regular Expressions to NFAs

Regular expressions only provide a way to define languages, while finite automata, including both DFA and NFA, allow us to determine whether some strings belong to a language or not. For any regular expression in Σ, it is possible to construct a corresponding NFA to recognize the same language defined by that regular expression [5]. There exists a syntax-directed algorithm that converts a regular expression to an NFA.

We start to show how to build an NFA from a regular expression without operators, which are base cases. Then we use inductive rules to construct a larger NFA accordingly.

Base cases:

- For the regular expression ϕ, its corresponding NFA is

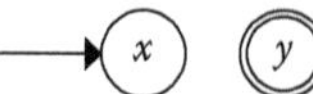

- For the regular expression ε, its corresponding NFA is

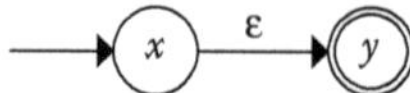

- For the regular expression a, where $a \in \Sigma$, its corresponding NFA is

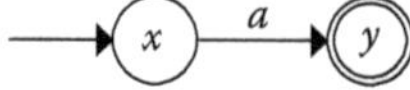

Inductive cases: Assuming s and t are regular expressions in Σ, and their NFAs are N(s) and N(t), respectively.

- For the regular expression $r = s|t$, its corresponding NFA N(r) is in the following, where the state x has two ε transitions pointing to the start states of N(s) and N(t) separately, and the accepting states of both N(s) and N(t) have ε transitions pointing to an accepting state of N(r).

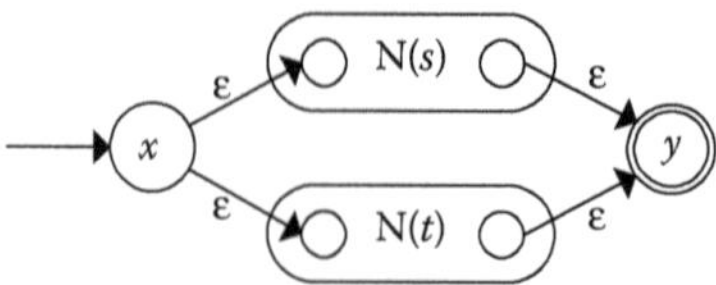

- For the regular expression $r = st$, its corresponding NFA N(r) is in the following, where the state x has ε transitions pointing to the start state of N(s), whose accepting state is connected to the start state of N(t) with an ε edge. Finally, we create a new accepting state y for N(r) and let the accepting state of N(t) to connect to it by an ε edge.

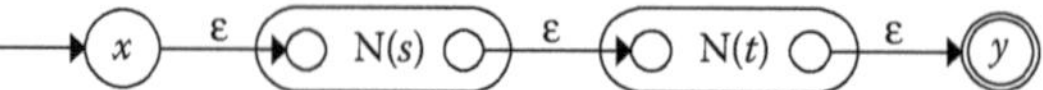

- For the regular expression $r = s^*$, its corresponding NFA N(r) is in the following, where we connect the accepting state of N(s) with its start state by an ε transition. However, the ε transition from x, the start state of N(r), to y, the accepting state of N(r), indicates r could be constituted from zero s.

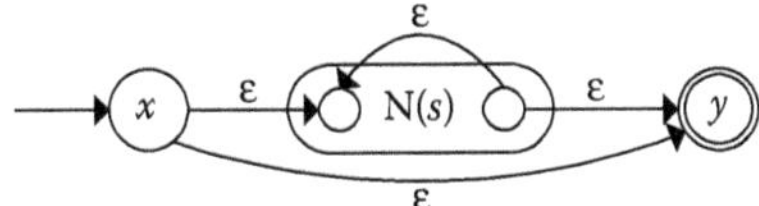

- It is obvious that a regular expression $r = (s)$ has the same NFA with its subexpression s.

Let us take this inductive algorithm to construct an NFA for a regular expression $r = 1^*0(0|1)^*$ in the following steps (some base steps are omitted):

1. Construct an NFA for 1^*:

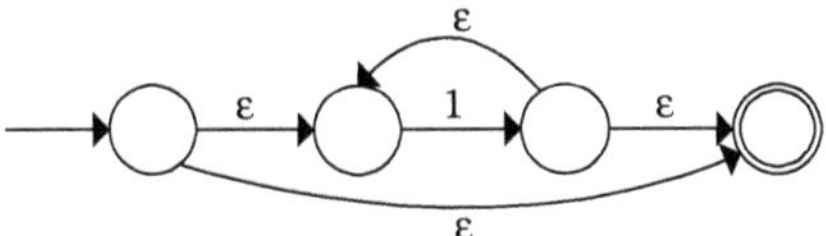

2. Construct an NFA for 1^*0:

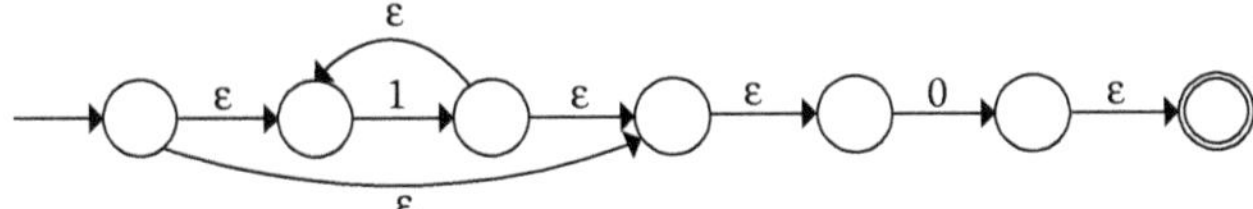

3. Construct an NFA for $0|1$:

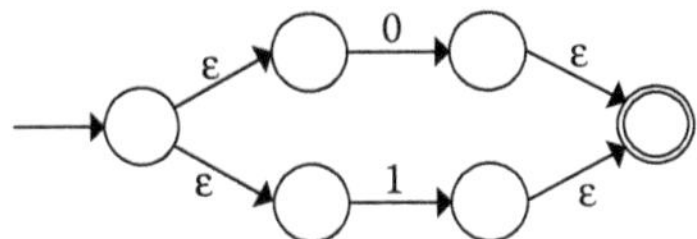

4. Construct an NFA for $(0|1)^*$:

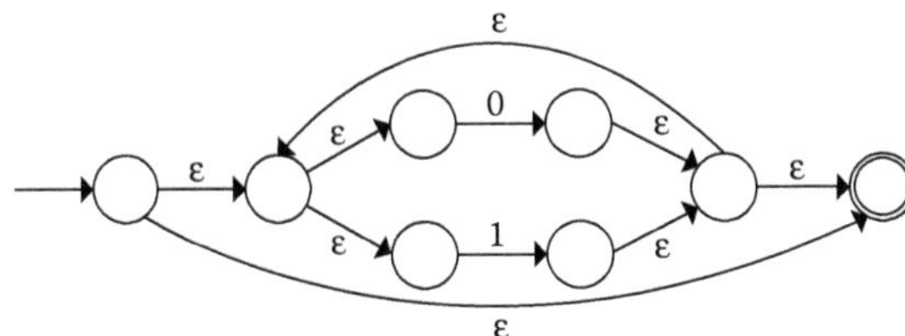

5. Finally, construct the NFA for $1^*0(0|1)^*$:

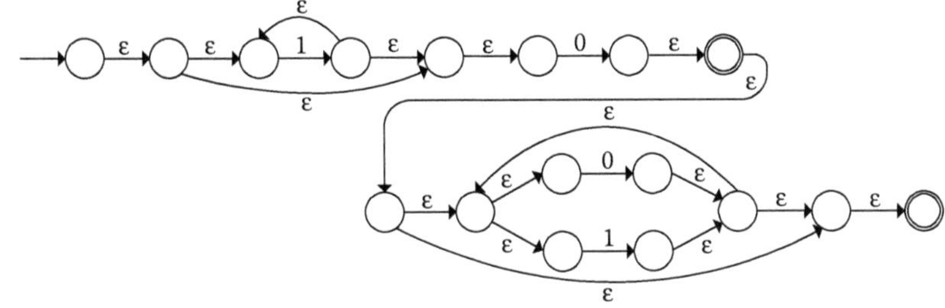

It is easy to find that many ε edges and corresponding states in this NFA could be removed without affecting its capability, but we should be very careful about this simplification transformation.

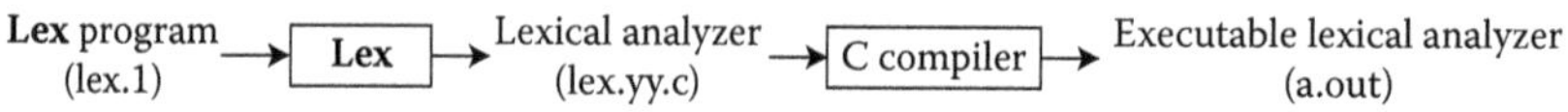

Lex program (lex.1) → [**Lex**] → Lexical analyzer (lex.yy.c) → [C compiler] → Executable lexical analyzer (a.out)

FIGURE 13.7 Generate a lexical analyzer using *Lex.*

13.2.5 Generator of the Lexical Analyzer

It has been mentioned that regular expressions are a powerful method to define or describe the structure of tokens. Then the conversion from a regular expression to an NFA and furthermore to a DFA is trivial. The DFA (or its minimized one) is just the recognizer for the language that described by the original regular expressions. There are quite a few ways to build a lexical analyzer based on this idea. **Lex** [6], as well as its recent version **Flex** [7], is a common tool to generate a lexical analyzer by specifying regular expressions to define patterns for tokens used in a programming language.

The **Lex** tool accepts a **Lex** program, including regular expressions and other descriptions, and produces the source program (usually C program) of a lexical analyzer [1]. After a regular compilation (by calling a C compiler), an executable lexical analyzer is then generated. Figure 13.7 clearly demonstrates this process.

13.3 Syntax Analysis

The syntax analysis (also called *parsing*) usually follows a lexical analysis in the flow of compilation. It accepts the output from a lexical analysis, a sequence of tokens, and decides whether they could constitute a legal program based upon predefined syntactic rules (grammars) [2]. The syntax analysis is the core phase that transforms an unstructured token sequence into a structured representation, usually a syntax tree (or parsing tree). For instance, given a context-free grammar G[E]* in Figure 13.8 and a token sequence "$i + i * i$" this syntax analysis could build a tree-like structure, as shown in Figure 13.9.

There are two approaches to do syntax analysis based on how a parsing tree is built: *top-down* and *bottom-up*. The bottom-up approach normally adopts a *shift–reduce* technique. Since modern programming languages, such as Java or C/C++, are more complicated than previous ones, it may not be feasible

	Production	Production id
G[E]:	$E \rightarrow E + T$	①
	$E \rightarrow T$	②
	$T \rightarrow T * F$	③
	$T \rightarrow F$	④
	$F \rightarrow (E)$	⑤
	$F \rightarrow i$	⑥

FIGURE 13.8 A context-free grammar.

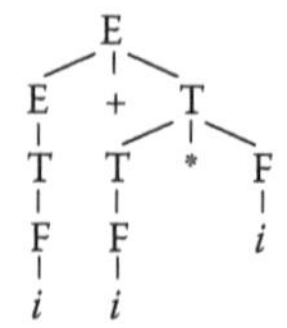

```
          E
       /  |  \
      E   +   T
      |     / | \
      T    T  *  F
      |    |     |
      F    F     i
      |    |
      i    i
```

FIGURE 13.9 A parsing tree for "$i + i * i$."

* Here, the "E" inside the square brackets is the start symbol of this grammar.

to build a syntax analyzer for them manually. Fortunately, there exist many tools to automatically generate the syntax analyzer from given language grammars.

In the following sections, we briefly introduce an LR(k) parsing originally proposed by Donald Knuth [8], which uses the finite state automata to decide either *shift* or *reduce* operation in the process of syntax analysis.

13.3.1 Framework of the LR Parsing

LR(k) parsing (or its variation) has been widely used in the syntax analysis. Here, the "L" indicates a left-to-right scanning of its input symbols, while the "R" means to construct a parsing tree according to the reverse steps of a rightmost derivation. The "k" is the number of symbols we attempt to look ahead to make a decision on choosing either *shift* or *reduce*. The rightmost derivation requires that, in each step, we always pick the rightmost nonterminal symbol which happens to be on the left-hand side of some production to be replaced by symbols on its right-hand side.

Let us take the grammar in Figure 13.8 for example. We start to derive a nonterminal symbol "E" (also the start symbol) and the rightmost derivation steps are given in Figure 13.10, where Figure 13.11 shows the corresponding process to construct a parsing tree. In each step, we always pick the rightmost nonterminal (underscored for clarity) to do derivation. The bottom-up LR parsing starts to sequentially read input symbols "$i + i * i$" from left to right and attempts to reduce them to a nonterminal "E" at last, which is exactly in the reversed order of its rightmost derivation. Therefore, if the input symbols could be ultimately reduced to the start symbol, a valid parsing tree is able to be constructed then.

An LR analyzer generally includes three parts: parsing driver, parsing table, and stacks [9]. The framework is demonstrated in Figure 13.12. The parsing driver reads input symbols sequentially, and then checks with the parsing table to make a decision of corresponding operations, either *shift*/*reduce* or *accept*/*reject* based on the current top element of the state stack. Hence, it is the parsing table that directs the process of parsing.

$$
\begin{array}{lll}
\mathrm{E} & \Rightarrow \mathrm{E} + \underline{\mathrm{T}} & \text{/* by production ① */} \\
 & \Rightarrow \mathrm{E} + \mathrm{T} * \underline{\mathrm{F}} & \text{/* by production ③ */} \\
 & \Rightarrow \mathrm{E} + \underline{\mathrm{T}} * i & \text{/* by production ⑥ */} \\
 & \Rightarrow \mathrm{E} + \underline{\mathrm{F}} * i & \text{/* by production ④ */} \\
 & \Rightarrow \underline{\mathrm{E}} + i * i & \text{/* by production ⑥ */} \\
 & \Rightarrow \underline{\mathrm{T}} + i * i & \text{/* by production ② */} \\
 & \Rightarrow \underline{\mathrm{F}} + i * i & \text{/* by production ④ */} \\
 & \Rightarrow i + i * i & \text{/* by production ⑥ */}
\end{array}
$$

FIGURE 13.10 The rightmost derivation.

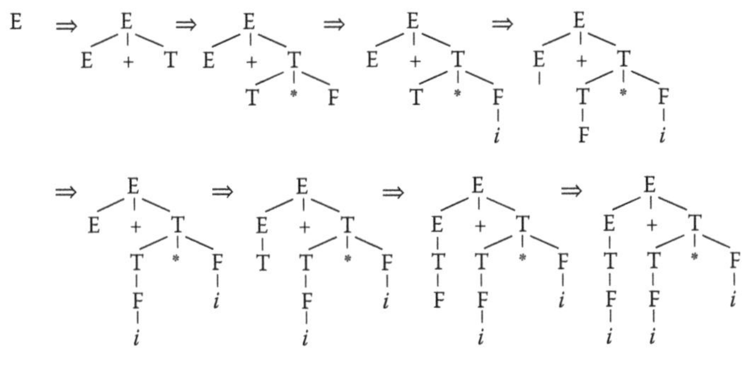

FIGURE 13.11 Building parsing tree for derivation.

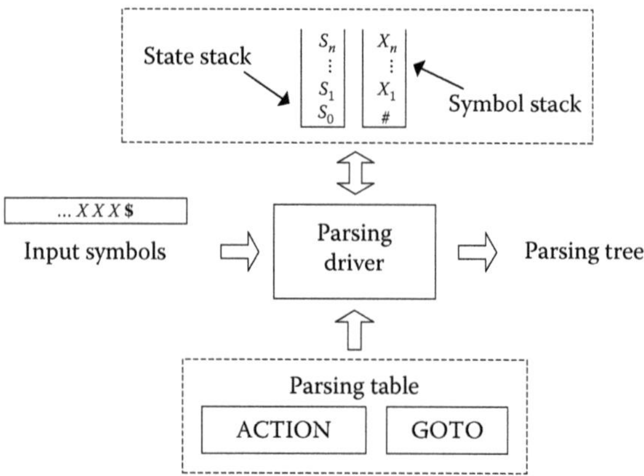

FIGURE 13.12 The framework of an LR analyzer.

TABLE 13.2 Parsing Table

	ACTION						GOTO		
State	+	*	(	)	*i*	#	E	T	F
0			S_4		S_5		1	2	3
1	S_6					acc			
2	r_2	S_7		r_2					
3	r_4	r_4		r_4					
4			S_4		S_5		8	2	3
5	r_6	r_6		r_6					
6			S_4		S_5			9	3
7			S_4		S_5				10
8	S_6			S_{11}					
9	r_1	S_7		r_1					
10	r_3	r_3		r_3					
11	r_5	r_5		r_5					

For instance, the parsing table for the grammar in Figure 13.8 is shown in Table 13.2 which has four kinds of operations:

- An "S_d" indicates a *shift* operation, such that the current input symbol and a state "*d*" are pushed into the symbol stack and the state stack, respectively.
- An "r_{id}" requires to do a *reduce* operation using the production numbered "id":

$$L \rightarrow \underbrace{X_1 \ldots\ldots X_n}_{n \text{ symbols}} \quad \text{------------------}\textit{production}\text{ id}$$

  It first pops out *n* elements from both the symbol stack and the state stack. Assuming the top of the state stack is S_{oldTop} after the popping, the driver will push the symbol L and state S_{newTop} into the symbol stack and the state stack, respectively, where S_{newTop} could be decided by checking the $\text{GOTO}[S_{\text{oldTop}}, L]$ in the parsing table.
- An "acc" indicates an accepting state if there is no more input symbols left. At this moment, the parsing driver accepts the whole input symbols. In other words, they pass through the syntax analysis and a valid parsing tree has been built successfully.

TABLE 13.3 Parsing Steps for "$i + i * i$"

Steps	State Stack	Symbol Stack	Input Symbols	Action	Goto
1	0		$i + i * i\#$	S_5	
2	0 5	i	$+i * i\#$	r_6	3
3	0 3	F	$+i * i\#$	r_4	2
4	0 2	T	$+i * i\#$	r_2	1
5	0 1	E	$+i * i\#$	S_6	
6	0 1 6	E+	$i^* i\#$	S_5	
7	0 1 6 5	E + i	$^* i\#$	r_6	3
8	0 1 6 3	E + F	$^* i\#$	r_4	9
9	0 1 6 9	E + T	$^* i\#$	S_7	
10	0 1 6 9 7	E + T*	$i\#$	S_5	
11	0 1 6 9 7 5	E + T*i	#	r_6	10
12	0 1 6 9 7 10	E + T*F	#	r_3	9
13	0 1 6 9	E + T	#	r_1	1
14	0 1	E	#	acc	

- Otherwise, the driver just rejects the input symbols which cannot fit for the grammars.

The parsing steps for the input symbols "$i + i * i$" are given in Table 13.3. A particular "#" is borrowed to indicate the end of input symbols and the state 0 is pushed into the state stack at the very beginning. For clarity, we attach the related *Action* and *Goto* operations in each parsing step. Since the last step comes up with an "acc" operation, the "$i + i * i$" is then accepted as a valid input satisfying the predefined grammars in Figure 13.8. The corresponding parsing tree is built automatically depending on all the parsing steps that perform the *reduce* operation.

It is obvious that the parsing table plays an important role in the parsing process because all operations, either *reduce* or *shift*, will be directed by the table. Actually, this table could be easily generated from the grammar based upon the technique of finite state automata.

13.3.2 Finite State Automata in the LR Parsing

Given the symbol stack and a current input symbol, how does the parser decide to execute *shift* or *reduce*? This decision is actually related to its current parsing state, where the theory of LR parsing uses the concept of "*item*" to formulate states.

An item of a grammar G is a production with a "dot" at somewhere in its right-hand side. For instance, the production "E → E + T" could correspond to the following items:

$$E \rightarrow \cdot E + T$$
$$E \rightarrow E \cdot + T$$
$$E \rightarrow E + \cdot T$$
$$E \rightarrow E + T \cdot$$

Roughly speaking, the "dot" in an item indicates some states in the parsing process, where at its left side are symbols that have been seen and at its right side are symbols that are supposed to come forth if this production will be used to do *reduce*. For instance, an item "E → E + ·T" tells us that the production will be used to reduce three successive symbols "E," "+," "T" to "E" once the former shows up at the top of the symbol stack, but for the current moment, we only obtain "E" and "+," and are waiting for the next "T" to come into the stack.

Since items relate to states in the parsing, they could be used to construct a deterministic finite automaton of the LR parsing. Before jumping into the automata, there are two related functions that need to be mentioned:

- Assuming I is a set of items, the CLOSURE(I) is constructed based upon the following requirements (recursively):
  1. $I \subseteq$ CLOSURE(I)
  2. If "$A \rightarrow \alpha \cdot B\beta$" $\in$ CLOSURE(I) and "$B \rightarrow \gamma$" is a production of the grammar, then "$B \rightarrow \cdot\gamma$" $\in$ CLOSURE(I).
- GO(I, X) is defined as the closure of the set of all items "$A \rightarrow \alpha X \cdot \beta$," such that "$A \rightarrow \alpha X \cdot \beta$" $\in$ I. For instance, assuming $I = \{$"$T \rightarrow T \cdot *F$"$\}$ in the grammar in Figure 13.8, GO(I,"*")={"$T \rightarrow T \cdot *F$," $F \rightarrow \cdot$ (E), $F \rightarrow \cdot$ i}, such that the last two items come from the computation of closure of {"$T \rightarrow T \cdot *F$"}.

To construct the automaton for a grammar G in Figure 13.8, we need to define an augmented grammar with adding a new start symbol G' as well as a production "G' $\rightarrow$ E," where E is the original start symbol [9]. For an LR automaton, its states are the sets of items and its transitions are given by the GO function. The start state of this LR automaton is CLOSURE({"G' $\rightarrow$ ·E"}), in this case, {"G' $\rightarrow$ ·E," "E $\rightarrow$ ·E + T," "E $\rightarrow$ ·T," "T $\rightarrow$ ·T * F," "T $\rightarrow$ ·F," "F $\rightarrow$ ·(E)," "F $\rightarrow$ ·i"}. Each rectangle in Figure 13.13, which is a set of items, represents a state in the LR automaton and we mark each "state j" as the set of items I_j.

It is the LR automaton that helps to build a parsing table as mentioned, including **ACTION** and **GOTO** parts. The **ACTION** $[j, X] = S_q$ indicates there is a transition labeled a terminal symbol X from state I_j to I_q, while **GOTO**$[j, X] = q$ comes from a transition labeled a nonterminal symbol X from state I_j to I_q. In addition, ACTION[1,#] = acc, because state I_1 in Figure 13.13 includes an item "G' $\rightarrow$ E·" which is the augmented production with the "dot" at the end of its right-hand side, indicating the start symbol "E" has been seen already. At this moment, if the parsing driver has finished reading all input symbols ("#" is used to mark the end of input), the parsing driver accepts the input symbols then.

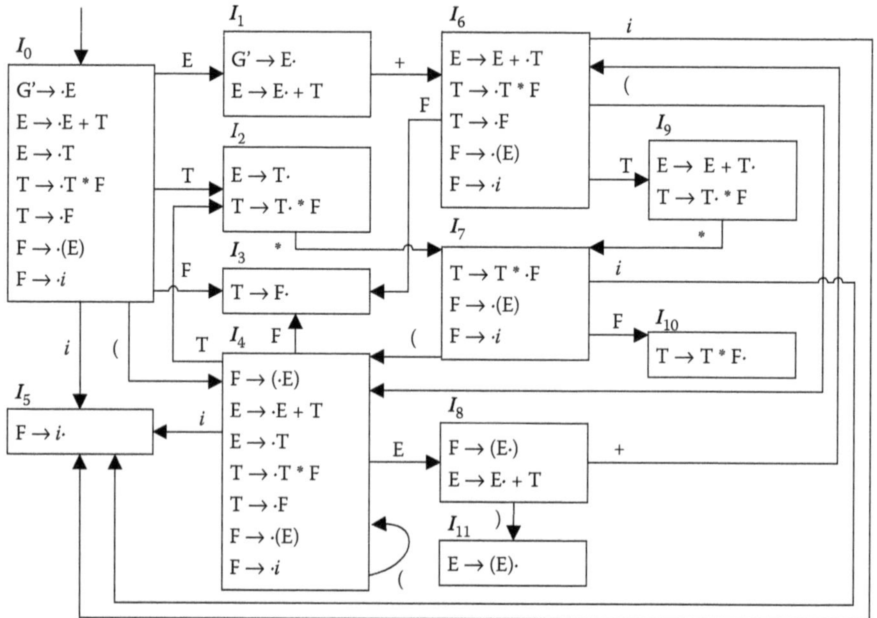

FIGURE 13.13 An LR automaton.

Given the LR automaton and an input symbols, the story behind the parsing table could be roughly explained as the following:

1. I_0 is the start state.
2. If the current state is I_1 and the input symbol is "#," then accept and exit, otherwise go to step (3).
3. Assuming the current state is I_j and the input symbol is X, there are two possibilities:
   a. I_j has items ending with a "dot," for instance, "$A \rightarrow \alpha\cdot$." In this case, back off $|\alpha|$ steps through the transition trace to a state I_c, which is treated as the new current state. Let A to be a new input symbol, then jump to step (2).
   b. I_j has items with a "dot" in front of X, for instance, "$A \rightarrow \alpha \cdot X\beta$." In this case, use the edge labeled X to transit to a new state I_c, which is treated as the new current state. Read the next input symbol and then jump to the beginning of step (2).

Steps (3.a) and (3.b) actually correspond to the *shift* and *reduce* operations, respectively. However, readers may find that they are not totally disjoint. For instance, assuming the current state is I_2 and the input is "∗," both *shift* and *reduce* operations are applicable, which is a *shift–reduce* conflict. In practice, some of these conflicts could be solved by computing a *FOLLOW* function* for all nonterminal symbols.

The *FOLLOW*(X) is defined to return a set of terminals that could appear immediately to the right of X in some sentential derivation form. For instance, *FOLLOW*(E) for the grammar in Figure 13.8 includes "+," but not "∗." Then, once the current state is I_2 and the input is "∗," it is inappropriate to do *reduce*, otherwise the result, which is "E," will be followed by a "∗." Therefore, the *shift–reduce* conflict in I_2 with a current input "*" could be solved and the correct operation at that moment is only *shift*.

Analogously, there is another category of conflicts: *reduce–reduce* conflicts, such that some states in the LR automaton have more than one item ending with the "dot." In that case, we could do *reduce*, but have multiple choices of available productions. Some of these conflicts are also solvable based on the *FOLLOW* idea. We summarize the solutions for *shift–reduce* and *reduce–reduce* conflicts below:

Assuming a state in the LR automaton is I, for instance,

$$I = \{\text{“}X \rightarrow \alpha \cdot b\beta,\text{”} \ \text{“}A \rightarrow \gamma \cdot,\text{”} \ \text{“}B \rightarrow \delta \cdot\text{”}\}$$

then both *shift–reduce* and *reduce–reduce* conflicts occur. However, if all of the following conditions are satisfied

$$FOLLOW(A) \cap FOLLOW(B) = \Phi$$
$$FOLLOW(A) \cap \{b\} = \Phi$$
$$FOLLOW(B) \cap \{b\} = \Phi$$

then we are able to deterministically decide the operations based on the current input symbol a:

- If $a = b$, then do *shift*; or
- If $a \in FOLLOW(A)$, then do *reduce* using the production "$A \rightarrow \gamma$"; or
- If $a \in FOLLOW(B)$, then do *reduce* using the production "$B \rightarrow \delta$"; otherwise
- Report an error and exit.

It is easy to extend the above statements to more general cases. If all *shift–reduce* and *reduce–reduce* conflicts in an LR automaton are solvable based upon this method, we call this corresponding grammar a Simple LR(1) or SLR(1) grammar [10]. Here, the "1" indicates we need to look ahead one symbol to make a decision.

As mentioned, the SLR(1) cannot solve all *shift–reduce* and *reduce–reduce* conflicts. The series of LR parsing still include other classic methods, such as LR(1), LALR(1) [11], and so on. They all heavily adopt the core technique of finite automata to construct their parsing table, but with some variation.

---

* The computation of *FOLLOW* for nonterminal symbols is trivial and the interested readers may refer to Ref. [1] for more information.

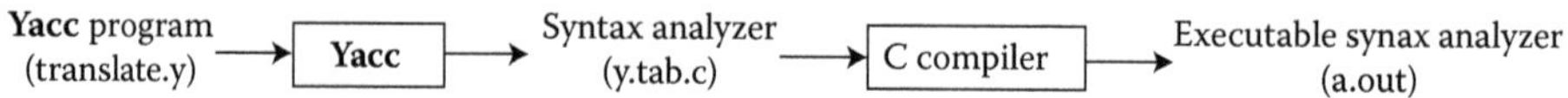

FIGURE 13.14 Generate a syntax analyzer using **Yacc**.

13.3.3 Generator of the Syntax Analyzer

Similar to the generator of the lexical analyzer as mentioned previously, there are corresponding tools, **Yacc** [12] and its open source version **Bison** [13], that are able to generate the syntax analyzer automatically with a predefined LALR(1) grammar. The working process, in Figure 13.14, is also analogical. A **Yacc** program, including a grammar as well as some user-defined behaviors, is fed into the **Yacc** tool, then a source program (usually C program) of the syntax analyzer is generated with a parsing driver as well as a parsing table, where the latter is constructed by the LR automaton technique. The executable version of a syntax analyzer comes after a regular C compilation.

13.4 Conclusions

In this chapter, we showed how to apply the technique of finite state automata to compilers. We mainly focused on two phases: lexical analysis and syntax analysis. The lexical analysis attempts to group neighboring input characters and recognize legal tokens, while the syntax analysis converts an unstructured token sequence into a structured parsing tree based on predefined language grammars.

We showed how to define the legal tokens by regular expressions and translate them into an NFA. After a simple conversion from the NFA to a DFA, the latter becomes a token recognizer indeed. Furthermore, an algorithm that converts an NFA to a DFA was also presented. The **Lex**, an automatic lexical analyzer generator, follows the idea to accept user-defined regular expressions, construct an NFA and then transform it to a DFA.

An LR(k) parsing is one of the most popular bottom-up syntax analysis approaches. It includes two main operations: *shift* and *reduce*. The operation *shift* continuously reads symbols from its input, while the operation *reduce* chooses a production to do reduction at some points. The decision of choosing either *shift* or *reduce*, therefore, becomes the core step in the parsing. Items are used to keep the current state of parsing, and *shift* or *reduce* decision totally depends on the current state in a finite automaton. Furthermore, *shift* makes the automaton do one-step transition forward, but *reduce* backs off several steps through previous transition path according to the chosen production in the language grammar. SLR(1) parsing handles some *shift–reduce* and *reduce–reduce* conflicts. The program **Yacc** is an automatic syntax analyzer generator that accepts user-defined grammars and adopts the automata technique to generate a syntax analyzer in the compiler.

References

1. A. V. Aho, M. S. Lam, R. Sethi, and J. Ullman. 2006. *Compilers: Principles, Techniques, and Tools.* 2nd ed. New Jersey: Prentice Hall.
2. J. E. Hopcroft, R. Motwani, and J. D. Ullman. 2000. *Introduction to Automata Theory, Languages, and Computation.* 2nd ed. Boston: Addison Wesley.
3. C. E. Shannon and J. McCarthy, eds. 1956. *Automata Studies,* Annals of Mathematics Studies, No. 34, pp. 43–98, Princeton: Princeton University Press.
4. M. O. Rabin and D. Scott. 1959. Finite automata and their decision problems. *IBM Journal of Research and Development* 3(2): 114–125.
5. K. Thompson. 1968. Regular expression search algorithm. *Communications of the ACM* 11(6): 419–422.

6. M. E. Lesk and E. Schmidt. 1990. Lex—A lexical analyzer generator. In UNIX Vol. II, A. G. Hume and M. D. McIlroy (Eds.), pp. 375–387 Philadelphia: W. B. Saunders Company.
7. *Flex: The Fast Lexical Analyzer.* Available at http://flex.sourceforge.net/.
8. D. E. Knuth. 1965. On the translation of languages from left to right. *Information and Control* 8:607–639.
9. S. Zhang, Y. Lv, W. Jiang, and G. Dai. 2007. *Compiler Principle* (In Chinese). 2nd ed. Beijing: Tsinghua University Press.
10. F. L. DeRemer. 1971. Simple LR(k) grammars. *Commun. ACM* 14(7): 453–460.
11. F. L. DeRemer. 1969. *Practical Translators for LR(k) Languages.* Technical Report. Massachusetts Institute of Technology, Cambridge, MA, USA.
12. S. C. Johnson. 1975. Yacc: Yet another compiler-compiler. In CS TR 32, Bell Labs. Available at http://dinosaur.compilertools.net/yacc/.
13. Bison—GNU parser generator. Available at http://www.gnu.org/software/bison/.

14

Finite State Models for XML Processing

Murali Mani
University of Michigan

14.1 Introduction

XML (eXtensible Markup Language) (W3C XML 2008) has become the de facto standard for information exchange over the World Wide Web. An XML document can define its own tags (and hence called extensible); this is unlike HyperText Markup Language (HTML), where the tags are predefined and specify formatting information. For instance, XML is used as the basis for Simple Object Access Protocol (SOAP) (W3C SOAP 2007) that is used in web services and allows applications to communicate over the Internet; XML is used in Mathematical Markup Language (MathML) (W3C MathML 2010) that allows mathematical notations and content to be processed by web applications.

XML documents can optionally specify their own schema. The schema defines what tags can be used in an XML document, and restricts the structure of documents. For instance, the schema for MathML enforces that in any MathML document, a fraction (<mfrac> element) consists of a numerator and a denominator (W3C MathML 2010).

Different languages have been proposed to specify schemas for XML documents. Some of the most widely used among these XML schema languages include Document Type Definition (DTD) (W3C XML 2008), W3C XML Schema (W3C XML Schema 2004), RELAX NG (RELAX NG 2001), and Schematron (Schematron 2006). In this chapter, we discuss how finite state models are used for validation of XML documents against the schema, where we check that the document satisfies all the structural constraints specified in the schema.

Another aspect of XML processing where finite state models is used is for answering queries over XML documents. XML queries specify patterns and finite state models are often used for determining document fragments that match the different patterns specified in a query. In this chapter, we outline how finite state models are used for XML query processing, considering both XPath (W3C XPath 2010) and XQuery (W3C XQuery 2010) expressions. We also describe some of the challenges in XML query processing, and how they are addressed by different state-of-the-art works.

Outline: The rest of this chapter is organized as follows. In Section 14.2, we describe how finite state models are used for validating XML documents against the schema. We present a simplified algorithm for XML document validation, and illustrate the algorithm with examples. In Section 14.3, we describe XPath and XQuery processing over XML documents. We discuss some of the issues in using finite state models for XPath processing, and describe how different approaches have tried to tackle the challenges differently. We describe a simplified algorithm for XPath query processing and illustrate it with examples. We then describe the challenges faced in XQuery processing, describe how different works have tackled XQuery processing, and describe an algorithm for XQuery processing using finite state models. In Section 14.4, we give our concluding remarks.

14.2 XML Validation

The formalism used to describe most XML schema languages is regular tree grammars (Comon et al. 2008). Schema languages such as DTD, XML Schemas, and RELAX NG can be described using this framework of regular tree grammars (Murata et al. 2005); however, there are schema languages such as Schematron (Schematron 2006) that are based on constraint languages, and not based on regular tree grammars and languages. In this section, we focus on XML schema languages that are based on regular tree grammars; variants of finite state automata are used for validating XML document instances against schemas in these languages.

Definition 14.1:

A regular tree grammar is a 4 tuple $G = (N, T, S, P)$, where

- N is a finite set of nonterminal symbols
- T is a finite set of terminal symbols
- $S \subseteq N$ is a set of start symbols
- P is a set of production rules of the form $X \rightarrow a\ r$, where $X \in N$, $a \in T$, and r is a regular expression over N. This rule says that X produces trees with a as the root and children that "satisfy" r. □

Example 14.1:

[Murata et al. 2005] The following grammar $G_1 = (N_1, T_1, S_1, P_1)$ is a regular tree grammar, where

N = *{Book, Chapter1, Chapter2, Pcdata}*
T = *{book, chapter, pcdata}*
S =*{Book}*
$P = \{Book \rightarrow book(Chapter1, Chapter2*), Chapter1 \rightarrow chapter(), Chapter2 \rightarrow chapter(Pcdata),$
$Pcdata \rightarrow pcdata()\}$

A document that is generated by the above grammar G_1 is shown in Figure 14.1.

Let us examine a simple validation algorithm for XML as in Murata et al. (2005) that checks whether a document is valid with respect to a given regular tree grammar. The algorithm is based on bottom-up tree automata (Comon et al. 2008), and works on the events in the document in the order in which they appear (see Figure 14.1a). The algorithm uses three stacks as follows:

1. Stack P, where each element in the stack is a set of symbols from N. The top of the stack indicates the symbols from N that can be used to match the next node in the document. Stack P is initialized to S (the set of start symbols in the grammar G).
2. Stack Y, where each element in the stack is a set of production rules from P. This set indicates the candidate set of production rules that can match the current node in the document. Stack Y is empty when we start the algorithm.

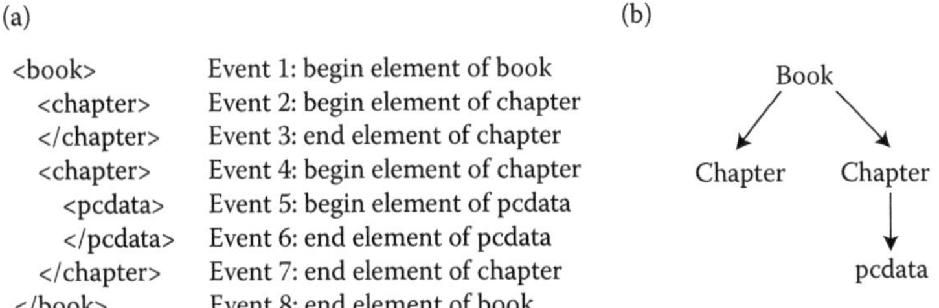
(a)
<book>
<chapter>
</chapter>
<chapter>
<pcdata>
</pcdata>
</chapter>
</book>
Event 1: begin element of book
Event 2: begin element of chapter
Event 3: end element of chapter
Event 4: begin element of chapter
Event 5: begin element of pcdata
Event 6: end element of pcdata
Event 7: end element of chapter
Event 8: end element of book
(b)
Book
Chapter
Chapter
pcdata

FIGURE 14.1 An XML document generated by the grammar G_1. (a) XML document, and the corresponding events. (b) Tree representation of the document in (a).

3. Stack S, where each element in the stack is a list of sets of symbols from N, where the first set of symbols indicate the symbols from N that match the first child of the current node (determined when we see the end element event for this first child), the second set of symbols indicate the symbols from N that match the second child of the current node (determined when we see the end element event for this second child) and so on. Stack Y is empty when we start the algorithm.

There are two classes of events that the algorithm primarily works on: start element events, and end element events. When we see a start element event for x, the algorithm does the following:

1. Find rules in P, where LHS is a symbol on the top of stack P. Let us denote these rules as $P' \subseteq P$.
2. Push set of rules from Step (1) (P') onto stack Y.
3. Push the set of symbols from N that appears in the RHS of rules in P' onto stack P.
4. Push an empty list onto stack S.

When we see the end element for x, the algorithm does the following:

1. Pop set of rules from stack Y. Let us denote these rules as P'.
2. Pop a list from stack S, and see which rules in P' "accept" this list. Let these rules be $P \subseteq P''$.
3. Append the set of symbols in N that appear in the LHS of the rules P'' as the next element of the list on top of stack S.
4. Pop set of symbols from stack P.

If for any start element event, Step (1) finds no rule then we report that the document is not valid. For an end element event, if Step (2) finds no rule in P'', then we report that the document is not valid. On the other hand, if we find that the stack S is empty, it implies that we have finished the document without any problems and we report that the document is valid.

Let us illustrate the above algorithm for the document in Figure 14.1, against the regular tree grammar $G_1 = (N_1, T_1, S_1, P_1)$ (Example 14.1). We will call the four rules in P_1 as Rule1, Rule2, Rule3, and Rule4. The three stacks before we see any events and after we see the start element event of book are shown below. Note that when we see start element event for book (<book>), Rule1 has LHS, Book, which matches the top of stack P. Therefore, {Rule1} forms P' and is pushed onto stack Y. The symbols that appear in the RHS of a rule in P' are {Chapter1, Chapter2} (they appear in the RHS of Rule1). Therefore, {Chapter1, Chapter2} is pushed onto stack P, and an empty list <> is pushed onto stack S.

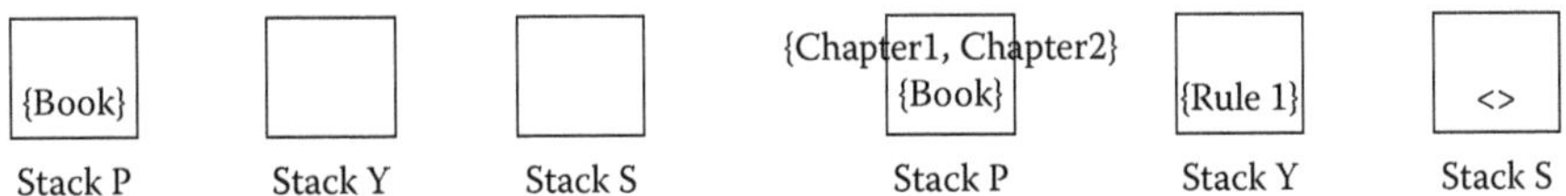
{Book}
Stack P
Stack Y
Stack S
{Chapter1, Chapter2}
{Book}
Stack P
{Rule 1}
Stack Y
<>
Stack S

The three stacks after we see events 2 and 3 are shown below. Event 2 is start element of chapter (<chapter>). Rules 2 and 3 have LHS as Chapter1 or Chapter2 (the symbols on top of stack P). Therefore, we push {Rule2, Rule3} onto stack Y. The symbols that appear in the RHS of Rules 2 and 3 {Pcdata} is pushed onto stack P, and an empty list <> is pushed onto stack S. Event 3 is end element of chapter (</chapter>). We pop from stack Y, and {Rule2, Rule3} becomes our P'. We pop from Stack S and we get the empty list <>. Only Rule2 "accepts" the empty list (Rule 3 accepts Pcdata). Therefore, $P'' = \{Rule2\}$. The LHS of Rule2 is Chapter1, and we append {Chapter1} to the top element of stack S, which now becomes <{Chapter1}>. We also pop an element from stack P.

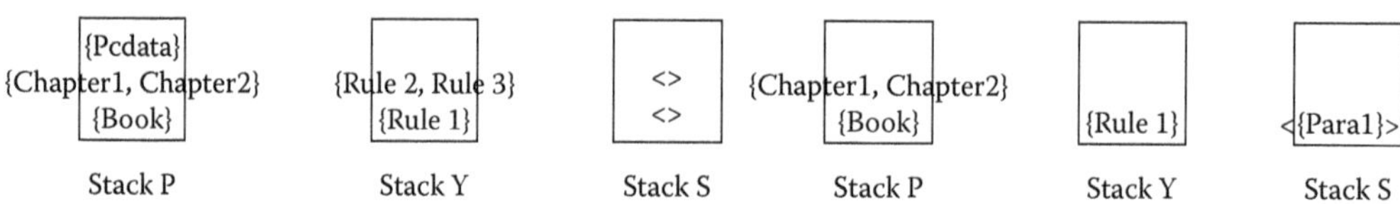

We continue this, and we illustrate the three stacks after events 7 and 8. Event 8 is end element of book (</book>). We pop from stack Y, and {Rule1} becomes our P'. We pop from Stack S and we get the list <{Chapter1}, {Chapter2}>. Rule1 "accepts" this list. Therefore, $P'' = \{\text{Rule1}\}$. Stack S is now empty; therefore, we need not do Step 3. Now we return that the document is valid.

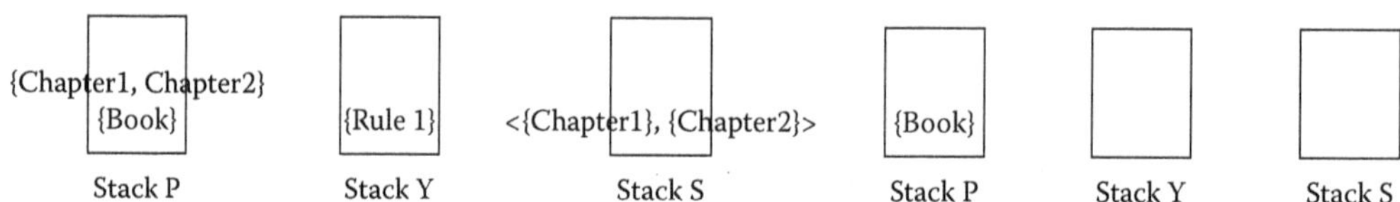

A feature worth noting is Step (2) when we see an end element event. Here, we need to match a list of set of symbols against a regular expression, and for this, we use a variant of a finite state automaton. The difference from a typical finite state automaton is that each "item" in our list is a set of symbols, and not one symbol as in typical finite state automaton. However, this variation requires a simple extension, and it is straightforward to construct a nondeterministic finite state automaton (NFA) to handle this.

To illustrate an invalid document, consider validating against G_1 the following document:

<book><chapter></chapter><chapter><pcdata></pcdata></chapter><chapter>
</chapter></book>

When the algorithm sees the end element event of book </book>, the top of stack S is the list <{Chapter1}, {Chapter2}, {Chapter1}> at the top, and stack Y has {Rule1} at the top. We see that Rule1 does not accept this list, and hence the document is reported to be not valid. We should mention that the algorithm we described is a simplified algorithm. Real validators should determine that the document is not valid when the last </chapter> is seen; this is needed for appropriate error messages. In Murata et al. (2005), we describe other algorithms used for XML validation, based on binary trees and based on derivatives. The chapter also describes hedge automata that can be used for XML validation (and also for XML processing).

Note that the algorithm that we described from Murata et al. (2005) in addition to checking whether the document is valid with respect to the schema can also check whether the document is "well formed" (a document is well formed if the opening and closing tags match). Another work that considers validation of XML documents is Segoufin and Vianu (2002), where the authors consider using finite state automata (without the stacks that we used) for validating XML documents. Finite state automata cannot be used for validation when the regular tree grammar is recursive (that is an element can have itself as a descendant). Therefore, the authors consider documents that are known to be well formed, and study when such documents can be validated using finite state automata. For instance, consider a grammar that produces any

number of <a> elements, nested within each other. Such a language can be described as (<a>^{n}</a>^{n}). Of course, we cannot validate the above using a finite state automaton. However, if we know that the document is well formed, then we need to check that the document is in the language (<a>*</a>*); we can use a finite state automaton for validating such a language. In Segoufin and Vianu (2002), the authors characterize for what grammars validation can be done without checking well-formedness using a finite state automaton.

14.3 XML Processing

XML Path Language (W3C XPath 2010) forms an essential component of XML query languages, such as XQuery (W3C XQuery 2010) and XML Transformation Languages (W3C XSLT 2007). XPath, by itself, is also often used as a query language. A query in XPath consists of multiple steps, where each step in turn consists of an axis, a node test, and a set of zero or more predicates. There are 13 different axes, including child, descendant, ancestor, parent, and attribute. The node test specifies the name of the node to match, and it can also be a wild card. Predicates are specified either as [path1] or as [path1 OPERATOR path2], where path1 (similarly path2) can either be XPath expressions or literal values; the predicate [path1] checks for the existence of the path specified; [path1 OPERATOR path2] checks that at least one of the values returned by path1 and at least one of the values returned by path2 satisfy the relationship (as specified by OPERATOR).

There are two different types of XML processing that are defined for XPath, as described in Bar-Yossef et al. (2005): (a) document filtering, that returns documents that contain the specified path, and (b) full-fledged query evaluation, that returns all the elements that match the specified path. Document filtering is an "easier" (smaller running time, and requires less memory) problem as compared to full-fledged query evaluation. To answer a document filtering query, we can perform full-fledged query evaluation, and return yes/no depending on whether the full-fledged query evaluation returns zero/at least one result. We will mostly explain full-fledged query evaluation in this chapter, though we will also remark on document filtering whenever appropriate.

An example XPath query is root/a[c/text()=5][d/text()=e/text()]/b. In this XPath query, there are three steps: the first step is looking for root element; the second step is looking for a child elements of the root element such that this a element has a c child element with text=5, and this a element has one d child element and one e child element whose text values match; and the third step is looking for b child elements of the a elements from the previous step.

Consider that the above XPath query is evaluated against the XML document shown in Figure 14.2 as a tree, which is streaming consisting of start element and end element events in order. We mainly consider streaming XML in this chapter because finite state models are especially useful for query evaluation over streaming XML documents. A portion of the XML document as a stream is: <root><a><c>5</c><d>3</d><e>4</e><e>3</e><b/></a>....</root>

In the above document, the root element has three a child elements; all these three a elements have a c child element which has the text 5. For the first a element, it has one d child element which has the text 3, and it has two e child elements, one of which has text 4, and another has text 3. As we have a d child

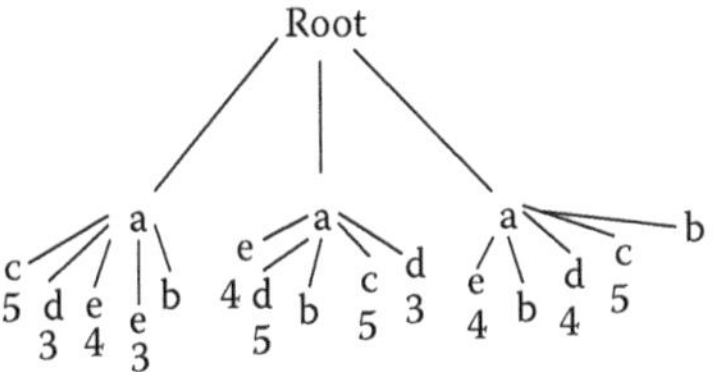

FIGURE 14.2 An example XML document represented as a tree.

element and an e child element whose values match, this a element matches all the predicates. Therefore, the b child elements of this a element are returned in the result.

For the second a element, none of the d child elements have text values (there are two d child elements with text values 5, 3) that match any of the e child elements (there is one e child element with text value 4). Therefore, this a element does not satisfy the predicate, and hence b child elements of this a element are not returned.

For the third a element, we have a d child element (with value 4) that matches a e child element (with value 4). Therefore, the two b child nodes of this a element are returned.

The result of the query is, therefore, the b child element of the first a element, and the two b child elements of the third a element; these three elements will be returned in the order in which they are seen in the document. Note that the above three b elements is the result of full-fledged query evaluation; on the other hand, if were to do document filtering, the result would be that the above document matches the specified XPath query (the specified XPath expression exists in the document).

This query also illustrates some of the key issues in XPath query answering. The first is that the predicates specified in the XPath query can be matched in any order for the different elements in the document. For instance, for the first a element, the predicate [c/text()=5] is matched first, then the predicate [d/text()=e/text()] is matched, and only after the predicates are matched, we get the b child elements. For the third a element, the first b child is seen before any predicate is matched, the predicate [d/text()=e/text()] is matched first, and then we match the predicate [c/text()=5], and after that we see the second b element. For the second a element, the b element is seen, then the predicate [c/text()=5] is satisfied, the predicate [d/text()=e/text()] is not satisfied.

Such ordering requires some nodes to be buffered, as very well studied and analyzed in Bar-Yossef et al. (2005). For the first a element, we need to buffer the value of d child element as 3, from the moment we see the d element till we see the e child element with value 3. We also need to buffer the value of the first e child element as 4, from the moment we see this e element till the predicate [d/text()=e/text()] is matched. Once we have seen the e child element with value 3, we can mark that this a element has all predicates satisfied, all the buffered d element values and e element values can be cleared, and the b child elements seen in future can be returned in the output. For the second a element, we need to buffer the value of e (4), the value of d (5), the b element which is "alive" (we may have to return this b element in the output, provided the predicates are satisfied), and another value of d (3). Only when we see the end of the a element, we know that this a element does not satisfy the predicates and hence all the buffered d values (3, 5), the buffered e value (4), and the buffered b element can be cleared. For the third a element, we need to buffer the value of e (4) till we see the value of d (4), we need to buffer the first b element till we see the value of c (5). Once we see the value of c element as 5, we know that the a element satisfies all the predicates, the buffered b element (the first b element) can now be output. Note that the second b element need not be buffered (as predicates are already satisfied when we see this element).

Another point worth noting in XPath processing is the storage needed for handling of multivariate predicates as described in Bar-Yossef et al. (2005), and the storage needed for storing "alive" events (that may be output provided some more predicates are satisfied). For instance, we see that for the second a element, the different values of e and the different values of d must be stored till we see the end of the a element, and we know that this a element will not satisfy all predicates. Similarly, the b child element of this a element also needs to be stored till we see the end of the a element. There are some works that utilize schema constraints (that specify order in which child elements appear) that may be available (Koch et al. 2004; Su et al. 2005).

Of the different works that we consider, XSQ (Peng and Chawathe 2005) and XPush (Gupta and Suciu 2003) use a single automaton with buffers to perform filtering of documents given an XPath query. Here they consider all the different possible orderings of predicates for a step in the XPath query, and construct one automaton that performs query evaluation. XPush considers evaluation of a large number of XPath expressions, they consider lazy construction of automaton, so that the high memory costs involved with storing the complete automaton corresponding to a large number of XPath queries is reduced. Both these works consider only univariate predicates of the form [path Op constant], and do not consider multivariate predicates of the form [path1 Op path2]. Because both these works study the problem of document filtering, and consider only univariate predicates, they do not buffer "alive" elements (elements that may be output in the result), or element values to be used for predicate evaluation (as needed for multivariate predicates).

There are other works that divide an XPath expression into multiple "linear" XPath expressions (Diao et al. 2003; Green et al. 2004). Linear path expressions can be very easily converted to finite state automata; the resulting matches from these linear path expressions should then be "joined" to determine the query results. In Green et al. (2004), the authors study primarily construction of deterministic finite state automata lazily for a large number of linear XPath expressions; the advantage of lazy construction of automata at run-time is that the size of the automaton in memory is much smaller than what would be if the automata were constructed at compile-time. In YFilter (Diao et al. 2003), the authors consider how to combine the results from different linear path expressions. Both Green et al. (2004) and Diao et al. (2003) do not consider multivariate predicates (of the form [d/text()=e/text()]).

Below, we illustrate how to break an XPath into multiple linear XPath expressions and combine the results from the different linear XPath expressions to form the complete results of the XPath expression, based on the example given above.

The XPath expression root/a[c/text()=5][d/text()=e/text()]/b can be broken into the following four linear XPath expressions:

Path 1: root/a/b
Path 2: root/a/c/text()
Path 3: root/a/d/text()
Path 4: root/a/e/text()

For ease of explanation, we assign ids (as subscripts) to the different elements in the document as shown in Figure 14.3.

Note that these IDs are given just for easier explanation, and are not present in the actual document. Also note that c1 has a text child with value 5.

When the different linear paths are evaluated against the document, we get the results shown in Figure 14.4.

The results of the XPath expression can be found by combining the above results. For instance, the a nodes that satisfy the predicate [c/text()=5] can be found based on results from Path 2 (Result1=SELECT DISTINCT a FROM Path2 WHERE text=5). The a nodes that satisfy the predicate [d/text()=e/text()] can be obtained based on results from Path 3 and Path 4 (Result 2=SELECT DISTINCT Path3.a FROM Path3, Path4 WHERE Path3.a=Path4.a AND Path3.text=Path4.text). We find that a1 and a3 satisfy this predicate. The b elements that are children of these a elements can be obtained from Path 1 as b1, b3, b4 (SELECT DISTINCT b FROM Path1 WHERE a IN Result1 AND a IN Result2).

Translation of linear XPath expressions into finite state automata is quite straightforward, and discussed very well in YFilter (Diao et al. 2003); here, the authors also describe how to combine different finite state automata for different linear path expressions into one automaton, so that at run-time only one

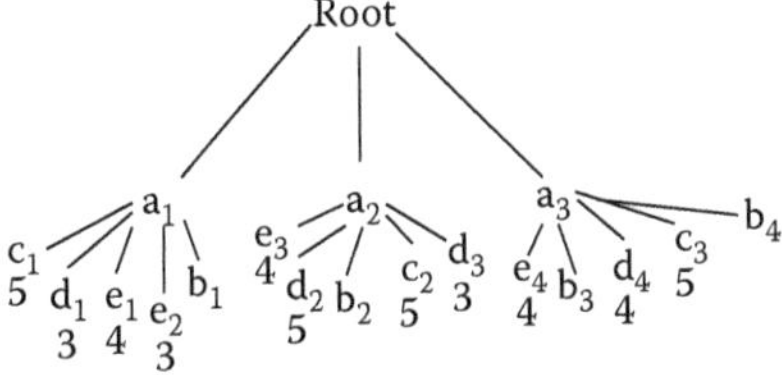

FIGURE 14.3 The XML document shown in Figure 14.2 with IDs for the different nodes.

Path 1		
Root	a	B
Root	a1	b1
Root	a2	b2
Root	a3	b3
Root	a3	b4

Path 2			
Root	a	C	Text
Root	a1	c1	5
Root	a2	c2	5
Root	a3	c3	5

Path 3			
Root	a	d	Text
Root	a1	d1	3
Root	a2	d2	5
Root	a2	d3	3
Root	a3	d4	4

Path 4			
Root	a	E	Text
Root	a1	e1	4
Root	a1	e2	3
Root	a2	e3	4
Root	a3	e4	4

FIGURE 14.4 Results from the different linear paths on the XML document shown in Figure 14.3.

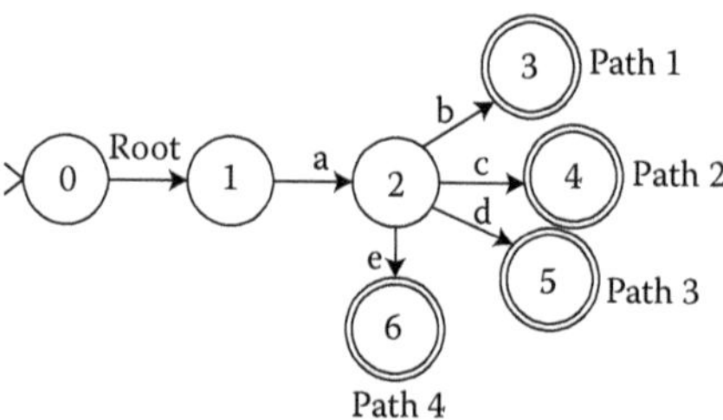

FIGURE 14.5 Combined automata for the four different linear path expressions.

automaton needs to be executed. The combined automaton corresponding to the four different linear path expressions is shown in Figure 14.5. While executing the automaton on the events corresponding to the XML document, a stack needs to be used to ensure correct nesting. For instance, suppose there was an x child of a, which the query is not interested in, and then we see a b child of this x element, we should not transition to state 3. Instead, we need to ensure that the end element of this x element is seen. When a final state is reached, the entire path from the root to that element is in the stack, and can be used to extract the results as shown in Figure 14.4.

Let us illustrate the execution of the automaton shown in Figure 14.5, as described in Raindrop (Wei et al. 2008) against a fragment from the XML document shown in Figure 14.3. In our document fragment, we will consider the first a child of the root:

<root><a1><c1>5</c1><d1>3</d1><e1>4</e1><e2>3</e2><b1></b1></a1> ....</root>.

In our stack, we will maintain the current set of states, and we use NULL to indicate that we are currently processing an element which is not part of the query. First, we have state 0 on the stack. When we see <root>, we look at the transitions from the states at the stack top (state 0), and we see that there is a transition to state 1, and push state 1 onto the stack. When we see <a1>, we push state 2 onto the stack. When we see <c1>, we push state 4 onto the stack; state 4 is a final state, and this matches Path 2; therefore, root, a1, c1, matches Path 2 (along with the child text nodes of c1, which is 5). Now, we see </c1>, which is an end element, and we pop one element (state 4) from the stack. Now we see d1, we push state 5 onto the stack; state 5 is a final state matching Path 3; therefore, root, a1, d1, 3 matches Path 3. Now we see </d1>, and pop state 5 from the stack. Now we see <e1>, we push state 6 onto the stack; state 6 is a final state matching Path 4; therefore, root, a1, e1, 4 matches

Path 4. Now we see </e1>, and pop state 6 from the stack. Now we see <e2>, we push state 6 onto the stack; state 6 is a final state matching Path 4; therefore, root, a1, e2, 3 matches Path 4. Now we see </e1>, and pop state 6 from the stack. Now we see <b1>, we push state 3 onto the stack; state 3 is a final state matching Path 1; therefore, root, a1, b1 matches Path 1. Now we see </b1>, and pop state 3 from the stack. Now we see </a1>, and pop state 2 from the stack. This process is continued for the remaining a elements also. Below we show the stack starting from the beginning of the document till we see </d1>.

At start of document	After <root>	After <a1>	After <c1>	After </c1>	After <d1>	After </d1>
			4, c1		5, d1	
		2, a1	2, a1	2, a1	2, a1	2, a1
	1, root	1, root	1, root	1, root	1, root	1, root
0	0	0	0	0	0	0

Suppose we execute the same automaton as shown in Figure 14.5, over a document fragment which includes elements that do not match the query as shown below.

<root><a1><x></x><c1>5</c1><d1>3</d1><e1>4</e1><e2>3</e2><b1>
</b1></a1>....</root>.

The stack at various points of time is shown below. The main point to note is that NULL is pushed onto the stack when we see <x>. After that, NULLs will continue to be pushed onto the stack for every start element till we pop all the NULLs out of the stack (when we see </x> for our example). After this, when we see <c1>, the execution will proceed as above.

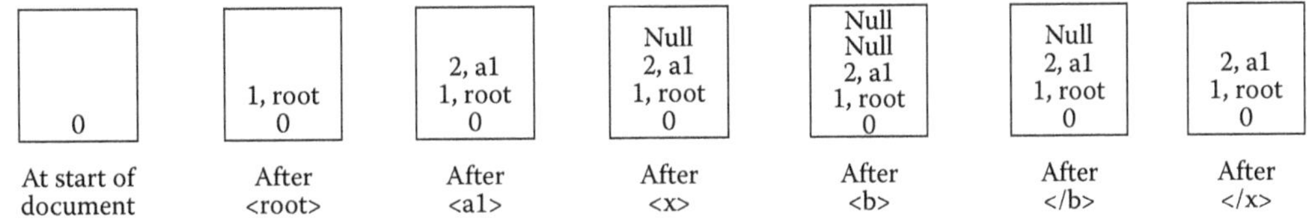

We examined how XPath queries can be evaluated using finite state automata, a stack and with buffers to hold temporary results. The buffers will then need to be joined to obtain the results for the query.

Now, let us examine how XQuery expressions, which consist of multiple XPath expressions along with other constructs, can be evaluated. There are different existing works which have studied XQuery evaluation, including XSM (Ludascher et al. 2002), Tukwila (Ives et al. 2002), and Raindrop (Wei et al. 2008). XQuery is motivated by SQL, and combines XPath expressions using FLWR (FOR-LET-WHERE-RETURN) expressions. The semantics for an XQuery expression is as follows. The FOR clause specifies multiple XPath expressions with variables. Each XPath expression variable will bind to an ordered set of nodes. For each combination of bindings for each of the variables, the remainder of the query is evaluated. The WHERE clause specifies predicates, and the bindings which do not satisfy the predicates are removed from further processing. The RETURN clause provides return structures (reorganization of elements can be specified and new elements can be formed in the RETURN clause) for each combination of bindings remaining after the WHERE clause predicates are evaluated. Note that each clause can in turn contain further nested FLWR expressions. The LET clause also specifies variables, but it binds one value to each combination of values in the FOR clause.

An example XQuery expression, an example document fragment, and the result of executing the XQuery expression on the document fragment are shown below (Wei et al. 2008).

Example XQuery expression:

```
for $a in open_auctions/auction[privacy]

let $e := $a/description/emph

where $e/text=''French Impressionism''

return <InterestedAuction> $a, $e </InterestedAuction>
```

Example Document fragment:

```
<open_auctions>

    <auction>

    <privacy>No</privacy>

    <description>

    Calendar of <emph>French Impressionism</emph>by<emph>
    Monet </emph>

    </description>

    <initial> $20 </initial>

</auction> ...
```

Result of evaluating the XQuery expression on the above document fragment:

```
<InterestedAuction>

  <auction>

    <privacy>No</privacy>

    <description>

    Calendar of <emph>French Impressionism</emph>by<emph>
    Monet</emph>

    </description>

    <initial> $20 </initial>

  </auction>

    <emph>French Impressionism</emph><emph>Monet </emph>

</InterestedAuction>
```

The above result is obtained as follows: the auction elements bind to $a variable; the two emph elements (with text "French Impressionism" and "Monet") bind to the $e variable as a single value; the predicate is checking whether one of the emph elements from the binding to $e has text = French Impressionism, this predicate is satisfied for this binding for $a. The RETURN clause specifies the return structure for this $a binding, which is to form an InterestedAuction element with child elements as the $a binding, and all the bindings for $e.

XSM (Ludascher et al. 2002) comes up with templates for different constructs in XQuery, including linear path expressions, for expressions, construction of elements; each template includes input buffers and output buffers. Tukwila (Ives et al. 2002) includes an X-scan operator that does the entire pattern matching, and there are other more traditional database operators which combine the different results of pattern matching returned by X-scan operator. Raindrop (Wei et al. 2008) is similar to Tukwila;

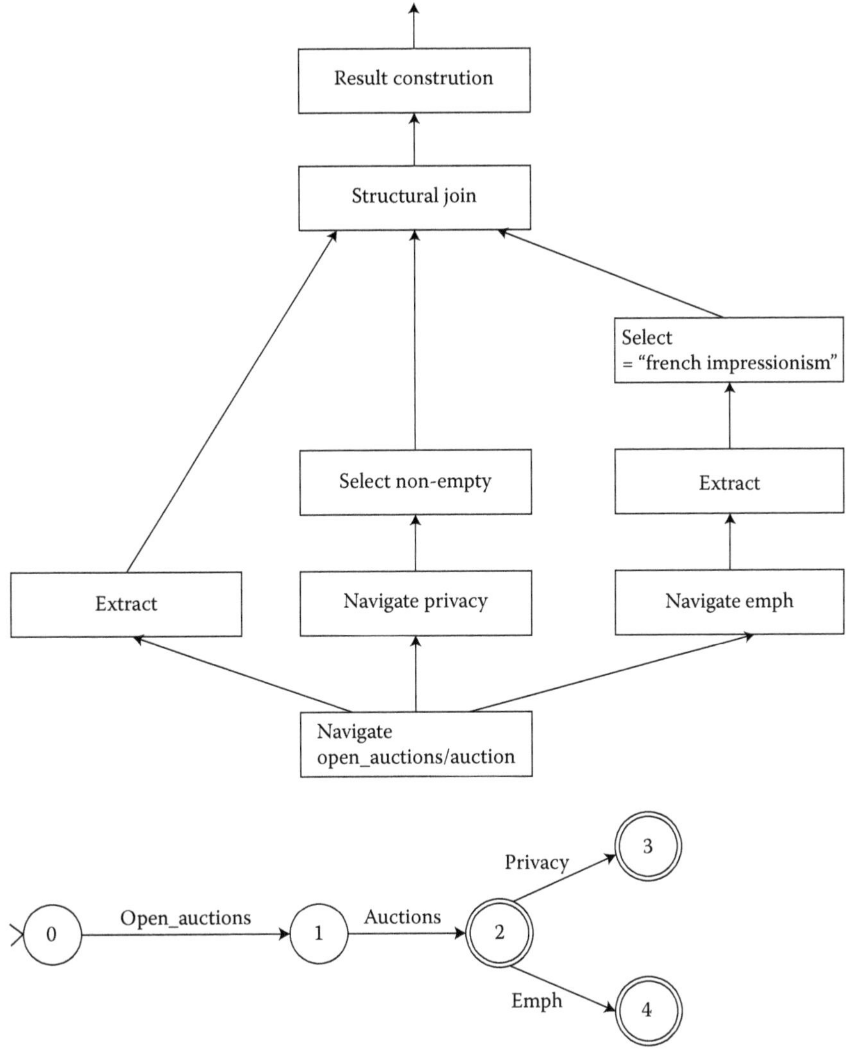

FIGURE 14.6 A plan for executing the example XQuery expression. The automaton shown below the plan is used to find the patterns, indicated by Navigate operators in the plan. The plan also includes Extract, Select, Join, and Result Construction operators.

however, instead of one X-scan operator that does all pattern matching, Raindrop considers different pattern matching as different operators. Furthermore, pattern extraction can be done using a finite state automaton, or can be done using document object model (DOM) (W3C DOM 2004) style operators after extraction of appropriate nodes. If a finite state model is used to extract patterns, then a push-model is used, and all elements that satisfy the subpattern are extracted. If a DOM style operator is used to extract subpatterns, then a pull-model is used, and only elements that satisfy the subpattern and also satisfy other predicates are extracted. Therefore, depending on the selectivity of different patterns, it may be more efficient to use a finite state automaton, or DOM-style operators for pattern extraction. A cost-model is developed to choose between the different pattern extraction methods based on the estimated costs of the different styles of pattern extraction (Su et al. 2008).

Let us illustrate the Raindrop approach using the example query given above. There are three patterns to be extracted:

$a=open_auctions/auction

$a/privacy

$e=$a/description/emph

The different patterns can be matched in the incoming XML document using finite state automata or using DOM style operators. Figure 14.6 illustrates where all the patterns are found using finite state automata. See that our automata is very similar to that used by YFilter (Diao et al. 2003) as discussed above. The patterns that are matched can further be manipulated using other operators similar to those used in relational algebra. The example plan in Figure 14.6 shows Extract operator for extracting the tokens and forming elements; Select operator for applying selection filters, and Structural Join operator that performs a structural join of the different inputs (in this case, the structural join ensures that all the different inputs are structurally descendants of the same auction element). There is also a result construction operator that can form output results.

14.4 Conclusions

In this chapter, we examined how finite state models are used for validating XML documents against schemas, and for XML query processing using XPath and XQuery expressions. We described simplified algorithms which should give a good idea on how finite state models can be used, and we described the main challenges with regard to XML processing, and how these challenges are handled. Finite state models play an important role in XML processing, and more work needs to be done on investigating how finite state models can be utilized for still more efficient XML processing.

Acknowledgments

Most of the algorithms that the author has published on XML validation and on XML query processing have been with other collaborators, including students. Acknowledgments are due to the different collaborators who have worked with the author in the different research projects.

References

Bar-Yossef, Z., Fontoura, M., and Josifovski. V. 2005. Buffering in query evaluation over XML streams. In *Proceedings of ACM PODS*, Baltimore, Maryland, pp. 216–227.

Comon, H., Dauchet, M., Gilleron, R., Jacquemard, F., Lugiez, D., Löding, C., Tison, S., and Tommasi, M. 2008. *Tree Automata Techniques and Applications*, Available at http://tata.gforge.inria.fr/.

Diao, Y., Altinel, M., Franklin, M. J., Zhang, H., and Fischer, P. M. 2003. Path sharing and predicate evaluation for high-performance XML filtering. *ACM Transactions on Database Systems* 28(4): 467–516.

Green, T. G., Gupta, A., Miklau, G., Onizuka, M., and Suciu, D. 2004. Processing XML streams with deterministic automata and stream indexes. *ACM Transactions on Database Systems* 29(4): 752–788.

Gupta, A. K. and Suciu, D. 2003. Stream processing of XPath queries with predicates. In *Proceedings of ACM SIGMOD*, San Diego, California, pp. 419–430.

Ives, Z. G., Halevy, A. Y., and Weld, D. S. 2002. An XML query engine for network-bound data. *VLDB Journal* 5(4): 380–402.

Koch, C., Scherzinger, S., Schweikardt, N., and Stegmaier, B. 2004. Flu XQuery: An optimizing XQuery processor for streaming XML data. In *Proceedings of VLDB*, Toronto, Canada, pp. 1309–1312.

Ludascher, B., Mukopadhyay, P., and Papakonstantinou, Y. 2002. A transducer-based XML query processor. In *Proceedings of VLDB*, Hong Kong, China, pp. 227–238.

Murata, M., Lee, D., Mani, M., and Kawaguchi, K. 2005. Taxonomy of XML schema languages using formal language theory. *ACM Transactions on Internet Technology* 5(4): 660–704.

Peng, F. and Chawathe, S. S. 2005. XSQ: A streaming XPath engine. *ACM Transactions on Database Systems* 30(2): 577–623.

RELAX NG. 2001. Relax NG Home Page. Available at http://www.relaxng.org/.

Schematron. 2006. Schematron: A language for making assertions about patterns found in XML documents. Available at http://www.schematron.com/.

Segoufin, L. and Vianu, V. 2002. Validating streaming XML documents. In *Proceedings of ACM PODS*, Madison, Wisconsin, pp. 53–64.

Su, H., Rundensteiner, E. A., and Mani, M. 2005. Semantic query optimization for XQuery over XML streams. In *Proceedings of VLDB*, Trondheim, Norway, pp. 277–288.

Su, H., Rundensteiner, E. A., and Mani, M. 2008. Automaton in or out: Run-time plan optimization for XML stream processing. In *International Workshop on Scalable Stream Processing Systems (SSPS)*, Nantes, France, pp. 38–47.

W3C DOM. 2004. Document Object Level (DOM) level 3 core specification. Available at http://www.w3.org/TR/DOM-Level-3-Core/.

W3C MathML. 2010. Mathematical Markup Language (MathML). Available at http://www.w3.org/TR/MathML3/.

W3C SOAP. 2007. Simple Object Access Protocol (SOAP). Available at http://www.w3.org/TR/soap/.

W3C XML. 2008. Extensible Markup Language (XML). Available at http://www.w3.org/TR/xml/.

W3C XML Schema. 2004. W3C XML Schema Home Page. Available at http://www.w3.org/XML/Schema.html.

W3C XPath. 2010. XML Path Language (XPath) 2.0. Available at http://www.w3.org/TR/xpath20/.

W3C XQuery. 2010. XQuery 1.0: An XML Query Language (2nd edition). Available at http://www.w3.org/TR/xquery/.

W3C XSLT. 2007. XSL Transformations (XSLT) version 2.0. Available at http://www.w3.org/TR/xslt20/.

Wei, M., Rundensteiner, E. A., and Mani, M. 2008. Processing recursive XQuery over XML streams: The Raindrop approach. *Document and Knowledge Engineering* 65(2): 243–265.

15

Petri Nets

Jiacun Wang
Monmouth University

15.1 Introduction

Petri nets were introduced in 1962 by Dr. Carl Adam Petri (Petri 1962). Petri nets are a powerful modeling formalism in computer science, system engineering, and many other disciplines. Petri nets combine a well-defined mathematical theory with a graphical representation of the dynamic behavior of systems. The theoretic aspect of Petri nets allows precise modeling and analysis of system behavior, while the graphical representation of Petri nets enables visualization of the modeled system state changes. This combination is the main reason for the great success of Petri nets. Consequently, Petri nets have been used to model various kinds of dynamic event-driven systems such as computers networks (Ajmone Marsan et al. 1986), communication systems (Merlin and Farber 1976; Wang 2006), manufacturing plants (Zhou and DiCesare 1989; Venkatesh et al. 1994; Desrochers and Ai-Jaar 1995), command and control systems (Andreadakis and Levis 1988), real-time computing systems (Tsai et al. 1995; Mandrioli et al. 1996), logistic networks (Landeghem and Bobeanu 2002), hybrid systems (Pettersson and Lennartson, 1995), biological systems (Kitakaze et al. 2005), and workflows (Aalst and Hee 2000; Lin et al. 2002) to mention only a few important examples. This wide spectrum of applications is accompanied by wide spectrum different aspects which have been considered in the research on Petri nets.

15.2 Graphic Representation

15.2.1 Graphic Representation

A Petri net is a particular kind of bipartite directed graphs populated by four types of objects. These objects are *places, transitions, directed arcs*, and *tokens*. Directed arcs connect places to transitions or transitions to places. In its simplest form, a Petri net can be represented by a transition together with an input place and an output place. This elementary net may be used to represent various aspects of the modeled systems. For example, a transition and its input place and output place can be used to represent a data processing event, its input data and output data, respectively, in a data processing system. In order to study the dynamic behavior of a Petri net modeled system in terms of its states and state changes, each place may potentially hold either none or a positive number of tokens. Tokens are a primitive concept for Petri nets in addition to places and transitions. The presence or absence of a token in a place may indicate whether a condition associated with this place is true or false, for instance (Peterson 1981).

Most theoretical work on Petri nets is based on the formal definition of Petri nets. However, a graphical representation of a Petri net is much more useful for illustrating the concepts of Petri net theory. A Petri net graph is a Petri net depicted as a bipartite directed multigraph. Corresponding to the definition of Petri nets, a Petri net graph has two types of nodes: a *circle* represents a place, and a *bar* or *box* represents a transition. Directed arcs (arrows) connect places and transitions, with some arcs directed from places to transitions and other arcs directed from transitions to places.

The Petri net shown in Figure 15.1 is composed of five places, namely p_1, p_2, p_3, p_4, and p_5, and four transitions, namely t_1, t_2, t_3, and t_4. p_1 is an input place to t_1 and t_2 because there is one directed arc contacting p_1 to t_1 and another one contacting p_1 to t_2. p_3 and p_4 are both output places to t_2 because there is an arc from t_2 to p_3 and an arc from t_2 to p_4. p_1 has one token, while all other places have 0 tokens. A directed arc can have a *weight*, which is labeled to the arc in the graphic representation. The weight of the arc from t_2 to p_3 and arc from p_3 to t_3 is 2. The weight of all other arcs is 1, by default.

The execution of a Petri net is controlled by the number and distribution of tokens in the Petri net. By changing distribution of tokens in places, which may reflect the occurrence of events or execution of operations, for instance, one can study the dynamic behavior of the modeled system. A Petri net is executed by *firing* transitions. However, only an *enabled* transition can fire.

A transition t is said to be *enabled* if each input place p of t contains at least the number of tokens equal to the weight of the directed arc connecting p to t. For example, both transitions t_1 and t_2 are enabled in the Petri net of Figure 15.1. The firing of an enabled transition t removes from each input place p the number of tokens equal to the weight of the arc from p to t, and deposits in each output place p the number of tokens equal to the weight of the arc from t to p. Figure 15.2 shows the new token distribution of the Petri net of Figure 15.1 after firing t_2.

Places are passive elements, while transitions are active elements. In a practical system model, places represent buffers, channels, geographical locations, conditions or states, transitions represent events, computation, transformation or transportation. Tokens represent objects (e.g., humans, goods and machines), information, conditions or states of objects. A system state is specified by token distribution in places,

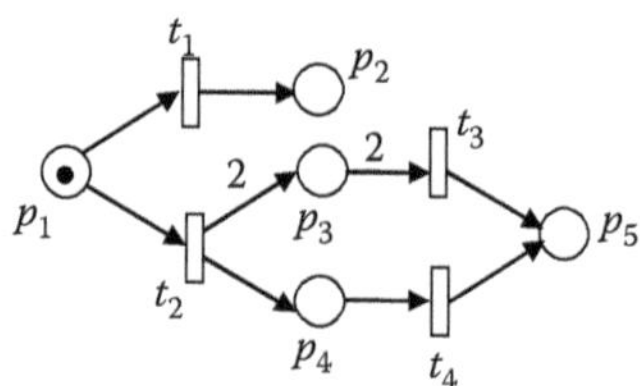

FIGURE 15.1 A Petri net with 5 places and 4 transitions.

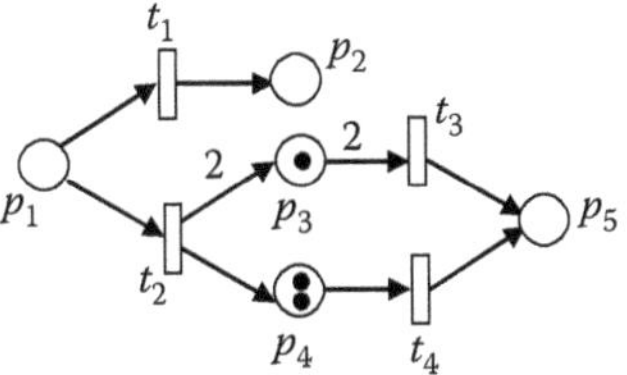

FIGURE 15.2 Firing of transition t_2.

which is called a *marking* in Petri net language. State transitions leading from one state to another are realized through transition firing.

Notice that since only enabled transitions can fire, the number of tokens in each place always remains nonnegative when a transition is fired. Firing a transition can never try to remove a token that is not there.

A transition without any input place is called a *source transition*, and one without any output place is called a *sink transition*. Note that a source transition is unconditionally enabled, and that the firing of a sink transition consumes tokens, but does not produce tokens.

A pair of a place p and a transition t is called a *self-loop*, if p is both an input place and an output place of t. A Petri net is said to be *pure* if it has no self-loops.

15.2.2 Modeling Example: A Reader–Writer System

Figure 15.3 shows the Petri net model of a classical reader–writer system. Initially, both the reader and writer are at rest (places *r_rest* and *w_rest* each contains a token). Then, the writer begins to write a mail (transition *write* fires, the token in *w_rest* is removed and a token deposited into place *mail*). Now transition *send* is enabled and can fire. Firing *send* removes the token from place *mail* and deposit a token to *w_rest* and a token to *mail_box*, meaning now the writer is at rest again and one mail is in the mail box.

Since both *mail_box* and *r_rest* have a token, transition *receive* is enabled. Firing *receive* removes the token from *mail_box* and *r_rest* and adds a token to place *received*. Now the *mail_box* is empty (no token in *mail_box*). After reading a received mail, the reader is waiting to receive new mails from the mail box (transition *read* fires and a token is added to place *r_rest*).

The above transition firing sequence *write-send-receive-read* brings the Petri net back to the initial marking, at which place *w_rest* and *r_rest* each has a token and all other places have no tokens. However, after the firing of *write* and *send*, these two transitions may fire again and again before the firing of *receive*. Each time *write* and *send* fire, one token that represents a mail is added to place *mail_box*. Therefore, the model of Figure 15.3 assumes the capacity of the mail box is unlimited. How to model a mail box that can hold up to a limited number of mails? Well, to this end, a common practice is to introduce a new place called *capacity*, which contains the number of tokens equal to the capacity of the mail box. Figure 15.4 shows a reader–writer system with a mailbox of size of two. In this model, each time *write* and *send* fire,

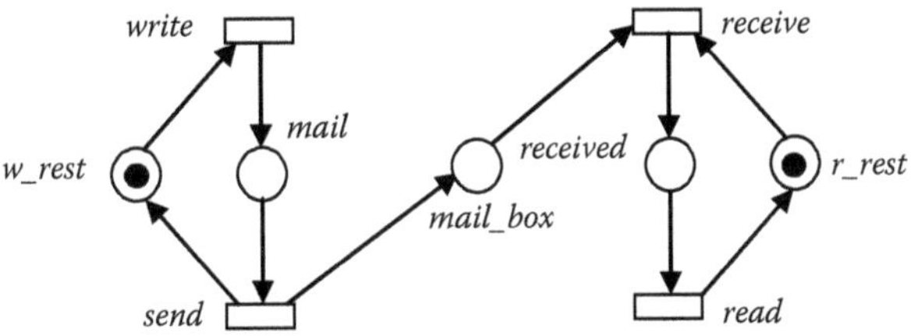

FIGURE 15.3 A reader–writer system.

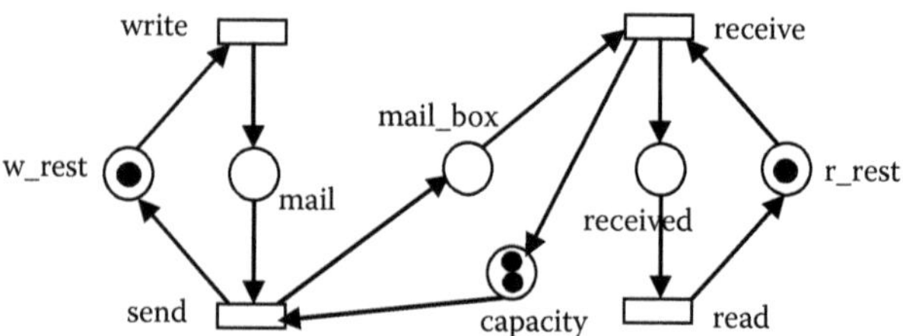

FIGURE 15.4 A reader–writer system in which the mailbox size is two.

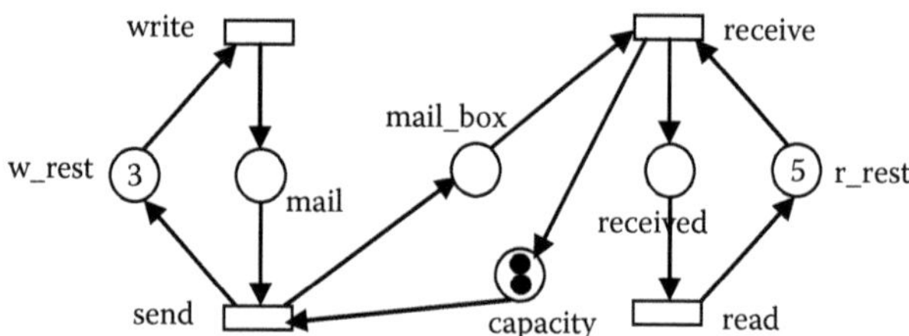

FIGURE 15.5 A reader–writer system with 5 readers and 3 writers.

a token is removed from place *capacity*. If *receive* does not fire, they can only fire a maximum of two times. But when *receive* fires, a token is returned to *capacity*, which allows *send* to fire again.

It is worth to mention that the number of tokens in *w_rest* and *r_rest* also model the number of writers and readers in the system, respectively. For example, if we want to model reader–writer system with three writers and five readers, then we only need to put three token in place *w_rest* and five tokens in *r_rest*. The new model is shown in Figure 15.5.

15.2.2.1 Modeling Example: A Manufacturing System

Consider a manufacturing system composed of 3 machines, 2 inspectors, 1 assembler, and 2 disassemblers. The system receives two types of parts (A and B) as inputs, and after processing the input parts, one A-part and two B-parts are assembled into a final product. The process is described as follows: raw parts arrive in pairs, A-parts are processed by machines 1 and 2 sequentially, while B-parts are processed by machine 3. Processed A- and B-parts are finally assembled by an assembler. Two inspectors are responsible for quality control. Inspector 1 examines A-parts after they are processed by machines 1 and 2. If they do not satisfy the quality requirements, they are sent back to be reprocessed by machines 2. Otherwise, they are sent for assembly. Inspector 2 examines the assembled products. If an assembled product satisfies the quality requirements, it is unloaded from the system as a final product; otherwise, it is disassembled either by disassembler 1 or 2 depending upon their status. Disassembler 1 generates A-parts to be sent back to machines 1 and 2 and two B-parts to assembler, respectively. Disassembler 2 generates A-parts to be sent back to machines 1 and 2 and two B-parts to machine 3. The flow is illustrated in Figure 15.6.

Figure 15.7 shows the Petri net model of this system. The legends of places and transitions in this model are as follows:

p_1: Parts ready
p_2: A-part
p_3: A-part after processed by machine 1
p_4: B-part after processed by machine 2
p_5: Inspection result of inspector 1
p_6: A-part for assembly

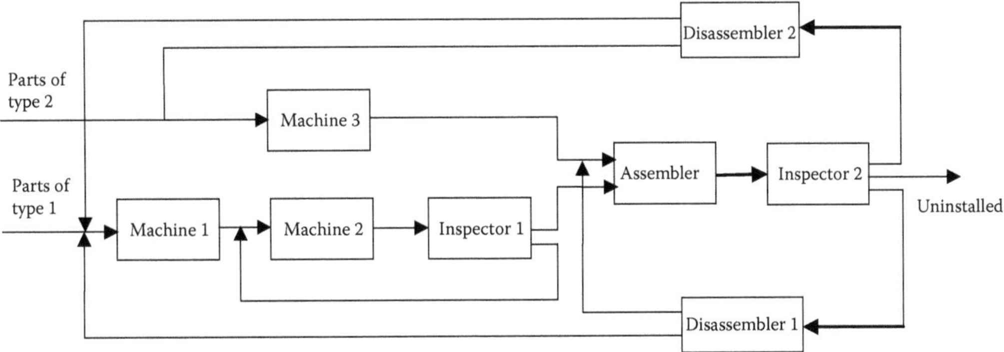

FIGURE 15.6 A manufacturing system.

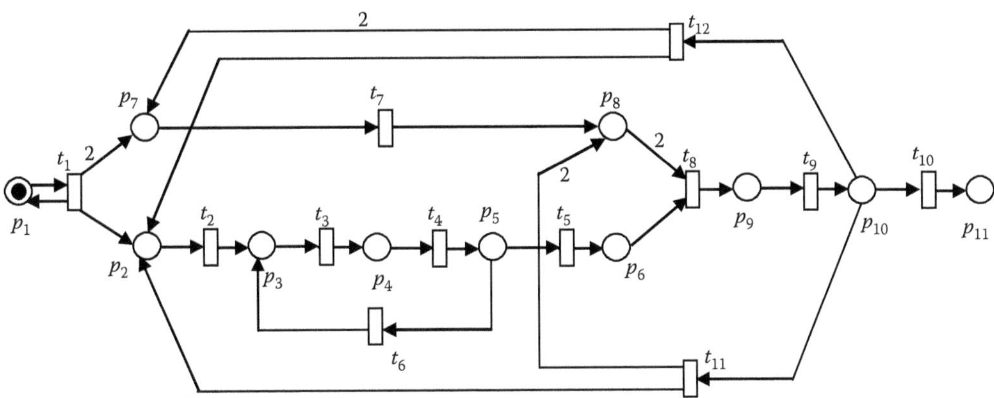

FIGURE 15.7 Petri net model of the manufacturing system in Figure 15.6.

p_7: B-part
p_8: B-part for assembly
p_9: Assembled product
p_{10}: Inspection result of inspector 2
p_{11}: Final product
t_1: One A-part and two B-parts arrive
t_2: Machine 1 works on an A-part
t_3: Machine 2 works on an A-part
t_4: Inspector 1 examines an A-part
t_5: A-part satisfies quality requirements
t_6: A-part does not satisfy quality requirements
t_7: Machine 3 works on B-part
t_8: Assembler works
t_9: Inspector 2 examines assembled product
t_{15}: Final product is unloaded
t_{11}: Disassembler 1 works
t_{12}: Disassembler 2 works

Notice that the two inspectors are modeled by t_4 and t_9. Their output places, namely p_5 and p_{10}, are decision points, each of which has multiple conflicting outgoing transitions. On the other hand, the transition t_9 models a merging (synchronization) action.

15.3 Formal Definition

A Petri net is formally defined as a 5-tuple $N = (P, T, I, O, M_0)$, where

1. $P = \{p_1, p_2, \ldots, p_m\}$ is a finite set of places
2. $T = \{t_1, t_2, \ldots, t_n\}$ is a finite set of transitions, $P \cup T \neq \emptyset$, and $P \cap T = \emptyset$
3. I: $T \times P \rightarrow \mathcal{N}$ is an *input function* that defines directed arcs from places to transitions, where $\mathcal{N}$ is a set of nonnegative integers
4. O: $T \times P \rightarrow \mathcal{N}$ is an *output function* that defines directed arcs from transitions to places
5. M_0: $P \rightarrow \mathcal{N}$ is the initial marking

A *marking* in a Petri net is an assignment of tokens to the places of a Petri net. An arc directed from a place p_j to a transition t_i defines p_j to be an *input place* of t_i, denoted by $I(t_i, p_j) = 1$. An arc directed from a transition t_i to a place p_j defines p_j to be an *output place* of t_i, denoted by $O(t_i, p_j) = 1$. If $I(t_i, p_j) = k$ (or $O(t_i, p_j) = k$), then there exist k directed (parallel) arcs connecting place p_j to transition t_i (or connecting transition t_i to place p_j).

The formal description of the Petri net in Figure 15.1 is as follows:

$$\begin{gathered}
P = \{p_1, p_2, p_3, p_4, p_5\};\\
T = \{t_1, t_2, t_3, t_4\};\\
I(t_1, p_1) = 1,\ \ I(t_1, p_i) = 0 \quad \text{for } i = 2, 3, 4, 5;\\
I(t_2, p_1) = 1,\ \ I(t_2, p_i) = 0 \quad \text{for } i = 3, 3, 4, 5;\\
I(t_3, p_3) = 2,\ \ I(t_3, p_i) = 0 \quad \text{for } i = 1, 2, 4, 5;\\
O(t_1, p_2) = 1,\ \ O(t_1, p_i) = 0 \quad \text{for } i = 1, 2, 4, 5;\\
O(t_2, p_3) = 2,\ \ O(t_2, p_4) = 1, O(t_2, p_i) = 0 \quad \text{for } i = 1, 3, 5;\\
O(t_3, p_5) = 1,\ \ O(t_3, p_i) = 0 \quad \text{for } i = 1, 2, 3, 4;\\
O(t_4, p_5) = 1,\ \ O(t_4, p_i) = 0 \quad \text{for } i = 1, 2, 3, 4;\\
M_0 = (1\,0\,0\,0\,0)^{\mathrm{T}}.
\end{gathered}$$

Mathematically, a transition t is said to be enabled iff

$$M(p) \geq I(t, p) \quad \text{for all } p \text{ in } P.$$

For example, the above condition holds for the Petri net of Figure 15.1 at the initial marking M_0, for transitions t_1 and t_2. Therefore, t_1 and t_2 are enabled and can fire. But for t_3, because $M_0(p_3) = 0$ while $I(t_3, p_3) = 2$, the condition does not hold. Therefore, t_3 is not enabled at M_0.

The new marking M' achieved by firing t at M can be calculated by

$$M'(p) = M(p) - I(t, p) + O(t, p) \quad \text{for all } p \text{ in } P.$$

Consider the Petri net shown in Figure 15.1 again. Under the initial marking, $M_0 = (1\,0\,0\,0\,0)^{\mathrm{T}}$, t_1 and t_2 are enabled. Firing of t_1 results in a new marking, say M_1. It follows from the firing rule that

$$M_1 = (0\,1\,0\,0\,0)^{\mathrm{T}}.$$

The Petri net is dead now because no transitions are enabled at t_2. If t_2 fires at M_0 instead of t_1, the new marking, say M_2, is

$$M_2 = (0\,0\,2\,1\,0)^{\mathrm{T}}.$$

At M_2, both transitions t_3 and t_4 are enabled. If t_3 fires, the new marking, say M_3, is

$$M_3 = (0\,0\,0\,1\,1)^{\mathrm{T}}.$$

If t_4 fires, the new marking, say M_4, is

$$M_4 = (0\,0\,2\,0\,1)^{\mathrm{T}}.$$

At M_3, t_4 can fire, while at M_4, t_3 can fire. In either case, a new marking, say M_5 is reached:

$$M_5 = (0\,0\,0\,0\,2)^{\mathrm{T}}.$$

M_5 is a dead marking.

15.4 Modeling Power

The typical characteristics exhibited by the activities in a dynamic event-driven system, such as concurrency, decision making, synchronization and priorities, can be modeled effectively by Petri nets.

1. *Sequential Execution.* In Figure 15.8a, transition t_2 can fire only after the firing of t_1 and transition t_3 can fire only after the firing of t_2. This imposes the precedence constraint "t_3 after t_2 and t_2 after t_1." Such precedence constraints are typical of the execution of the parts in a dynamic system. Also, this Petri net construct models the causal relationship among activities. A practical example is transitions t_2, t_3, and t_4 in Figure 15.4, which model events "Machine 1 works on an A-part," "Machine 2 works on an A-part," and "Inspector 1 examines an A-part," respectively. These three events occur sequentially in the system.
2. *Conflict.* Transitions t_1, t_2, and t_3 are in conflict with each other in Figure 15.8b. All are enabled but the firing of any transition leads to the disabling of the other two transitions. Such a situation will arise, for example, when a machine has to choose among part types or a part has to choose among several machines. The resulting conflict may be resolved in a purely nondeterministic way or in a probabilistic way, by assigning appropriate probabilities to the conflicting transitions.
3. *Concurrency.* In Figure 15.8c, the transitions t_1, t_2, and t_3 are concurrent. Concurrency is an important attribute of system interactions. In the manufacturing system example of Figure 15.7, t_2 and t_7 are concurrent, because after t_1 fires, both p_2 and p_6 get a token. t_2 and t_7 model events "Machine 1 works on an A-part" and "Machine 3 works on a B-part," respectively, which can happen at the same time.
4. *Synchronization.* It is quite normal in a dynamic system that an event requires multiple resources. The resulting synchronization of resources can be captured by transitions of the type shown in Figure 15.8d. Here, t_1 is enabled only when each of p_1, p_2, and p_3 receives a token. The arrival of a token into each of the three places could be the result of a possibly complex sequence of operations elsewhere in the rest of the Petri net model. Essentially, transition t_1 models the joining operation.
5. *Mutually exclusive.* Two processes are mutually exclusive if they cannot be performed at the same time due to constraints on the usage of shared resources. Figure 15.8e shows this structure. For example, a robot may be shared by two machines for loading and unloading. Two such structures are parallel mutual exclusion and sequential mutual exclusion.
6. *Priorities.* The classical Petri nets discussed so far have no mechanism to represent priorities. Such a modeling power can be achieved by introducing an *inhibitor arc*. The inhibitor arc connects an

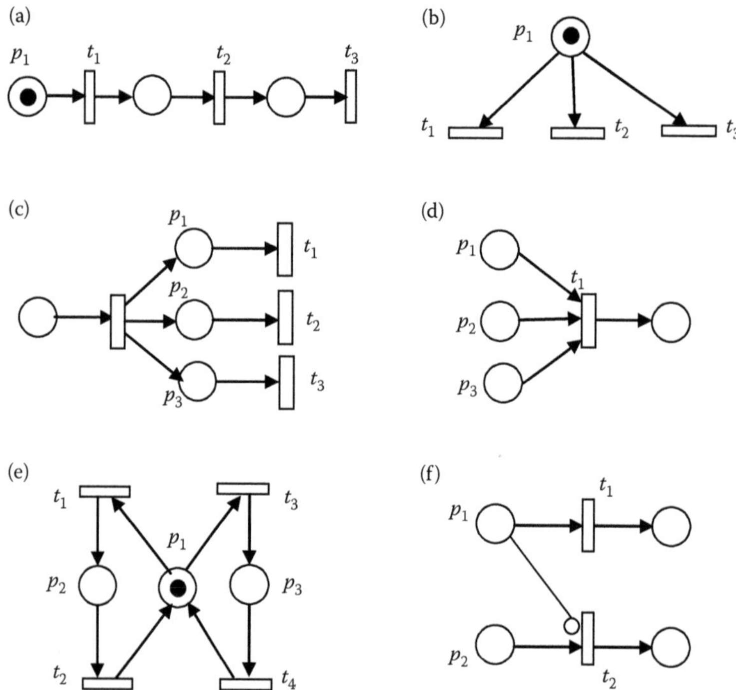

FIGURE 15.8 Typical Petri net structures. (a) sequential, (b) conflict, (c) concurrent, (d) synchronization, (e) mutual exclusive, and (f) priority.

input place to a transition, and is pictorially represented by an arc terminated with a small circle. The presence of an inhibitor arc connecting an input place to a transition changes the transition enabling conditions. In the presence of the inhibitor arc, a transition is regarded as enabled if each input place, connected to the transition by a normal arc (an arc terminated with an arrow), contains at least the number of tokens equal to the weight of the arc, and no tokens are present on each input place connected to the transition by the inhibitor arc. The transition firing rule is the same for normally connected places. The firing, however, does not change the marking in the inhibitor arc connected places. A Petri net with an inhibitor arc is shown in Figure 15.8f. t_1 is enabled if p_1 contains a token, while t_2 is enabled if p_2 contains a token and p_1 has no token. This gives priority to t_1 over t_2: In a marking in which both p_1 and p_2 have a token, t_2 would not be able to fire until t_1 is fired.

15.5 Petri Net Analysis

We have introduced the modeling power of Petri nets in the previous sections. However, modeling by itself is of little use. It is necessary to *analyze* the modeled system. This analysis will hopefully lead to important insights into the behavior of the modeled system.

There are four common approaches to Petri net analysis: (1) reachability analysis, (2) the matrix-equation approach, (3) invariant analysis, and (4) simulation. The first approach involves the enumeration of all reachable markings, but it suffers from the state-space explosion issue. The matrix equations technique is powerful but in many cases it is applicable only to special subclasses of Petri nets or special situations. The invariant analysis determines sets of places or transitions with special features, as token conservation or cyclical behavior. For complex Petri net models, discrete-event simulation is an option to check the system properties.

15.5.1 Reachability Analysis

Reachability analysis is conducted through the construction of reachability tree if the net is bounded. Given a Petri net N, from its initial marking M_0, we can obtain as many "new" markings as the number of the enabled transitions. From each new marking, we can again reach more markings. Repeating the procedure over and over results in a tree representation of the markings. Nodes represent markings generated from M_0 and its successors, and each arc represents a transition firing, which transforms one marking to another.

The above tree representation, however, will grow infinitely large if the net is unbounded. To keep the tree finite, we introduce a special symbol ω, which can be thought of as "infinity." It has the properties that for each integer n, $\omega > n$, $\omega + n = \omega$, and $\omega \geq \omega$. Generally speaking, we do not know if a Petri net is bounded or not before we perform the reachability analysis. However, we can construct a *coverability tree* if the net is unbounded or a reachability tree if the net is bounded according to the following general algorithm:

1. Label the initial marking M_0 as the root and tag it "new."
2. For every new marking M:
   - 2.1 If M is identical to a marking already appeared in the tree, then tag M "old" and go to another new marking.
   - 2.2 If no transitions are enabled at M, tag M "dead-end" and go to another new marking.
   - 2.3 While there exist enabled transitions at M, do the following for each enabled transition t at M:
     - 2.3.1 Obtain the marking M'' that results from firing t at M.
     - 2.3.2 On the path from the root to M if there exists a marking M' such that $M'(p) \geq M''(p)$ for each place p and $M' \neq M''$, that is, M'' is coverable, then replace $M''(p)$ by ω for each p such that $M'(p) > M''(p)$.
     - 2.3.3 Introduce M' as a node, draw an arc with label t from M to M', and tag M' "new."

If ω appears in a marking, then the net is unbounded and the tree is a coverability tree, otherwise, the net is bounded and the tree is a reachability tree. Merging the same nodes in a coverability tree (reachability tree) results in a coverability graph (*reachability graph*).

Example 15.1: (Reachability Analysis)

Consider the Petri net shown in Figure 15.1. The analysis in Section 15.3 indicates six reachable markings:

$$M_0 = (1\,0\,0\,0\,0)^{\mathrm{T}}, \quad M_1 = (0\,1\,0\,0\,0)^{\mathrm{T}}, \quad M_2 = (0\,0\,2\,1\,0)^{\mathrm{T}},$$
$$M_3 = (0\,0\,0\,1\,1)^{\mathrm{T}}, \quad M_4 = (0\,0\,2\,0\,1)^{\mathrm{T}}, \quad M_5 = (0\,0\,0\,0\,2)^{\mathrm{T}}.$$

The reachability tree of this Petri net is shown in Figure 15.9a, and the reachability graph is shown in Figure 15.9b.

15.5.2 Incidence Matrix and State Equation

For a Petri net with n transitions and m places, the incidence matrix $A = [a_{ij}]$ is an $n \times m$ matrix of integers and its entry is given by

$$a_{ij} = O(t_i, p_j) - I(t_i, p_j).$$

In writing matrix equations, we write a marking M_k as an $m \times 1$ column vector. The jth entry of M_k denotes the number of tokens in place j immediately after the kth firing in some firing sequence. The

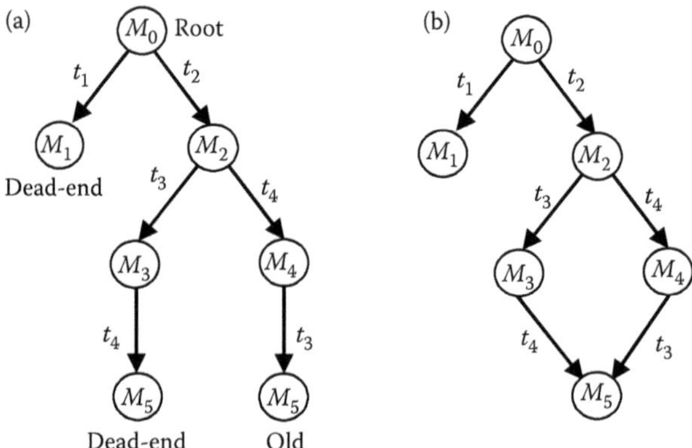

FIGURE 15.9 (a) Reachability tree. (b) Reachability graph.

kth firing or *control vector* u_k is an $n \times 1$ column vector of $n-1$ 0's and one nonzero entry, a 1 in the ith position indicating that transition i fires at the kth firing. Since the ith row of the incidence matrix A denotes the change of the marking as the result of firing transition i, we can write the following state equation for a Petri net:

$$M_k = M_{k-1} + A^{\mathrm{T}} u_k, \quad k = 1, 2, \ldots$$

Suppose that a destination marking M_{d} is reachable from M_0 through a firing sequence $\{u_1, u_2 \ldots, u_d\}$. Writing the state equation for $k = 1, 2, \ldots, d$ and summing them, we obtain

$$M_{\mathrm{d}} = M_0 + A^{\mathrm{T}} \sum_{k=1}^{\mathrm{d}} u_k$$

15.5.3 Software Tools

For complex Petri net models, simulation is another way to check the system properties. The idea is simple, that is, using the execution algorithm to run the net. Simulation is an expensive and time-consuming technique. It can show the presence of undesirable properties but cannot prove the correctness of the model in general case. Despite this, Petri net simulation is indeed a convenient and straightforward yet effective approach for engineers to validate the desired properties of a discrete event system. For this reason, many software tools have been developed by Petri net researchers. A list of Petri net simulation tools along with feature descriptions can be found in the following *Petri Nets World* website: http://www.informatik.uni-hamburg.de/TGI/PetriNets/

15.6 Petri Net Properties

As a mathematical tool, Petri nets possess a number of properties. These properties, when interpreted in the context of the modeled system, allow system designer to identify the presence or absence of the application domain specific functional properties of the system under design. Two types of properties can be distinguished, behavioral and structural ones. The behavioral properties are those which depend on the initial state or marking of a Petri net. The structural properties, on the other hand, do not depend on the initial marking of a Petri net. They depend on the topology, or net structure, of a Petri net. In this section, we briefly introduce three behavioral properties: reachability, safeness, and liveness, and two structural properties: invariants, and siphons and traps.

15.6.1 Reachability

An important issue in designing event-driven systems is whether a system can reach a specific state, or exhibit a particular functional behavior. In general, the question is whether the system modeled with a Petri net exhibits all desirable properties as specified in the requirement specification, and no undesirable ones.

In order to find out whether the modeled system can reach a specific state as a result of a required functional behavior, it is necessary to find such a transition firing sequence that would transform its Petri net model from the initial marking M_0 to the desired marking M_j, where M_j represents the specific state, and the firing sequence represents the required functional behavior. In general, a marking M_j is said to be *reachable* from a marking M_i if there exists a sequence of transitions firings which transforms M_i to M_j. A marking M_j is said to be *immediately reachable* from M_i if firing an enabled transition in M_i results in M_j. The set of all markings reachable from marking M is denoted by $R(M)$.

For example, as shown in Figure 15.9, the Petri net of Figure 15.1 has five reachable markings, in which M_1 and M_2 are immediately reachable from the initial marking M_0.

15.6.2 Safeness

In a Petri net, places are often used to represent information storage areas in communication and computer systems, product and tool storage areas in manufacturing systems, and so on. It is important to be able to determine whether proposed control strategies prevent from the overflows of these storage areas. The Petri net property which helps to identify the existence of overflows in the modeled system is the concept of *boundedness*.

A place p is said to be *k-bounded* if the number of tokens in p is always less than or equal to k (k is a nonnegative integer number) for every marking M reachable from the initial marking M_0, that is, $M \in R(M_0)$. It is *safe* if it is 1-bounded.

A Petri net $N = (P, T, I, O, M_0)$ is k-bounded (safe) if each place in P is k-bounded (safe). It is *unbounded* if k is infinitely large. For example, the Petri net of Figure 15.1 is 2-bounded, but the net of Figure 15.3 is unbounded, because if transitions *write* and *send* keep firing, the place *mail_box* will get infinite number of tokens.

15.6.3 Liveness

The concept of liveness is closely related to the *deadlock* situation, which has been situated extensively in the context of computer operating systems.

A Petri net modeling a deadlock-free system must be *live*. This implies that for any reachable marking M, it is ultimately possible to fire any transition in the net by progressing through some firing sequence. This requirement, however, might be too strict to represent some real systems or scenarios that exhibit deadlock-free behavior. For instance, the initialization of a system can be modeled by a transition (or a set of transitions) which fire a finite number of times. After initialization, the system may exhibit a deadlock-free behavior, although the Petri net representing this system is no longer live as specified above. For this reason, different levels of liveness are defined. Denote by $L(M_0)$ the set of all possible firing sequences starting from M_0. A transition t in a Petri net is said to be:

1. $L0$-live (or dead) if there is no firing sequence in $L(M_0)$ in which t can fire.
2. $L1$-live (potentially firable) if t can be fired at least once in some firing sequence in $L(M_0)$.
3. $L2$-live if t can be fired at least k times in some firing sequence in $L(M_0)$ given any positive integer k.
4. $L3$-live if t can be fired infinitely often in some firing sequence in $L(M_0)$.
5. $L4$-live (or live) if t is $L1$-live (potentially firable) in every marking in $R(M_0)$ (Murata 1989).

For example, all the three transitions in the net of Figure 15.1 are $L1$-live because all the four transitions can only fire once. However, all transitions in the net of Figure 15.4 are $L4$-live, because they are all $L1$-live in every reachable marking.

15.6.4 Invariants

In a Petri net, arcs describe the relationships among places and transitions. They are represented by two matrices I and O. By examining the linear equations based on the execution rule and the matrices, one can find subsets of places over which the sum of the tokens remains unchanged. One may also find a transition firing sequence brings the marking back to the same one. The concepts of *S-invariant* and *T-invariant* are introduced to reflect these properties.

Mathematically, an S-invariant is an integer solution y of the homogeneous equation

$$Ay = 0,$$

and a T-invariant is an integer solution x of the homogeneous equation

$$A^{\mathrm{T}}x = 0.$$

The nonzero entries in an S-invariant represent weights associated with the corresponding places so that the weighted sum of tokens on these places is constant for all markings reachable from an initial marking. These places are said to be covered by an S-invariant. The nonzero entries in a T-invariant represent the firing counts of the corresponding transitions which belongs to a firing sequence transforming a marking M_0 back to M_0. Although a T-invariant states the transitions comprising the firing sequence transforming a marking M_0 back to M_0, and the number of times these transitions appear in this sequence, it does not specify the order of the transition firings.

Invariant findings may help in analysis for some Petri net properties. For example, if each place in a net is covered by an S-invariant, then it is bounded. However, this approach is of limited use since invariant analysis does not include all the information of a general Petri net.

The set of places (transitions) corresponding to nonzero entries in an S-invariant $y \geq 0$ (T-invariant $x \geq 0$) is called the *support of an invariant* and is denoted by $\|x\|$ ($\|y\|$). A support is said to be *minimal* if there is no other invariant y_1 such that $y_1(p) \leq y(p)$ for all p. Given a minimal support of an invariant, there is a unique minimal invariant corresponding to the minimal support. We call such an invariant a *minimal-support invariant.* The set of all possible minimal-support invariants can serve as a generator of invariants. That is, any invariant can be written as a linear combination of minimal-support invariants.

Example 15.2:

Figure 15.10 shows a simple manufacturing system with a single machine and a buffer. The capacity of the buffer is 1. A raw part can enter the buffer only when it is empty, otherwise it is rejected. As soon as the part residing in the buffer it gets processed, the buffer is released and can accept another coming part. Fault may occur in the machine when it is processing a part. After being repaired, the machine continues to process the uncompleted part. The places and transitions in this Petri net are as follows:

p_1: The buffer (size of two) is empty
p_2: Parts are in buffer
p_3: The machine is free
p_4: The machine is processing a part
p_5: The machine is failed
t_1: A part arrives
t_2: The machine starts processing a part
t_3: The machine ends processing a part
t_4: The machine fails
t_5: Repair the machine

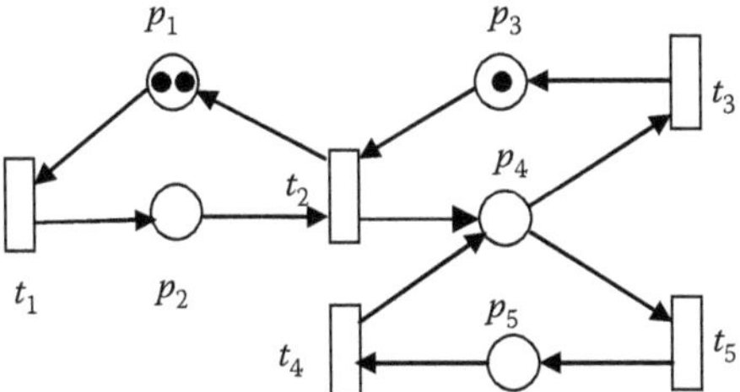

FIGURE 15.10 A simple manufacturing system with a single machine and a buffer.

We are going to find the S-invariants of the net and use them to see how this simple manufacturing system works. First we obtain the incidence matrix directly from the model:

$$A = \begin{bmatrix} -1 & 1 & 0 & 0 & 0 \\ 1 & -1 & -1 & 1 & 0 \\ 0 & 0 & 1 & -1 & 0 \\ 0 & 0 & 0 & -1 & 1 \\ 0 & 0 & 0 & 1 & -1 \end{bmatrix}.$$

Then, by solving $Ay = 0$ we get two minimal-support S-invariants, $y_1 = (1\,1\,0\,0\,0)^{\mathrm{T}}$ and $y_2 = (0\,0\,1\,1\,1)^{\mathrm{T}}$ where $\|y_1\| = \{p_1, p_2\}$ and $\|y_2\| = \{p_3, p_4, p_4\}$ are corresponding minimal supports. Since the initial marking is $M_0 = (2\,0\,1\,0\,0)^{\mathrm{T}}$, then $M_0^{\mathrm{T}} y_1 = 1$ and this leads to the fact that

$$M(p_1) + M(p_2) = 2.$$

It shows that the buffer is either free or one slot occupied or both slots occupied. Similarly, it results from $M_0^{\mathrm{T}} y_2 = 1$ that

$$M(p_3) + M(p_4) + M(p_5) = 1.$$

It shows how the machine spends its time. It is either up and waiting, or up and working, or down.

15.6.5 Siphons and Traps

Let $^*S$ and S^* denote the sets of input and output transitions of place set S, respectively. Given a PN $P = (P, T, I, O, M_0)$, a subset of places S of P is called a *siphon* if $^*S \subseteq S^*$. Intuitively, if a transition is going to deposit a token to a place in a siphon S, the transition must also remove a token from S. On the other hand, a subset of places S of P is called a trap if $S^* \subseteq {}^*S$. Intuitively, a trap S represents a set of places in which every transition consuming a token from S must also deposit a token back into S.

For example, the set $S = \{w\_rest, mail, mail\_box\}$ in the Petri net of Figure 15.3 is a siphon, because $^*S = \{write, send\}$, $S^* = \{write, send, receive\}$ and thus $^*S \subset S^*$. The set $S = \{mail\_box, r\_rest, received\}$ in the same Petri net is a siphon, because $S^* = \{receive, read\}$, $S^* = \{send, receive, read\}$ and thus $S^* \subset {}^*S$.

15.7 Colored Petri Nets

In a standard Petri net, tokens are indistinguishable. Because of this, Petri nets have the distinct disadvantage of producing very large and unstructured specifications for the systems being modeled. To tackle this issue, high-level Petri nets were developed to allow compact system representation. Colored Petri nets (Jensen 1981) and Predicate/Transition (Pr/T) nets (Genrich and Lautenbach 1981) are among the most popular high-level Petri nets. We will introduce colored Petri nets in this section.

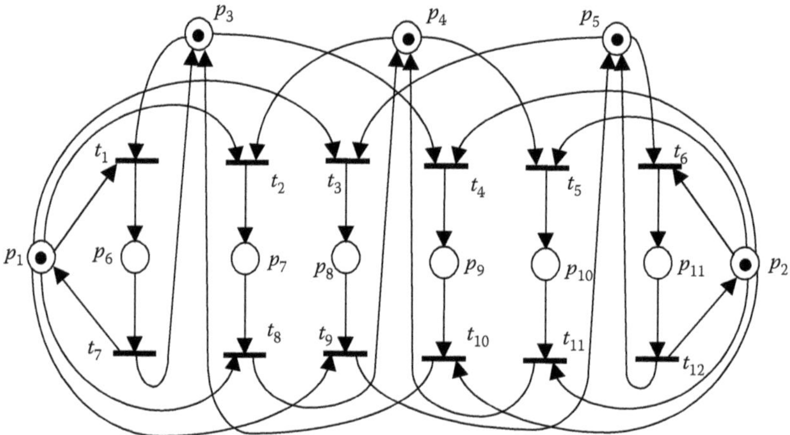

FIGURE 15.11 Petri net model of a simple manufacturing system.

Introduced by Kurt Jensen in, a *Colored Petri Net* (CPN) has its each token attached with a color, indicating the identity of the token. Moreover, each place and each transition has attached a set of colors. A transition can fire with respect to each of its colors. By firing a transition, tokens are removed from the input places and added to the output places in the same way as that in original Petri nets, except that a functional dependency is specified between the color of the transition firing and the colors of the involved tokens. The color attached to a token may be changed by a transition firing and it often represents a complex data-value. CPNs lead to compact net models by using of the concept of colors. This is illustrated by Example 15.2.

Example 15.3: (A Manufacturing System)

Consider a simple manufacturing system comprising two machines M1 and M2, which process three different types of raw parts. Each type of parts goes through one stage of operation, which can be performed on either M1 or M2. After the completion of processing of a part, the part is unloaded from the system and a fresh part of the same type is loaded into the system. Figure 15.11 shows the (uncolored) Petri net model of the system. The places and transitions in the model are as follows:

p_1 (p_2): Machine M1 (M2) available
p_3 (p_4, p_5): A raw part of type 1 (type 2, type 3) available
p_6 (p_7, p_8): M1 processing a raw part of type 1 (type 2, type 3)
p_9 (p_{10}, p_{11}): M2 processing a raw part of type 1 (type 2, type 3)
t_1 (t_2, t_3): M1 begins processing a raw part of type 1 (type 2, type 3)
t_4 (t_5, t_6): M2 begins processing a raw part of type 1 (type 2, type 3)
t_7 (t_8, t_9): M1 ends processing a raw part of type 1 (type 2, type 3)
t_{15} (t_{11}, t_{12}): M2 ends processing a raw part of type 1 (type 2, type 3)

Now let us take a look at the CPN model of this manufacturing system, which is shown in Figure 15.12. As we can see, there are only 3 places and 2 transitions in the CPN model, compared at 11 places and 12 transitions in Figure 15.11. In this CPN model, p_1 means machines are available (corresponding to places p_1 and p_2 in Figure 15.11), p_2 means parts available (corresponding to places p_3–p_5 in Figure 15.11), p_3 means processing in progress (corresponding to places p_6–p_{11} in Figure 15.11), t_1

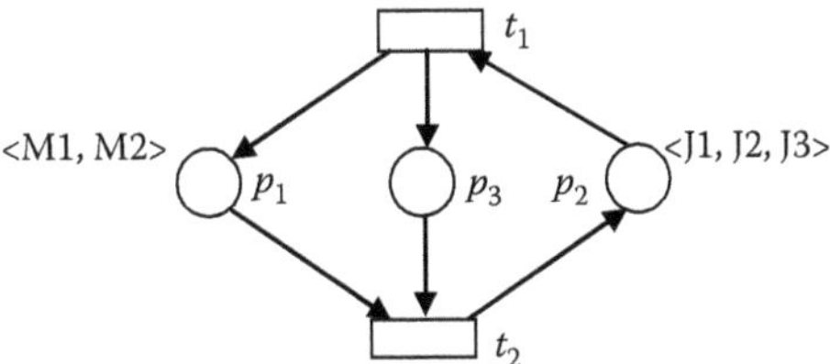

FIGURE 15.12 Colored Petri net model of the manufacturing system.

means processing starts (corresponding to transitions t_1-t_6 in Figure 15.11), and t_2 means processing ends (corresponding to transitions t_7-t_{12} in Figure 15.11). There are three color sets: *SM*, *SP*, and $SM \times SP$, where $SM = \{\mathrm{M1, M2}\}$, $SP = \{\mathrm{J1, J2, J3}\}$. The color of each node is as follows:

$$\begin{aligned} \mathrm{C}(p_1) &= \{\mathrm{M1, M2}\}, \\ \mathrm{C}(p_2) &= \{\mathrm{J1, J2, J3}\}, \\ \mathrm{C}(p_3) &= \mathrm{SM} \times \mathrm{SP}, \\ \mathrm{C}(t_1) &= C(t_2) = \mathrm{SM} \times \mathrm{SP}. \end{aligned}$$

CPN models can be analyzed through reachability analysis. As for ordinary Petri nets, the basic idea behind reachability analysis is to construct a reachability graph. Obviously, such a graph may become very large, even for small CPNs. However, it can be constructed and analyzed totally automatically, and there exist techniques which makes it possible to work with condensed occurrence graphs without losing analytic power. These techniques build upon equivalence classes. Another option to the CPN model analysis is simulation. Readers are referred to Jensen (1997) for a detailed description of the concepts, analysis methods and practical use of colored Petri nets.

15.8 Timed Petri Nets

The need for including timing variables in the models of various types of dynamic systems is apparent since these systems are real time in nature. In the real world, almost every event is time related. When a Petri net contains a time variable, it becomes a *Timed Petri Net* (Wang 1998). The definition of a timed Petri net consists of three specifications:

- The topological structure
- The labeling of the structure
- Firing rules

The topological structure of a timed Petri net generally takes the form that is used in a conventional Petri net. The labeling of a timed Petri net consists of assigning numerical values to one or more of the following things:

- Transitions
- Places
- Arcs connecting the places and transitions

The firing rules are defined differently depending on the way the Petri net is labeled with time variables. The firing rules defined for a timed Petri net control the process of moving the tokens around.

The above variations lead to several different types of timed Petri nets. Among them, deterministic time Petri nets (Ramchandani 1974) and stochastic timed Petri nets (Molloy 1982; Bause and Kritzinger 2002), in which time variables are associated with transitions, are the two most widely used extended Petri nets.

For more theory and applications of stochastic Petri nets, refer to (Hass 2002) and (Lindemann 1998).

15.8.1 Deterministic Timed Petri Nets

The introduction of deterministic time labels into Petri nets was first attempted by Ramchandani (1974). In his approach, the time labels were placed at each transition, denoting the fact that transitions are often used to represent actions, and actions take time to complete. The obtained extended Petri nets are called *Deterministic Timed Petri Nets* (DTPNs for short). Ramamoorthy and Ho (1980) used such an extended model to analyze system performance. The method is applicable to a restricted class of systems called *decision-free nets.* This class of nets involves neither decisions nor nondeterminism. In structural terms, each place is connected to the input of no more than one transition, and to the output of no more than one transition.

A deterministic timed transitions Petri net (DTPN) is a 6-tuples (P,T,I,O,M_0,τ), where (P,T,I,O,M_0) is a Petri net, $\tau : T \rightarrow R^+$ is a function that associates transitions with deterministic time delays.

A transition t_i in a DTPN can fire at time τ if and only if

1. For any input place p of this transition, there have been the number of tokens equal to the weight of the directed arc connecting p to t_i in the input place continuously for the time interval $[\tau - \tau_i, \tau]$, where τ_i is the associated firing time of transition t_i.
2. After the transition fires, each of its output places, p, will receive the number of tokens equal to the weight of the directed arc connecting t_i to p at time τ.

An important application of DTPN is to calculate the cycle time of a Petri net model. For a decision-free Petri net where every place has exactly one input arc and one output arc, the minimum cycle time (maximum performance) C is given by

$$C = \max\left\{\frac{T_k}{N_k} : k = 1,2,\ldots,q\right\}$$

where

$T_k = \sum_{t_i \in L_k} \tau_i =$ sum of the execution times of the transitions in circuit k;

$N_k = \sum_{p_i \in L_k} M(p_i) =$ total number of tokens in the places in circuit k;

$q \; =$ number of circuits in the net.

Example 15.4: (A Communication Protocol)

Consider the communication protocol between two processes, one indicated as the sender, and the other as the receiver. The sender sends messages to a buffer, while the receiver picks up messages from the buffer. When it gets a message, the receiver sends an ACK back to the sender. After receiving the ACK from the receiver, the sender begins processing and sending a new message. Suppose that the sender takes 1 time unit to send a message to the buffer, 1 time unit to receive the ACK, and 3 time units to process a new message. The receiver takes 1 time unit to get the messages from the buffer, 1 time unit to send back an ACK to the buffer, and 4 time units to process a received message. The DTPN model of this protocol is shown in Figure 15.13. The legends of places and transitions and timing properties are as follows:

p_1: The sender ready
p_2: Message in the buffer
p_3: The sender waiting for ACK
p_4: Message received
p_5: The receiver ready
p_6: ACK sent

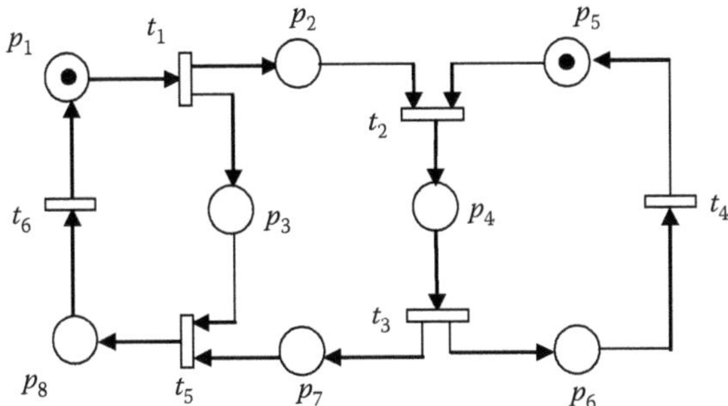

FIGURE 15.13 Petri net model of a simple communication protocol.

p_7: ACK in the buffer
p_8: ACK received
t_1: The sender sends a message to the buffer. Time delay: 1 time unit
t_2: The receiver gets the messages from the buffer. Time delay: 1 time unit
t_3: The receiver sends back an ACK to the buffer. Time delay: 1 time unit
t_4: The receiver processes the message. Time delay: 4 time units
t_5: The sender receivers the ACK. Time delay: 1 time unit
t_6: The sender processes a new message. Time delay: 3 time units

---

There are three circuits in the model. The cycle time of each circuit is calculated as follows:

$$\text{circuit } p_1t_1p_3t_5p_8t_6p_1 : C_1 = \frac{T_1}{N_1} = \frac{1+1+3}{1} = 5,$$

$$\text{circuit } p_1t_1p_2t_2p_4t_3p_7t_5p_8t_6p_1 : C_2 = \frac{T_2}{N_2} = \frac{1+1+1+1+3}{1} = 7,$$

$$\text{circuit } p_5t_2p_4t_3p_6t_4p_5 : C_3 = \frac{T_3}{N_3} = \frac{1+1+4}{1} = 6.$$

After enumerating all circuits in the net, we know the minimum cycle time of the protocol between the two processes is 7 time units.

15.8.2 Stochastic Timed Petri Nets

Stochastic timed Petri nets (STPNs) are Petri nets in which stochastic firing times are associated with transitions. An STPN is essentially a high-level model that generates a stochastic process. STPN-based performance evaluation basically comprises modeling the given system by an STPN and automatically generating the stochastic process that governs the system behavior. This stochastic process is then analyzed using known techniques. STPNs are a graphical model and offer great convenience to a modeler in arriving at a credible, high-level model of a system.

The simplest choice for the individual distributions of transition firing times is negative exponential distribution. Because of the memoryless property of this distribution, the stochastic process associated with the STPN is a continuous time homogeneous Markov chain with state space in one to one correspondences to with marking in $R(M_0)$, the set of all reachable markings. The transition rate matrix of the Markov chain can be easily constructed from the reachability graph given the firing rates of the transitions of the STPN. Exponential timed stochastic Petri nets, often called *Stochastic Petri Nets* (SPNs), were independently proposed by Natkin (1980) and Molloy (1981), and their capabilities in the performance analysis of real systems have been investigated by many authors.

An SPN is a six-tuple $(P, T, I, O, M_0, \Lambda)$ in which (P, T, I, O, M_0) is a Petri net and $\Lambda : T \rightarrow R$ is a set of firing rates whose entry λ_k is the rate of the exponential individual firing time distribution associated with transition t_k. Natkin and Molloy have shown that the marking process of an SPN is a continuous time Markov chain. The state space of the Markov chain is the reachable set $R(M_0)$. Suppose there are $s$markings in $R(M_0)$, and the underlying Markov chain is ergodic, then the steady state probability distribution $\Pi = (\pi_0, \pi_1, \ldots, \pi_s)$ can be obtained by resolving the following linear system:

$$\Pi Q = 0$$

$$\sum_j \pi_j = 1, \quad \text{where } \pi_j \geq 0, \; j = 0, 1, 2, \ldots$$

where Q is a transition rate matrix whose elements outside the main diagonal are the rates of the exponential distributions associated with the transitions from state, while the elements on the main diagonal make the sum of the elements of each row equal to zero. Denote by $E(M_i)$ the set of all enabled transition at marking M_i, and T_{ij} the set of enabled transitions at marking M_i whose firings lead the SPN to another marking M_j. Then Q is determined as follows:

$$q_{ij} = \sum_{t_k \in T_{ij}} \lambda_k,$$

$$q_i = q_{ii} = -\sum_{t_k \in E(M_i)} \lambda_k.$$

The probability of marking M_i changing to M_j is the same as the probability that one of the transitions in the set T_{ij} fires before any of the transitions in the set $T \backslash T_{ij}$. Since the firing times in an SPN are mutually independent exponential random variables, it follows that the required probability has the specific value given by

$$\alpha_{ij} = \frac{q_\iota \phi}{q_\iota}.$$

In the expression for α_{ij} deduced above, note that the numerator is the sum of the rates of those enabled transitions in M_i, the firing of any of which changes the marking from M_i to M_j; whereas the denominator is the sum of the rates of all the enabled transitions in M_i. Also note that $\alpha_{ij} = 1$ if and only if $T_{ij} = E(M_i)$.

Example 15.5: **(A Stochastic Petri Net)**

Figure 15.14 shows a simple SPN model with its reachable markings and its reachable graph. The linear system of steady-state probabilities is

$$\pi_0 + \pi_1 + \pi_2 + \pi_3 + \pi_4 = 1$$

Let $\Lambda = (1\,1\,1\,1)$, then solution to this system is

$$\pi_0 = \pi_4 = 2/7, \quad \pi_1 = \pi_2 = \pi_3 = 1/7$$

The analysis of an SPN model is usually aimed at the computation of more aggregate performance indices than the probabilities of individual markings. Several kinds of aggregate results can easily be obtained from the steady-state distribution over reachable markings, such as the probability of an event defined through place markings, the average number of tokens in a place, the frequency of firing a transition, that is, the average number of times the transition fires in unit time, and the average delay of a token in traversing a subnet in steady-state conditions (Ajmone Marsan 1990).

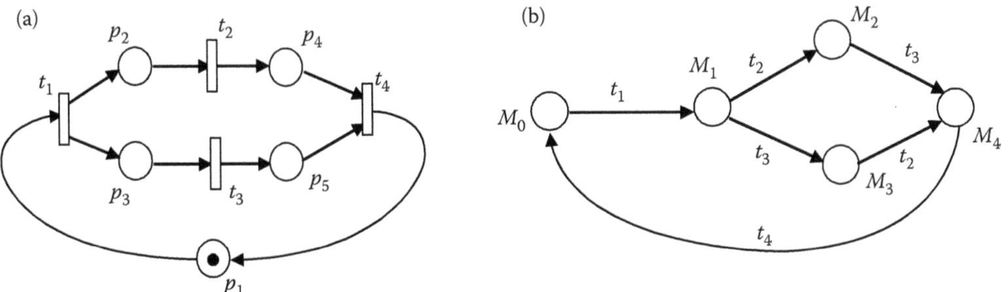

FIGURE 15.14 (a) SPN model. (b) Reachability graph.

15.9 Continuous Petri Nets and Hybrid Petri Nets

The Petri nets we have discussed so far are all discrete Petri nets, because transitions fire at certain instants with zero duration. A *continuous Petri net* has the same structure as a discrete Petri net. The main difference is that a transition in a continuous Petri net fires continuously with a flow that may be externally generated by an input signal and may also depend on the continuous making. A *continuous place* can hold a nonnegative real number as its content. A *continuous transition* fires continuously at the speed of a parameter assigned at the continuous transition. Continuous Petri nets are suitable for modeling systems in which continuous or batch material flows go through a number of stages of processing (Pettersson and Lennartson, 1995).

Hybrid Petri nets (Alla and David, 1998) are an extension to traditional Petri nets with introduction of continuous places and continuous transitions, or a mixture of discrete Petri nets and continuous Petri nets. Therefore, a hybrid Petri net has a discrete part and a continuous part. The discrete part models the logic functioning of a system and the continuous part models a system with continuous flows.

15.10 Concluding Remarks

Petri nets have been proven to be a powerful modeling tool for various types of dynamic event-driven systems. Since Petri nets were introduced in 1962, numerous research papers have been published. The research has been conducted on many branches, with each branch exploring a promising application aspect of this formalism. Given the rich research results from the Petri net society, it is hard to cover all of them in such a short book chapter. Therefore, this chapter only aims at briefly introducing the most basic concepts, theory and applications of ordinary Petri nets and a few of most popular extended Petri nets.

References

Ajmone Marsan, M. 1990. Stochastic Petri nets: An elementary introduction. *Advances in Petri Nets, LNCS* 424.

Ajmone Marsan, M., M.G. Balbo, and G. Conte. 1986. *Performance Models of Multiprocessor Systems*, MA: The MIT Press.

Alla, H. and R. David. 1998. Continuous and hybrid Petri nets. *J Circuits Syst Comput*, 8, 159–188.

Andreadakis, S.K., and A.H. Levis. 1988. Synthesis of distributed command and control for the outer air battle. *Proceedings of the 1988 Symposium on C² Research*. SAIC, McLean, VA.

Bause, F., and P. Kritzinger. 2002. *Stochastic Petri Nets—An Introduction to the Theory*. Germany: Vieweg-Verlag.

Desrochers, A., and R. Ai-Jaar. 1995. *Applications of Petri Nets in Manufacturing Systems: Modeling, Control and Performance Analysis*. New York, NY: IEEE Press.

Genrich, J.H., and K. Lautenbach. 1981. System modeling with high-level Petri nets. *Theoretical Computer Science* 13:109–136.

Haas, P. 2002. *Stochastic Petri Nets: Modelling, Stability, Simulation*. New York: Springer-Verlag.

Jensen, K. 1981. Colored Petri nets and the invariant-method. *Theoretical Computer Science* 14:317–336.

Jensen, K. 1997. *Coloured Petri Nets. Basic Concepts, Analysis Methods and Practical Use*, 3 Vols. London: Springer-Verlag.

Kitakaze, H., H. Matsuno, and N. Ikeda. 2005. Prediction of debacle points for robustness of biological pathways by using recurrent neural networks. *Genome Informatics*, 16(1): 192–202.

Lin, C., L. Tian, and Yaya Wei. 2002. Performance equivalent analysis of workflow systems. *Journal of Software* 13(8): 1472–1480.

Lindemann, C. 1998. *Performance Modelling with Deterministic and Stochastic Petri Nets*. Hoboken, NJ: John Wiley & Sons.

Mandrioli, D., A. Morzenti, M. Pezze, P. San Pietro and S. Silva. 1996. A Petri net and logic approach to the specification and verification of real time systems. In: *Formal Methods for Real Time Computing* (C. Heitmeyer and D. Mandrioli eds), Hoboken, NJ: John Wiley & Sons Ltd.

Merlin, P., and D. Farber. 1976. Recoverability of communication protocols—Implication of a theoretical study. *IEEE Transactions on Communications* 24: 1036–1543.

Molloy, M. 1981. *On the Integration of Delay and Throughput Measures in Distributed Processing Models*. PhD thesis, UCLA, Los Angeles, CA.

Molloy, M. 1982. Performance analysis using stochastic Petri nets. *IEEE Transactions on Computers* 31(9): 913–917.

Murata, T. 1989. Petri nets: Properties, analysis and applications. *Proceedings of the IEEE* 77(4): 541–580.

Natkin, S. 1980. *Les Reseaux de Petri Stochastiques et Leur Application a l'evaluation des Systemes Informatiques*. These de Docteur Ingegneur, Cnam, Paris, France.

Peterson, J.L. 1981. *Petri Net Theory and the Modeling of Systems*. NJ: Prentice-Hall.

Petri, C.A. 1962. *Kommunikation mit Automaten*. English Translation, 1966: *Communication with Automata*, Technical Report RADC-TR-65-377, Rome Air Dev. Center, New York.

Pettersson, S., and B. Lennartson. 1995. Hybrid modelling focused on Hybrid Petri nets. *2nd European Workshop on Real-time and Hybrid systems*, Grenoble, France, 303–309.

Ramamoorthy, C., and G. Ho. 1980. Performance evaluation of asynchronous concurrent systems using Petri nets. *IEEE Transaction on Software Engineering* 6(5): 440–449.

Ramchandani, C. 1974. *Analysis of Asynchronous Concurrent Systems by Petri Nets*. Project MAC, TR-120, MIT, Cambridge, MA.

Tsai, J., S. Yang, and Y. Chang. 1995. Timing constraint Petri nets and their application to schedulability analysis of real-time system specifications. *IEEE Transactions on Software Engineering* 21(1): 32–49.

van der Aalst, W., and K. van Hee. 2000. *Workflow Management: Models, Methods, and Systems*. MA: MIT Press.

van Landeghem, R., and C.-V. Bobeanu. 2002. Formal modeling of supply chain: An incremental approach using Petri nets. *14th European Simulations Symposium and Exhibition*. Dresden, Germany.

Venkatesh, K., M. C. Zhou, and R. Caudill. 1994. Comparing ladder logic diagrams and Petri nets for sequence controller design through a discrete manufacturing system. *IEEE Trans. on Industrial Electronics* 41(6): 611–619.

Wang, J. 1998. *Timed Petri Nets, Theory and Application*. Boston: Kluwer Academic Publishers.

Wang, J. 2006. Charging information collection modeling and analysis of GPRS networks. *IEEE Transactions on Systems, Man and Cybernetics, Part C* 36(6): 473–481.

Zhou, M. C., and F. DiCesare. 1989. Adaptive design of Petri net controllers for error recovery in automated manufacturing systems. *IEEE Trans. on Systems, Man, and Cybernetics* 19(5): 963–973.

16 Statecharts

Hanlin Lu and Sheng Yu
University of Western Ontario

16.1 Introduction

Statecharts (Harel 1987) was introduced by David Harel in 1987 as one of the most popular state-based visual/graphical formalism for designing reactive systems. With hierarchy, concurrency, and many other features, statecharts were more practical and expressive than the classical state diagrams. They have been adopted by a number of object modeling techniques and languages, including the Object Modeling Technique (OMT) (Rumbaugh et al. 1991), where they are called the "State Transition Diagrams," and the Unified Modeling Language (UML) (Rumbaugh 2004; OMG 2007), where they are known as the "State Machines," for modeling the dynamic behaviors of objects. Besides, there are many formalisms for statecharts supported by software such as STATEMATE (Harel et al. 1990), Rhapsody (Harel and Kugler 2001), Fujaba (Geiger and Albert 2005), and so on.

An accurate and comprehensive description of statecharts will benefit their applications greatly. However, statecharts' semantics was not defined precisely in Harel's original paper (Harel 1987) (as described in (von der Beeek 1994)). The study for the semantics of statecharts has began by Harel et al. (1987) right after the publication of Harel (1987). Since then, many related papers have been published, for example, Pnueli and Shalev (1991) and Huizing et al. (1988). In 1994, von der Beeck listed 19 issues concerning statecharts' semantics in von der Beeek (1994). In 1996 in Harel and Naamad (1996), introduced the semantics of statecharts implemented in STATEMATE (Harel et al. 1990) and discussed how they deal with the issues described in von der Beeek (1994).

During the last 20 years, it has attracted much attention to formalize statecharts' semantics, especially for the UML State machines, and various kinds of formal approaches have been introduced. Modeling the state machines using Petri-net has been studied in Baresi and Pezzè (2001), King and Pooley (1999), Merseguer et al. (2002), Bernardi et al. (2002), Saldhana and Shatz (2000), Hu and Shatz (2004), Hu and Shatz (2006); logic based modeling for statecharts has been studied in Graw et al. (1999), Rossi et al. (2004), Lano (1998), Lavazza et al. (2001), Drusinsky (2006); models based on the graph transformation systems have been described in Kong et al. (2009), Engels et al. (2000), Kuske (2001), Kuske et al. (2002), Gogolla and Presicce (1998), Gogolla et al. (2002), Ermel et al. (2005), Höscher et al. (2006), Ziemann et al. (2004a), Ziemann (2004b), Maggiolo-Schettini and Peron (1994), Maggiolo-Schettini and Peron (1996), Varró (2002); statecharts are described by extended hierarchical finite automata in Latella et al. (1999); a template-based approach Niu et al. (2003) is used to describe the state-machine semantics in Taleghani and Atlee (2006), and so on.

Apparently, we have listed only a few approaches that have been studied toward statecharts formalization. In the survey by Crane and Dingel (2005), the semantics for the UML state machines were categorized into 26 kinds based on the formalization approach. In Lund et al. (2010), the authors surveyed 12 different types of formalization for UML state machines based on the type of semantics.

We find that although so many formal models have been applied to describe statecharts, the computation power of statecharts was seldom considered. And not much research on the computation power of the models that has been used to formalize statecharts has been done, since Drusinsky and Harel (1994) and Hirst and Harel (1994). On the contrary, we can find descriptions like "statecharts are extended finite automata" in many places to equate the computation power of statecharts to finite automata. It has been shown in Lu and Yu (2009) clearly that they are very different. We show in this chapter that the computation power of statecharts is more appropriately modeled by Interaction Machines (Wegner 1997).

Interaction Machines (IMs) were introduced by Peter Wegner (Wegner 1997) more than 10 years ago, which are considered to be a further development of TMs (TM) models. Wegner claimed that Tms shut out the world during the computation and stop after generating outputs, thus, they do not model interactions (Wegner 1997; 1998; Goldin and Wegner 2008). In Wegner (1997), he introduced IMs by extending TMs with dynamic streams, which record histories of interactions with the environment. We consider that IMs are more accurate models than TMs for statecharts since they are better models for interactions (Wegner 1997).

Based on Wegner's IMs, Sheng Yu defined Interactive Turing Machines (ITMs) in Yu (2001). ITMs are variant models of IMs that extend Multi-tape TMs with interaction tapes. Sequential ITMs (SITMs) provide detailed relations between TMs' computing and sequential interactions. In this chapter, we use SITMs as accurate models for statecharts.

In Section 16.2, we review the basic elements of statecharts. In Section 16.2.7, we mention statecharts in UML. In Section 16.3, we consider the computation power of statecharts and we provide a specific example of statechart in Section 16.3.2. In Section 16.3.5.2, we study the linkage in detail between statecharts and interaction-based models. We conclude our chapter in Section 16.4.

16.2 Basic Elements of Statecharts

Reactive systems (Harel 1987) are event-driven systems that continuously interact with both the outside and inside environments, such as embedded systems on cell phones, object-based software systems, and so on. They may contain multiple components that operate concurrently and interaction histories may have impact on their future behaviors.

As claimed by David Harel, the description of large and complex reactive systems may require unmanageable and exponentially large number of states and complex interactions, which are not well expressed by notations from classic state diagrams (Conway 1963; Booth 1967; Parnas 1969; Feyock 1977; Hopcroft and Ullman 1979).

Classic state diagrams use a finite set of vertices and edges denoting the states and transitions, and a finite set of symbols associated with the edges denoting the inputs and outputs. Statecharts inherits all elements from classic state diagram, moreover, it includes new features, such as *composite states*, *orthogonal regions*, *transition guards*, and so on. In this section, we review the basic elements of statecharts.

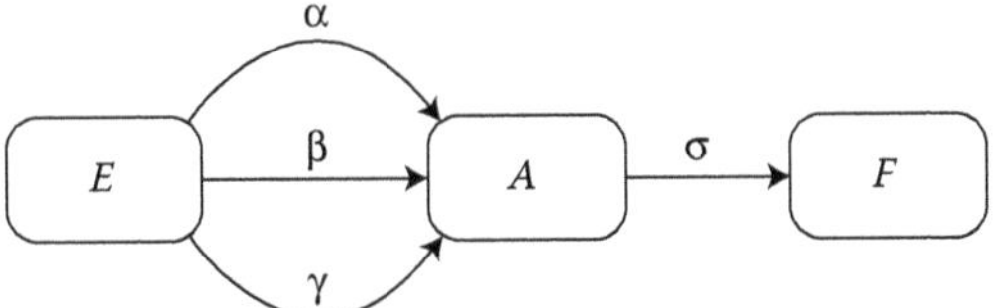

FIGURE 16.1 A statechart with three simple states.

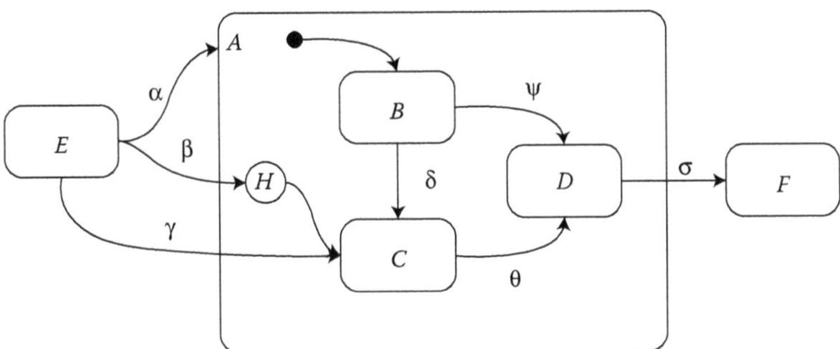

FIGURE 16.2 A statechart containing a composite state.

16.2.1 States

The states represent the system's (finite) memory of interactions. Each state models a situation in which the system has a distinguishing reaction behavior. When it is used to model software objects, a state is also representing a set of object's variables values.

In statecharts, we can depict states in different level of details, a state can be refined by a set of substates and lower-level transitions, a group of states can also form a superstate. For example, the set of states, $\{A, E, F\}$, in Figure 16.1 are simple states. We can refine state A into a composite state as in Figure 16.2. The UML state machines has a third kind, the submachine states, which are essentially composite states.

16.2.2 Regions

In case, a reactive system has several orthogonal components that operate concurrently and independently, their states can be described separately by using multiple *regions*, where each region represents one component. For example, in Figure 16.3, the composite state M has two regions, X and Y. Compared with enumerating the combinations of sub-states from every orthogonal parts, regions are more intuitive and smaller in the size of description.

One region can receive outputs from other regions within the same state. All states and transitions in one region are exclusively included within that region. In Figure 16.3, X and Y are two regions of state M, a transition from state A cannot target at any states of region Y, for example $\{D, E, F\}$.

Note that all regions within the same composite state must be activated or disabled simultaneously. Entering/leaving any region of a composite state will activate/deactivate all other regions of that state.

16.2.3 Events, Actions, Conditions, and Activities

An *event* is an input that triggers a state transition, such as a method invocation on an object. The process of an event, however, can be right away, deferred, or rejected.

An *action* is an output that can be attached to a transition or a state, as in Figure 16.4. When attached to a transition, it is carried out during the transition (before entering its targeting state), for example, "Beep". If attached to a state, the action will need to specify whether it is carried out upon entering or

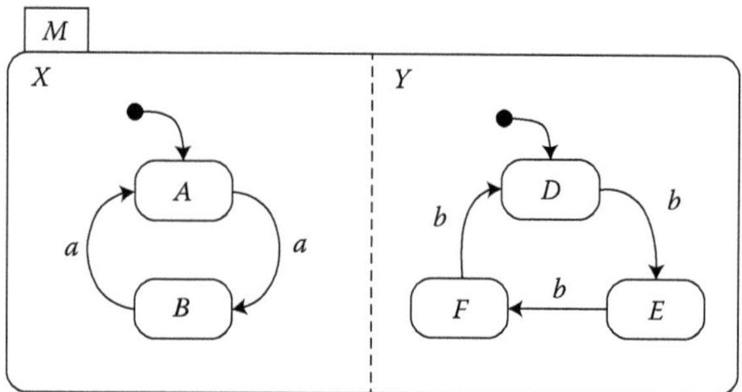

FIGURE 16.3 A composite state M with 2 regions.

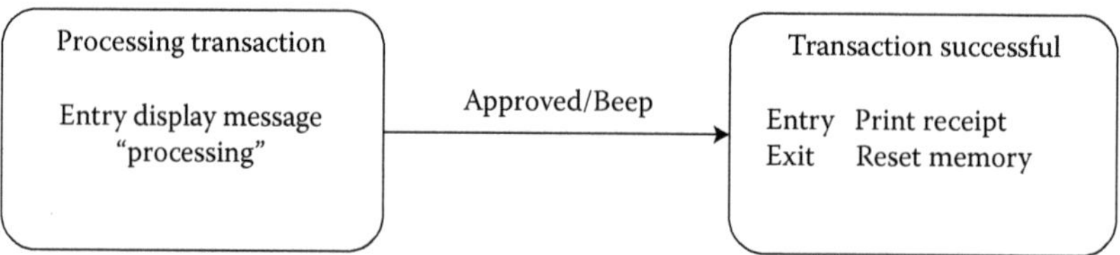

FIGURE 16.4 Actions attached to states and transitions.

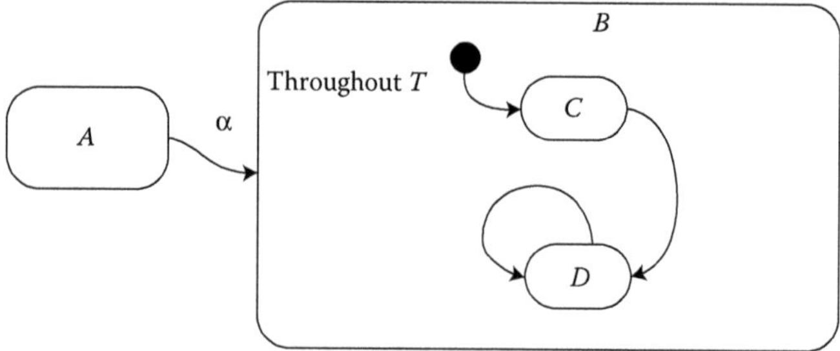

FIGURE 16.5 An activity T with its associated state B.

leaving the state, for example "Print Receipt" and "Reset Memory" in state "Transaction Successful". In statecharts, an *action* is "instantaneous" (Harel 1987), which means we do not consider time during its occurrence, either it is a function computation or updating a variable's value, and so on.

Different from *actions*, an *activity* describes an output by a sequence of actions, which means the time (steps of computation) is considered during such occurrence. The description is associated with a state and it specify the entry and exit actions for that state. As shown in Figure 16.5, an activity T is associated with state B. An action $start(T)$ will be generated when entering B and an action $stop(T)$ will be generated upon leaving B.

16.2.4 Transitions

The *transitions* are specified by their source and target states, triggering events, actions, and guard conditions. Basically, a transition can have at least the source and target states.

The guard conditions are boolean valued expressions without side effects, which are evaluated before the transitions take place. The transitions are enabled only when their conditions have true values. (In Harel's statecharts, the expressions are placed within a pair of parentheses, while in UML state machines,

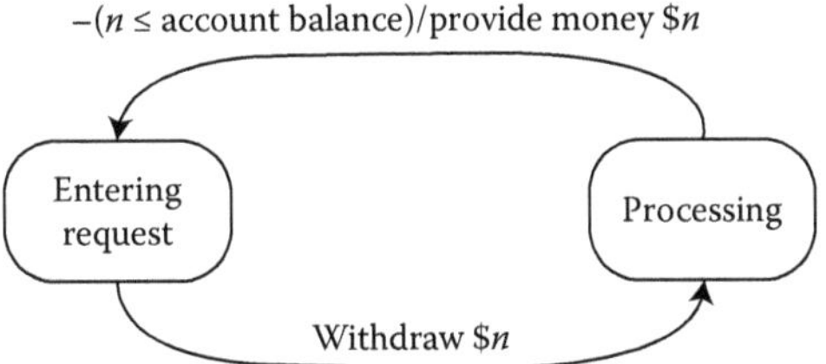

FIGURE 16.6 A transition with a guard condition.

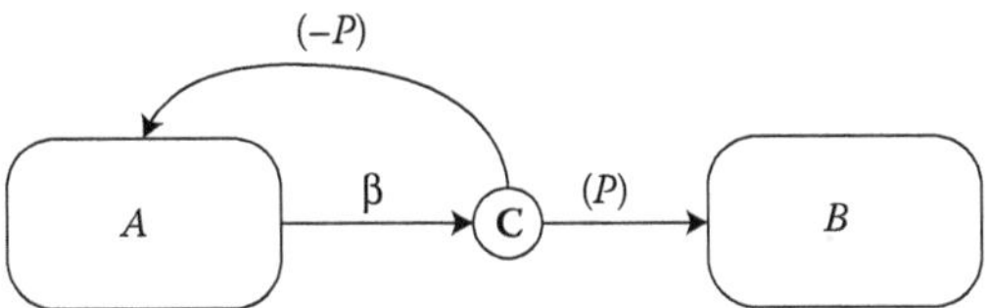

FIGURE 16.7 Transitions connected by a conditional pseudo state.

they are placed within brackets.) For example, an ATM machine will offer money only if the client has a sufficient balance, as shown in Figure 16.6.

Except targeting at a set of orthogonal regions, every transition can have only one deterministic target when it is triggered. However, for a group of transitions, if they originate from the same source with the same event but different targets and condition guards, only one of these guards can be true at a time such that the path will be chosen dynamically during the process. In this case, we use a conditional pseudo state (a circle with "C" in Figure 16.7) for a simplified description. In Figure 16.7, state *A* transits to *B* on event *e* when condition *P* is true, otherwise it will reenter *A*. In UML state machines, the circle with "C" is replaced by a diamond symbol, as in Figure 16.12.

16.2.5 Pseudo States

Generally, the *pseudo states*, which we have shown in previous figures, are used as connection points for transitions, such as *initial pseudo states*, *terminate pseudo states*, *conditional pseudo states*, and so on. An *initial pseudo state* indicates the default starting state or substate of a statechart or a composite state and it is usually required. The *terminate pseudo state* is an optional elements in UML state machines (see Figure 16.12), however, is not included in the original statecharts. In fact, the computation of a reactive system does not have to stop after it is shut down, for example you may not want to reset an operating system every time you turn it off.

There are also pseudo states that are not used for connecting transitions, such as the *history pseudo states*. When a transition targets at a *history pseudo state* of a composite state *A*, the most recently visited substate of *A* will be activated. More specifically, there are two kinds of *history pseudo state*: *shallow* and *deep*, where the first represents a substate of the first level, the second represents a substate of the lowest level. Note that the history pseudo state needs to specify a default substate to activate in case the composite state has never been visited. In Figure 16.8, when *a* triggers the transition from *D* to *A*, the system will choose either *B* or *C*. While in Figure 16.9, the transition from *D* to *A* will have the targeting state among *E*, *F*, *G*, *I*, or *J*.

16.2.6 Variables (Storage)

Variables represent the storage components in a system that carry computation histories. Using them to model reaction behaviors can help reduce the description size of a diagram, such as build up guard

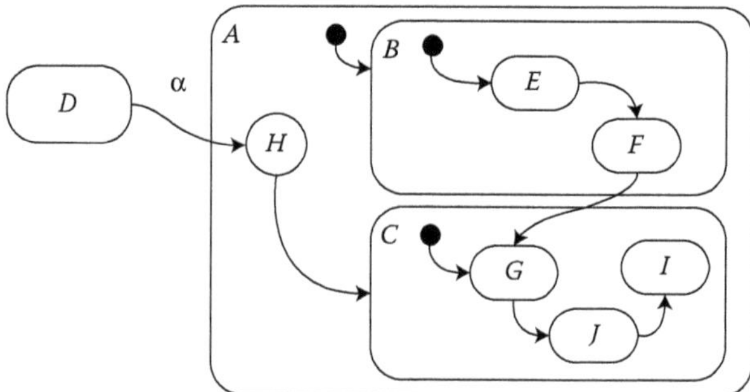

FIGURE 16.8 A composite state with a shallow history pseudo state.

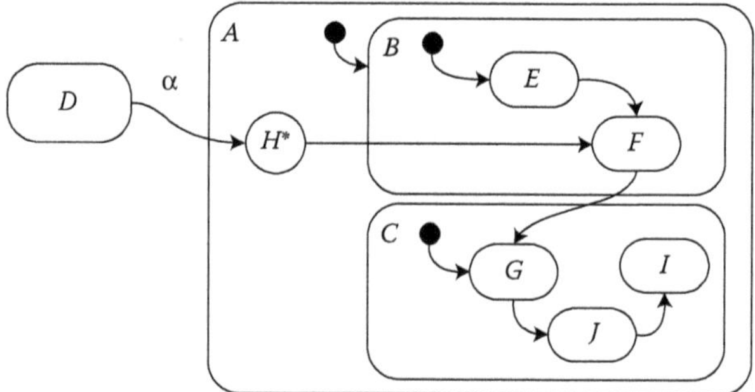

FIGURE 16.9 A composite state with a deep history pseudo state.

expressions, events, or actions. The set of variable values, which will depend on usage in statecharts, are not necessarily finite.

Variables were not considered as essential elements in many literatures on statecharts modeling, for example Harel (1987), Douglass (1999), Drusinsky (2006), and von der Beeek (1994) though they represent essential behaviors in many reactive systems. An example can be found in Harel (1987, page 251–252), where the original statecharts were introduced. In the wristwatch statechart, a transition from state "Displays" to "Alarm-beeps" is triggered by "T(current time) hits T_1(alarm setting)." The wristwatch can memorize an arbitrary alarm setting "T_1" and will activate an alarm when current time hits it. This behavior is modeled by variables rather than states.

An important question concerning statechart variables is whether they are bounded, that is, whether they present infinite storage. Since variables have not been formally included in the definition of statecharts before, such a question has not been discussed. It is clear that if the variables in statecharts are actually used as infinite storage, statecharts are not finite automata (FAs). Then, the question whether statecharts have similar computing power as finite automata is solved right away. However, we feel that it is important to show that, as in Lu and Yu (2009), such cases do exist commonly and, thus, the behaviors of such statecharts cannot properly described by FAs.

A shallow/deep history pseudo state that represents a substate of a composite state can be considered as a bounded variable. When using the history pseudo state, we do not have to draw all possible subsequent transition paths separately. The difference between whether we use a shallow history pseudo state is shown in Figures 16.10 and 16.11. In Figure 16.10, state B contains a shallow history pseudo state, which represents the last visited substate, M or N, when leaving B. The transition from D to B will activate either M or N, based on its reaction history, as a target substate of B. The statechart in Figure 16.11 is equivalent to that in Figure 16.10. Instead of including any history pseudo state, in Figure 16.11, we can enumerate all possible computing paths at the cost of more states and transitions.

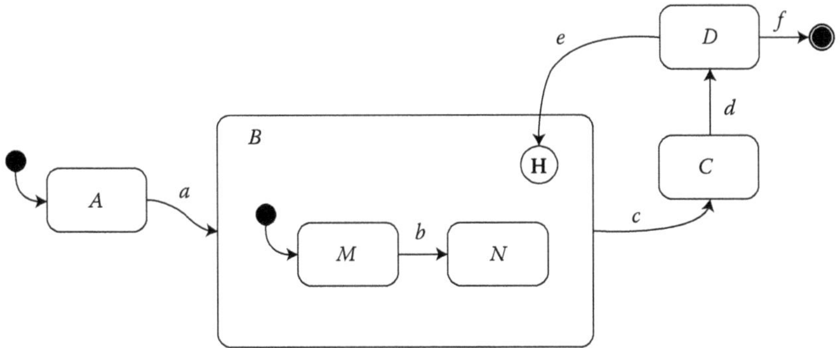

FIGURE 16.10 A statechart using a history pseudo state.

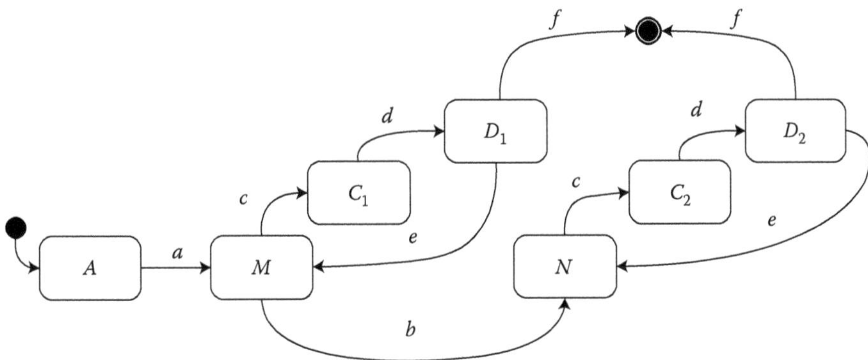

FIGURE 16.11 A statechart without any history pseudo state.

However, for complex reaction behaviors, variables are not necessarily finite. We may consider a statechart of an *ATM* machine, where a variable is used to represent the account balance of a client. The current account balance affects the *ATM*'s interaction behaviors with the client, for example, "approve a withdraw request," and its value is unbounded. Also, variables that do not directly affect state transitions may be included. For example, suppose our *ATM* statechart has a variable containing a string of welcome message. The value is only used for "one time" display, but not for controlling any state transitions.

Note that variables of statecharts may present more than one type of usage, for example, a data member of an object stands as both the states and variables.

16.2.7 The UML State Machines

Unified Modeling Language (UML) (Rumbaugh 2004; OMG 2007) is one of the most widely used modeling language in Object Orient software engineering. In UML, "State Machines," which is based on David Harel's statecharts, serve as primary ways to depict event-driven behaviors of reactive systems. There are two types of state machines in UML: protocol state machines and behavior state machines, where the second type is more close to David Harel's statecharts.

Protocol state machine is a simple transition diagram which models only states, transitions, events, and conditions. This type of state machines focus on the sequence of inputs for objects to respond, without showing the implementation details or complex reaction behaviors.

Different from the protocol state machine, behavior state machines allow to use more expressive notations, for example, actions associated to transitions or states. This type of diagrams inherited most of the features from Harel's statecharts, though they slightly change some of the notations from Harel (1987).

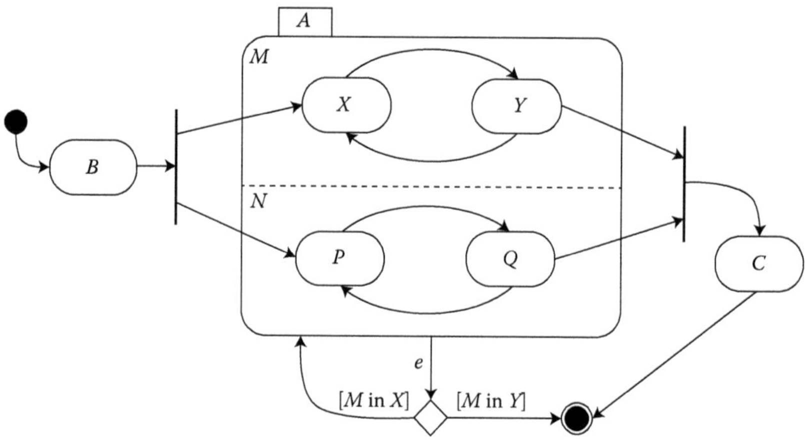

FIGURE 16.12 A statechart with compound transitions.

For example, as in Figure 16.12, we can use connection points called fork/joint instead of drawing multiple transitions to/from orthogonal regions; a conditional pseudo state is presented by a diamond symbol, and so on. Also, UML state machines provide more specified semantics, although there are still semantics issues (Reggio and Wieringa 1999; Fecher et al. 2005).

16.3 The Computing Power of Statecharts

The computation power of statecharts cannot be described by *FAs* in general, however, most descriptions in literature consider that statecharts are similar to finite state machines (*FSMs*) or finite automata (*FAs*) although no accurate arguments or proofs have been provided. We can prove this by analyzing the statechart of an *Automatic Teller Machine (ATM)* model as in Section 16.3.2.

Clearly, many statecharts can be modeled by finite automata. However, those facts do not imply that the computation power of statecharts in general is limited to that of finite automata. In this section, we show that there are commonly used statecharts that cannot be described by any finite automaton. Their computation power is clearly beyond what finite automata can do.

Instead of FAs, we claim that IMs, introduced more than 10 years ago by Peter Wegner (Wegner 1997), are much more accurate theoretical models than TMs for statecharts, since they describe interaction behaviors (Wegner 1997). We study the linkage in detail between statecharts and interaction-based models.

16.3.1 A Formal Definition of Statecharts

There are already many models for formalizing statecharts' semantics. In this chapter, however, we intend to study their computing power rather than any specific definition of semantics. Hence, we need a brief yet powerful definition for statecharts, which includes only the essential elements.

Definition 16.1: (Statecharts)

Formally, a statechart is a 9-tuple,

$$(Q, v, E, A, C, \delta, \eta, s, T),$$

where Q is the nonempty finite set of states; v is an unbounded variable; E is the nonempty finite set of events; A is the finite set of actions, which is split into two subsets: the set of external actions A_{ex}, which is a finite set of symbols, and the set of internal actions A_{in}, which is a finite set of functions $p: v \to v$; C is the finite set of conditions, where each condition is a boolean-valued function of the form $q: v \to \{0, 1\}$; $\delta: Q \times E \times C^* \to Q$ is the transition function; $\eta: \delta \to A^*$ is the action function; s is the initial state; and $T \subseteq Q$ is the set of terminate states.

We use only one unbounded variable v, for simplicity, to represent all variables used in instances of statecharts. Each state $q \in Q$ can be defined recursively as a statechart S_q of the next level. However, a multilevel statechart can always be transformed into an equivalent one-level statechart although the number of states can increase significantly.

This definition is not intended to include all the features of statecharts. For example, the "activities" (Harel 1987) can be included in the definition, which are sets of actions associated with states; the history pseudo states can be described by variables, and so on.

16.3.2 An Example of Statecharts

We provide an ATM model, as in Figure 16.13, to illustrate how statecharts describe reactive systems. In Figure 16.13, the ATM can collect, dispense money to identified clients, and maintain their information, for example banking card and account balance. The client, after being identified, can place any deposit/withdraw requests before choosing "*Exit.*" For security, withdrawing more money than the client's balance will never be approved and the client will have to input the PIN again.

We model the ATM example by statechart S, which is a 9-tuple

$$(Q, v, E, A, C, \delta, \eta, s, T).$$

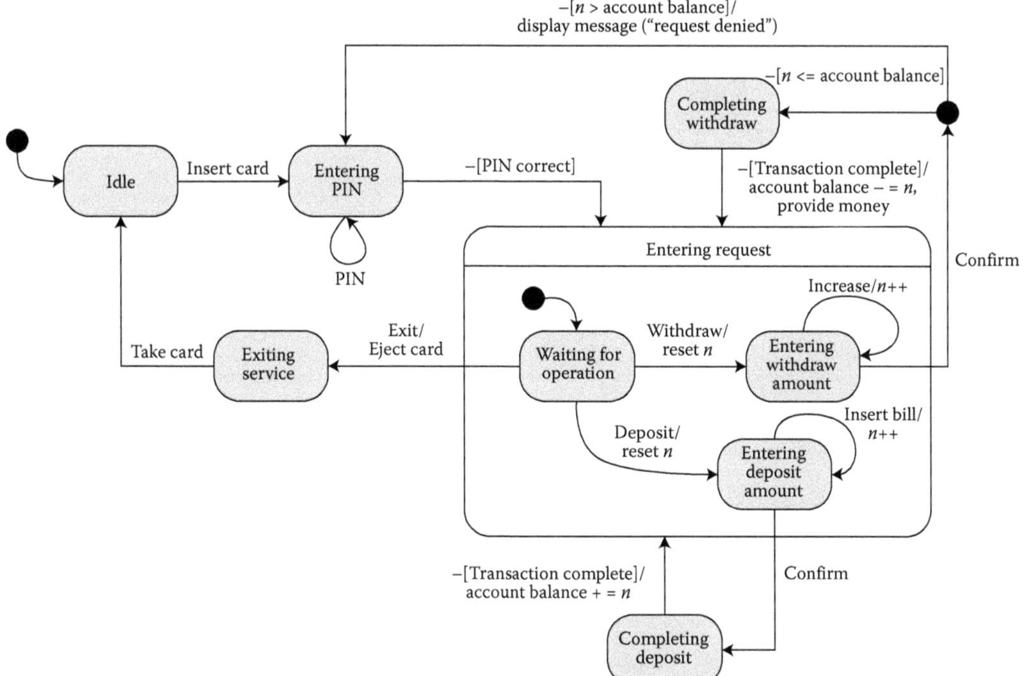

FIGURE 16.13 A statechart of an ATM model.

The set of states Q of S consists of {*Idle, Entering PIN, Entering request, Completing withdraw, Completing deposit, Exiting service*}, where the "Entering request" is a composite state containing three substates, {*Waiting for operation, Entering withdraw amount, Entering deposit amount*}. "*Idle*" is the starting state as well as the terminate state, for example $s = Idle$ and $T = \{Idle\}$. The variables {*n, account balance*} are represented by v of S.

The event set E of S consists of {*Insert card, Take card, Enter PIN, Withdraw, Deposit, Increase, Insert bill, Confirm, Exit*}, which are abbreviated to i, t, p, w, d, a, b, c, and e, respectively, in the following. Especially, "a" is represent event "Increase" and b is for "Insert bill."

Transitions without trigger events are attached by "-" instead and conditions expressions are placed within "[ ]"s, such as "$[n \leqslant accountbalance]$". The set of external actions are {*Display Message, Provide money, Eject card*}, the internal actions are {reset n, $n{+}{+}$, $account\ balance += n$, $account\ balance -= n$}.

16.3.3 Are Statecharts Finite Automata

Now we can consider the computing power of our statechart S in Section 16.3.2. In order to analyze S's language, we consider "*Idle*" to be the accepting state, although S does not "stop" after serving each client. The acceptable strings of S should be the successful input scenarios leading from the "*Idle*" state to itself, which are the event sequences beginning with the symbol "i" and ending with "t."

We consider S's language L_S to be the set of event sequences that a client has been involved from the very first session, instead of only the current session. For example, the event sequence of a client's first session is "ipdbbet" and the second is "ipwaet," then we apply the word "ipdbbet" to the first and "ipdbbetipwaet" to the second. Note that L_S is the set of all valid sequences of events, where each word may or may not be applied by the client. Intuitively, every word $w \in L_S$ satisfies that the number of withdrawals ("a"s) is no more that of deposits ("b"s) in any prefix of w.

We define a morphism

$$h : E^* \rightarrow E^*$$

such that

$$h(a) = a, \quad h(b) = b, \quad \text{and}$$
$$h(i) = h(t) = h(p) = h(e) = h(w) = h(d) = \varepsilon.$$

Then we have,

$$h(L_S) = \{x \mid |y|_a \leq |y|_b \text{ for any prefix } y \text{ of } x\}.$$

The language $h(L_S)$ is clearly not regular. Since the family of regular languages is closed under morphisms (Hopcroft and Ullman 1979; Yu 1997), then we have the following result as an immediate consequence.

Theorem 16.1:

L_S is not a regular language.

The above arguments appear to be obvious. Clearly, the existence of variables in statecharts separate them from finite automata. However, variables were not included as an essential part of statecharts in their original definitions (Harel 1987; Harel and Politi 1998). As we have mentioned in Section 16.2.6, the variables are one of the important components of statecharts and should be included explicitly in their formal definition.

Our improved definition makes it more clear what statecharts can do. The example in this section clearly separates statecharts, in general, from finite automata. This shows that the "common sense" that statecharts are finite automata which has been claimed in many writings is clearly untrue.

However, we still have the question what the computation power of statecharts exactly is, that is, what theoretical model accurately characterizes statecharts.

16.3.4 Interaction Machines

Peter Wegner claimed that TMs shut out the world during the computation and stop after generating outputs; thus, they can not model interaction behaviors (Wegner 1997; 1998; Goldin and Wegner 2008). He introduced IMs (Wegner 1997), which are considered to be a further development of TMs.

IMs extend TMs with dynamic streams to record interaction histories. The output of an IM will depend on both the current input and its previous interactions. Based on Wegner's IMs, Sheng Yu defined Interactive Turing Machines (ITMs) (Yu 2001), which extend Multi-tape Turing machines with interaction tapes. Sequential ITMs (SITMs) (Yu 2001) provide detailed relations between TMs' computing and sequential interactions. In this chapter, we use SITMs as the models for statecharts.

For simplicity, we define an *SITM* with an interaction tape and only one work tape. It interacts with any other ITM through the interaction tape, which is shared by the two machines. When the interaction tape is being written by one, it is read-only to the other and it uses two special characters as the delimiters to indicate the end of each SITM's output. Read/wirte heads of the two SITMs on the interaction tape can only move toward the same side or stay.

Formally, a *Sequential Interactive Turing machine (SITM)* M is an 11-tuple

$$(I, O, Q, \Gamma, \gamma, \omega, s, \#, \$, B, T),$$

where I is the finite input alphabet; O is the finite output alphabet; Q is the nonempty finite set of states; $\#, \$, B$ are three special symbols, where # is the delimiter preceding an input string and at the end of an output string, \$ is the other delimiter that is ending an input string and preceding an output string, B is the blank symbol of the interaction tape and work tape; Γ is the finite work-tape alphabet;

$$\gamma : Q \times (I \cup \{\#, \$\}) \times \Gamma \rightarrow Q \times \{R, S\} \times (\Gamma \times \{L, R, S\})$$

is the transition function of M in the reading phase (when the head on the interaction tape is only allowed to read), where $\{R, S\}$ are movement directions of the head on the interaction tape and $\{L, R, S\}$ are movement directions of the working tape's head;

$$\omega : Q \times \{B\} \times \Gamma \rightarrow Q \times (O \cup \{\#\} \times \{R, S\}) \times (\Gamma \times \{L, R, S\})$$

is the transition function of M in the writing phase; $s \in Q$ is the starting state of M; $T \subseteq Q$ is the set of terminating states.

16.3.5 Linkage between Statecharts and Interactive Turing Machines

In this section, we will use a general notion to study the linkage between statecharts and Interactive Turing Machines.

16.3.5.1 Labeled Transition Systems

Basically, both statecharts and Interactive Turing Machines satisfy the general notion, Labeled transition systems (LTSs) (Plotkin 1981) consist only of the states, transitions, and labels. Note that, in LTSs, none of the three elements are necessarily finite.

Definition 16.2: **(Labeled transition systems (LTS))**

A Labeled Transition System (LTS) is a 3-tuple $(S, T, \{\xrightarrow{t} | t \in T\})$, where S is the set of states, T is the set of labels, $\xrightarrow{t} \subseteq S \times S$, where $t \in T$, is the set of labeled transitions.

Example 16.1:

A statechart $S = (Q_s, v_s, E_s, A_s, C_s, \delta_s, \eta_s, s_s, T_s)$ can be expressed as $L_s = ((Q_s, v_s), E_s, \xrightarrow{E_s} \subseteq (Q_s, v_s) \times (Q_s, v_s))$ in terms of an LTS. The set of states of L_s is the pair formed by S's states Q_s and the variable v_s, to avoid confusion with the *states* of S, Q_s, we will call this pair S's *configuration* in the following: the labels is the set of events E_s of S; the transitions are obtained by applying S's transition function δ_s to its action functions η_s, though those external actions do not directly affect statechart's behavior; thus, they are not considered here.

Example 16.2:

Let an SITM M be the tuple

$$(I_m, O_m, Q_m, \Gamma_m, \gamma_m, \omega_m, s_m, \#, \$, B, T_m).$$

We can write an LTS L_m as the following, which is exactly M,

$$((Q_m, \Gamma^*, P), I_m \cup \{\#, \$, B\} \cup \{\epsilon\}, \xrightarrow{I_m \cup \{\#,\$,B\} \cup \{\epsilon\}} \subseteq (Q_m, \Gamma^*, P) \times (Q_m, \Gamma^*, P)).$$

The state set of L_m is the tuple formed by M's state Q_m, the set of possible strings on M's work tape Γ^* and the set of all positions of the head, P, on work tape, to avoid confusion with M's states Q_m, we call this tuple M's *configuration* in the following: the labels of L_m contains all possible symbols on M's interaction tape, $I_m \cup \{\#, \$, B\}$, and the empty symbol ϵ, since the head on the interaction tape is allowed to stay after each transition which means no new input will be read at the next step; the set of transitions is isomorphic to the set $\gamma_m \cup \omega_m$, though every transition does not contain the output that M writes on the interaction tapes.

To express a sequence of state transition behaviors, we will use the following definition.

Definition 16.3:

Let $L = (S, T, \{\xrightarrow{t} \mid t \in T\})$ be an LTS. If $\{a_1, \ldots, a_n\} \in T$, $\{P_0, \ldots, P_n\} \in S$, we write

$$P_0 \xrightarrow{a_1} \ldots \xrightarrow{a_n} P_n,$$

if

$$P_0 \xrightarrow{a_1} P_1 \xrightarrow{a_2} \ldots \xrightarrow{a_n} P_n.$$

We also write

$$P(\xrightarrow{a_1})^* P',$$

if there is an arbitrary number of transitions labeled by a_1 such that

$$P \xrightarrow{a_1} \ldots \xrightarrow{a_1} P'.$$

16.3.5.2 Bisimulation and Observation Equivalence

Our method to show statecharts are Interaction Machines is based on bisimulation, which has been studied by Park (1981) and Milner (1989).

Bisimulation is an equivalence relation describing whether two agents behave in the same way based on observation. The "agents" are computing systems that are identified by their *states*, for example, statecharts and SITMs. The relation is also known as the *weak bisimulation*, since agents' internal interactions that cannot be detected from outside are not required to match exactly.

We may notice from Example 16.2, there are transitions in the SITM that are triggered by nothing (an empty input) from the outside such that they execute without awareness. Such input is called a *silent action* in Milner (1989).

Definition 16.4: (Bisimulation)

A binary relation $R \subseteq \mathcal{P} \times \mathcal{P}$ over agents is a (weak) *bisimulation* if $(P, Q) \in R$ implies, for all $\alpha \in Act$,

1. Whenever $P \overset{\alpha}{\rightarrow} P'$ then, for some Q', $Q(\overset{\epsilon}{\rightarrow})^* \overset{\hat{\alpha}}{\rightarrow} (\overset{\epsilon}{\rightarrow})^* Q'$ and $(P', Q') \in R$
2. Whenever $Q \overset{\alpha}{\rightarrow} Q'$ then, for some P', $P(\overset{\epsilon}{\rightarrow})^* \overset{\hat{\alpha}}{\rightarrow} (\overset{\epsilon}{\rightarrow})^* P'$ and $(P', Q') \in R$

where P is the set of agents; Act is the set of labels; $\hat{\alpha} = \alpha$ if α is nonsilent, for example, $\alpha \in Act \backslash \{\epsilon\}$, otherwise $\hat{\alpha} = \epsilon$.

Based on the above definitions, we can claim the following:

Theorem 16.2:

For any Statechart S, there is a Sequential Interactive Turing Machine M such that M and S satisfy the bisimulation relation.

Proof. We prove this by providing a method to construct an Sequential Interactive Turing Machine (SITM) M for any given statechart S such that M bisimulate S. We may consider the statechart S in Example 16.1.

Let M's input alphabet I_m formed by the all the elements of the events E_s of S such that $I_m = E_s$. Since M's work tape and S's variables are essentially equivalent storage models, there exists a bijection $f : \Gamma^* \rightarrow v_s$ between the string on the work tape and the value of the variable. However, at each computing step, the work tape is allowed to read/write only one symbol from a finite set, which is different from operating a variable, M may require a finite number of steps to modify its work tape in order to simulate the change of value of v_s in S.

Let M have the same number of states as S such that there is a bijection between Q_m and Q_s. Note that Q_m will be expanded later. Let the initial configuration of M, (s_m, ϵ, p_0) simulate that of S (s_s, val_0), where $\epsilon \in \Gamma^*$ represents the value of the blank work tape and p_0 is the starting position of the work tape's head, $val_0 \in v_s$ is the initial value of v_s.

For every configuration $(q_s, val) \in (Q_s, v_s)$ of S, if and only if $\exists i \in E_s$ such that

$$(q_s, val) \overset{i}{\rightarrow} (q_s', val'),$$

we construct a (finite sequence of) transition(s)

$$(q_m, t_m, p)(\overset{\epsilon}{\rightarrow})^* \overset{i}{\rightarrow} (\overset{\epsilon}{\rightarrow})^*(q_m', t_m', p'),$$

in M such that ϵ is the "silent action" of M, (q_m, t_m, p) simulates (q_s, val) and (q'_m, t'_m, p') simulates (q'_s, val'), where

$$(q_m, t_m, p), (q'_m, t'_m, p') \in (Q_m, \Gamma^*, P)$$

and

$$(q_s, val), (q'_s, val') \in (Q_s, v_s).$$

The set of states Q_m may be expanded after the construction of transitions.

Since all the configurations of M are constructed to simulate that of S, it can be verified that for every configuration of M, $(q_m, t_m, p) \in (Q_m, \Gamma^*, P)$, if there exists a transition

$$(q_m, t_m, p) \xrightarrow{i} (q'_m, t'_m, p'),$$

where $i \in I_m \cup \{\epsilon\}$ and (q_m, t_m, p) is simulating $(q_s, v_s) \in (Q_s, V_s)$, then

1. If $i \neq \epsilon$, S has the configuration (q'_s, v'_s) such that

$$(q_s, v_s) \xrightarrow{i} (q'_s, v'_s),$$

where $e_s \in E_s$ is identical to i_m, and (q'_s, v'_s) is simulating (q'_m, t'_m, p');
2. If $i = \epsilon$, S does not have any corresponding transition and (q_s, v_s) is simulating (q'_m, t'_m, p'). □

Theorem 16.3:

For any SITM N, there is a Statechart T such that T and N satisfy the bisimulation relation.

The proof for Theorem 16.2 has been omitted since it is similar to the proof for Theorem 16.1. Now we have shown that statecharts and SITMs are equivalent models.

16.3.5.3 An Example of Bisimulation

We will investigate in our example in Section 16.3.2 to see how an SITM bisimulates a statechart. First, we flatten the statechart S by N, as shown in Figure 16.14, where the "Entering Request" state is replaced by its substate, {*Waiting for operation, Entering withdraw amount, Entering deposit amount*}.

We redraw N further by N', in Figure 16.15, where all the elements have shorter names. For example, state "Z" and "Y" substitute state "Entering deposit amount" and "Entering withdraw amount"; the event "Insert card" is replaced by "i", as in Section 16.3.2, which triggers a transition from state "A" to "H"; the symbols in the "[ ]"s are transition conditions correspond to those of statechart N, for example, "[#]" (refers to "$amount > accountbalance$"), "[\$ or B]" (replaces "$amount = accountbalance$ or $amount < accountbalance$") and "[pc]" (equals to "PIN Correct"). For simplicity, we assume that our ATM model processes requests from only one client (with card "i") and it uses symbol "p" as PIN, and so on. (In fact the model can be extended without any problem.) Also, since the external actions, such as "Provide money," do not directly affect N's interaction behavior, we remove them in N'. It is obvious that N' is essentially the same as N as well as the original S.

Follow the method in Section 16.3.5.2, we can build an SITM M bisimulating N', as shown in Figure 16.16. The set of input I of M is the same as the event set E of N'. Each state of M bisimulates the state with the "same name" of N', except that states $\{E_1, E_1\}$ of M bisimulate state E of N' and $\{F_1, F_2\}$ bisimulate state F of N'. The labels placed within "[ ]"s are the symbols that M read from its working tape. Those "{}"s placed after "/" contain M's internal writing operations on its working tape, though $\{\leftarrow, \downarrow, \rightarrow\}$ replace $\{L, S, R\}$ to avoid confusion with tape symbols. The working tape alphabet is the set $\{\$, \#, B\}$, where "$B$" is the blank symbol. The interaction tape alphabet is the set $\{i, p, w, d, a, b, c, e, t\}$. At

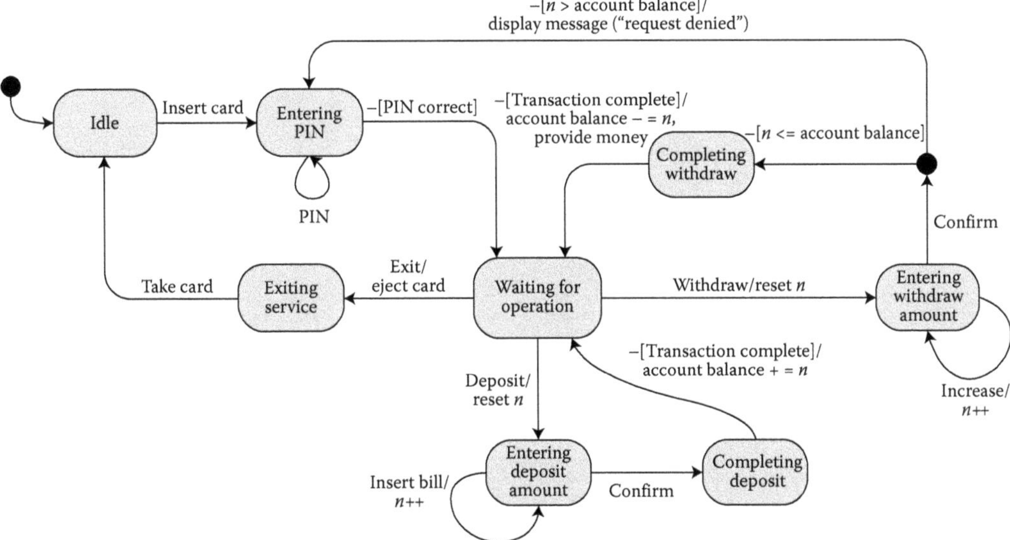

FIGURE 16.14 The flat statechart N.

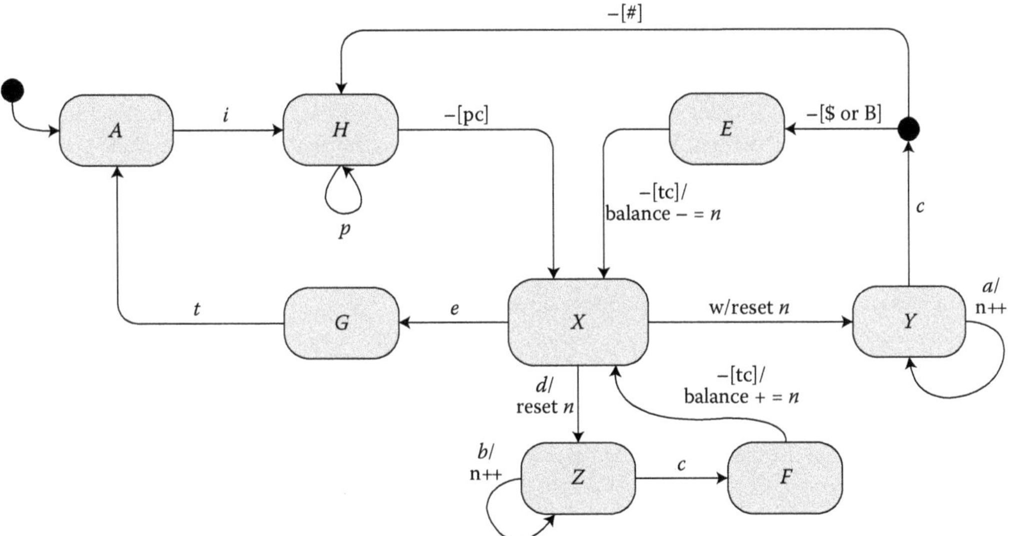

FIGURE 16.15 The simplified statechart N′.

the beginning, the working tape contains user's account balance, which is a string of "$"s, and its working head starts from the right most symbol. Since we remove the outputs from N', M does not need to write on its interaction tape for simulation. Since bisimulation does not require internal actions of the two agents be the same, it can be verified that M bisimulates the statechart N', and therefore M bisimulates N as well as our original statechart S.

16.4 Conclusion

In this chapter, we have reviewed the basic features of Harel's statecharts and, through a few examples, we have explained how they model reactive systems.

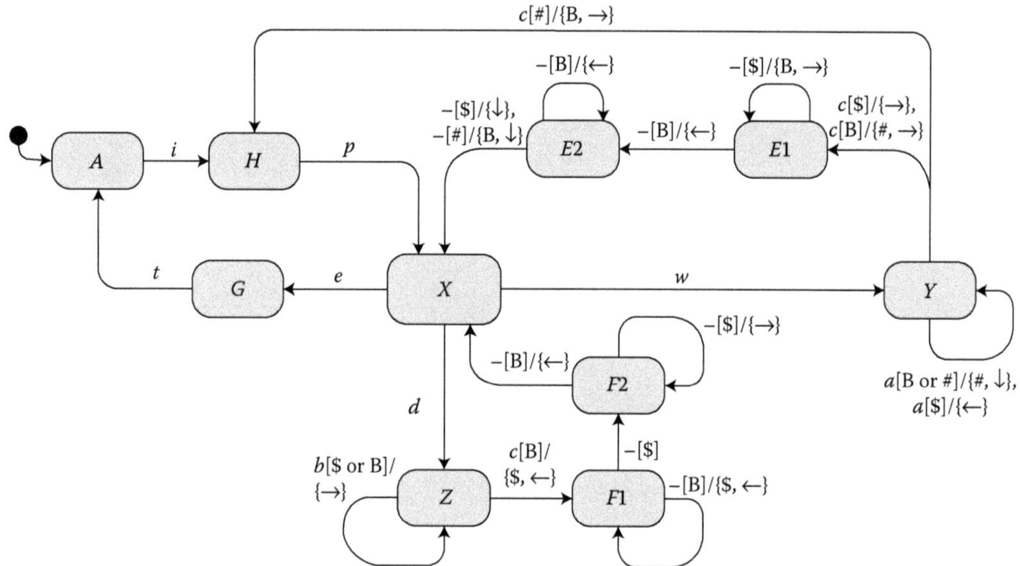

FIGURE 16.16 The Sequential ITM M.

Although much research have been done to achieve a precise semantics for statecharts, the issues on the computation power of statecharts have not gained much attention. In this chapter, we prove that statecharts are not Finite Automata in general. We use an approach which is different from other popular formalisms, for example, Petri-net, Temporal Logic, and Graph transformation. We claim that Wegner's Interaction Machines are more precise models for statecharts, and we have shown that statecharts and Interactive Turing Machines (Yu 2001) are equivalent models based on bisimulation.

We believe that an accurate and comprehensive description of statecharts will benefit the applications of statecharts greatly. Since statecharts are considered an important model for objects, it also has the implication that a more accurate description for objects is given.

References

Parnas, D. L. 1969. On the use of transition diagrams in the design of a user interface for an interactive computer system. In *Proceedings of the ACM Conference* 1969, 379–185, ACM.

Feyock, S. 1977. Transition diagram-based CAI/HELP systems. *International Journal of Man-Machine Studies* 9: 399–413.

Conway, M. E. 1963. Design of a separable transition-diagram compiler. *Communications of the ACM* 6(7):396–408.

Booth, T. L. 1967. *Sequential Machines and Automata Theory.* New York: John Wiley & Sons.

Douglass, B. Powel. 1999. Uml statecharts. Technical report, I - Logix. http://www.embedded.com/1999/9901/9901feat1.htm.

Harel, D. 1987. Statecharts: A visual formulation for complex systems. *Science of Computer Programming,* 8(3):231–274.

Harel, D., A. Pnueli, J. P. Schmidt, and S. Sherman. 1987. On the formal semantics of statecharts. In *Proceedings of the Second IEEE Symposium on Logic in Computation*, 54–64, IEEE.

Pnueli, A. and M. Shalev. 1991. What is in a step: On the semantics of statecharts. In *TACS '91: Proceedings of the International Conference on Theoretical Aspects of Computer Software*, 244–264, Berlin, Heidelberg: Springer-Verlag.

Huizing, C., R. Gerth, and W. P. de Roever. 1988. Modeling statecharts behaviour in a fully abstract way. In *Proceedings of the 13th Colloquium on Trees in Algebra and Programming*, 271–294, London: Springer-Verlag.

Rumbaugh, J., M. Blaha, W. Premerlani, F. Eddy, and W. Lorensen. 1991. *Object-Oriented Modeling and Design*. Upper Saddle River, NJ: Prentice Hall, Inc.

Rumbaugh, J., I. Jacobson, and G. Booch. 2004. *Unified Modeling Language Reference Manual, The (2nd Edition)*. Boston, MA: Pearson Education, Inc.

Object Modeling Group. 2007 Uml 2.0 superstructure specification. Technical report, OMG. http://www.omg.org/technology/documents/formal/uml.htm

Harel, D. 2002. Synthesizing state-based object systems from LSC specifications. *International Journal of Foundations of Computer Science.*, 13(1):5–51.

Harel, D. and M. Politi. 1998. *Modeling Reactive Systems with Statecharts*, New York: McGraw-Hill.

Hopcroft, J. E. and J. D. Ullman. 1979. *Introduction to Automata Theory, Languages, and Computation*. Boston, MA: Addison-Wesley.

Booch, G., R. A. Maksimchuk, and M. W. Engle. 2007. *Object-Oriented Analysis and Design with Applications*. Boston, MA: Addison-Wesley.

Yu, S. 1997. Regular languages. In *Handbook of Formal Languages*, Vol. 1, edited by G. Rozenberg and A. Salomaa, New York: Springer-Verlag.

Rich, E. 2008. *Automata, Computability and Complexity: Theory and Applications*. Upper Saddle River, NJ: Prentice Hall.

Milner, R. 1971. An algebraic definition of simulation between programs. in *Proceeding of 2nd IJCAI*, 481–489, San Francisco: Morgan Kaufmann Publishers Inc.

Milner, R. 1989. *Communication and Concurrency*, 88–128, Upper Saddle River, NJ: Prentice Hall.

Park, D. 1981. *Concurrency and Automata on Infinite Sequences*. Theoretical Computer Science, LNCS 104, 167–183, London: Springer-Verlag.

Aho, A. V., J. E. Hopcroft, and J. D. Ullman. 1974. *The Design and Analysis of Computer Algorithms*. Boston, MA: Addison Wesley.

de Roever, W.-P. and K. Engelhardt. 1998. *Data Refinement: Model-Oriented Proof Methods and Their Comparison*. Cambridge Tracts in Theoretical Computer Science 47, New York: Cambridge University Press.

von der Beeek, M. 1994. A comparison of statecharts variants. in *Formal Techniques in Real-Time and Fault-Tolerant Systems (FTRTFT'94)*, LNCS 863, 128–148, London: Springer.

Crane, M. L. and J. Dingel. 2005. On the semantics of UML State machines: Categorization and Comparison Technical Report, Queen's University.

Lund, M. S., A. Refsdal, and K. Stølen. 2010. Semantics of UML models for dynamic behavior: a survey of different approaches, In *Proceedings of the 2007 International Dagstuhl Conference on Model-Based Engineering of Embedded Real-Time Systems*, LNCS 6100, 77–103, Berlin/Heidelberg, Germany: Springer-Verlag.

Crane, M. L. and J. Dingel. 2005. UML Vs. Classical Vs. Rhapsody Statecharts: Not all models are created equal. In *Proceedings of 8th International Conference on Model Driven Engineering Languages and Systems (MoDELS'05)*, LNCS 3713, 97–112, Berlin/Heidelberg, Germany: Springer-Verlag.

von der Beeck, M. 2001. Formalization of UML-statecharts. In *Proceedings of the 4th International Conference on The Unified Modeling Language, Modeling Languages, Concepts, and Tools*, 406–421, London: Springer-Verlag.

Börger, E., A. Cavarra and E. Riccobene. 2000. Modeling the dynamics of UML State Mahcines. In *Proceedings of the International Workshop on Abstract State Machines, Theory and Applications*, LNCS 1912, 223–241, Berlin/Heidelberg, Germany: Springer-Verlag.

Huizing, C. and R. Gerth. 1992. *Semantics of Reactive Systems in Abstract Time*. LNCS 600, 291–314, Berlin/Heidelberg, Germany: Springer-Verlag.

Reggio, G. and R. Wieringa. 1999. Thirty one problems in the semantics of UML 1.3 dynamics. In *OOPSLA99 Workshop, Rigorous Modelling and Analysis of the UML: Challenges and Limitations*, Denver, Colorado.

Fecher, H., J. Schönborn, M. Kya, and W.-P. de Roever. 2005. 29 new unclarities in the semantics of UML 2.0 State Machines. in *Proceedings of International Conference on Formal Engineering Methods*, LNCS 3785, 52–65, Berlin/Heidelberg, Germany: Springer-Verlag.

Fecher, H. and J. Schönborn. 2007 *UML 2.0 State Machines: Complete Formal Semantics via Core State Machine.* in *Proceeding of FMICS 2006 and PDMC 2006*, LNCS 4346, 244–260, Berlin/Heidelberg, Germany: Springer-Verlag.

Harel, D. 1997. Some thoughts on statecharts, 13 years later. In *Proceedings of the 9th International Conference on Computer Aided Verifcation (CAV'97)*, LNCS 1254, 226–231, Berlin/Heidelberg, Germany: Springer-Verlag.

Baresi, L. and M. Pezzè. 2001. On formalizing UML with high-level Petri nets. In *Proceedings of the Concurrent Object-Orient Programming and Petri Nets*, LNCS 2001, 271–300, Berlin/Heidelberg, Germany: Springer-Verlag.

King, P. and R. Pooley. 1999. Using UML to derive stochastic Petri net models. In *Proceedings of the 15th UK Performance Engineering Workshop (UKPEW'99)*, 45–56, Bristol, UK.

Merseguer, J., J. Campos, S. Bernardi, and S. Donatelli. 2002. A compostional semantics for UML state machines aimed at performance evaluation. In *Proceedings of the 6th International Workshop on Discrete Event Systems*, 295–302, Washington, DC: IEEE Computer Society Press.

Bernardi, S., S. Donatelli, and J. Merseguer. 2002. From UML sequence diagrams and statecharts to analysable Petri net models. In *Proceedings of the 3rd International Workshop on Software and Performance (WOSP'02)*, 35–45, New York: ACM.

Saldhana, J. A. and S. M. Shatz. 2000. UML diagrams to object Petri net models: An approach for modeling and analysis. In *Proceedings of the International Conference of Software and Knowledge Engineering SEKE00*, 103–110, Chicago, IL.

Hu, Z. and S. M. Shatz. 2004. Mapping UML diagrams to a Petri net notation to exploit tool support for system simulation. In *Proceedings of the International Conference on Software Engineering and Knowledge Engineering (SEKE04)*, 213–219, Banff Alberta, Canada.

Hu, Z. and S. M. Shatz. 2006. Explicit modeling of semantics associated with composite states in UML statecharts In*Automated Software Engineering* 13(4): 423–467.

Graw, G., P. Herrmann and H. Krumm. 1999. Composing objectoriented specificactions and verifications with cTLA. In *Proceedings of Workshop on Semantics of Objects and Processes (SOAP99)*, 7–22, London: Springer-Verlag.

Rossi, C., M. Enciso, and I. P. de Guzmán. 2004. Formalization of UML statemachines using temporal logic. *Software and Systems Modeling*, 3(1): 31–54.

Drusinsky, D. 2006. *Modeling and Verification Using UML Statecharts.* Oxford: Elsevier Newnes.

Lano, K. 1998. Logical specification of reactive and real-time systems. *Journal of Logic and Computation*, 8(5): 679–711.

Lavazza, L., G. Quaroni, and M. Venturelli. 2001. Combining UML and formal notations for modelling real-time systems. *ACM SIGSOFT Software Engineering Notes* 26(5): 196–206.

Kong, J., K. Zhang, J. Dong, and D. Xu. 2009. Specifying behavioral semantics of UML diagrams through graph transformations. *The Journal of Systems and Software*, 82: 292–306.

Engels, G., J. H. Hausmann, R. Heckel, and S. Sauer. 2000. Dynamic meta modeling: a graphical approach to the operational semantics of behavioral diagrams in UML. In *Proceedings of the UML 2000*, LNCS 1939, 323–337, Berlin/Heidelberg, Germany: Springer-Verlag.

Kuske, S. 2001. A formal semantics of UML state machines based on structured graph transformation, In *Proceedings of the 4th International Conference on The Unified Modeling Language, Modeling Languages, Concepts, and Tools*, 241–256, London: Springer-Verlag.

Kuske, S., M. Gogolla, R. Kollmann, and H.-J. Kreowski. 2002. An integrated semantics for UML class, object and state diagrams based on graph transformation. In *Proceedings of the 3rd International Conference on Integrated Formal Methods*, LNCS 2335, 11–28, London: Springer-Verlag.

Gogolla, M. and F. P. Presicce. 1998. State diagrams in UML: A formal semantics using graph transformations. In *Proceedings of the ICSE98 Workshop Precise Semantics of Modeling Techniques*. Technical report, TUM-I9803, 55–72, University of Munich.

Gogolla, M., P. Ziemann, and S. Kuske. 2009. Towards an integrated graph based semantics for UML. *Software and System Modeling* 8(3):403–422.

Ermel, C., K. Hülscher, S. Kuske, and P. Ziemann. 2005. Animated simulation of integrated UML behavioral models based on graph transformation. In *Proceedings of the IEEE Symposium on Visual Languages and Human-Centric Computing*, 125–133, Washington, DC: IEEE Computer Society.

Höscher, K., P. Ziemann, and M. Gogolla. 2006. On translating UML models into graph transformation systems. *Journal of Visual Languages and Computing*, 17(1): 78–105.

Ziemann, P., K. Höscher, and M. Gogolla. 2004. From UML models to graph transformation systems. In *Proceedings of the Workshop on Visual Languages and Formal Methods, Electronic Notes in Theoretical Computer Science (ENTCS)*, 127(4): 17–33.

Ziemann, P., K. Höscher, and M. Gogolla. 2004. Coherently explaining UML statechart and collaboration diagrams by graph transformations. In *Proceedings of the Brazilian Symposium on Formal Methods, Electronic Notes in Theoretical Computer Science (ENTCS)*, Vol. 130, 263–280.

Maggiolo-Schettini, A. and A. Peron. 1994. Semantics of full statecharts based on graph rewriting. In *Proceeding of the Graph Transformation in Computer Science*, LNCS 776, 265–279, Berlin/Heidelberg, Germany: Springer-Verlag.

Maggiolo-Schettini, A. and A. Peron. 1996. A graph rewriting framework for statecharts semantics. In *Proceedings of the 5th International Workshop on Graph Grammars and their Application to Computer Science*, LNCS 1073, 107–121, Berlin/Heidelberg, Germany: Springer-Verlag.

Varró, D. 2002. A formal semantics of UML statecharts by model transition systems, In *Proceeding of the 1st International Conference on Graph Transformation (ICGT 2002)*, LNCS 2505, 378–392, London: Springer-Verlag.

Jin, Y., R. Esser, and J. W. Janneck. 2004. A method for describing the syntax and semantics of UML statecharts. *Journal of Software and System Modeling*, 3: 150–163.

Taleghani, A. and J. M. Atlee. 2006. Semantic variations among UML stateMachines. *Model Driven Engineering Languages and Systems*, LNCS 4199, 245–259, Berlin/Heidelberg, Germany: Springer-Verlag.

Niu, J., J. M. Atlee, and N. Day. 2003. Template semantics for model-based notations. *IEEE Transactions on Software Engineering*, 29(10): 866–882.

Latella, D., I. Majzik, and M. Massink. 1999. Towards a formal operational semantics of UML statechart diagrams. In *Proceedings of the 3rd International Conference on Formal Methods for Open Object-Oriented Distributed Systems*, 331–347, Deventer: Kluwer.

Geiger, L. and A. Zündorf. 2005. Statechart modeling with Fujaba. In *Proceedings of the International Workshop on Graph-Based Tools (GraBaTs'04)*, Electronic Notes in Theoretical Computer Science, 127(1): 37–49.

Harel, D. and A. Naamad. 1996. The STATEMATE semantics of statecharts. *ACM Transactions on Software Engineering and Methodology*, 5(4): 293–333.

Harel, D., H. Lachover, A. Naamad et al. 1990. STATEMATE: A working environment for the development of complex reactive systems. *IEEE Transactions on Software Engineering*, 16(4): 403–414.

Harel, D. and H. Kugler. 2001. The rhapsody semantics of statecharts (or, on the executable core of the UML) In *Integration of Software Specification Techniques for Application in Engineering*, LNCS 3147, 325–354, Berlin/Heidelberg, Germany: Springer-Verlag.

Drusinsky, D. and D. Harel. 1994. On the power of bounded concurrency I: Finite automata. *Journal of the ACM*, 41: 517–539.

Hirst, T. and D. Harel. 1994. On the power of bounded concurrency II: pushdown automata. *Journal of the ACM*, 41: 540–554.

Wegner, P. 1997. Why interaction is more powerful than algorithms. *Communications of the ACM*, 40(5):80–91.

Wegner, P. 1998. Interactive foundations of computing. *Theoretical Computer Science* 192(2):315–351.

Goldin, D. and P. Wegner. 2008. The interactive nature of computing: Refuting the strong Church-Turing thesis. *Minds and Machines*, 18(1):17–38.

Yu, S. 2001. The time dimension of computation models. *Where Mathematics, Computer Science, Linguistics and Biology Meet*, Chapter 14, 161–172, Norwell: Kluwer Academic Publishers.

Lu, H. and S. Yu. 2009. Are statecharts finite automata. In *Proceedings of the 14th International Conference on Implementation and Application of Automata*, LNCS 5642, 258–261, Berlin/Heidelberg, Germany: Springer-Verlag.

Plotkin, G. D. 1981. *A Structural Approach to Operational Semantics*, Report DAIMI-FN-19, Computer Science Dept, Århus University, Denmark.

17 Model Checking

Zhenhua Duan and Cong Tian
Xidian University

17.1 Principle of Model Checking

The term model checking was coined by Clarke and Emerson [1], and Sifakis and Queille independently [2]. Model checking is a general approach and is applied in areas like hardware and software verifications. Due to its success in several projects and the high degree of support by tools, there is an increasing interest in industry in model checking—various companies have started building their own research groups and sometimes have developed their own in-house model checkers.

The basic idea of model checking is to use algorithms, executed by computer tools, to verify the correctness of systems. As illustrated in Figure 17.1, the user inputs a description of a finite model of the system (the possible behavior) specified by a Kripke structure, automaton or Labeled Transition Systems (LTS), and so on, and a description of the requirements specification (the desirable behavior) defined by a formalism such as temporal logic formula and leaves the verification up to the machine. If an error is recognized, the tool provides a counterexample showing under which circumstances the error can be generated. The counterexample consists of a scenario in which the model behaves in an undesired way.

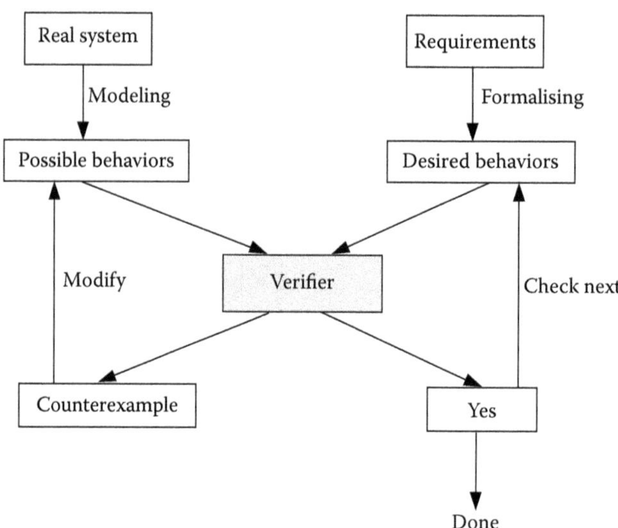

FIGURE 17.1 Principle of model checking.

Thus, the counterexample provides evidence that the model is faulty and needs to be revised. This allows the user to locate the error and to repair the model specification before continuing. If no errors are found, the user can refine its model description (e.g., by taking more design decisions into account, so that the model becomes more concrete/realistic) and can restart the verification process.

17.2 Temporal Logics

Logics extended by operators that allow the expression of properties about executions, in particular those that can express properties about the relative order between events, are called temporal logics (TL). Such logics are rather well established and have their origins in other fields decades ago. The introduction of these logics into computer science is due to Pnueli [3]. Temporal logic is a well-accepted and commonly used specification technique for expressing properties of computations of reactive systems at a rather high level of abstraction.

There are number of temporal logics which can be divided into two categories, linear-time and branching-time logics. Linear-time logics are concerned with properties of paths while branching-time logics describe properties depending on the branching of computational tree structures.

17.2.1 Linear Temporal Logic

Given a nonempty finite set AP of atomic propositions, the set of all Propositional Linear TL (PLTL) formulas over AP is the set of all formulas built from elements from AP using negation ($\neg$), disjunction ($\vee$), next operator ($\bigcirc$), and the until operator ($\cup$). The syntax is given below:

$$\phi ::= P \mid \neg\phi \mid \varphi \vee \psi \mid \bigcirc\varphi \mid \varphi \cup \phi$$

PLTL formulas are interpreted in linear-time structures. A linear-time structure over AP is an infinite sequence $x = x(0), x(1), \ldots$ where each $x(i)$ is a valuation $\mathrm{AP} \rightarrow \{true, false\}$. Whether a formula ϕ holds in a linear-time structure x at a position i, denoted $x, i \models \phi$ is defined according to the following rules:

- $x, i \models P$ if $x(i)(P) = true$, for $P \in \mathrm{AP}$
- $x, i \models \neg\phi$ if $x, i \not\models \phi$

- $x, i \models \varphi \vee \psi$ if $x, i \models \varphi$ or $x, i \models \psi$
- $x, i \models \bigcirc\varphi$ if $x, i+1 \models \varphi$
- $x, i \models \varphi \cup \psi$ if there exists $j \geq i$ such that $x, i' \models \varphi$ for $i' \in \{i, i+1, \ldots, j-1\}$ and $x, j \models \psi$

The abbreviations *true*, *false*, $\wedge$, $\rightarrow$, and $\leftrightarrow$ are defined as usual. In particular, $true \overset{\text{def}}{=} \phi \vee \neg\phi$ and $false \overset{\text{def}}{=} \phi \wedge \neg\phi$ for any formula ϕ. In addition, eventually ($\Diamond$) and always ($\Box$) temporal operators can be derived by $\Diamond\phi \overset{\text{def}}{=} true \cup \phi$ and $\Box\phi \overset{\text{def}}{=} \neg\Diamond\neg\phi$, respectively.

17.2.2 Computation Tree Logic

The formulas of Computation Tree Logic (CTL) are built over set AP of atomic propositions, defined by the following grammar:

$$\phi ::= P \mid \neg\phi \mid \phi \vee \phi \mid EX\phi \mid E(\phi \cup \varphi) \mid A(\phi \cup \varphi)$$

Let $p \in \text{AP}$, and $M = (S, R, L)$ be a Kripke structure, $s \in S$, and ϕ, φ be CTL formulas. The set of paths starting in state s of model M is defined by

$$P_M(s) = \{\sigma \in S^{\omega} \mid \sigma[0] = s\}$$

The satisfaction relation $\models$ is defined by:

$$\begin{array}{lll}
s \models p & \texttt{iff} & p \in \text{L}(s) \\
s \models \neg\phi & \texttt{iff} & s \not\models \phi \\
s \models \phi \vee \varphi & \texttt{iff} & s \models \phi \text{ or } s \models \varphi \\
s \models EX\phi & \texttt{iff} & \exists\sigma \in P_M(s) \cdot \sigma[1] \models \phi \\
s \models E(\phi \cup \varphi) & \texttt{iff} & \exists\sigma \in P_M(s) \cdot P_M(s) \cdot (\exists j \geq 0 \cdot \sigma[j] \models \varphi \wedge (\forall 0 \leq k < j \cdot \sigma[k] \models \phi)) \\
s \models A(\phi \cup \varphi) & \texttt{iff} & \forall\sigma \in P_M(s) \cdot P_M(s) \cdot (\exists j \geq 0 \cdot \sigma[j] \models \varphi \wedge (\forall 0 \leq k < j \cdot \sigma[k] \models \phi))
\end{array}$$

Let $F\phi \overset{\text{def}}{=} true \cup \phi$ and $G\phi \overset{\text{def}}{=} \neg F \neg\phi$, we define the following abbreviations:

$$\begin{array}{lcl}
EF\phi & \overset{\text{def}}{=} & E(true \cup \phi) \\
AF\phi & \overset{\text{def}}{=} & A(true \cup \phi) \\
EG\phi & \overset{\text{def}}{=} & \neg AF\neg\phi \\
AG\phi & \overset{\text{def}}{=} & \neg EF\neg\phi \\
AX\phi & \overset{\text{def}}{=} & \neg EX\neg\phi
\end{array}$$

17.2.3 Propositional Projection Temporal Logic

Propositional Projection Temporal Logic (PPTL) [7,8] is a kind of linear temporal logics based on intervals rather than points. Let *Prop* be a countable set of atomic propositions. The formula P of PPTL is given by the following grammar:

$$P ::= p \mid \bigcirc P \mid \neg P \mid P_1 \vee P_2 \mid (P_1, \ldots, P_m)\ prj\ P$$

where $p \in Prop$, $P_1, \ldots, P_m$ and P are all well-formed PPTL formulas. $\bigcirc$ (next) and *prj* (projection) are basic temporal operators.

Following the definition of Kripke's structure [9], we define a state s over *Prop* to be a mapping from *Prop* to $B = \{true, false\}$, $s : Prop \longrightarrow B$. We will use $s[p]$ to denote the valuation of p at state s. An interval σ is a nonempty sequence of states, which can be finite or infinite. The length, $|\sigma|$, of σ is ω if σ is infinite, and the number of states minus 1 if σ is finite. To have a uniform notation for both finite and infinite intervals, we will use extended integers as indices. That is, we consider the set N_0 of nonnegative integers and ω, $N_\omega = N_0 \cup \{\omega\}$, and extend the comparison operators, $=, <, \leq$, to N_ω by considering $\omega = \omega$, and for all $i \in N_0$, $i < \omega$. Moreover, we define $\preceq$ as $\leq -\{(\omega, \omega)\}$. To simplify definitions, we will denote σ by $< s_0, \ldots, s_{|\sigma|} >$, where $s_{|\sigma|}$ is undefined if σ is infinite. With such a notation, $\sigma_{(i..j)}$ $(0 \leq i \preceq j \leq |\sigma|)$ denotes the subinterval $< s_i, \ldots, s_j >$, $\sigma^{(k)}$ $(0 \leq k \preceq |\sigma|)$ denotes $< s_k, \ldots, s_{|\sigma|} >$, i.e., kth suffix of σ, and σ^k $(0 \leq k \preceq |\sigma|)$ denotes $< s_0, \ldots, s_k >$, i.e., kth prefix of σ. Note that for an infinite interval σ, the ωth prefix σ^ω and suffix $\sigma^{(\omega)}$ are undefined and denoted by $\perp$. Further, the concatenation of a finite σ with another interval (or empty string) σ' denoted by $\sigma \cdot \sigma'$ is defined as follows.

$$\sigma_1 \cdot \sigma_2 = \begin{cases} \sigma_1 & \text{if } \sigma_2 = \epsilon \\ \sigma_2 & \text{if } \sigma_1 = \epsilon \\ < s_0, \ldots, s_i, s_{i+1}, \ldots > & \text{if } \sigma_1 = < s_0, \ldots, s_i > \text{ and } \sigma_2 = < s_{i+1}, \ldots > \text{ and } i \in N_0 \\ \perp & \text{otherwise} \end{cases}$$

Let $\sigma = < s_0, s_1, \ldots, s_{|\sigma|} >$ be an interval and $r_1, \ldots, r_h$ be integers $(h \geq 1)$ such that $0 \leq r_1 \leq r_2 \leq \ldots \leq r_h \preceq |\sigma|$. The projection of σ onto $r_1, \ldots, r_h$ is the interval (namely projected interval)

$$\sigma \downarrow (r_1, \ldots, r_h) = < s_{t_1}, s_{t_2}, \ldots, s_{t_l} >$$

where $t_1, \ldots, t_l$ is obtained from $r_1, \ldots, r_h$ by deleting all duplicates. That is, $t_1, \ldots, t_l$ is the longest strictly increasing subsequence of $r_1, \ldots, r_h$. For instance, $<s_0, s_1, s_2, s_3, s_4> \downarrow (0, 0, 2, 2, 2, 3) = <s_0, s_2, s_3>$. This is convenient to define an interval obtained by taking the endpoints (rendezvous points) of the intervals over which $P_1, \ldots, P_m$ are interpreted in the projection construct.

An interpretation is a tuple $\mathcal{I} = (\sigma, k, j)$, where σ is an interval, k is an integer, and j an integer or ω such that $k \preceq j \leq |\sigma|$. We use the notation $(\sigma, k, j) \models P$ to denote that formula P is interpreted and satisfied over the subinterval $< s_k, \ldots, s_j >$ of σ with the current state being s_k. The satisfaction relation $(\models)$ is inductively defined as follows:

`I-prop`	$\mathcal{I} \models p$ iff $s_k[p] = true$, for any given proposition p
`I-not`	$\mathcal{I} \models \neg P$ iff $\mathcal{I} \not\models P$
`I-or`	$\mathcal{I} \models P \vee Q$ iff $\mathcal{I} \models P$ or $\mathcal{I} \models Q$
`I-next`	$\mathcal{I} \models \bigcirc P$ iff $k < j$ and $(\sigma, k+1, j) \models P$
`I-prj`	$\mathcal{I} \models (P_1, \ldots, P_m)\ prj\ Q$ if there exist integers $k = r_0 \leq r_1 \leq \ldots \preceq r_m = j$ such that $(\sigma, r_{l-1}, r_l) \models P_l, 1 \leq l \leq m$, $(\sigma', 0, \lvert\sigma'\rvert) \models Q$ for one of the following σ': (a) $r_m < j$ and $\sigma' = \sigma \downarrow (r_0, \ldots, r_m) \cdot \sigma_{(r_m+1..j)}$ or (b) $r_m = j$ and $\sigma' = \sigma \downarrow (r_0, \ldots, r_h)$ for some $0 \leq h \leq m$

The abbreviations *true*, *false*, $\wedge$, $\rightarrow$, and $\leftrightarrow$ are defined as usual. In particular, $true \stackrel{\text{def}}{=} P \vee \neg P$ and $false \stackrel{\text{def}}{=} P \wedge \neg P$. Also we have the following derived formulas:

EMPTY	$empty$	$\stackrel{\text{def}}{=}$	$\neg \bigcirc true$	MORE	$more$	$\stackrel{\text{def}}{=}$	$\neg empty$
0-NEXT	$\bigcirc^0 P$	$\stackrel{\text{def}}{=}$	P	N-NEXT	$\bigcirc^n P$	$\stackrel{\text{def}}{=}$	$\bigcirc(\bigcirc^{n-1} P)$
0-LEN	$len(0)$	$\stackrel{\text{def}}{=}$	$empty$	N-LEN	$len(n)$	$\stackrel{\text{def}}{=}$	$\bigcirc^n empty$
SKIP	$skip$	$\stackrel{\text{def}}{=}$	$len(1)$	W-NEXT	$\odot P$	$\stackrel{\text{def}}{=}$	$empty \vee \bigcirc P$
CHOP	$P; Q$	$\stackrel{\text{def}}{=}$	$(P, Q)\ prj\ empty$	SOM	$\Diamond P$	$\stackrel{\text{def}}{=}$	$true; P$
ALWAYS	$\Box P$	$\stackrel{\text{def}}{=}$	$\neg \Diamond \neg P$	KEEP	$keep(P)$	$\stackrel{\text{def}}{=}$	$\Box(\neg empty \rightarrow P)$
HALT	$halt(P)$	$\stackrel{\text{def}}{=}$	$\Box(empty \leftrightarrow P)$	FIN	$fin(P)$	$\stackrel{\text{def}}{=}$	$\Box(empty \rightarrow P)$

where $n \geq 1$, $\odot$ (weak next), $\Box$ (always), $\Diamond$ (sometimes), and ; (chop); *empty* denotes an interval with zero length, and *more* means the current state is not the final one over an interval; *halt*(P) is true over

TABLE 17.1 Precedence Rules

Precedence	Operators	Precedence	Operators
1	$\neg$	2	$\bigcirc$, $\odot$, $\Diamond$, $\Box$
3	$\wedge$, $\vee$	4	$\rightarrow$, $\leftrightarrow$
5	*prj*, ;		

TABLE 17.2 Logic Laws for PPTL

Law1	$\Box(P \wedge Q) \equiv \Box P \wedge \Box Q$
Law2	$\Diamond(P \vee Q) \equiv \Diamond P \vee \Diamond Q$
Law3	$\bigcirc(P \vee Q) \equiv \bigcirc P \vee \bigcirc Q$
Law4	$\bigcirc(P \wedge Q) \equiv \bigcirc P \wedge \bigcirc Q$
Law5	$R; (P \vee Q) \equiv (R; P) \vee (R; Q)$
Law6	$(P \vee Q); R \equiv (P; R) \vee (Q; R)$
Law7	$\Diamond P \equiv P \vee \bigcirc \Diamond P$
Law8	$\Box P \equiv P \wedge \odot \Box P$
Law9	$more \wedge \neg \bigcirc P \equiv more \wedge \bigcirc \neg P$
Law10	$\neg \odot P \equiv \bigcirc \neg P$
Law11	$\bigcirc P; Q \equiv \bigcirc(P; Q)$
Law12	$w \wedge (P; Q) \equiv (w \wedge P); Q$
Law13	$Q\ prj\ empty \equiv Q$
Law14	$empty\ prj\ Q \equiv Q$
Law15	$(P_1, \ldots P_m)\ prj\ empty \equiv P_1; \ldots; P_m$
Law16	$(P, empty)\ prj\ Q \equiv (P \wedge \Diamond empty)\ prj\ Q$
Law17	$(P_1, \ldots, w \wedge empty, P_{t+1}, \ldots, P_m)\ prj\ Q \equiv (P_1, \ldots, w \wedge P_{t+1}, \ldots, P_m) prj\ Q$
Law18	$(P_1, \ldots, (P_i \vee P_i'), \ldots, P_m)\ prj\ Q \equiv ((P_1, \ldots, P_i, \ldots, P_m)\ prj\ Q) \vee$ $((P_1, \ldots, P_i', \ldots, P_m)\ prj\ Q)$
Law19	$(P_1, \ldots, P_m)\ prj\ (P \vee Q) \equiv ((P_1, \ldots, P_m)\ prj\ Q) \vee ((P_1, \ldots, P_m)\ prj\ Q)$
Law20	$(\bigcirc P_1, \ldots, P_m)\ prj\ \bigcirc Q \equiv \bigcirc(P_1; (P_2, \ldots, P_m)\ prj\ Q)$
Law21	$(w \wedge P_1, \ldots, P_m)\ prj\ q \equiv w \wedge ((P_1, \ldots, P_m)\ prj\ Q)$
Law22	$(P_1, \ldots, P_m)\ prj\ (w \wedge Q) \equiv w \wedge ((P_1, \ldots, P_m)\ prj\ Q)$

an interval if and only if P is true at the final state, $fin(P)$ is true as long as P is true at the final state and $keep(P)$ is true if P is true at every state ignoring the final one.

A formula P is satisfied by an interval σ, denoted by $\sigma \models P$, if $(\sigma, 0, |\sigma|) \models P$. A formula P is called satisfiable if $\sigma \models P$ for some σ. A formula P is valid, denoted by $\models P$, if $\sigma \models P$ for all σ.

In order to avoid an excessive number of parentheses, the following precedence rules are used as shown in Table 17.1, where 1 = highest and 5 = lowest.

Table 17.2 shows some useful logic laws. Note that w is a state formula. The proofs of the logic laws can be found in Refs. [7,8].

17.2.4 Interval Temporal Logic

Propositional Interval Temporal Logic (PITL) is also a kind of interval based temporal logics. It was originally defined over finite intervals [4]. After the introduction of PPTL interpreted over finite and infinite intervals [6] in 1994, PITL was further extended to infinite models [5] in 1995.

The formulas of PITL are given by the following grammar:

$$P ::= p \mid \bigcirc P \mid \neg P \mid P_1 \vee P_2 \mid P; Q \mid P\ proj\ Q$$

where $p \in Prop$, the set of atomic propositions, P_1, P_2, and Q are all well-formed PITL formulas. The next ($\bigcirc$), chop (;), and projection (*proj*) are basic temporal operators.

The semantics of PITL can be defined in the similar way as PPTL. The satisfaction relation ($\models$) is inductively defined as follows:

`I-prop` $\mathcal{I} \models p$ iff $s_k[p] = true$, for any given proposition p
`I-not` $\mathcal{I} \models \neg P$ iff $\mathcal{I} \not\models P$
`I-or` $\mathcal{I} \models P \vee Q$ iff $\mathcal{I} \models P$ or $\mathcal{I} \models Q$
`I-next` $\mathcal{I} \models \bigcirc P$ iff $k < j$ and $(\sigma, k+1, j) \models P$
`I-chop` $\mathcal{I} \models P;Q$ iff there exists r such that $k \leq r \leq j$ and $(\sigma, k, r) \models P$ and $(\sigma, r, j) \models Q$
`I-proj` $\mathcal{I} \models P\ proj\ Q$ if there exist finitely many integers $k = r_0 < r_1 < \ldots < r_m = j$, such that $(\sigma, r_{l-1}, r_l) \models P, 1 \leq l \leq m$, and $(\sigma', 0, |\sigma'|) \models Q$, where $\sigma' = \sigma \downarrow (r_0, \ldots, r_m)$

Here, $\sigma \downarrow (r_0, \ldots, r_m) =< s_{r_0}, \ldots, s_{r_m} >$. Note that P^* can be derived from projection construct by $P^* \stackrel{\text{def}}{=} P\ proj\ true$.

With the infinite models, $P\ proj\ Q$ cannot be defined. Therefore, formulas of PITL with infinite models can be given by the following grammar:

$$P ::= p \mid \bigcirc P \mid \neg P \mid P_1 \vee P_2 \mid P;Q \mid P^*$$

where $p \in Prop$, P_1, P_2 and Q are all well-formed PITL formulas. next ($\bigcirc$), chop (;) and chop star ($*$) are basic temporal operators.

For atomic proposition, negation, disjunction, next and chop constructs, the satisfaction relation is defined the same as in finite case. In addition, the satisfaction relation for chop star is defined as follows:

`I-chop star` $\mathcal{I} \models P^*$ iff there are integers $k = r_0 \leq r_1 \leq \ldots \leq r_{n-1} \preceq r_n = j$ such that $(\sigma, r_{l-1}, r_1) \models P$, for all $1 \leq l \leq n$

A formula P is satisfied by an interval σ, denoted by $\sigma \models P$, if $(\sigma, 0, |\sigma|) \models P$. A formula P is called satisfiable if $\sigma \models P$ for some σ. A formula P is valid, denoted by $\models P$, if $\sigma \models P$ for all σ.

17.2.5 Expressiveness

Figure 17.2 depicts the relationship among CTL, PLTL, and PPTL (with star). It follows that PPTL is more expressive than both PLTL and CTL, whereas PLTL and CTL are incomparable.

In PLTL, but not in CTL, for example, $A(F(p \wedge Xp))$ is a PLTL formula for which there does not exist an equivalent formula in CTL. In CTL, but not in PLTL, the formula $AGEFp$ is a CTL formula for which there does not exist an equivalent formula in PLTL. Factually, PLTL has the same expressive power as

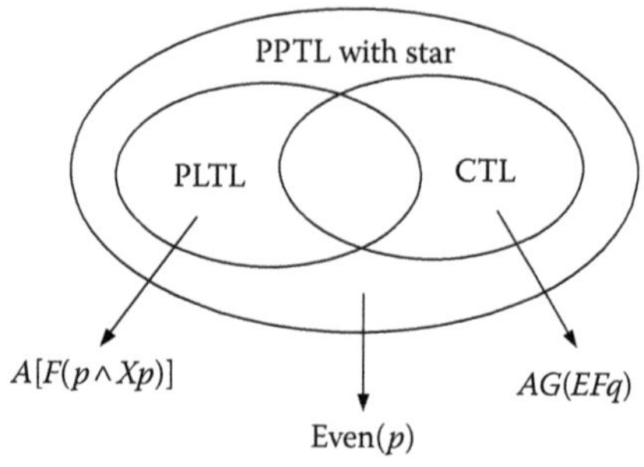

FIGURE 17.2 Relationship between CTL*, CTL, and PLTL.

star-free regular expression. For CTL, the exact expressiveness still remains open, even though it has been proved to be less expressive than full regular expressions. For example, *even*(p), that is, p must hold at every even state without considering other states, cannot be expressed with neither PLTL nor CTL. However, PPTL has the same expressiveness with full regular expressions. That is PPTL is powerful than both PLTL and CTL.

17.3 Basic Model Checking Scheme for PLTL

Thanks to Wolper et al. [10], any PLTL formula can be equivalently transformed to a Büchi automaton, the basic scheme for model checking PLTL formulas is obtained. First, Büchi automata A_ϕ and A_{sys} are constructed for defining the desired property ϕ and specifying the model *sys* of the system, respectively. The accepting runs of A_ϕ correspond to the desired behavior of the model. If all possible behavior is desirable, that is, when $L_\omega(A_{sys}) \subseteq L_\omega(A_\phi)$, then we can conclude that the model satisfies ϕ. This seems to be a plausible approach. However, the problem to decide language inclusion of Büchi automata is PSPACE-complete.

Observe that

$$L_\omega(A_{sys}) \subseteq L_\omega(A_\phi) \Leftrightarrow (L_\omega(A_{sys} \cap L_\omega(\bar{A_\phi})) = \emptyset)$$

where $\bar{A}$ is the complement of A that accept $\Sigma^\omega \setminus L_\omega(A)$ as a language. The construction of $\bar{A}$ for the Büchi automaton A, is however, quadratically exponential: if A has n states then $\bar{A}$ has c^{n^2} states, for some constant $c > 1$. This means that the resulting automaton is very large. Using the observation that the complement automaton of A_ϕ is equal to the automaton for the negation of ϕ:

$$L_\omega(\bar{A_\phi}) = L_\omega(A_{\neg\phi})$$

we arrive at the following more efficient method for model checking PLTL which is usually more efficient. This is shown in Figure 17.3. The idea is to construct a Büchi automaton for the negation of the desired property ϕ. Thus, the automaton $A_{\neg\phi}$ models the undesired computations of the model that we want to check. If A_{sys} has a certain accepting run that is also an accepting run of $A_{\neg\phi}$ then this is an example run that violates ϕ, so we conclude that ϕ is not a property of A_{sys}. If there is no such common run, then ϕ is satisfied. Emerson and Lei [11] have proven that the third step is decidable in a linear time, and thus falls outside the class of NPC-problems.

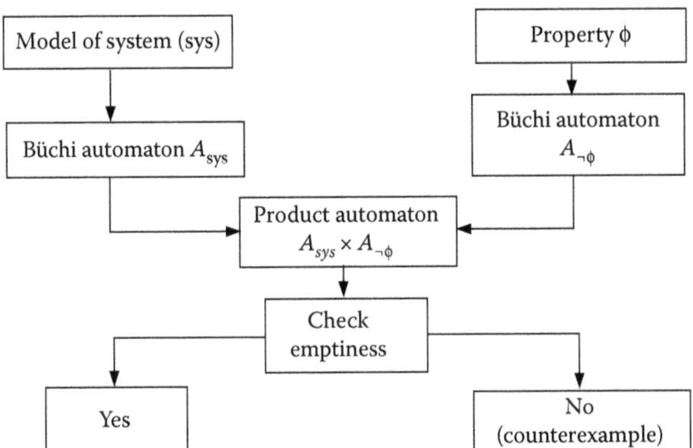

FIGURE 17.3 Model checking PLTL.

17.4 Model Checking Scheme for CTL

We use transition system $T = \{S, R, L, I\}$ to denote the model of the system to be verified, where S is the set of states, R the transition relation, L the labeling function, and I the set of initial states. $[\varphi]_T$ means the set of states s of T such that $T, s \models \varphi$. The model checking problem can then be rephrased as deciding whether or not $I \subseteq [\varphi]_T = \emptyset$. The satisfaction sets $[\varphi]_T$ can be computed by induction on the structure of ϕ, as follows:

$$
\begin{aligned}
[v]_T &= \{s : v \in s(V)\} \quad (\text{for } v \in V) \\
[\neg\varphi]_T &= S \setminus [\varphi]_T \\
[\varphi_1 \vee \varphi_2]_T &= [\varphi_1]_T \cup [\varphi_2]_T \\
[EX\varphi]_T &= \delta^{-1}([\varphi]_T) = \{s : t \in [\varphi]_T \text{ for some } A, t \text{ s.t. } (s, A, t) \in \delta\} \\
[EG\varphi]_T &= \mathit{gfp}(\lambda X \cdot [\varphi]_T \cap \delta^{-1}(X)) \\
[\varphi_1 EU \varphi_2]_T &= \mathit{lfp}(\lambda X \cdot [\varphi_2]_T \cup ([\varphi_1]_T \cap \delta^{-1}(X)))
\end{aligned}
$$

where $\mathit{lfp}(f)$ and $\mathit{gfp}(f)$, for a function $f : 2^S \to 2^S$, denote the least and greatest fixed points of f. The clauses for the EG and EU connectives are justified from the recursive characterizations:

$$
\begin{aligned}
EG\varphi &\equiv \varphi \wedge EXEG\varphi \\
\varphi_1 EU \varphi_2 &\equiv \varphi_2 \vee (\varphi_1 \wedge (EX(\varphi_1 EU \varphi_2))
\end{aligned}
$$

17.5 Model Checking Scheme for PITL and PPTL

Model checking PITL and PPTL is mainly based on the transformation from logic formulas to Büchi automata where normal form and Normal Form Graph (NFG) are important intermediates.

17.5.1 Normal Form of PPTL Formulas

Normal form is an important notation for constructing NFGs of PPTL formulas. In the following, we briefly give its definition and relevant concepts. The detailed explanation can be found in Ref. [24].

Definition 17.1: (Normal Form)

Let Q_p be the set of atomic propositions appearing in a PPTL formula Q. The normal form of Q can be defined by,

$$Q \equiv \bigvee_{j=0}^{n_0} (Q_{ej} \wedge \mathit{empty}) \vee \bigvee_{i=0}^{n_1} (Q_{ci} \wedge \bigcirc Q'_i)$$

where $Q_{ej} \equiv \bigwedge_{k=1}^{m_0} \dot{q}_{jk}$, $Q_{ci} \equiv \bigwedge_{h=1}^{m} \dot{q}_{ih}$, $q_{jk}, q_{ih} \in Q_p$, for any $r \in Q_p$, $\dot{r}$ denotes r or $\neg r$; Q'_i is a PPTL formula without "$\vee$" being the main operator. □

According to the definition, in a normal form, $p \wedge q \wedge \bigcirc(\Box p \vee q)$ must be written as $p \wedge q \wedge \bigcirc(\Box p) \vee p \wedge q \wedge \bigcirc q$ since "$\vee$" is the main operator of $\Box p \vee q$. Implicitly, Q'_i is also not permitted to be the form of $\neg \bigwedge_{k=1}^{n \geq 2} P_k$. Further, for convenience, we call $\bigvee_{j=0}^{n_0}(Q_{ej} \wedge \mathit{empty})$ the terminating part whereas $\bigvee_{i=0}^{n_1}(Q_{ci} \wedge \bigcirc Q'_i)$ the non-terminating (or future) part of the normal form. In some circumstances,

$$Q \equiv \underbrace{\bigvee_{j=0}^{n_0}(Q_{ej} \wedge empty)}_{\text{Terminating part}} \vee \underbrace{\bigvee_{j=0}^{n}(Q_{ci} \wedge \bigcirc Q_i')}_{\text{Future part}}$$

FIGURE 17.4 Normal form of PPTL formulas.

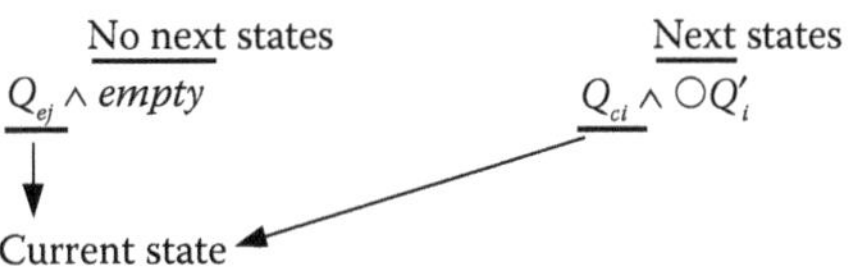

FIGURE 17.5 Basic components of a normal form.

for convenience, we write $Q_e \wedge empty$ instead of $\bigvee_{j=0}^{n_0}(Q_{ej} \wedge empty)$ and $\bigvee_{i=0}^{r}(Q_i \wedge \bigcirc Q_i')$ instead of $\bigvee_{i=0}^{n_1}(Q_{ci} \wedge \bigcirc Q_i')$. Thus,

$$Q \equiv (Q_e \wedge empty) \vee \bigvee_{i=0}^{r}(Q_i \wedge \bigcirc Q_i')$$

where Q_e and Q_i are state formulas.

The normal form of a formula divides the formula into two parts: the terminating part and the future part as shown in Figure 17.4. The terminating part is a terminating formula, since $\bigvee_{j=0}^{n_0}(Q_{ej} \wedge empty) \equiv (\bigvee_{j=0}^{n_0}(Q_{ej} \wedge empty)) \wedge empty$ while the future part is a non-local formula because of $\bigvee_{i=0}^{n}(Q_{ci} \wedge \bigcirc Q_i') \equiv \bigvee_{i=0}^{n}(Q_{ci} \wedge \bigcirc Q_i') \wedge more$. In addition, in the normal form, $Q_{ej} \wedge empty$ or $Q_{ci} \wedge \bigcirc Q_i'$, is called a basic component of the normal form. As shown in Figure 17.5, for each basic component in the terminating part, $Q_{ej} \equiv \bigwedge_{k=1}^{m_0} \dot{q}_{jk}$ means that each proposition or its negation in $\{\dot{q}_{j1}, \ldots, \dot{q}_{jk}\}$ holds at the current state while *empty* means that the current interval ending point is reached; for each basic component in the future part, similarly, $Q_{ci} \equiv \bigwedge_{h=1}^{m} \dot{q}_{ih}$ indicates that each proposition or its negation in $\{\dot{q}_{i1}, \ldots, \dot{q}_{ih}\}$ is true at the current state and $\bigcirc Q_i'$ means that Q_i' holds over the interval starting from the next state.

Definition 17.2: (Complete Normal Form)

Let Q_p be the set of atomic propositions appearing in a PPTL formula Q. The complete normal form of Q is defined by

$$Q \equiv \bigvee_{j=0}^{n_0}(Q_{ej} \wedge empty) \vee \bigvee_{i=0}^{n_1}(Q_{ci} \wedge \bigcirc Q_i')$$

where like in normal form, $Q_{ej} \equiv \bigwedge_{k=1}^{m_0} \dot{q}_{jk}$, $Q_{ci} \equiv \bigwedge_{h=1}^{m} \dot{q}_{ih}$, $q_{jk}, q_{ih} \in Q_p$, for any $r \in Q_p$, $\dot{r}$ denotes r or $\neg r$; further $\bigvee_i Q_{ci} \equiv true$ and $\bigvee_{i \neq j}(Q_{ci} \wedge Q_{cj}) \equiv false$; Q_i' is an arbitrary PPTL formula. □

Note that a complete normal form may be not a normal form, since "$\vee$" is possibly the main operator of Q_i'. It has been proved that any PPTL formula in normal form can be further transformed into its complete normal form [24] which plays an important role for transforming a formula in the negation form into its normal form.

We now give the algorithm Nf for transforming a PPTL formula into its normal form. Given a formula R as input, we first do some preprocesses by Pre(R) to eliminate all implications, equivalences, double negations, *skip*, *more*, *len*(n), and $\Diamond$ in R by replacing all subformulas of the form $P \rightarrow Q$ by $\neg P \vee Q$, $P \leftrightarrow Q$ by $\neg P \wedge \neg Q \vee P \wedge Q$, $\neg\neg P$ by P, *skip* by $\bigcirc empty$, *more* by $\bigcirc true$, *len n* by $\bigcirc^n empty$, and $\Diamond P$

TABLE 17.3 Algorithm for Translating a PPTL Formula to Its Normal Form

```
Function NF(R)
/* precondition: R is a PPTL formula and in PRE*/
/* postcondition: NF(R) computes an equivalent normal form for formula R*/
begin function
    case
        R is true: return empty ∨ ○true;
        R is false: return false;
        R is a state formula: return NFDNF(DNF(R));
        R is μ ∧ ○P: return R;
        R is μ ∧ ○^n P: return μ ∧ ○(○^(n−1))P;
        R is μ ∧ empty: return R;
        R is P ∨ Q: return NF(P)∨NF(Q);
        R is (P_1, ..., P_m) prj Q: return PRJ(R);
        R is P*: return empty∨ CHOP((NF(P)∧more); P*);
        R is ¬P: return NEG(CONF(NF(P)));
        R is □P: return NF(P ∧ empty)∨ NF(P ∧ ○□P)
        R is P ∧ Q: return AND(NF(P),NF(Q))
        R is P; Q: return CHOP(R);
    end case
end function
```

by $true; P$. Algorithm DNF is used to translate a state formula into its disjunctive normal form. And algorithm NFDNF will transform a disjunction normal form by replacing all basic products $\bigwedge_{j=1}^{m} q_{ij}$ by $\bigwedge_{j=1}^{m} \dot{q}_{ij} \wedge empty \vee \bigwedge_{j=1}^{m} \dot{q}_{ij} \wedge \bigcirc true$. We say that a formula P is in PRE if P has been processed by PRE(P). In the following algorithms, w denotes a state formula and μ a basic product. We can now write pseudo codes for Algorithm NF as shown in Table 17.3.

In algorithm NF(R), the basic constructs such as atomic propositions, the next, disjunction, negation, and projection constructs need to be considered. However, to improve the efficiency, other constructs such as always, and chop are also taken into account. Intuitively, if R is *true*, *false* or a state formula, the transformation is straightforward; if R is $\mu \wedge \bigcirc P$ or $\mu \wedge empty$, R is already in its normal form; if R is $P \vee Q$ the algorithm only needs to transform P and Q into the normal form; if R is $(P_1, \ldots, P_m)\ prj\ Q$ (resp. $P; Q$) the algorithm calls function PRJ (resp. function CHOP) to transform it; in addition, $\square P \equiv P \wedge empty \vee P \wedge \bigcirc\square P$, so the algorithm needs to transform $P \wedge empty$, $P \wedge \bigcirc\square P$ into the normal forms; for $P \wedge Q$, the algorithm transform P and Q into the normal forms at first, then transforms $P \wedge Q$ into its normal form by Algorithm AND.

Algorithm PRJ and CHOP are straightforward. It considers all possible projection constructs (resp. chop constructs) and uses the corresponding logic laws to transform them as shown in Tables 17.4 and 17.5, respectively.

For formula $\neg P$, we first transform formula P into its normal form, then we convert it into its complete normal form by algorithm CONF. Accordingly, $\neg P$ can be transformed to its normal form. The ideas are shown in algorithm CONF and NEG respectively (see Tables 17.6 and 17.7).

In algorithm AND, we assume that P and Q are in the normal forms. So P (resp. Q) is in the form of $P_1 \vee P_2$, $P_e \wedge empty$, or $P_i \wedge \bigcirc P_i'$. If P is in the form of $P_1 \vee P_2$, then algorithm AND deals with $P_1 \wedge Q \vee P_2 \wedge Q$ and transforms $P_1 \wedge Q$ and $P_2 \wedge Q$ by calling itself. If $P \equiv P_e \wedge empty$, $Q \equiv Q_e \wedge empty$ or $P \equiv P_i \wedge \bigcirc P_i'$, $Q \equiv Q_i \wedge \bigcirc Q_i'$ then the transforming is straightforward. If $P \equiv P_e \wedge empty$, $Q \equiv Q_i \wedge \bigcirc Q_i'$ or $P \equiv P_i \wedge \bigcirc P_i'$, $Q \equiv Q_e \wedge empty$ then $P \wedge Q \equiv false$. The algorithm in pseudo code is illustrated in Table 17.8.

TABLE 17.4 Algorithm for Translating Projection Constructs to Normal Form

```
Function PRJ(R)
/* precondition: R is (P_1, ..., P_m) prj Q*/
/* postcondition: PRJ(R) computes an equivalent NF for formula R*/
begin function
    case
      R is empty prj empty: return empty;
      R is empty prj Q: return NF(Q);
      R is P prj empty: return NF(P);
      R is ○P prj ○Q: return ○(P; Q);
      R is (○P_1, ..., P_m) prj ○Q: return ○(P_1; ((P_2, ..., P_m) prj Q);
      R is (empty, P_2, ..., P_m) prj Q: return PRJ((P_2, ..., P_m) prj Q);
      R is (P_1, ..., P_m) prj empty: return (P_1; (P_2; (...; P_m)));
      R is (w ∧ P_1, ..., P_m) prj Q: return NF(w∧ PRJ((P_1, ..., P_m) prj Q));
      R is (P_1, ..., P_m) prj (w ∧ Q): return NF(w∧ PRJ((P_1, ..., P_m) prj Q));
      R is (P_1 ∨ P'_1, ..., P_m) prj Q: return PRJ((P_1, ..., P_m) prj Q)∨PRJ((P'_1, ..., P_m) prj Q);
      R is (P_1, ..., P_m) prj (Q ∨ Q'): return PRJ((P_1, ..., P_m) prj Q)∨PRJ((P_1, ..., P_m) prj Q');
      otherwise: return PRJ((NF(P_1), ..., P_m) prj NF(Q));
    end case
end function
```

TABLE 17.5 Algorithm for Translating Chop Constructs to Normal Form

```
Function CHOP(R)
/* precondition: R is P; Q*/
/* postcondition: CHOP(R) computes an equivalent NF for formula R*/
begin function
    case
      R is empty; Q: return NF(Q);
      R is ○P; Q: return ○(P; Q);
      R is w ∧ P; Q: return NF(w∧CHOP(P; Q));
      R is P_1 ∨ P_2; Q: return CHOP(P_1; Q)∨CHOP(P_2; Q);
      otherwise: return CHOP(NF(P); Q);
    end case
end function
```

Obviously, the normal form enables us to transform the formula into two parts: the present and future ones. The present part is a state formula while the future part is either *empty* or a next formula. This standard generic form inspires us to construct a graph for describing the models of a formula.

17.5.2 Normal Form Graph

For a PPTL formula P, NFG of P is a directed graph, $G = (CL(P), EL(P), V_0)$, where $CL(P)$ denotes the set of nodes, $EL(P)$ denotes the set of edges in the graph, and $V_0 \subseteq CL(P)$ means the set of root nodes. In $CL(P)$, each node is specified by a formula in PPTL, while in $EL(P)$, each edge is a directed arc, labeled with a state formula Q_e, from node Q to node R and identified by a triple, (Q, Q_e, R). Accordingly, for

TABLE 17.6 Algorithm for Transforming a Normal Form to its Complete Normal Form

```
Function CONF(R)
/* precondition: R is in normal form, m_i and M_i are defined as in Lemma 3.1*/
/* postcondition: CONF(R) computes an equivalent complete normal form for R*/
begin function
  case
    R is P ∧ ○P': return P ∧ ○P' ∨ ¬P ∧ ○false;
    R is P ∧ ○P' ∨ ⋁_{i=1}^n (m_j ∧ ○M'_j): return ⋁_{i=1}^n (P ∧ m_i ∧ ○(P' ∨ M'_i))∨
                                          ⋁_{i=1}^n (¬P ∧ m_i ∧ ○M'_i)
    R is P_e ∧ empty ∨ ⋁_{i=1}^n (P_i ∧ ○P'_i): return P_e ∧ empty∨ CONF(P_1 ∧ P'_1 ∨
                                          CONF(⋁_{i=2}^n (P_i ∧ ○P'_i)))
  end case
end function
```

TABLE 17.7 Algorithm for Transforming the Negation of a Complete Normal Form

```
Function NEG(R)
/* precondition: R ≡ R_e ∧ empty ∨ ⋁_{i=1}^n (m_i ∧ ○M'_i), is in complete normal form*/
/* postcondition: NEG(R) computes an equivalent complete normal form for ¬R*/
begin function
    return ¬R_e ∧ empty ∨ ⋁_{i=1}^n (m_i ∧ ○¬M'_i)
end function
```

TABLE 17.8 Algorithm for Transforming *and Construct* to the Normal Form

```
Function AND(P, Q)
/* precondition: P and Q are in normal form*/
/* postcondition: AND(P, Q) computes an equivalent NF for P ∧ Q*/
begin function
  case
    P is P_1 ∨ P_2: return AND(P_1, Q)∨AND(P_2, Q);
    Q is Q_1 ∨ Q_2: return AND(P, Q_1)∨AND(P, Q_2);
    P is P_e ∧ empty and Q is Q_e ∧ empty: return P_e ∧ Q_e ∧ empty;
    P is P_i ∧ ○P'_i and Q_j ∧ ○Q'_j: return P_i ∧ Q_j ∧ ○(P'_i ∧ Q'_j)
    P is P_e ∧ empty, Q is Q_j ∧ ○Q'_j or P is P_i ∧ ○P'_i, Q is Q_e ∧ empty: return false
  end case
end function
```

convenience sometimes, a node in an NFG or LNFG in the next section is called a formula. NFG of a PPTL formula is inductively defined in Definition 17.3.

Definition 17.3: (Normal Form Graph, NFG)

For a PPTL formula P, the set $CL(P)$ of nodes and $EL(P)$ of edges connecting nodes in $CL(P)$ are inductively defined as follows:

1. If P is not in the form of $\bigvee_i P_i$, $P \in CL(P)$, $V_0 = \{P\}$, otherwise, $P_i \in CL(P)$, $P_i \in V_0$ for each i

2. For all $Q \in CL(P) \setminus \{\varepsilon, false\}$, if Q is transformed into its normal form $\bigvee_{j=0}^{h}(Q_{ej} \wedge empty) \vee \bigvee_{i=0}^{k}(Q_{ci} \wedge \bigcirc Q_i')$, then $\varepsilon \in CL(P)$, $(Q, Q_{ej}, \varepsilon) \in EL(P)$ for each j, $1 \le j \le h$; $Q_i' \in CL(P)$, $(Q, Q_{ci}, Q_i') \in EL(P)$ for all i, $1 \le i \le k$

The NFG of formula P is the directed graph $G = (CL(P), EL(P), V_0)$. □

In an NFG, any root node in V_0 is denoted by a circle with an input edge originating from none nodes, ε node is marked by a small black dot, and each of other nodes by a single circle. Each edge is denoted by a directed arc connecting two nodes. A finite path is a finite alternating sequence of nodes and edges, $\pi =< n_0, e_0, n_1, e_1, \ldots, \varepsilon >$ from a root node to the ε node, while an infinite path is an infinite alternating sequence of nodes and edges, $\pi =< n_0, e_0, n_1, e_1 \ldots, n_i, e_i \ldots, n_j, e_j, n_i, e_i, \ldots, n_j, e_j, \ldots >$ departing from the root node with some nodes, for example, $n_i, \ldots, n_j$, occurring for infinitely many times. For convenience, we use $\texttt{Inf}(\pi)$ to denote the set of nodes which infinitely often occur in the infinite path π. In some circumstances, in a path of NFG of formula Q, a node n_i can be replaced by a formula $Q_i \in CL(Q)$ and an edge e_i can be replaced by a state formula $Q_{ie} \in EL(Q)$.

An algorithm for constructing NFGs of a formula is given in Table 17.9. The algorithm uses *mark*[] to indicate whether or not a node needs to be decomposed. If *mark*[P] = *0* (unmarked), P needs further to be decomposed, otherwise if *mark*[P] = *1* (marked), P has been decomposed or needs not to be

TABLE 17.9 Algorithm for Constructing NFG of a PPTL Formula

Function NFG(P)
/* precondition: P is a PPTL formula*/
/* postcondition: NFG(P) computes NFG of P, $G = (CL(P), EL(P), V_0,)$*/
begin function
  case
    P is $\bigvee_i P_i$: $CL(P) = \{P_i | P_i$ appears in $\bigvee_i P_i\}$; $Mark[P_i] = 0$ for each i;
        $V_0 = \{P_i | P_i$ appears in $\bigvee_i P_i\}$;
    P is not $\bigvee_i P_i$: $CL(P) = \{P\}$; $Mark[P] = 0$; $V_0 = \{P\}$;
  end case
  $EL(P) = \emptyset$; AddE = AddN = 0;
  while there exists $R \in CL(P) \setminus \{\varepsilon, false\}$, and *mark*[$R$] ==0
    mark[R] =1; /*marking R is decomposed*/
    Q =NF(R);
    case
      Q is $\bigvee_{j=1}^{h} Q_{ej} \wedge empty$: AddE=1; /*first part of NF needs added*/
      Q is $\bigvee_{i=1}^{k} Q_i \wedge \bigcirc Q_i'$: AddN=1; /*second part of NF needs added*/
      Q is $\bigvee_{j=1}^{h} Q_{ej} \wedge empty \vee \bigvee_{i=1}^{k} Q_i \wedge \bigcirc Q_i'$: AddE=AddN=1; /*both parts need added*/
    end case
    if AddE == 1 **then** /*add first part of NF*/
      $CL(P) = CL(P) \cup \{\varepsilon\}$; $EL(P) = EL(P) \cup \bigcup_{j=1}^{h}\{(R, Q_{ej}, \varepsilon)\}$; AddE=0;
    if AddN == 1 **then for** i = 1 **to** k, /*add second part of NF*/
      if $Q_i' \notin CL(P)$,$CL(P) = CL(P) \cup \bigcup_{i=1}^{k}\{Q_i'\}$;
        if Q_i' is not *false*, *mark*[Q_i']=0; **else** *mark*[Q_i']=1; $EL(P) = EL(P) \cup \bigcup_{i=1}^{k}\{(R, Q_i, Q_i')\}$; AddN=0;
  end while
  return G;
End function

TABLE 17.10 Algorithm for Simplifying an NFG

```
Function SIMPLIFY(G)
/* precondition: G = (CL(P), EL(P), V_0) is an NFG of PPTL formula P*/
/* postcondition: SIMPLIFY(G) computes an NFG of P, G' = (CL'(P), EL'(P),
V'_0), which contains no redundant nodes*/
begin function
   CL'(P) = CL(P); EL'(P) = EL(P); V'_0 = V_0;
   while ∃ R ∈ CL'(P) and R is not ε and has no edges departing from
     do (CL'(P) = CL'(P) \ {R}; EL'(P) = EL'(P) \ ⋃_i(R_i, R_e, R);)
                    /* ⋃_i(R_i, R_e, R) denotes the set of edges connecting to node R*/
     if R ∈ V'_0, V'_0 = V'_0 \ {R};
   end while
   return G';
end function
```

done. Algorithm NF is used to transform a formula into its normal form. Further, in the algorithm, two global boolean variables AddE and AddN are employed to indicate whether or not terminating part and non-terminating part of the normal form is encountered, respectively.

In the NFG of a PPTL formula P generated by algorithm NFG, the set $CL(P)$ of nodes and the set $EL(P)$ of edges are inductively produced by repeatedly transforming the unmarked nodes into their normal forms. So, one question we have to answer is whether or not the transforming process terminates. Fortunately, we can prove that, for any PPTL formula P, the number of nodes in $CL(P)$ is finite, that is, $|CL(P)| = k \in N_0$ [24]. This guarantees the termination of algorithm NFG.

Further, in an NFG constructed by algorithm NFG, some nodes might have no successors (e.g., $P \wedge \neg P$). These nodes are redundant and can be removed. Algorithm SIMPLIFY is useful for eliminating redundant nodes of an NFG as shown in Table 17.10.

In the following, an example is given to show how the NFG is constructed and simplified.

Example 17.1:

Constructing NFG of formula,

$$P \equiv \bigcirc^2(\Box \bigcirc p\,; q) \vee (p\,; \Box \bigcirc q)$$

Step 1. Build the NFG of formula P.

As depicted in Step 1 of Figure 17.6, initially, the root node is created and named with formula P. To get the successor nodes, formula $P \equiv \bigcirc^2(\Box \bigcirc p\,; q) \vee (p\,; \Box \bigcirc q)$ is transformed to its normal form $\bigcirc(\bigcirc(\Box \bigcirc p; q)) \vee p \wedge \bigcirc(q \wedge \Box \bigcirc q) \vee p \wedge \bigcirc(true; \Box \bigcirc q)$. Thus, nodes $\bigcirc(\Box \bigcirc p; q)$, $q \wedge \Box \bigcirc q$ and $true; \Box \bigcirc q$ are generated from the root and connected by directed arcs labeled with *true*, p and p, respectively (Step 2 of Figure 17.6). Subsequently, node $\Box \bigcirc p; q$ is created from node $\bigcirc(\Box \bigcirc p; q)$ as shown in Step 3 of Figure 17.6. Since the normal form of $q \wedge \Box \bigcirc q$ is $q \wedge \bigcirc(q \wedge \Box \bigcirc q)$, an edge labeled with q is emanated from node $q \wedge \Box \bigcirc q$ to itself (see Step 4 of Figure 17.6). Further, by transforming $true; \Box \bigcirc q$ to its normal form, $true; \Box \bigcirc q \equiv \bigcirc(q \wedge \Box \bigcirc q) \vee \bigcirc(true; \Box \bigcirc q)$, two edges both labeled by *true* are created from node $true; \Box \bigcirc q$ to node $q \wedge \Box \bigcirc q$ and node $true; \Box \bigcirc q$ itself, respectively (see Steps 5 and 6 of Figure 17.6). Then chop formula $\Box \bigcirc p; q$ is encountered. We transform it into normal form $\Box \bigcirc p; q \equiv \bigcirc(p \wedge \Box \bigcirc p; q)$. So node $(p \wedge \Box \bigcirc p; q)$ is created as shown in Step 7 of Figure 17.6.

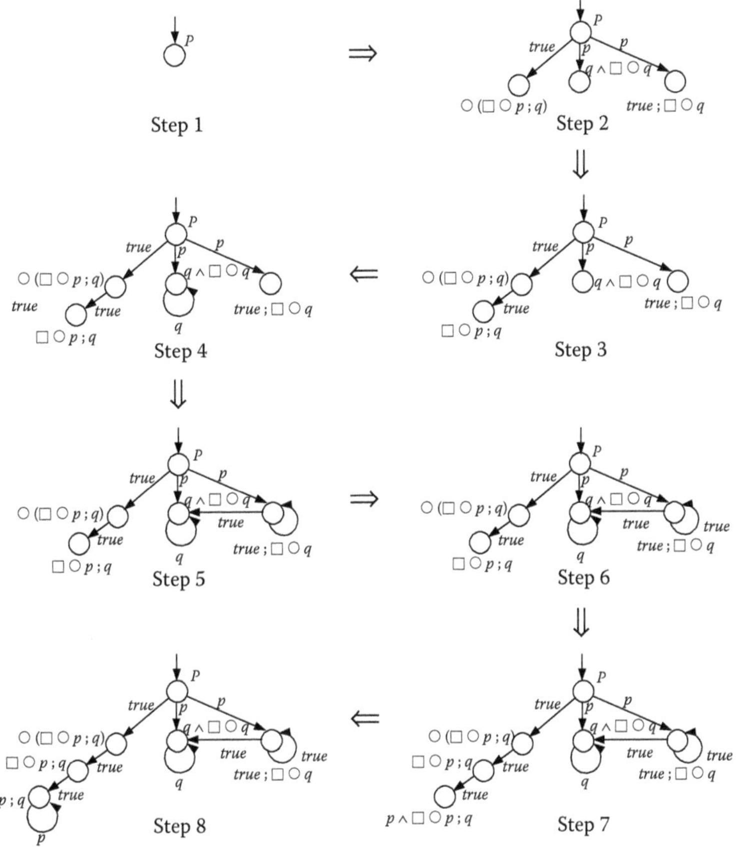

FIGURE 17.6 Constructing the NFG of P.

Finally, since $p \wedge \Box \bigcirc p; q \equiv p \wedge \bigcirc(p \wedge \Box \bigcirc p; q)$, a self loop with label p is added to node $p \wedge \Box \bigcirc p; q$ as illustrated in Step 8 of Figure 17.6.

Step 2. Obtain simplified NFG of formula P.

According to Algorithm Simplify, no nodes need to be removed. □

17.5.3 Satisfiability Based on NFGs

Intuitively, all models of a formula are implicitly contained in its NFG. For easily expressing the relationship between models of formula Q and paths of NFG G of Q, functions $P2M(\pi, G)$ and $M2P(\sigma, G)$ are formally defined below. Given a path $\pi = < Q, Q_{0e}, Q_1, Q_{1e}, \ldots >$ in G, an interval σ_π can be obtained by

$$\sigma_\pi = P2M(\pi, G) = \begin{cases} < s_0, s_1, \ldots, s_n >, \quad \text{for} 1 \leq i \leq n \\ \quad s_i[q] = \mathit{true} \quad \text{if } q \in Q_{ie}, \text{ and } s_i[q] = \mathit{false} \quad \text{if } \neg q \in Q_{ie}, \\ \quad \text{if } \pi =< Q, Q_{0e}, Q_1, Q_{1e}, \ldots, Q_{ne}, \varepsilon > \text{ and } \pi \in G \\ < s_0, s_1, \ldots, (s_i, \ldots, s_j)^\omega >, \quad \text{for } 1 \leq k \leq j \\ \quad s_k[q] = \mathit{true} \quad \text{if } q \in Q_{ke}, \text{ and } s_k[q] = \mathit{false} \quad \text{if } \neg q \in Q_{ke}, \\ \quad \text{if } \pi =< Q, Q_{0e}, Q_1, Q_{1e}, \ldots, (Q_i, Q_{ie}, \ldots, Q_j, Q_{je})^\omega > \quad \text{and } \pi \in G \end{cases}$$

Correspondingly, for a model $\sigma =< s_0, s_1, \ldots >\models Q$, a path π_σ over the NFG G of Q can be obtained by

$$\pi_\sigma = M2P(\sigma, G) = \begin{cases} < Q, Q_{0e}, Q_1, Q_{1e}, \ldots, Q_{ne}, \varepsilon >, \quad \text{for } 1 \le i \le n \\ \quad Q_{ie} = \bigwedge_k \dot{q}_k, \dot{q}_k \equiv q_k, \quad \text{if } s_i[q_k] = true, \\ \quad \text{and } \dot{q}_k \equiv \neg q_k, \quad \text{if } s_i[q_k] = false, Q_i \in CL(Q) \\ \quad \text{if } \sigma =< s_0, s_1, \ldots, s_n > \\ < Q, Q_{0e}, Q_1, Q_{1e}, \ldots, (Q_i, Q_{ie}, \ldots, Q_j, Q_{je})^\omega >, \quad \text{for } 1 \le h \le j \\ \quad Q_{he} = \bigwedge_k \dot{q}_k, \dot{q}_k \equiv q_k, \quad \text{if } s_h[q_k] = true, \\ \quad \text{and } \dot{q}_k \equiv \neg q_k, \quad \text{if } s_h[q_k] = false, Q_h \in CL(Q) \\ \quad \text{if } \sigma =< s_0, s_1, \ldots, (s_i, \ldots, s_j)^\omega > \end{cases}$$

For finite models, the following theorem is easily proved [24].

Theorem 17.1:

Finite paths in the NFG of PPTL formula P precisely characterize finite models of P. □

For infinite models, it is much more intricate because of the involvement of chop and projection operators. In fact, not all of infinite paths in the NFG of a formula are infinite models of the corresponding formula. We investigate the case carefully in the following.

A formula R is called a chop formula if $R \equiv P \wedge Q$ and $P \equiv P_1; P_2$, where P_1, P_2 and Q are any PPTL formulas. Further, $P_1; P_2$ is called a chop component of chop formula R. Formally, a chop formula R_c can be defined as follows

$$R_c ::= P; Q \mid R_c \wedge R_c \mid R_c \wedge R$$

where P, Q and R are any PPTL formulas.

For instance, $\bigcirc p; q$, $(\bigcirc p; q) \wedge \bigcirc r$ are chop formulas while $\bigcirc(\bigcirc p; q)$ and $\neg(\bigcirc p; q)$ are not, where p, q, r are atomic propositions. Note that $P; Q$ is a chop formula but $\neg(P; Q)$ is not. This will be formally analyzed later.

For chop construct $P; Q$, an infinite model $\sigma =< s_0, s_1, \ldots, s_k, \ldots >\models P; Q$ if and only if there exists $i \in N_0$, such that $\sigma^i \models P$ and $\sigma^{(i)} \models Q$. This implies that if infinite model $\sigma =< s_0, s_1, \ldots, s_k, \ldots >\models P$ and there is no finite prefix $\sigma^i \models P$ ($i \in N_0$), it fails to satisfy $P; Q$. Note that in the weak version of chop construct, $P; Q$ is still satisfied. For convenience, we say finiteness of P in strong chop construct $P; Q$ (FSC_Property for short) is satisfiable over an infinite model σ, denoted by predicate $FSC(\sigma, P; Q) = true$, if there exists $i \in N_0$ such that $\sigma^i \models P$ and $\sigma^{(i)} \models Q$. Actually, an infinite path of the NFG of $P; Q$ presents an infinite model of either $P; Q$ or P. So, FSC_Property needs to be considered if a node (formula) is a chop formula.

Correspondingly, in NFGs, when constructing NFG of $P; Q$, initially, $P; Q$ is transformed into its normal form,

$$\begin{aligned} P; Q &\equiv (\bigvee_{j=0}^{n_0} P_{ej} \wedge empty \vee \bigvee_{i=0}^{n_1} (P_{ci} \wedge \bigcirc P'_i)); Q \\ &\equiv \bigvee_{j=0}^{n_0} \underline{(P_{ej} \wedge empty; Q)} \vee \bigvee_{i=0}^{n_1} P_{ci} \wedge \bigcirc(P'_i; Q) \\ &\equiv \bigvee_{j=0}^{n_0} (P_{ej} \wedge Q) \vee \bigvee_{i=0}^{n_1} P_{ci} \wedge \bigcirc(P'_i; Q) \end{aligned}$$

Subsequently, the new generated formula $P_i'; Q$ needs to be repeatedly transformed into its normal form in the same way. Whenever a final state of P is reached, that is, $P_e \wedge empty; Q$ is encountered, where P_e is a state formula, FSC_Property of $P; Q$ is satisfied over the path π departing from the root node.

Formally, let $\pi =< R, R_{0e}, R_1, R_{1e}, \ldots >$ be an infinite path in the NFG of formula R, and $\pi^{(k)} = < R_k, R_{ke}, \ldots >$ be the kth suffix of π. Whether or not FSC_Property of R is satisfiable over π, denoted by predicate $FSC(\pi, R)$, can be defined as follows.

1. $fsc(\pi, R) = true$ if R is a non-chop formula
2. $fsc(\pi, P; Q) = true$ if there exists $i \in N_0$ such that $R_i \equiv P_{ie} \wedge \varepsilon; Q$
3. $fsc(\pi, P \wedge Q) = true$ if $fsc(\pi, P) = true$ and $fsc(\pi, Q) = true$
4. $fsc(\pi, P \vee Q) = true$ if $fsc(\pi, P) = true$ or $fsc(\pi, Q) = true$
5. $FSC(\pi, R) = true$ if $fsc(\pi^{(k)}, R_k) = true$ for all $k \in N_0$

Based on this, it can be proved that in the NFG of PPTL formula R, an infinite path π with $FSC(\pi, R) = true$ precisely characterizes infinite models of R.

17.5.4 Decision Procedure Based on LNFG

To explicitly display whether or not the FSC_Property of a chop formula is satisfied, extra propositions $\mathtt{l}_k$, $k \in N_0$ and $k > 0$, are introduced. Let $Prop_l = \{\mathtt{l}_1, \mathtt{l}_2, \ldots\}$ be the set of extra propositions. $Prop \cap Prop_l = \emptyset$. Note that these extra propositions are merely employed to mark nodes and are not allowed to appear in a PPTL formula. When constructing NFGs by normal forms, for any chop formula $P; Q$, we equivalently transform it as $P \wedge fin(\mathtt{l}_k); Q$. Since $fin(\mathtt{l}_k)$ can be recursively expressed by

$$fin(\mathtt{l}_k) \equiv \mathtt{l}_k \wedge empty \vee \bigcirc fin(\mathtt{l}_k)$$

we have

$$\begin{aligned} & P \wedge fin(\mathtt{l}_k); Q \\ & \equiv (\bigvee_{j=0}^{n_0} P_{ej} \wedge empty \vee \bigvee_{i=0}^{n_1} (P_{ci} \wedge \bigcirc P_i')) \wedge (\mathtt{l}_k \wedge empty \vee \bigcirc fin(\mathtt{l}_k)); Q \\ & \equiv (\bigvee_{j=0}^{n_0} P_{ej} \wedge \mathtt{l}_k \wedge empty \vee \bigvee_{i=0}^{n_1} (P_{ci} \wedge \bigcirc (P_i' \wedge fin(\mathtt{l}_k)))); Q \\ & \equiv \bigvee_{j=0}^{n_0} (\underline{P_{ej} \wedge \mathtt{l}_k \wedge empty; Q}) \vee \bigvee_{i=0}^{n_1} (P_{ci} \wedge \bigcirc (P_i' \wedge fin(\mathtt{l}_k); Q)) \\ & \equiv \bigvee_{j=0}^{n_0} (\underline{P_{ej} \wedge \mathtt{l}_k \wedge Q}) \vee \bigvee_{i=0}^{n_1} (P_{ci} \wedge \bigcirc (P_i' \wedge fin(\mathtt{l}_k); Q)) \end{aligned}$$

Thus, by using $fin(\mathtt{l}_k)$, FSC_Property of $P; Q$ is satisfied if there exists an edge where $\mathtt{l}_k$ holds. And $fin(\mathtt{l}_k)$ occurring in a node $P \wedge fin(\mathtt{l}_k); Q$ means that FSC_Property of $P; Q$ has not been satisfied at this node. For convenience, for a node in the form of $P \wedge fin(\mathtt{l}_k); Q$ or $\bigwedge_{i=1}^{n} R_i$ with some $R_i \equiv P_i \wedge fin(\mathtt{l}_k); Q_i$, we add an extra label $\tilde{\mathtt{l}}_k$ in this node to mean that the finiteness of some chop formula has not been satisfied at this node.

Accordingly, Labeled Normal Form Graph (LNFG) is defined based on NFG with the usage of $\mathtt{l}_k$ propositions.

TABLE 17.11 Algorithm for Constructing LNFG of a PPTL Formula

```
Function LNFG(P)
/* precondition: P is a PPTL formula*/
/* postcondition: LNFG(P) computes LNFG of P, G = (CL(P), EL(P), V0, L = {L1,...,Lm})*/
begin function
  case
    P is ∨_i Pi: CL(P) = {Pi|Pi appears in ∨_i Pi}; Mark[Pi] = 0 for each i;
                 V0 = {Pi|Pi appears in ∨_i Pi};
    P is not ∨_i Pi: CL(P) = {P}; Mark[P] = 0; V0 = {P};
  end case
  EL(P) = ∅; AddE = AddN = 0; k = 0; L = ∅;
  while there exists R ∈ CL(P) \ {ε, false}, and mark[R] ==0
    if R ≡ P; Q or R ≡ ∧_{i=1}^n Ri and ∃Ri ≡ Pi; Qi and no fin(k) has been added in the chop construct
      k=k+1; Rewrite R as P ∧ fin(1k); Q or ∧_{i=1}^n Ri with some Ri being Pi ∧ fin(1k); Qi; Lk = {R}; L = L ∪ {Lk};
      if there exist P ∧ fin(1s); Q or ∧_{i=1}^n R'i with some R'i ≡ P ∧ fin(1s); Q,
        and P ∧ fin(1x); Q or ∧_{i=1}^n R'i with some R'i ≡ P ∧ fin(1x); Q, 1 ≤ l < x < k
        delete the node P ∧ fin(1k); Q or ∧_{i=1}^n R'i with some R'i ≡ P ∧ fin(1s); Q;
        re-adding the edges to P ∧ fin(1s); Q or ∧_{i=1}^n R'i; continue;
    Q =NF(R); mark[R] =1;                                          /*marking R is decomposed*/
    case
      Q is ∨_{j=1}^h Qej ∧ empty: AddE=1;                          /*first part of NF needs added*/
      Q is ∨_{i=1}^k Qi ∧ ○Q'i : AddN=1;                           /*second part of NF needs added*/
      Q is ∨_{j=1}^h Qej ∧ empty ∨ ∨_{i=1}^k Qi ∧ ○Q'i: AddE=AddN=1;    /*both parts need added*/
    end case
    if AddE == 1 then                                              /*add first part of NF*/
      CL(P) = CL(P) ∪ {ε}; EL(P) = EL(P) ∪ ∪_{j=1}^h {(R, Qej, ε)}; AddE=0;
    if AddN == 1 then for i = 1 to k,                              /*add second part of NF*/
      if Q'i ∉ CL(P),CL(P) = CL(P) ∪ ∪_{i=1}^k {Q'i};
        if Q'i is not false, mark[Q'i]=0; else mark[Q'i]=1;
      EL(P) = EL(P) ∪ ∪_{i=1}^k {(R, Qi, Q'i)}; AddN=0;
  end while
  return G;
End function
```

Definition 17.4: (Labeled Normal Form Graph, LNFG)

For a PPTL formula P, its LNFG is a tuple $G = (CL(P), EL(P), V_0, \mathbb{L} = \{\mathbb{L}_1, \ldots, \mathbb{L}_m\})$, where $CL(P)$, $EL(P)$ and V_0 are identical to the ones in NFG, each $\mathbb{L}_k \subseteq CL(P)$, $1 \leq k \leq m$, is the set of nodes with $\tilde{1_k}$ labels. □

Algorithm LNFG based on Algorithm NFG is given in Table 17.11 by further rewriting the chop component $P; Q$ as $P \wedge fin(1_k); Q$ whenever a new chop formula is created. And anytime if a formula $\bigwedge_i R_i$ with some R_i being $P \wedge fin(k); Q$ is created, and there are already two previous nodes in the branch to $\bigwedge_i R_i$ which are the same to R_i excepting for the label index k, instead of creating the new node $\bigwedge_i R_i$, only an edge is added back to the node $\bigwedge_i R_i$ with the smaller label index.

TABLE 17.12 Algorithm for Checking Whether or Not PPTL P Is Satisfiable

```
Function CHECK(P)
/* precondition: P is a PPTL formula*/
/* postcondition: CHECK(P) checks whether formula P is satisfiable or not.*/
begin function
    G = LNFG(P);
    if there exists ε node in CL(P),
    return P is satisfiable with finite models;
    if there exists infinite path π with Inf(π) ⊈ 𝕃_i, for all 1 ≤ i ≤ m
    return P is satisfiable with infinite models;
    else return unsatisfiable;
end function
```

Also, by the construction of LNFG, especially the usage of $\tilde{1_k}$ labels, a node $n \in \mathbb{L}_k$ means that FSC_Property of the chop formula with $\tilde{1_k}$ label has not been satisfied currently, while $n \notin \mathbb{L}_k$, for any k, indicates that FSC_Property of all chop formulas have been satisfied at this state. Accordingly, for an infinite path with $\mathtt{Inf}(\pi) \not\subseteq \mathbb{L}_k$, the FSC_Property of the chop formula with $\tilde{1_k}$ label is satisfiable since at least a node without $\tilde{1_k}$ exists in $\mathtt{Inf}(\pi)$. Further, for any chop formulas over the path, their FSC_Properties are satisfiable, that is, $FSC(\pi, Q) = true$, iff $\mathtt{Inf}(\pi) \not\subseteq \mathbb{L}_i$ for any $1 \leq i \leq m$. Therefore, it has that in the LNFG of a formula P, finite paths precisely characterize finite models of P; infinite paths with $\mathtt{Inf}(\pi) \not\subseteq \mathbb{L}_i$, for all $1 \leq i \leq m$ precisely characterize infinite models of P.

Consequently, a decision procedure for checking the satisfiability of a PPTL formula P can be constructed based on the LNFG of P. In the following, a sketch of the procedure, Algorithm CHECK in pseudo code, is given in Table 17.12.

In order to use our approach, we have developed a tool in C++. With this tool, all the algorithms introduced above have been implemented. For any PPTL formula, the tool can automatically transform it to its normal form, and LNFG, then check whether the formula is satisfiable or not.

Example 17.2:

Checking the satisfiability of formula $P \equiv (p \wedge \Box \bigcirc p; \bigcirc \Box q) \wedge (\Box r; \bigcirc \Box q)$.

By Algorithm LNFG, LNFG $G = (CL(P), EL(P), V_0, \mathbb{L} = \{\mathbb{L}_1, \ldots, \mathbb{L}_m\})$ of formula $(p \wedge \Box \bigcirc p; \bigcirc \Box q) \wedge (\Box r; \bigcirc \Box q)$ is constructed as depicted in Figure 17.7, where $CL(P) = \{n_0, n_1\}$, $EL(P) = \{(n_0, p \wedge r, n_0), (n_0, p \wedge r, n_1), (n_1, p, n_1)\}$, $V_0 = \{n_0\}$, $\mathbb{L} = \{\mathbb{L}_1, \mathbb{L}_2\}$, $\mathbb{L}_1 = \{n_0, n_1\}$ and $\mathbb{L}_2 = \{n_0\}$. The formula is unsatisfiable since there exists no node without $\tilde{1_1}$ label which is reachable from n_0 and n_1.

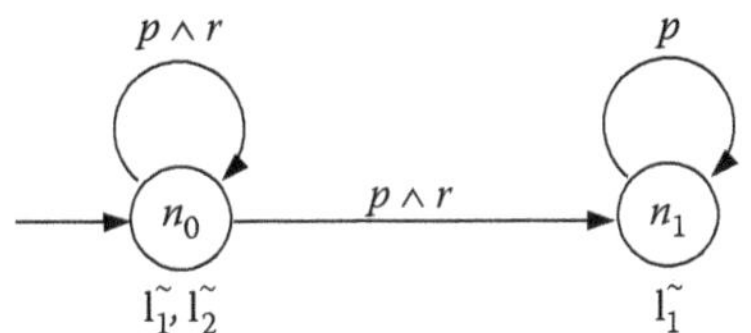

$n_0 : (p \wedge \Box \bigcirc p \wedge fin\,(l_1); \bigcirc \Box q)$
$\wedge (\Box r \wedge fin\,(l_2); \bigcirc \Box q)$

$n_1 : (p \wedge \Box \bigcirc p \wedge fin\,(l_1); \bigcirc \Box q)$
$\wedge (\Box q)$

FIGURE 17.7 LNFG of $(p \wedge \Box \bigcirc p; \bigcirc \Box q) \wedge (\Box r; \bigcirc \Box q)$.

17.5.5 From LNFGs to Büchi Automata

Now we focus on how to transform an LNFG to a Büchi automaton. We first transform an LNFG to a Generalized Büchi Automaton (GBA) [26]. Factually, an LNFG contains all the information of the corresponding GBA. The set of nodes is in fact the set of locations in the corresponding GBA; each edge (v_i, Q_e, v_j) forms a transition; there exists only one initial location, the root node; the set of accepting locations consists of ε node and the nodes which can appear in infinite paths for infinitely many times. Given an LNFG $G = (CL(P), EL(P), V_0, \mathbb{L} = \{\mathbb{L}_1, \ldots, \mathbb{L}_m\})$ of formula P, an GBA, $B = (Q, \Sigma, I, \delta, F = \{F_1, \ldots, F_m\})$, over an alphabet Σ can be constructed as follows:

- Sets of the locations Q and the initial locations I: $Q = V$, and $I = \{v_0\}$.
- Transition δ: Let $\dot{q}_k$ be an atomic proposition or its negation, and we define a function $atom(\bigwedge_{k=1}^{m_0} \dot{q}_k)$ for picking up atomic propositions or their negations appearing in $\bigwedge_{k=1}^{m_0} \dot{q}_k$ as follows,

$$atom(true) = true$$

$$atom(\dot{q}_k) = \begin{cases} \{q_k\}, & \text{if } \dot{q}_k \equiv q_k \ 1 \le k \le l \\ \{\neg q_k\}, & \text{otherwise} \end{cases}$$

$$atom\left(\bigwedge_{k=1}^{m_0} \dot{q}_k\right) = atom(\dot{q}_1) \cup atom\left(\bigwedge_{k=2}^{m_0} \dot{q}_k\right)$$

For each $e_i = (v_i, Q_e, v_{i+1}) \in E$, there exists $v_{i+1} \in \delta(v_i, atom(Q_e))$. For node ε, $\delta(\varepsilon, \epsilon) = \{\varepsilon\}$.
- Accepting set $F = \{F_1, \ldots, F_m\}$: It has been proved that infinite paths with $\texttt{Inf}(\pi) \not\subseteq \mathbb{L}_i$ for all $1 \le i \le m$ precisely characterize infinite models of P. This can be equivalently expressed by "infinite paths with $\texttt{Inf}(\pi) \cap \overline{\mathbb{L}_i} \neq \emptyset$ for all $1 \le i \le m$ precisely characterize infinite models of P", where $\overline{\mathbb{L}_i}$ denotes $CL(P) \setminus \mathbb{L}_i$. So, $F_i = \overline{\mathbb{L}_i}$ for each i. In addition, by employing the stutter extension rule, $\{\varepsilon\}$ is also an accepting set.

Formally, algorithm LNFG-GBA shown in Table 17.13 is used for transforming an LNFG to a Büchi automaton. Consequently, the algorithm presented in [26] for transforming a GBA to a Büchi automaton can be applied.

17.5.6 Model Checking PPTL

With our model checking algorithm, the system to be verified is modeled as a Büchi automaton A_s, while the property is specified by a PPTL or PITL formula P. To check whether or not the system satisfies P, $\neg P$ is transformed into an LNFG, and further a Büchi automaton A_p. The system can be verified by computing the product automaton of A_s and A_p, and then checking whether the words accepted by the product automaton is empty or not as shown in Figure 17.8. If the words accepted by the product automaton is empty, the system can satisfy the property otherwise the system cannot satisfy the property, and a counterexample can be found.

Within SPIN, a translator is realized to automatically transform a formula in linear temporal logic to Never Claim. To implement model checking PPTL in SPIN, we also provide a translator from PPTL formulas to Never Claims as shown in Figure 17.9. We have realized the translator in C++ according to the algorithms presented in the previous chapters. Further, the translator has successfully been integrated within the original SPIN.

17.5.7 Complexity and Discussions

In Ref. [16], we have proved that the complexity for the satisfiability of PPTL is nonelementary by reducing the empty problem of star-free expressions [17], which is proved to be nonelementary, to the satisfiability

TABLE 17.13 Algorithm for Obtaining a GBA from an LNFG

```
Function LNFG-GBA(G)
/* precondition: G = (CL(P), EL(P), V_0, 𝕃 = {𝕃_1, ..., 𝕃_m}) is the LNFG of PPTL formula P*/
/* postcondition: LNFG-GBA(G) computes an GBA B = (Q, Σ, I, δ, F = {F_1, ..., F_n}) from G*/
begin function
    Q = ∅; F = {F_1, ..., F_m}; F_i = ∅, 1 ≤ i ≤ m; I = ∅;
    for each node v_i ∈ V,
        add a state q_i to Q, Q = Q ∪ {q_i};
    end for
    for each node v_i ∈ overline(𝕃_i),
        add a state q_i to F_i, F_i = F_i ∪ {q_i};
    end for
    if v_i is ε, F = F ∪ {q_i}; δ(q_i, ϵ) = {q_i};
    if q_0 ∈ V_0, I = I ∪ {q_0};
    for each edge e = (v_i, P_e, v_j) ∈ E,
        q_j ∈ δ(q_i, atom(P_e));
    end for
    return B = (Q, Σ, I, δ, F)
end function
```

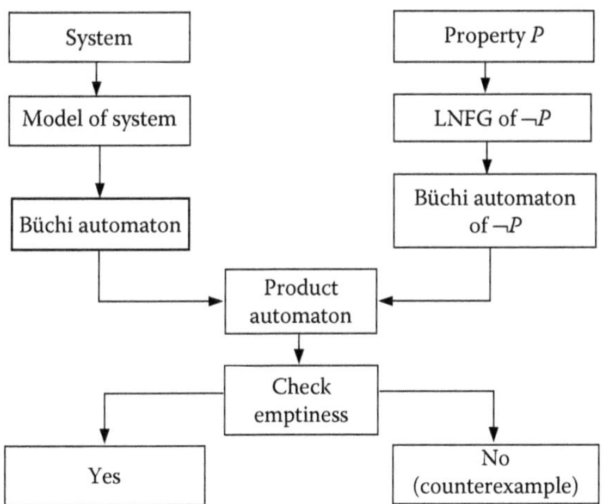

FIGURE 17.8 Automata based model checking PPTL.

of PPTL formulas. Here, the model checking algorithm is mainly based on the decision algorithm by further transforming LNFGs to Büchi automata. Accordingly, the complexity for model checking PPTL is also nonelementary. Essentially, the non-elementary complexity of PPTL is caused by chop construct, since negations cannot pushed in front of atomic propositions for a formula with chop operators. Given a PPTL formula P with the length being n and the maximal nesting depth of chop operators being d, the exact complexity will be d-exponential. However, in practice, the length of the formula specifying the property of the system is not too long and nesting depth of chop operators is limited. Thus, the approach is tolerable.

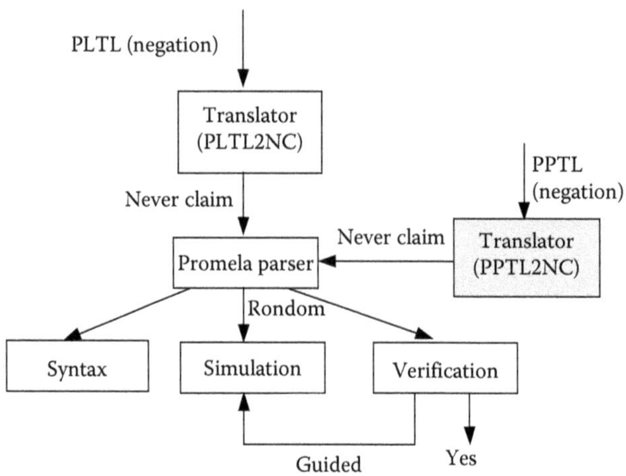

FIGURE 17.9 Model checking PPTL based on SPIN.

17.5.8 Model Checking Leaping Frogs Problem

Problem Description: A pond has a line of lily pads from left to right. To start with, each brown frog is on its own lily pad on the left side of the pond, each green frog is on its own lily pad on the right side of the pond, and there is an empty lily pad between the two colors of frogs, that is, `BBB_GGG` where `B` and `G` present a brown and green frog, respectively. The problem below requires the brown frogs to end up on the right side of the pond and the green frogs to end up on the left side of the pond, that is, `GGG_BBB`, by following these rules: (1) only one frog can jump at one time (during one move); (2) frogs can either jump onto the next lily pad (if it is empty) or they can jump over a different-colored frog (e.g., green frogs cannot jump over green frogs); (3) when they jump over a frog they must land on the empty lily pad right beside that frog; (4) you can never have more than one frog on a lily pad; (5) frogs can only move forward (brown frogs move to the right and green frogs move to the left).

Now we use PPTL model checking to check whether or not there is a method for the brown frogs to end up on the right side of the pond and the green frogs to end up on the left side of the pond by the above rules. First, the leaping frogs problems (3 green frogs and 3 brown frogs) is modeled in PROMELA. The desired property is when no frogs can leap, the brown frogs are on the right side of the pond and the green frogs are on the left side of the pond, that is, `GGG_BBB`. To this end, three propositions p, q, and r are defined to express that "brown frogs are on the left hand side and green frogs are on the right hand side", "no frogs can leap forever" and "green frogs are on the left hand side and brown frogs are on the right hand side," respectively. Accordingly, the property is specified by $p; (q \wedge r)$. When implemented by our model checker, it outputs an leaping method for the frogs to change positions as illustrated in Figure 17.10.

17.6 Unified Model Checking with PTL

This section presents a unified model checking approach with Projection TL (PTL) based on SAT. To this end, a Modeling, Simulation and Verification Language (MSVL) is defined based on PTL. Further, normal forms and NFGs for MSVL programs are defined. The finiteness of NFGs for MSVL programs is proved in details. Moreover, by modeling a system with an MSVL program p, and specifying the desirable property of the system with a PPTL formula ϕ, whether or not the system satisfies the property (whether or not $p \rightarrow \phi$ is valid) can equivalently be checked by evaluating whether or not $\neg(p \rightarrow \phi) \equiv p \wedge \neg\phi$ is

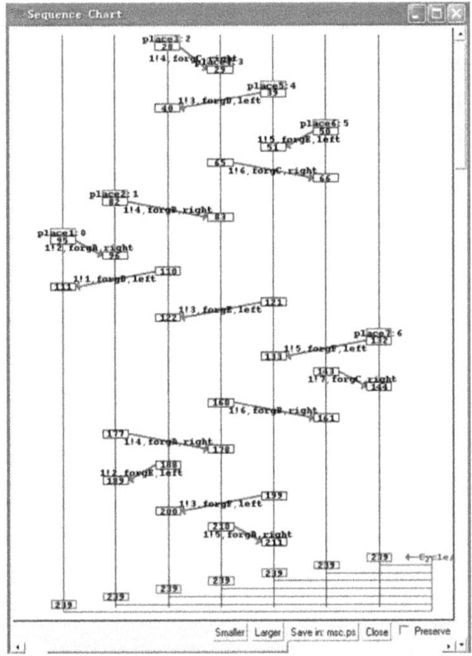

FIGURE 17.10 Verification result.

unsatisfiable. Finally, the satisfiability of a formula in the form of $p \wedge \neg\phi$ is checked by constructing the NFG of $p \wedge \neg\phi$, and then inspecting whether or not there exist paths acceptable in the LNFG.

17.6.1 Modeling, Simulation, and Verification Language

The Language MSVL is a subset of Projection Temporal Logic with framing technique, and an extension of Framed Tempura [18–20]. It can be be used for the purpose of modeling, simulation, and verification of software and hardware systems.

17.6.2 Framing

Framing is concerned with the persistence of the values of variables from one state to another. Intuitively, the framing operation on variable x, denoted by $frame(x)$, means that variable x always keeps its old value over an interval if no assignment to x is encountered. For the definition of *frame* operator, a new assignment called *a positive immediate assignment* is defined as

$$x \Leftarrow e \stackrel{\text{def}}{=} x = e \wedge p_x$$

where p_x is an atomic proposition associated with state (dynamic) variable x, and notice that p_x cannot be used for other purpose. To identify an occurrence of an assignment to a variable, say x, we make use of a flag called the assignment flag, denoted by a predicate $af(x)$; it is true whenever an assignment of a value to x is encountered, and false otherwise. The definition of the assignment flag is $af(x) \stackrel{\text{def}}{=} p_x$, for every variable x. There are state framing (lbf) and interval framing ($frame$) operators. Intuitively, when a variable is framed at a state, its value remains unchanged if no assignment is encountered at that state.

A variable is framed over an interval if it is framed at every state over the interval.

$$lbf(x) \stackrel{\text{def}}{=} \neg af(x) \rightarrow \exists b : (\odot x = b \wedge x = b)$$
$$frame(x) \stackrel{\text{def}}{=} \Box(more \rightarrow \bigcirc lbf(x))$$

where b is a static variable.

17.6.3 The MSVL Language

The arithmetic expression e and boolean expression b of MSVL are inductively defined as follows:

$$e ::= n \mid x \mid \bigcirc x \mid \odot x \mid e_0 \; op \; e_1 (op ::= + \mid - \mid * \mid \backslash \mid mod)$$
$$b ::= true \mid false \mid e_0 = e_1 \mid e_0 < e_1 \mid \neg b \mid b_0 \wedge b_1$$

where n is an integer and x is a variable. The elementary statements in MSVL are defined as follows:

TERMINATION:	$empty$
ASSIGNMENT:	$x = e$
P-I-ASSIGNMENT:	$x \Leftarrow e$
STATE FRAME:	$lbf(x)$
INTERVAL FRAME:	$frame(x)$
CONJUNCTION:	$p \wedge q$
SELECTION:	$p \vee q$
NEXT:	$\bigcirc p$
ALWAYS:	$\Box p$
CONDITIONAL:	$if \; b \; then \; p \; else \; q \stackrel{\text{def}}{=} (b \rightarrow p) \wedge (\neg b \rightarrow q)$
EXISTS:	$\exists x : p$
PROJECTION:	$(p_1, \ldots, p_m) \; prj \; p$
SEQUENCE:	$p ; q$
WHILE:	$while \; b \; do \; p \stackrel{\text{def}}{=} (p \wedge b)^* \wedge \Box(empty \rightarrow \neg b)$
PARALLEL:	$p \parallel q \stackrel{\text{def}}{=} (p \wedge (q; true)) \vee (q \wedge (p; true))$
AWAIT:	$await(b) \stackrel{\text{def}}{=} (frame(x_1) \wedge \ldots \wedge frame(x_h)) \wedge \Box(empty \leftrightarrow b)$ where $x_i \in V_b = \{x \mid x \text{ appears in } b\}$

where x denotes a variable, e stands for an arbitrary arithmetic expression, b a boolean expression, and $p_1, \ldots, p_m$, p and q stand for programs of MSVL. The assignment $x = e$, positive immediate assignment $x \Leftarrow e$, $empty$, $lbf(x)$, and $frame(x)$ are basic statements and the others are composite ones.

The assignment $x = e$ means that the value of variable x is equal to the value of expression e. Positive immediate assignment $x \Leftarrow e$ indicates that the value of x is equal to the value of e and the assignment flag for variable x, p_x, is $true$. Statements of $if \; b \; then \; p \; else \; q$ and $while \; b \; do \; p$ are the same as that in the conventional imperative languages. The next statement $\bigcirc p$ means that p holds at the next state while $\Box p$ means that p holds at all the states over the whole interval from now. $p \wedge q$ means that p and q are executed concurrently and share all the variables during the mutual execution. $p \vee q$ means p or q are executed. $empty$ is the termination statement meaning that the current state is the final state of the interval over which the program is executed. The sequence statement $p; q$ means that p is executed from the current state to its termination while q will hold at the final state of p and be executed from that state. The existential quantification $\exists x : p$ intends to hide the variable x within the process p. $lbx(x)$ means the value of x in the current state equals to value of x in the previous state if no assignment to x occurs, while $frame(x)$ indicates that the value of variable x always keeps its old value over an interval if no assignment to x is encountered. Different from the conjunction statement, the parallel statement allows both the processes to specify their own intervals. For example, $len(2)\|len(3)$ holds but $len(2) \wedge len(3)$ is obviously false. Projection can be thought of as a special parallel computation which is executed on different time scales. The projection $(p_1, \ldots, p_m)\ prj\ q$ means that q is executed in parallel with $p_1, \ldots, p_m$ over an interval obtained by taking the endpoints of the intervals over which the $p_i's$ are executed. In particular, the sequence of $p_i's$ and q may terminate at different time points. Finally, *await b* does not change any variable, but waits until the condition b becomes *true*, at which point it terminates.

Further, the following derived statements are useful in practice.

MULTIPLE SELECTION:	$OR_{k=1}^{n} \stackrel{\text{def}}{=} p_1 \vee p_2 \vee \ldots \vee p_n$
CONDITIONAL:	$if\ b\ do\ p \stackrel{\text{def}}{=} if\ b\ do\ p\ else\ empty$
WHEN:	$when\ b\ do\ p \stackrel{\text{def}}{=} await(b); p$
GUARDED COMMAND:	$b_1 \rightarrow p_1 \Box \ldots \Box b_n \rightarrow p_n \stackrel{\text{def}}{=} OR_{k=1}^{n}(when\ b_k\ do\ p_k)$
REPEAT:	$repeat\ p\ until\ c \stackrel{\text{def}}{=} p; while\ \neg c\ do\ p$

17.6.4 Normal Forms and NFGs of MSVL

Definition 17.5:

A program q in MSVL is in normal form if

$$q \equiv \bigvee_{i=1}^{l} q_{ei} \wedge empty \vee \bigvee_{j=1}^{t} q_{cj} \wedge \bigcirc q_{fj}$$

where $0 \leq l \leq 1$, $t > 0$, and $l + t \geq 1$. For $1 \leq j \leq t$, q_{fj} is a general MSVS program; whereas q_{ei} $(i = 1)$ and q_{cj} $(1 \leq j \leq t)$ are *true* or all are state formulas of the form:

$$(x_1 = e_1) \wedge \ldots \wedge (x_l = e_l) \wedge \dot{p_{x_1}} \wedge \ldots \wedge \dot{p_{x_l}}$$

where $e_k \in D$, data domain, $(1 \leq k \leq l)$. □

Theorem 17.2:

Any MSVL program q can be transformed into its normal form.

The proof for transforming most of the statements in MSVL into normal form can be found in Refs. [7,19]. The other statements of MSVL can be transformed in a similar way. □

Modeling a system with an MSVL program (formula in PTL) p, according to the normal form, we can construct a graph, namely normal form graph (NFG), which explicitly illustrates the state space of the system. Actually, the NFG also presents the models satisfying formula p. For an MSVL program p, the NFG of p is a directed graph, $G = (CL(p), EL(p))$, where $CL(p)$ denotes the set of nodes and $EL(p)$ denotes the set of edges in the graph. In $CL(p)$, each node is specified by a program in MSVL, while in $EL(p)$, each edge is a directed arc labeled with a state formula p_e from node q to node r and identified by a triple, (q, p_e, r). $CL(p)$ and $EL(p)$ of G can be inductively defined as in Definition 17.6.

Definition 17.6:

For a program p, the set $CL(p)$ of nodes and the set $EL(p)$ of edges connecting nodes in $CL(p)$ are inductively defined as follows:

1. $p \in CL(p)$;
2. For all $q \in CL(p) \setminus \{\varepsilon, false\}$, if $q \equiv \bigvee_{i=1}^{l} q_{ei} \wedge empty \vee \bigvee_{j=1}^{t} q_{cj} \wedge \bigcirc q_{fj}$, then $\varepsilon \in CL(p)$, $(q, q_{ei}, \varepsilon) \in EL(p)$ for each i; $q_{fj} \in CL(p)$, $(q, q_{cj}, q_{fj}) \in EL(p)$ for all j;

The NFG of formula p is the directed graph $G = (CL(p), EL(p))$. □

Definition 17.6 implies an algorithm for constructing NFGs of MSVL programs. In the NFG of a program p generated by Definition 17.6, the set $CL(p)$ of nodes and the set $EL(p)$ of edges are inductively produced by repeatedly transforming the new created nodes into their normal forms. So, one question we have to answer is whether or not the transforming process terminates. Fortunately, we can prove that, for any MSVL program p, the number of nodes in $CL(p)$ is finite.

Let $D = \{d_1, \ldots, d_n\}$ be a finite set of data, $V = \{x_1, \ldots, x_m\}$ a finite set of variables, and *Prop* a countable set of atomic propositions. To prove the finiteness of NFGs of MSVL programs, we first prove that, for any MSVL program, it can be equivalently expressed by a PPTL formula.

Theorem 17.3:

Any program p in MSVL can be equivalently expressed by a formula $\Phi(p)$ in PPTL.

The proof of the theorem proceeds by induction on structures of programs in MSVL which can be found in Ref. [25]. □

In the previous section, we have proved the finiteness of NFGs of PPTL formulas. Hence, the conclusion also holds for MSVL programs since any MSVL program can be equivalently expressed by a PPTL formula.

Example 17.3:

NFG of MSVL program $frame(x) \wedge (x = 2 \vee x = 3) \wedge if\ (x = 2)\ then\ len(2)\ else\ \{len(3)\}$ can be constructed as shown in Figure 17.11. □

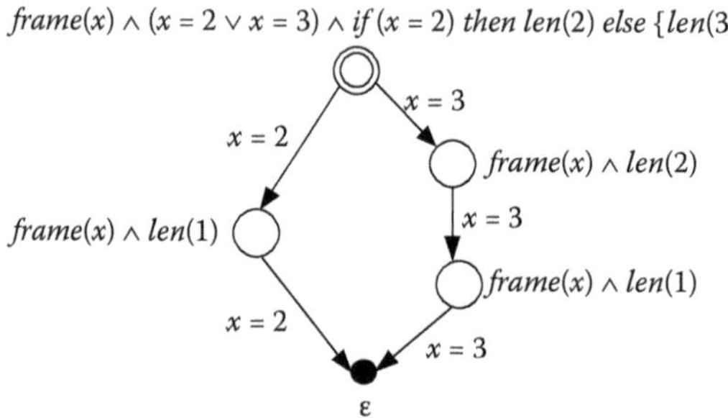

FIGURE 17.11 NFG of MSVL program $frame(x) \wedge (x = 2 \vee x = 3) \wedge if\ (x = 2)\ then\ len(2)\ else\ \{len(3)\}$.

17.6.5 Model Checking MSVL Based on SAT

17.6.5.1 Basic Approach

Modeling the system to be verified by an MSVL program p, and specifying the desirable property of the system by a PPTL formula ϕ, to check whether or not the system satisfies the property, we need to prove the validation of

$$p \rightarrow \phi$$

If $p \rightarrow \phi$ valid, the system satisfies the property, otherwise the system violates the property. Equivalently, we can check the satisfiability of

$$\neg(p \rightarrow \phi) \equiv p \wedge \neg\phi$$

If $p \wedge \neg\phi$ is unsatisfiable ($p \rightarrow \phi$ is valid), the system satisfies the property, otherwise the system fails to satisfy the property, and for each $\sigma \models p \wedge \neg\phi$, σ determines a counterexample that the system violates the property. Accordingly, our model checking approach can be translated to the satisfiability of PTL formulas of the form $p \wedge \neg\phi$, where p is an MSVL program and ϕ is a formula in PPTL. Since both model p and property ϕ are formulas in PTL, we call this model checking a unified approach.

To check the satisfiability of PTL formula $p \wedge \neg\phi$, we still construct the NFG (as well as LNFG) of $p \wedge \neg\phi$. As depicted in Figure 17.12, initially, we create the root node $p \wedge \neg\phi$, then we transform p and $\neg\phi$ into their normal forms, respectively. By computing the conjunction of normal forms of p and $\neg\phi$, new nodes ε and $p_{fj} \wedge \neg\phi_{fs}$, and edges $(p \wedge \neg\phi, p_{ei} \wedge \neg\phi_{ck}, \varepsilon)$ from node $p \wedge \neg\phi$ to ε, $(p \wedge \neg\phi, p_{cj} \wedge \neg\phi_{cs}, p_{fj} \wedge \neg\phi_{fs})$ from $p \wedge \neg\phi$ to $p_{fj} \wedge \neg\phi_{fs}$ are created. Further, by dealing with each new created nodes $p_{fj} \wedge \neg\phi_{fs}$ using the same methods as the root nodes $p \wedge \neg\phi$ repeatedly, the NFG of $p \wedge \neg\phi$ can be produced. Thus, it is apparent that each node in the NFG of $p \wedge \neg\phi$ is in the form of $p' \wedge \neg\phi'$, where p' and ϕ' are nodes in the NFGs of p and $\neg\phi$, respectively.

Considering the FSC_Property of chop formulas, LNFGs are needed to be constructed for the satisfiability of formulas in the form of $p \wedge \neg\phi$. Based on Algorithm LNFG in Chapter 3, a recursive Algorithm U-LNFG in Table 17.14 is formalized for constructing LNFG of PTL formulas in the form of $p \wedge \neg\phi$. In the algorithm, another function NF(p) is called to produce the normal form of a PPTL formula or

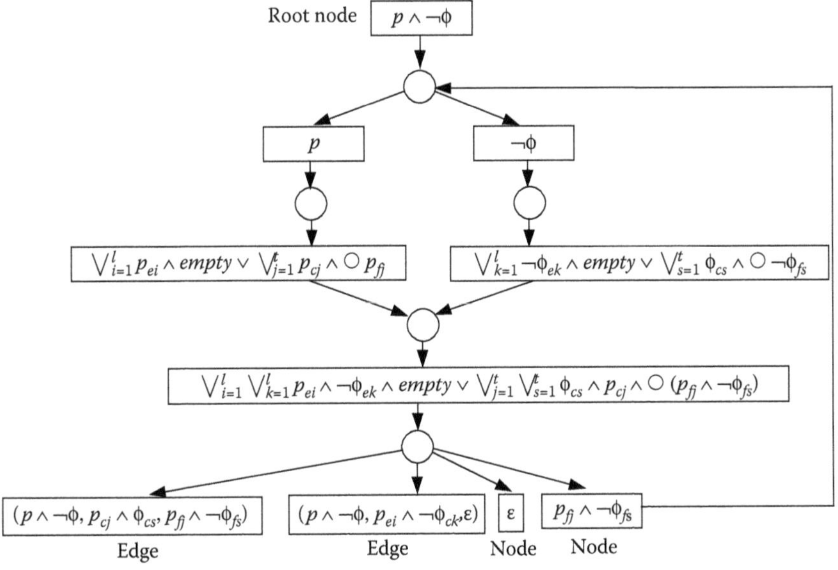

FIGURE 17.12 Constructing NFG of $p \wedge \neg\phi$.

TABLE 17.14 Algorithm for Constructing LNFG of a $p \wedge \neg\phi$

```
Function U-LNFG(p ∧ ¬φ)
/* precondition: p is a program in MSVL, ¬φ is a formula in PPTL*/
/* postcondition: U-LNFG(p ∧ ¬φ) computes LNFG of p ∧ ¬φ, G=(CL(p ∧ ¬φ), EL(p ∧ ¬φ), V0, 𝕃 = {𝕃1,...,𝕃m})*/
begin function
  CL(p ∧ ¬φ) = {p ∧ ¬φ}; EL(p ∧ ¬φ) = ∅; mark[p ∧ ¬φ] = 0; AddE = AddN = 0;
  k = 0; 𝕃 = ∅;
  while there exists R ≡ r ∧ ¬φ ∈ CL(p ∧ ¬φ) \ {ε}, and mark[r ∧ ¬φ] == 0
     if R ≡ ⋀_{i=1}^n R_i and ∃R_i ≡ P_i; Q_i
       if ∃ ⋀_{i=1}^n R'_i with some R'_i ≡ P ∧ fin(l_S); Q in 𝕃_s,
          Transform R as ⋀_{i=1}^n R_i with some R_i being P_i ∧ fin(l_S); Q_i; 𝕃_s = 𝕃_s ∪ {R};
     else k=k+1; Transform R as ⋀_{i=1}^n R_i with some R_i being
          P_i ∧ fin(l_K); Q_i;
          𝕃_k = {R}; 𝕃 = 𝕃 ∪ {𝕃_k};
    do mark[r ∧ ¬φ] =1;                                        /*marking r ∧ ¬φ is decomposed*/
    Q =NF(r)∧NF(¬φ);
     case
       Q is ⋁_{j=1}^h Q_ej ∧ empty: AddE=1;                      /*first part of NF needs added*/
       Q is ⋁_{i=1}^k Q_i ∧ ○Q'_i : AddN=1;                       /*second part of NF needs added*/
       Q is ⋁_{j=1}^h Q_ej ∧ empty ∨ ⋁_{i=1}^k Q_i ∧ ○Q'_i: AddE=AddN=1;   /*both parts need added*/
     end case
     if AddE == 1 then                                          /*add first part of NF*/
       CL(p ∧ ¬φ) = CL(p ∧ ¬φ) ∪ {ε}; EL(p ∧ ¬φ) = EL(p ∧ ¬φ) ∪ ⋃_{j=1}^h {(R, Q_ej, ε)};
       AddE=0;
     if AddN == 1 then for i = 1 to k,                          /*add second part of NF*/
       if Q'_i ∉ CL(p ∧ ¬φ),CL(p ∧ ¬φ) = CL(p ∧ ¬φ) ∪ ⋃_{i=1}^k {Q'_i};
         if Q'_i is not false, mark[Q'_i]=0; else mark[Q'_i]=1;
       EL(p ∧ ¬φ) = EL(p ∧ ¬φ) ∪ ⋃_{i=1}^k {(R, Q_i, Q'_i)}; AddN=0;
   end while
   return G;
End function
```

an MSVL program p. Finally, satisfiability of $p \wedge \neg\phi$ based on LNFG can be referred to Chapter 4. For the complexity of the algorithm, roughly speaking, if $|cl(p)| = O(n)$ and $|cl(\neg\phi)| = O(m)$, at most, $|cl(p \wedge \neg\phi)| = O(n \times m)$.

17.6.5.2 Model Checker

We have developed a model checking tool (prototype) based on our model checking algorithm. Generally, the prototype can work in three modes: modeling, simulation and verification. With the modeling mode, given the MSVL program p of a system, the state space of the system can implicitly be given as an NFG of p. In the simulation mode, an execution path of the NFG of the system is output according to minimal model semantics of MSVL [19]. Under the verification mode, given a system model described by an MSVL program, and a property specified by a PPTL formula, it can automatically be checked whether the system satisfies the property or not, and the counterexample can be given if the system does not satisfy the property.

17.6.5.3 Example

As an example, consider the mutual exclusion problem of two processes competing for a shared resource which is often analyzed in literatures. Pseudo code for this example can be given as shown in Figure 17.13. We assume that the processes are executed in one time unit in an interleaving manner. The wait statement makes a process into sleep. When all processes are asleep the scheduler tries to find a process satisfying waiting condition and reactivates the corresponding process. If all of the waiting conditions are false the system stalls. This mutual exclusion problem can be coded in MSVL as shown below. Notice that the underlined code can be ignored with the current part since it is for the purpose of making a counterexample later on.

	process A
	forever
$A \cdot pc = 0$	**wait for** $B{\cdot}pc = 0$
$A \cdot pc = 1$	*access shared resource*
	end forever
	end process

	process B
	forever
$B \cdot pc = 0$	**wait for** $A \cdot pc = 0$
$B \cdot pc = 1$	*access shared resource*
	end forever
	end process

FIGURE 17.13 Pseudo code for two processes A and B competing for a shared resource.

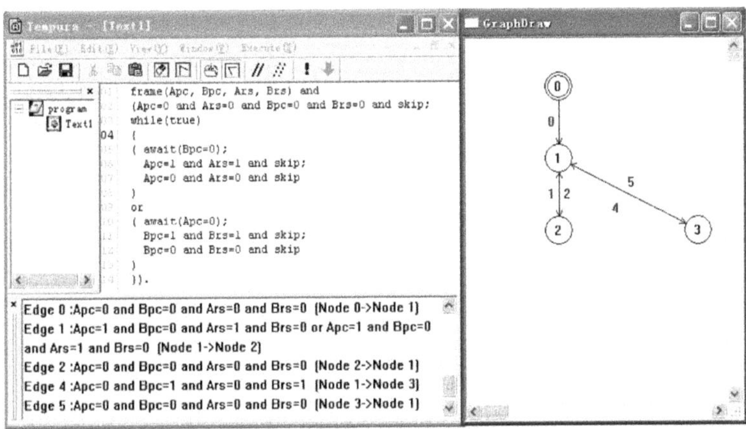

FIGURE 17.14 NFG of the mutual exclusion problem.

frame(Apc, Bpc, Ars, Brs) and
(Apc=0 and Ars=0 and Bpc=0 and Brs=0 and skip;
while(true){
(await(Bpc=0);
Apc=1 and Ars=1 and skip;
Apc=0 and Ars=0 and skip)
or
(Apc=1 and Ars=1 and Bpc=1 and Brs=1 and skip;
Apc=0 and Ars=0 and Bpc=0 and Brs=0 and skip)
or
(await(Apc=0);
Bpc=1 and Brs=1 and skip;
Bpc=0 and Brs=0 and skip) }).

In the MSVL program, $Ars = 1$ ($Brs = 1$) means processes A (B) is in the shared resource, while $Ars = 0$ ($Brs = 0$) means processes A (B) has released the shared resource. With the modeling mode of MSVL, the state space of the mutual exclusion problem can be created and presented as an NFG as shown in Figure 17.14. In the NFG, edge 0 indicates that neither process A nor B is in the shared resource; edge 1 (from 1 to 2) indicates that process A is in the shared resource and B is not; edge 2 (from 2 to 1) indicates that neither process A nor B is in the shared resource; edge 4 (from 1 to 3) indicates that process B is in the shared resource and A is not; edge 5 (from 3 to 1) indicates that neither process A nor B is in the shared resource.

As a result, the property, "processes A and B will never be in the shared resource in the same time," should hold. That is, *Ars* and *Brs* will never be assigned with 1 at the same time. By employing propositions p and q to denote $Ars = 1$ and $Brs = 1$, respectively, this property can be specified by $\Box(\neg(p \wedge q))$ in PPTL.

FIGURE 17.15 Verification result.

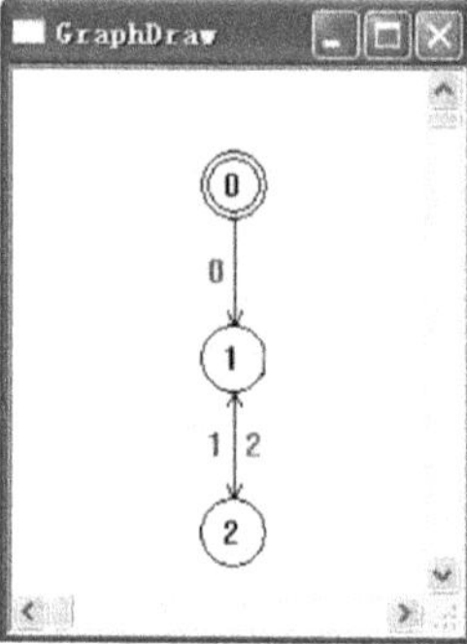

FIGURE 17.16 Verification result.

FIGURE 17.17 Verification result.

With the verification mode of MSVL, we add the following code

</define p:Ars=1; define q:Brs=1; always(~(p and q))/>

to the beginning of the MSVL code for the mutual exclusion problem, then run the code with model checker, an empty NFG with no edges is produced as shown in Figure 17.15. This means that the formula is unsatisfiable, and the system satisfies the property.

Suppose that, when A is in the shared resource, B is possible in the shared resource. To model it, we add the code with underline to the previous MSVL code of the system. Now we check whether or not the system satisfies $\Box(\neg(p \wedge q))$, and the resulting NFG is produced as shown in Figure 17.16. Obviously, there exist infinite paths where node 1 and node 2 appear infinitely often. Thus, the property cannot be satisfied.

As you can see, MSVL and PPTL can be used to verify properties of programs in a similar way as kripke structures (or automata) and PLTL (or CTL) do. However, the expressive power of PLTL and CTL is limited. They cannot express regular properties such as "a property Q holds at even states ignoring odd ones over an interval (or computation run)" [22]. This type of property can be specified and verified by PPTL. In the following, we further verify such a property of the mutual exclusion problem.

The mutual exclusion problem has a special property: "neither A nor B is in the shared resource, immediately after A or B released the shared resource; and when A or B is in the shared resource, it releases the resource at the next state." Basically, this property is a regular property. It can be specified by, $\neg p \wedge \neg q \wedge (\bigcirc^2 \neg p \wedge \neg q \wedge empty)^*$ in PPTL. Thus, we can add *</define p:Ars=1; define q:Brs=1; ~p and ~q and (next(next(~p and ~q and empty)))# />* to the beginning of the MSVL code for the mutual exclusion problem, then run the code with model checker, an empty NFG with no edges is produced as shown in Figure 17.17. Hence, the formula is unsatisfiable, and the system satisfies the property.

17.7 State Space Explosion Problem

The exponential growth of the state space is referred to as the state-space explosion problem. This problem not only appears for model checking, but also for various other problems like performance analysis that are strongly based on state space exploration. In order to combat this problem in the setting of model checking various approaches have been considered.

17.7.1 Symbolic Model Checking

Various methods for model checking, like model checking linear and branching temporal logics, are all based on a system description in terms of (a kind of) finite state-transition diagram. The most obvious way to store this automaton in memory is to use a representation that explicitly stores the states and the transitions between them. Typically data structures such as linked lists and adjacency matrices are used. The idea of the BDD (Binary Decision Diagram) technique is to replace the explicit state-space representation by a compact symbolic representation. The state space, being a Büchi automaton, CTL-model, or a region automaton, is represented using binary decision diagrams, models that are used for the compact representation of boolean functions. A BDD [23] is similar to a binary decision tree, except that its structure is a directed acyclic graph, and there is a total order on the occurrences of variables that determine the order of vertices when traversing the graph from root to one of its leaves (i.e. end-point). Once the state space is encoded using BDDs, the model checking algorithm at hand is carried out using these symbolic representations. This means that all manipulations performed by these algorithms must be carried out, as efficiently as possible, on BDDs.

An important selling point of BDD is that when the variables are in a specific order, the size of the BDD can be exponentially reduced. However, the BDD size can vary greatly for different variable orderings. The problem of finding an optimal variable ordering is NP-hard. Even though BDD cannot guarantee the avoidance of the state space explosion in all cases, it provides a compact representation of several systems, allowing these systems to be verified that would be impossible to handle using explicit state enumeration methods. It turns out that in particular the operations needed for CTL model checking can be carried out efficiently on BDDs.

17.7.2 Partial Order Reduction

Systems consisting of a set of components that cooperatively solve a certain task are quite common in practice, for example, hardware systems, communications protocols, distributed systems, and so on. Typically, such systems are specified as the parallel composition of n processes P_i, for $0 < i \leq n$. The state space of this specification equals in worst case $\Pi_{i=1}^{n} |P_i|$, the product of the state spaces of the components. A major cause of this state space explosion is the representation of parallelism by means of interleaving. Interleaving is based on the principle that a system run is a totally ordered sequence of actions. In order to represent all possible runs of the system, all possible interleaving of actions of components need to be represented and, consequently, parallel composition of state spaces amounts to building the Cartesian product of the state spaces of the components. For n components, thus $n!$ different orderings (with $2n$ states) are explicitly represented. In a traditional state space exploration all these orderings are considered. For checking a large class of properties, however, it is sufficient to check only some (in most of the cases even just one) representative of all these interleaving. For example, if two processes both increment an integer variable in successive steps, the end result is the same regardless of the order in which these assignments occur. The underlying idea of this approach is to reduce the interleaving representation into a partial-order representation. System runs are now no longer totally ordered sequences, but partially ordered sequences. In the best case, the state space of the parallel composition of n processes P_i now reduces to $\Sigma_{i=1}^{n} |P_i|$ and only a single ordering of $n+1$ states needs to be examined.

17.7.3 Bounded Model Checking

The original motivation of bounded model checking was to leverage the success of SAT in solving Boolean formulas to model checking. Bounded model checking relies on the observation that state sequences of fixed length, say k, can be represented using k copies of the variables used to represent a single state. The set of fixed-length sequences that represent termination or looping runs of a given finite state transition system T can, therefore, be encoded by formulas of propositional logic, as well as the semantics of PLTL formula φ over such sequences. For any given length k, the existence of a state sequence of length k that represents a run of T satisfying φ can thus be reduced to the satisfiability of a certain propositional formula, which can be decided using efficient SAT solvers. During the last few years there has been a tremendous increase in reasoning power of SAT solvers. They can now handle instances with hundreds of thousands of variables and millions of clauses. Symbolic model checkers with BDDs, on the other hand, can check systems with no more than a few hundred latches. Though clearly the number of latches and the number of variables cannot be compared directly, it seemed plausible that solving model checking with SAT could benefit the former. Experiments have shown that it can solve many cases that cannot be solved by BDD-based techniques. The inverse is also true: there are problems that are better solved by BDD-based techniques but cannot by BMC. BMC joins the arsenal of automatic verification tools but does not replace any of them.

17.7.4 Abstraction

Although techniques such as symbolic model checking and partial-order reduction attempt to battle the infamous state explosion problem, the size of systems that can be analysed using model checking remains relatively limited: even astronomical numbers such as 10^{100} states are generated by systems with a few hundred bits, which is a far cry from realistic hardware or software systems. Model checking must, therefore, be performed on rather abstract models. When the analysis of big models cannot be avoided, it is rarely necessary to consider them in full detail in order to verify or falsify some given property. This idea can be formalized as an abstraction function (or relation) that induces some abstract system model such that the property holds of the original, concrete model if it can be proven for the abstract model.

Unfortunately, the abstract model usually does not satisfy exactly the same properties as the original one so that a process of refinement may be necessary. Generally, one requires the abstraction to be sound (the properties proved on the abstraction are true of the original system); However, most often, the abstraction is not complete (not all true properties of the original system are true of the abstraction). An example of abstraction is, on a program, to ignore the values of non boolean variables and to only consider boolean variables and the control flow of the program; such an abstraction, though it may appear coarse, may in fact be sufficient to prove, for example, properties of mutual exclusion. Counter-example guided abstraction refinement (CEGAR) begins checking with a coarse (imprecise) abstraction and iteratively refines it. When a violation (counterexample) is found, the tool analyzes it for feasibility (i.e., is the violation genuine or the result of an incomplete abstraction?). If the violation is feasible, it is reported to the user; if it is not, the proof of infeasibility is used to refine the abstraction and checking begins again.

References

1. E.M. Clarke and E.A. Emerson. Design and syntesis of synchronization skeletons using branching time temporal logic. In *Logic of Programs 81*, LNCS 131, Springer, New York City, NY, 1981.
2. J.P. Quielle and J. Sifakis. Specification and verification of concurrent systems in CESAR. In *Proceedings of the 5th International Symposium on Programming*, Denmark, pp. 337–350, 1981.
3. A. Pnueli. The temporal logic of programs. In *Proceedings of the 18th IEEE Symposium on Foundations of Computer Science*. IEEE, New York, 46–67.2, 1977.
4. B.Moszkowski. Reasoning about Digital Circuits. Ph.D thesis, Department of Computer Science, Stanford University. TRSTAN-CS-83–970, 1983.

5. B.C. Moszkowski. Compositional reasoning about projected and infinite time. In *Proceeding of the First IEEE Int'l Conf. on Enginneering of Complex Computer Systems (ICECCS'95)*, Japan, pp. 238–245. IEEE Computer Society Press, 1995.
6. Z. Duan, M. Koutny and C. Holt. Projection in Temporal Logic Programming. Logic Programming and Automated Reasoning, LNCS 822, 1994.
7. Z. Duan. An Extended Interval Temporal Logic and A Framing Technique for Temporal Logic Programming. PhD thesis, University of Newcastle Upon Tyne, May 1996.
8. Z. Duan. *Temporal Logic and Temproal Logic Programming.* ISBN 7-03-016651-5/TP.3158, Science Press, China, 2006.
9. S.A. Kripke. Semantical analysis of modal logic I: Normal propositional calculi. *Z. Math. Logik Grund. Math.* 9, 67–96, 1963.
10. P. Wolper, M. Vardi, and A.P. Sistla. Reasoning about infinite computation paths. *Proceedings 24th IEEE Symposium on Foundations of Computer Science*, Tucson, AZ, pp. 185–194, 1983.
11. E.A. Emerson and C.L. Lei. Modalities for model checking: Branching time logic strikes back. In *Proceedings 12th ACM Symposium on Principles of Programming Languages*, New Orleans, LA, pp. 84–96, 1985.
12. S. Kono. A combination of clausal and non-clausal temporal logic programs. In *Lecture Notes in Artificial Intelligence*, vol. 897, pp. 40–57. Springer Verlag, New York, NY, 1995.
13. H. Bowman and S. Thompson. *A Tableau Method for Interval Temporal Logic with Projection.* Springer-Verlag, Berlin H. de Swart (ed.):TABLEAUX98, LNAI 1397, 1998.
14. H. Bowman and S. Thompson. A decision procedure and complete axiomatization of interval temporal logic with projection. *Journal of Logic and Computation* 13(2), 195–239, 2003.
15. A. Tarski. A lattice-theoretical fixpoint theorem and its applications. *Pacific Journal of Math*, 5, 285–309, 1955.
16. C. Tian and Z. Duan. Complexity of propositional projection temporal logic with star. *Mathematical Structure in Computer Science* 19(1): 73–100, 2009.
17. L. Stockmeyer. The Complexity of Decision Problems in Automata Theory and Logic, PhD thesis, MIT, Available as Project MAC Technical Report 133, 1974.
18. Z. Duan, X. Yang and M. Kounty. Semantics of Framed Temporal Logic Programs. *Proceedings of ICLP 2005*, Barcelona, Spain, LNCS 3668, pp. 256–270, Oct. 2005.
19. Z. Duan, X. Yang and M. Koutny. Framed Temporal Logic Programming. *Science of Computer Programming* 70, 31–61, Elsevier Science, 2008.
20. X. Yang and Z. Duan. Operational semantics of framed tempora. *Journal of Logic and Algebraic Programming*, 78(1), 22–51, Nov./Dec. 2008.
21. Glynn Winskel. *The Formal Semantics of Programming Languages.* Foundations of Computing. Cambridge, Massachusetts: The MIT Press.
22. Gerard J. Holzmann. The model checker spin. *IEEE Transactions on Software Engineering*, 23(5), 279–295, May 1997.
23. S.B. Akers. On a theory of boolean functions. *Journal of Society Industrial Math*, 7(4), 487–497, 1959.
24. Z. Duan, C. Tian, and Li Zhang. A decision procedure for propositional projection temporal logic with infinite models. *Acta Informatica*, 45(1), 43–78, 2008.
25. Z. Duan and C. Tian. *A Unified Model Checking Approach with Projection Temporal Logic.* ICFEM2008, LNCS5256, pp. 167–186, Springer Verlag, October, 2008.
26. Joost-Pieter Katoen. *Concepts, Algorithms, and Tools for Model Checking.* Lecture Notes of the Course Mechanised Validation of Parrel Systems, 1999.

18

System Modeling with UML State Machines

Omar El Ariss
Penn State University

Dianxiang Xu
Dakota State University

18.1 Introduction

State machine diagrams are part of the Object Management Group Unified Modeling Language (OMG UML) (Object Management Group 2010). UML state machine diagrams, which were previously called statechart diagrams in UML 1, are an object-based variation of Harel's statecharts (Harel 1987) with the addition of activity diagram notations. David Harel introduced statecharts as an extension to state machines (Hopcroft et al. 2006). The aim was to represent the behavior of complex systems in a clear and understandable form without suffering from explosion in the number of states and edges. The main originality of statecharts was in the introduction of orthogonality and hierarchy.

UML state machine diagrams have two different types of state machines:

- *Behavioral state machines*: are graphical models used to depict the functional specifications of a system. These models are used to represent the dynamic behavior of the system. The behavior of the system is modeled through the interaction of the system's subcomponents with each other and with the system's environment. Behavioral state machine models are usually constructed during the design phase of the software development lifecycle and are used for the analysis, verification, and correctness of the design. They are also used to guide software developers through the implementation process.
- *Protocol state machines*: are used to model the sequence of events a UML classifier can respond to, and under which conditions, without representing the classifier's behavior. Therefore, protocol state machines are subset of behavioral state machines without the use of behavioral notations. These

state machines identify the events that can be called by specifying their usage mode of operation (under which state and precondition they can be called) and their expected results (through the use of post conditions). The states of a protocol state machine cannot have deep history, shallow history, or actions. Protocol transitions, on the other hand, only focus on the operations and not on the actions.

In what follows, we are going to concentrate on how to use UML state machine diagrams to model the dynamic behavior of computer systems. That means, we are going to only focus on behavioral state machines and not on protocol state machines. The rest of this chapter further discuss the elements of a state machine and how they are used in order to model the behavior of systems. The concepts are defined then described through the use of three running examples: a light controller, a phone and a microwave system. We will also give a brief summary on recent research and real applications of UML state machines.

18.2 UML Behavioral State Machines

Behavioral state machines are useful to depict the behavior of reactive systems. These are systems that change their behavior and output in response to external (either the user or the environment) interactions or from internal interactions such as communication between the system components or its subcomponents. In addition, state machines are capable of representing complex behavior by simplifying the system into smaller parts. This type of representation supports modular design and allows scalability, inheritance, and reusability of system components.

State machines express the behavior of the system by showing the relation between the system's current state with its past states. This allows a detailed analysis of the system behavior, and the understating of the operations of the system from a particular state. It also shows how a particular behavior is not only a direct cause of the system's current input but also caused by its history of past interactions.

A state machine diagram contains the following graphic elements:

- *States*: A state is a mode, situation or a stage of interest in the behavior of a system. States are either dynamic or static in nature, and can be used to represent hierarchy and orthogonality.
- *Transitions*: A transition is a relationship between two states, the source state and the target state. Transitions depict the system and subsystem interactions.
- *Events*: An event is an occurrence that causes change in the behavior of the system. In other words, an event is a trigger that causes a transition from one state to another.
- *Pseudostates*: are various types of graphical elements that are neither states nor transitions. They are used to enrich the descriptiveness and preciseness of both transitions and the behavior of the system as a total.

In this chapter, we use the following naming convention to distinguish between states and events: state names start with an uppercase character while event names start with a lowercase character.

18.3 States

A state is modeled using a circular rectangle with the state's name being represented either inside the triangle or outside the triangle using a name tab. Figure 18.1 shows three different ways to represent a state called *Off*:

1. The first representation has the state name inside the rectangle. This type of representation is usually used to model simple static states.
2. The second representation has the state name inside a name tab. This type of state representation is useful when modeling orthogonal composite states.

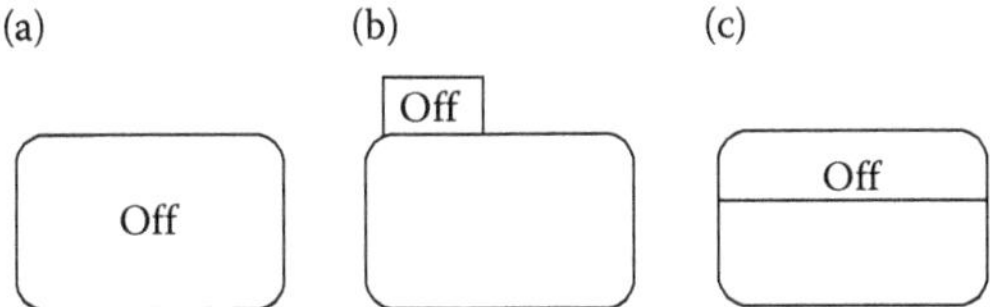

FIGURE 18.1 Giving names to states. Three different ways to give names to states: (a) name is directly inside the rectangle, (b) name is inside a tab, (c) name is inside a separate compartment.

3. The third representation divides the rectangle into two different compartments. The state name is given inside the first compartment. This type of representation is usually useful when modeling simple dynamic states or hierarchical composite states.

There are three different types of states:

1. *Simple state*: represents a basic situation. A simple state can either be dynamic or static in behavior. Static states are states that have no internal behavior and only wait for a certain event (condition) to occur. Consider as an example the behavioral representation of a light switch, which is shown in Figure 18.2. This model has two states: a *Lights Off* state and a *Lights On* state, where the *Lights Off* state indicates the light is turned off while the *Lights On* state indicates that the lights are on. Both of these states are simple because they are representing a basic mode of behavior: light presence or light absence. In addition, both of these states are static in their behavior; the moment one of these states becomes active then all it does is to wait for a particular event to occur.

   Dynamic states on the other hand, are states that perform some type of action while they are active. At the time of execution of an action, the state will not accept any event until the action is completed. UML state machines support three types of behavior inside a state:
   - *Entry*: the specified behavior is executed at the moment the state becomes active.
   - *Exit*: the specified behavior is executed at the moment the state becomes inactive.
   - *Do*: the specified behavior continues to execute as long as the state is active.
2. *Composite state*: is a state that supports hierarchy or orthogonality. In case of hierarchy, the composite state is composed of one region. As for orthogonality, a composite state encloses two or more regions, where each region executes concurrently. A region in a hierarchical or orthogonal state contains states and transitions. The states inside a region in turn can also be simple states, composite states, or submachine states.

Hierarchy is introduced by allowing states to encompass other states. Hierarchical states represent abstraction of the software behavior, where lower level abstractions are depicted as states that are enclosed inside other states. To demonstrate this point, let us consider the behavior of a phone. Figure 18.3 shows a partial representation of the phone. Two simple states are used:

- *Off state*: to depict that the phone is turned off and is not being used.
- *On state*: to depict that the phone is turned on and is currently being used.

This model depicts the phone behavior at a high level of abstraction. To further analyze and understand the behavior of the phone while it is being used requires a more detailed representation, or a lower level abstraction, of what is happening while the phone is on. This in depth representation of the operational

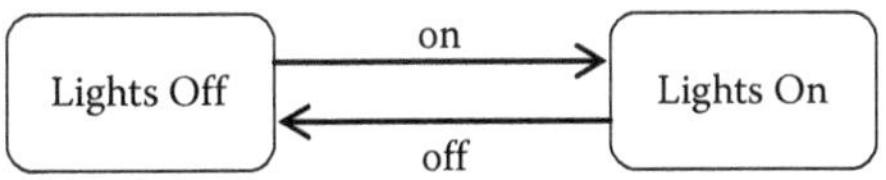

FIGURE 18.2 State machine for a light switch.

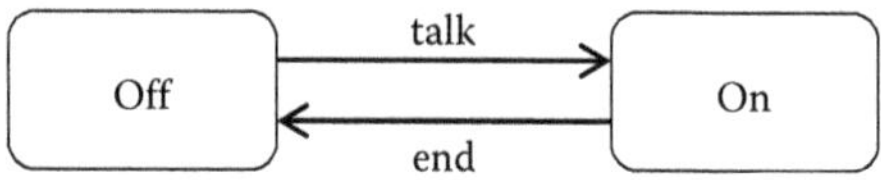

FIGURE 18.3 Partial state machine for a handheld phone.

phone behavior should include multiple modes of interest, such as calling and dialing. Therefore, there is a need to replace the static *On* state with a hierarchical composite state.

Figure 18.4 shows the representation of the phone behavior while it is on as a composite state that supports hierarchy. This representation is not intended to be complete, but was partially constructed to explain the difference between simple states and hierarchical states. As we further discuss the rest of the state machine components, we are going to return to this phone model and improve on it.

This composite state consists of three simple states and two transitions. Two of these states include dynamic behavior. For instance, the *Ready* state indicates that the phone is ready and can be used to make a call. The action of having a dial tone is performed immediately upon entry to this state. The moment this state exits, due to a *press digit* event, the dial tone should be stopped.

Moreover, through the use of hierarchy it is possible to introduce lower level behavior with ease and without worrying about intricate alterations to the model. For instance, suppose that we want to show the detailed behavior of the dialing process. In that case, the *Dialing* state of Figure 18.4 can be replaced by a composite state that describes the dialing behavior at a lower level of abstraction.

Orthogonal states represent concurrent behavior. This allows the representation of system functionalities that should be present and executing at the same time. Orthogonality is supported through the use of parallel states separated by dashed lines (and-states). Each dashed line will divide the composite state into two separate regions. These orthogonal regions will run concurrently. In other words, the software system or subsystem must be in all its regions the moment the composite state becomes active. Communication between orthogonal parts is done through the use of actions and through broadcasting of events. This notion of orthogonality is very helpful in representing subcomponents of a system while pertaining modularity in the behavioral representation. The notion of communication between orthogonal components will be discussed in Section 18.8.

Consider again the light switch example but now with the support of a dimmer. This light controller allows the user to switch the lights on and off and also to control the brightness of the bulbs. The dimmer allows the user to select three level of brightness: dim, medium, and bright. Therefore, this light controller has two main functionalities or modes of operation: a switch and a dimmer. These operations can run simultaneously. Thus, the light controller state machine should support orthogonality.

Figure 18.5 shows the representation of the light controller behavior. This composite state supports concurrency of two functionalities, and is, therefore, divided into two regions. The first region represents the behavior of the switch. The second region represents the behavior of the dimmer. The dimmer representation is composed of three simple states that depict the mode of brightness the light is in. Concurrency means that the orthogonal regions are both active at the same time. Consequently, the

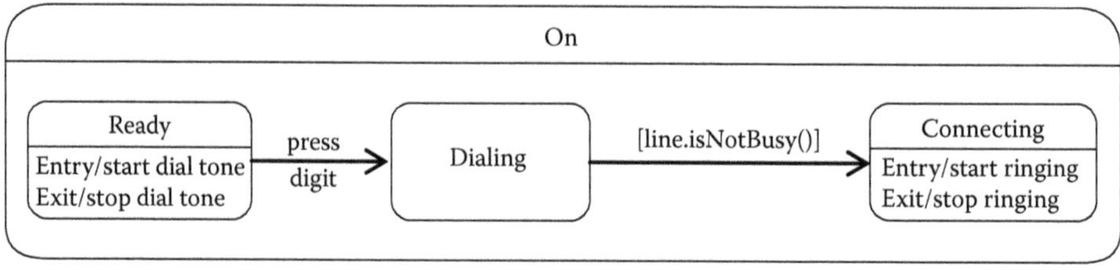

FIGURE 18.4 Partial representation of the On hierarchical composite state.

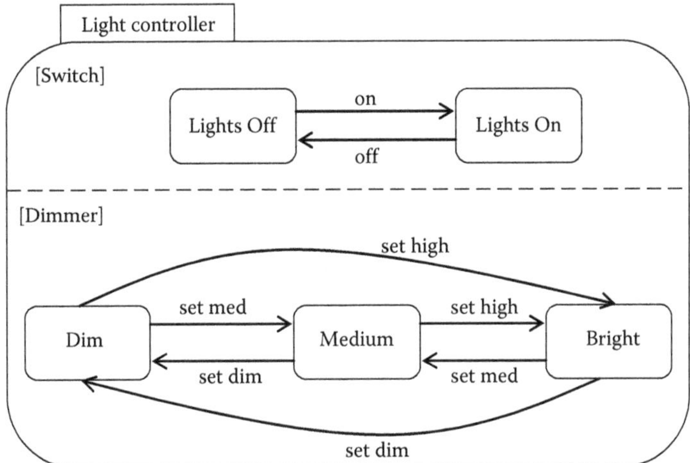

FIGURE 18.5 Partial representation of a light controller.

behavior of each region can change simultaneously, where the light can be on or off as the level of brightness is being modified.

3. *Submachine state*: a submachine state is semantically equivalent to a composite state. All that is discussed previously about composite states apply to submachine states. The main advantages of using submachine states are reusability and factoring of common behavior. These advantages will be apparent as we discuss the pseudostates, mainly speaking the entry and exit points.

18.4 Transitions

States represent the modes of operation that we are interested in while modeling the behavior of an object. Transitions are used to depict the object's dynamic behavior, hence the occurrence of change in behavior. A transition is graphically represented as a directed arc from a source state to a target state. Figure 18.6 shows a general representation of a transition. A transition can be optionally labeled by the following elements:

- *Event-list*: an event or list of events that cause the transition to be triggered or fired. In case there is a list of events, then the occurrence of one event from this list is enough to trigger the transition.
- *Guard*: is a conditional statement that is evaluated to determine whether the transition should be triggered or not. Figure 18.4 has two transitions, the first is triggered by a *press digit* event while the second is triggered only when the guard (having the destination number not to be busy) is to true.
- *Action-list*: an operation or set of operations that will execute when the transition is triggered.

During the operational behavior or execution of a system, a particular state can be active or inactive. The occurrence of a transition indicates the following:

- The source state was the active state before the triggering of the transition.
- The source state exits because of the transition occurrence and becomes inactive.

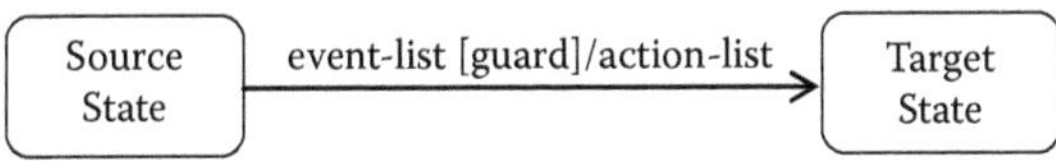

FIGURE 18.6 Graphical representation for a transition.

- The target state becomes the active state.

In case, the active state is a composite state then all the regions inside the composite state are also active, where each region should have one and only one active state. Let us consider the situation, where the light controller state of Figure 18.5 becomes active. That means both the switch and the dimmer regions are also active, where one state from each region is active. For example, the light controller can be in both *Lights On* and *Medium* states thus indicating that the lights are switched on with a medium brightness. In case of the switch region, it is impossible to have both the *Lights On* state and the *Lights Off* state to be active at the same time. Only one of these states can be active at a time.

18.5 Events

Events are internal or external interactions that cause transitions in the behavior of an object or a system. The response of the system to an event depends on its current state and the type of events this state accepts. If an event occurs while the system is currently at a state that cannot handle this type of event, then the system behavior will not be changed and the event will be ignored. For example, Figure 18.3 has two events: a *talk* event that turns on the phone and an *end* event that turns off the phone. Let us consider that the current state of our modeled phone is the *Off* state, then the system will only react to one event which is pressing on the talk button. The phone will ignore any other event it receives, such as pressing on the end button or on the dial numbers, and will remain in its current state. The only event that the *Off* state accepts is the *talk* event, which will cause the phone's current active state to be the *On* state.

The behavior of some systems might require time dependency. UML state machines support time through the use of timed events. This type of event keeps track of the time a state is active and compares it to a boundary value. The event happens when the boundary value is exceeded. The general format for this type of event is: after(time), where time can either be absolute or relative.

The *Ready* state of Figure 18.4 indicates that the phone is turned on, has a dial tone and is ready to be used for making a call. In conventional phones, the presence of the dial tone indicates that the phone can currently be used. This dial tone will disappear after a certain amount of time and will be replaced by another tone (called the partial dial tone) that informs the user to hang up or press the flash button. In order to include this behavior in our phone model, we have to use a timed event. Figure 18.7 shows the newly added behavior to the *On* state. The behavior now indicates that the phone will be ready to be used, and if the user did not attempt any call during 10 s then a timed event will occur causing a transition to the *Not Ready* state. The user from the *Not Ready* state can either press on the flash button to return to the *Ready* state or can exit the *On* state by pressing the end button, as shown in Figure 18.3.

18.6 Pseudostates

In addition to states, events, and transitions, UML state machines use pseudostates to enrich and precisely define the dynamic behavior of a system. The latest version of UML introduced three additional graphic elements: entry point, exit point, and terminate pseudostates. Currently, there are 10 different types of pseudostate, where each type has its own representation and usage. This section discuss seven different types of pseudostates. The complete specifications for the pseudostates are described and defined in the OMG Unified Modeling Language Superstructure (Object Management Group 2010).

18.6.1 Initial Pseudostate

When dealing with the activation of composite states, there is a need to clearly specify which of the enclosed states will become active. Looking at Figure 18.5, specifically the switch region, it is hard to tell which of the two states is the default state. Although, our own experience that comes from dealing with light switches tells us that the default state should be *Lights Off*, the model does not explicitly depict that.

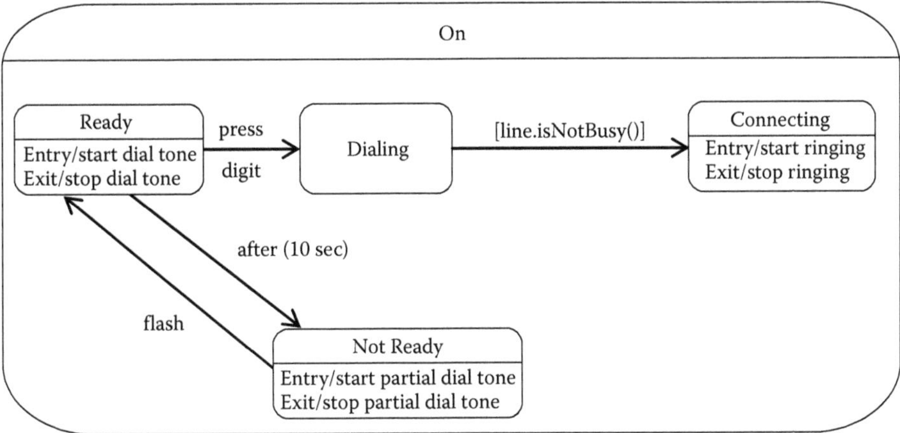

FIGURE 18.7 Partial representation of the On state with a timed event.

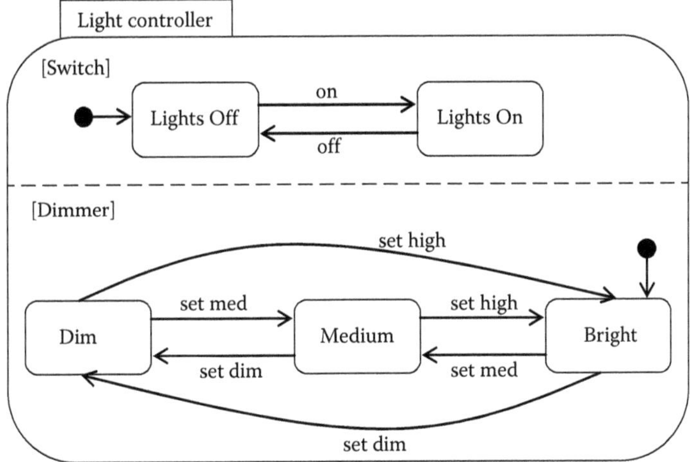

FIGURE 18.8 State machine for a light controller.

Similarly, examining the dimmer region does not inform us which of the three states is initially activated. Therefore, there is a need for a graphic notation that clearly identifies the default state inside a region. Each region should have at most one initial pseudostate.

Figure 18.8 now clearly defines the default states for the two regions of the light controller state machine. An initial pseudostate is represented as a black filled circle with a transition emanating from it toward the default state. Therefore, there are two default states, the *Lights Off* state and the *Bright* state for the switch region and the dimmer region, respectively. It is clear now from looking at Figure 18.8 that the default behavior for the light is being off while set to the highest brightness.

18.6.2 Choice Pseudostate

A choice pseudostate is used to depict the behavior of dynamic conditional branching. A choice pseudostate is represented as a diamond-shaped symbol that receives one transition without a trigger as an incoming source. From this pseudostate, decision choices are represented as two or more outgoing transitions each having a distinct guard. It is important that the choice representation of the guard conditions

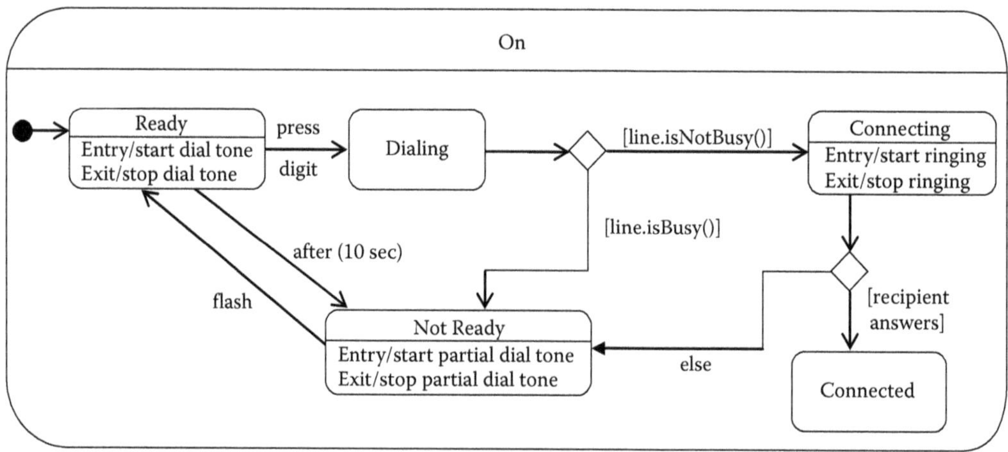

FIGURE 18.9 Partial representation of the On state with choice pseudostates.

should be well defined in order to cover all the decision situations. UML state machines also support static conditional branching through the use of a junction pseudostate.

Going back to the phone example, the modeled representation of the *On* state does not yet cover the behavior after the call is getting connected. What happens next depends on the behavior of the person getting this call. In case the recipient answers, then the phone call gets connected. On the other hand, if nobody answers then the ringing tone should be replaced by the partial dial tone indicating that the phone cannot currently be used. Figure 18.9 shows the modified behavior of the *On* state. The decision making process is shown in the figure through a transition from the *Connecting* state to the diamond symbol (the choice pseudostate). At the point of a transition occurrence, the guards of the decision pseudostate are evaluated dynamically to determine which path is to be chosen. In this example, there is an if–else decision-making process, where there is one guard and an *else* notation. In case the guard evaluates to true, then a transition to the *Connected* state occurs. If the *[recipient answers]* guard is false then the else path is chosen and the *Not Ready* state becomes the active state. The use of the *else* notation is very beneficial when dealing with choice pseudostates because it makes sure that the decision making process is well defined.

Figure 18.9 also shows another choice pseudostate that evaluates whether the callee's line is busy or not. In case the line is not busy, then the call can be made. This is shown as a transition to the *Connecting* state. If the targeted line is busy then the call cannot be made and the phone notifies the user through a partial dial tone. This is shown in the figure as a transition to the *Not Ready* state.

18.6.3 History Pseudostates

In some systems, it is relevant to remember the last current internal state of a composite state or a submachine state when this composite state turns inactive. Examples of this type of situation are systems that remember their last configuration before they are turned off. UML state machines support this type of behavior through the use of shallow and deep history pseudostates. A shallow history pseudostate is used to remember the last active internal state of a composite state but not the substates of this internal state. This saved configuration will be used as the starting configuration when the composite state becomes active again. Deep history pseudostate on the other hand, keeps track of the recent configuration of both the internal state and its substates.

Consider the simplified behavior of a microwave system. The system can be used for cooking or heating food for a specified amount of time. At any moment while the microwave is being operational, the user is allowed to cancel the current operation or to pause it by either pressing on the start button or by opening

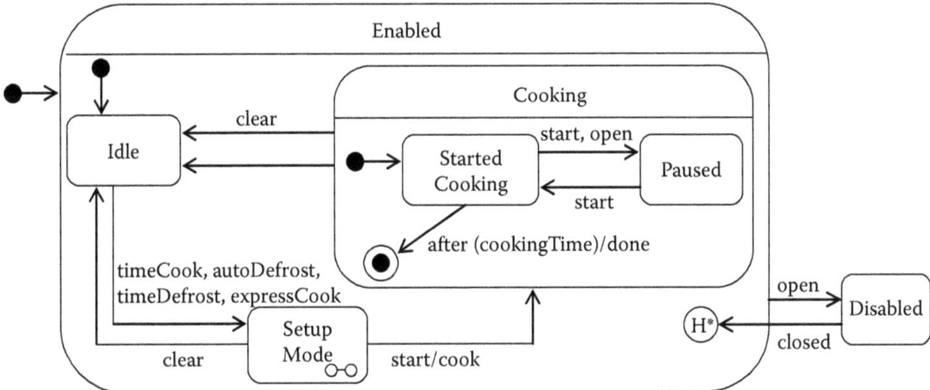

FIGURE 18.10 State machine representing the cooking behavior of a microwave.

the microwave door. In case the microwave door is opened, then the cooking operation will only resume when the door is closed again followed by a press on the start button. Figure 18.10 shows the state machine for the cooking behavior of a microwave.

At the high level of operation, the microwave model has two states:

- An enabled state, where the microwave can be used.
- A disabled state, where the microwave cannot be used because the door is open.

The initial state for the microwave system is the *Enabled* state. The user through this state can select one from the many cooking functions and can start, stop, or pause the cooking operation. Let us consider as an example that the user opens the microwave door while its current state is the cooking state. In this case, a transition from the *Started Cooking* state to the *Paused* state takes place, the *Enabled* state becomes inactive and then the *Disabled* state becomes the active state. In case the door gets closed, the *Enabled* state becomes active again and should return to the last internal state configuration it was in before the door was first opened, which is the *Paused* state.

In order to model that behavior using state machine representations, we should use a deep history pseudostate. A shallow history pseudostate in this case will not be sufficient because it will only remember that the microwave was in the *Cooking* state without keeping track which of its internal states was last active. This will cause the cooking operation to start again from the beginning. The use of a deep history pseudostate, shown in Figure 18.10 as a circle with "H*" enclosed in it, will make sure that the *Cooking* state is enabled, and its currently activated internal state is the *Paused* state.

18.6.4 Entry and Exit Points

Before the introduction of entry and exit points, UML state machines had no restrictions at all into how to enter or exit a composite state. A submachine or a composite state can be entered either through a transition to its boundary or through a direct transition to one of its internal states. A transition to the boundary of a submachine state is semantically equivalent to entering its region through its initial state. Similarly, a submachine can be exited either through a transition from its boundary (called a high-level transition) or through a transition from any of its internal states. A high-level transition indicates that the composite state and all its enclosed hierarchical or orthogonal states become inactive. An example of a high-level transition is the *end* transition, as shown in Figure 18.11. The occurrence of an *end* event while the phone is on, will cause a transition from the *On* state to the *Off* state form which ever internal state it was currently in.

The problem with permitting entrance to and exiting from internal states makes the representation of the composite state to be dependent on these internal transitions. Thus, it will be difficult to reuse

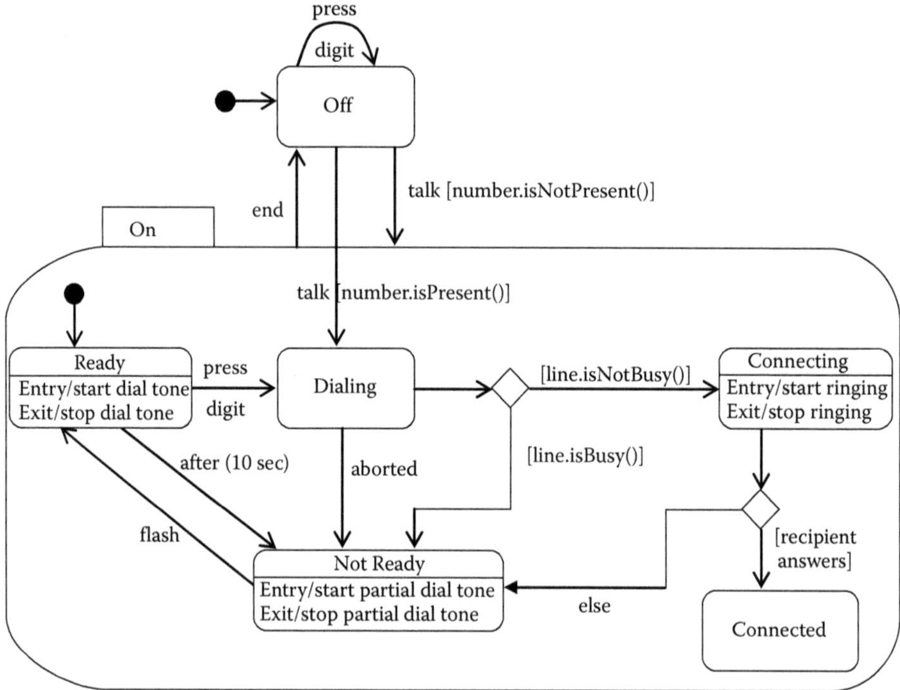

FIGURE 18.11 The phone state machine.

these composite states in other state machines. The following example will demonstrate this concept. Figure 18.11 shows the state machine representation for the phone system. The representation introduces new behavior into our phone model. Now the user can enter the phone number while the phone is still turned off and then press on the talk button. This new feature will cause a direct transition from the *Off* state to the *Dialing* state.

We can see from Figure 18.11 that there are two transitions that have the *Off* state as the source state and the *On* state as a target state. The occurrence of the *talk [number.isNotPresent()]* transition activates the composite state through its boundary. That means the On state will be entered through its default state, which is the *Ready* state. This transition models the behavior of turning the phone on, then getting the dial tone signal. The occurrence of the *talk [number.isPresent()]* transition enters the composite state not through its default state but directly to its *Dialing* state. This transition will only occur when the guard is true, indicating that the user has first entered the number while the phone is still off then pressed on the talk button. The representation of the *On* state is now dependent on this transition, and, therefore, it has hard to reuse this component.

Through the use of entry and exit points, it is possible to hide the internal structure of the submachine and limit the dependability of the component on external transitions. This will improve the modularity of the component and its ability to be reused. An entry point is represented as a small circle while an exit point is represented as a small circle with a cross inside of it. Entry and exit points can be placed either on the border of or within a submachine state. Figure 18.12 shows the state machine representation for the phone, where the internal representation for the *On* state is hidden. The small icon of two connected circles inside the On state is optionally used to indicate that this state is a composite one and has an expanded behavior. The *On* state has one entry point called *direct call*. The use of the entry point hides the direct transition to the internal state and removes the dependence of behavior between the *On* and *Off* states.

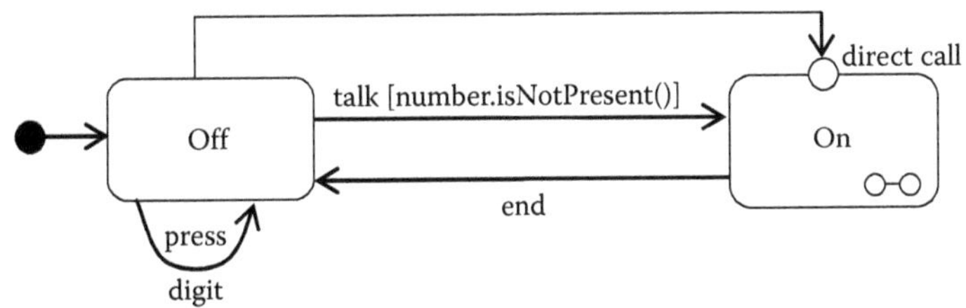

FIGURE 18.12 The phone state machine at a high level of abstraction.

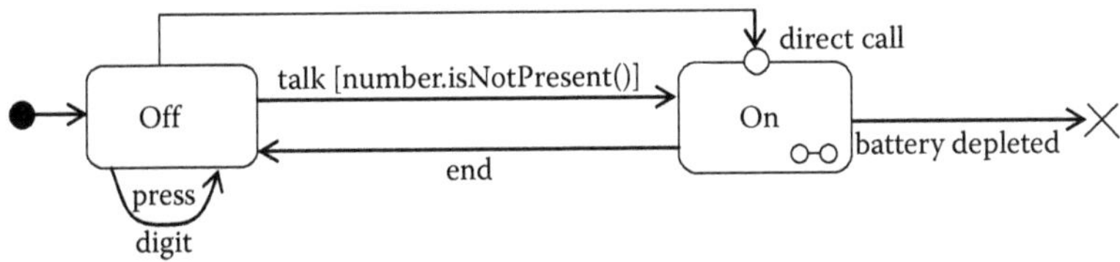

FIGURE 18.13 The phone state machine with a terminate pseudostate.

18.6.5 Terminate Pseudostate

A terminate pseudostate is used to represent a complete stoppage of the behavior of a state machine. This implies that the execution of the system has been terminated. At the moment of termination, the state machine cannot respond to events anymore and as a result cannot change its behavior. An example of a complete stoppage to the phone behavior is having an empty battery while the phone is being used. Figure 18.13 shows the modified behavior to our running example. The terminate pseudostate is shown in the figure as a cross symbol.

18.7 Final State

The behavior of a system can either execute repeatedly without completion or can execute for a certain set of operations then stop. This difference in behavior depends on the type of system or the type of operation to be modeled. UML state machines are capable of representing both types of behavior. Through the use of a final state, a state diagram can depict the completion of behavior. A final state is a special type of a UML state and is represented as a black circle enclosed within a small circle. A transition to a final state indicates that the encompassing region has completed its execution. The completion of the behavior in all the orthogonal regions of a composite state implies that the behavior of this composite state has also been completed. Multiple final states can be used in a single region to depict that there are more than one way to complete its behavior. The completion of a composite state will trigger a completion event.

Until now all we have modeled are reactive systems. These systems encompass a behavior that continuously waits for events, reacts to them and then waits again. The modeling of this type of behavior, which is always waiting for an event occurrence, does not require the use of a final state. A good example for a completion behavior is the detailed representation of a dialing process. Until now we have represented the phone's dialing behavior as a simple state.

Figure 18.14 shows the lower level behavior of the dialing process, where a composite state is used instead of a simple state. This hierarchical state becomes active as soon the user starts pressing on any of the digit keys. The *Dialing* composite state has two states: a *Partial Dial* default state and a final state. The default state indicates that the dial process has not been completed yet, while the final state indicates the dialing process has been completed successfully. The model shows that for the transition to the final state to occur, a valid number should be entered followed by three seconds of idle time from the time the last

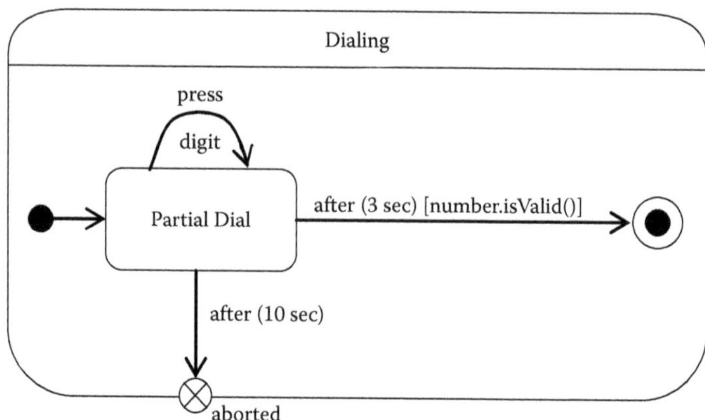

FIGURE 18.14 The Dialing composite state.

digit has been dialed. The completion of the *Dialing* state will trigger a completion event, thus causing a transition from the *Dialing* state in Figure 18.11 to the choice pseudostate.

It is relevant to mention that the composite state of Figure 18.14 makes use of an exit point to represent the behavior of leaving this state without completing its behavior. This situation will either occur when:

- The user quits in the middle of the dialing process.
- The number of digits entered is less than or more than the valid number of digits.

18.8 System Design and Component Communication

The modeling of system behavior using traditional finite state machines usually commences by first coming up with the states the system goes in. In other words, the first step in designing a system is to identify the modes of operations that are of interest. Next, transitions and events are added to connect these operations or states together. UML state machines add hierarchy and orthogonality to the model, and, therefore, the design process commences in a different direction. The representation should first take into consideration the components and subcomponents of a system and how these components and subcomponents will interact with each other. Accordingly, the design process starts by breaking down the behavior into separate parts following a top down hierarchical specification. This modularity in design allows software practitioners to focus on designing each component on its own, then later work on its communication with the other system parts. In this section, we are going to briefly describe these concepts using the microwave system as an example.

A conventional microwave system is composed of the following main functional components:

- *Display*: provides the user with the current time or the current status of the cooking operation.
- *Door Sensor*: is responsible for checking the state of the microwave door; whether it is currently opened or closed.
- *Light*: depicts the light behavior inside the microwave compartment. The light should be turned on every time the microwave door is opened or when the microwave is operational.
- *Microwave controller*: allows the user to control the operational behavior of a microwave. The behavior of this component has already been discussed and modeled, where Figure 18.10 shows its state machine representation. One thing to notice in this representation is that the *Setup Mode* state is a composite state; notice the composite icon. This preference of representation was chosen to

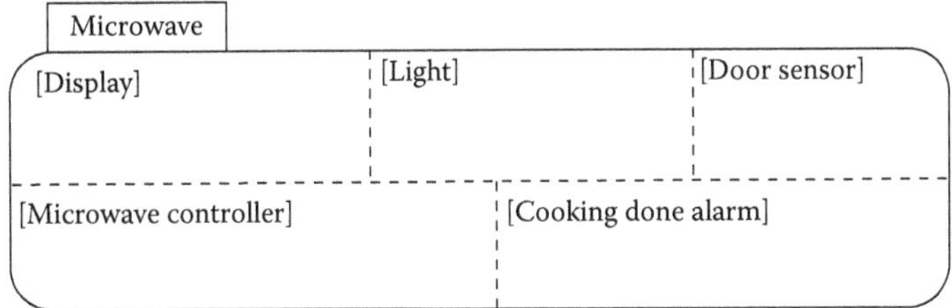

FIGURE 18.15 A partial microwave state machine representation.

hide the behavioral decomposition of the *Setup Mode* in order to decrease the size and to simplify the behavior of the controller component.

- *Cooking Done Alarm*: notifies the user using an audible beep that the cooking operation has been completed. This alarm will sound every one minute until the user opens the microwave door.

After coming up with the system components then understating their functionalities, the next step is to check their mode of execution. Some of the questions to consider during this process are: which components will run in parallel, which will run sequentially, and which will encompass other components. In case of the microwave system, all of the five components should run concurrently. Figure 18.15 shows the partial representation of the whole system, where each component is modeled as an orthogonal region.

The next thing to do is to model each orthogonal region (system component) one at a time. The process begins by identifying the states and substates of interest. Events, transitions and pseudostates are then introduced in order to model the behavior between these states. Figure 18.16 shows the complete behavior for the microwave.

Communication between orthogonal regions occurs through the use of:

Simultaneous transitions: Some transitions in a state machine might share the same event. In that case, the occurrence of this event might cause the trigger of multiple transitions. This happens if and only if the target source for each of these transitions is currently active. An example is the *clear* event. Consider that the microwave is in its cooking state, then the current substates for the *Display*, *Light*, *Cooking Done Alarm*, *Door Sensor*, and *Microwave Controller* regions are the *Counter*, *On*, *Off*, *Closed*, and *Cooking* states, respectively. Three of these active states share the *clear* event. Therefore, an occurrence of a *clear* event will trigger three simultaneous transitions causing the configuration of active substates to be the following: *Clock*, *Off*, *Off*, *Closed*, and *Idle*.

Generation of an event: This is done through the use of the '/' symbol followed by the action (event to be generated). This is referred to broadcasting of events, where one part of a submachine generates an event that causes one or more transitions in the whole state machine. Consider the microwave at its initial configuration when the status of the door sensor changes from false to true. This will cause the *[sensor.isEnabled()]*guard to be true and will generate an *open* action. This action will trigger the following transitions:

- *Light region*: a transition from the *Off* state to the *On* state.
- *Microwave Controller region*: a transition from the *Enabled* state to the *Disabled* state.

18.9 Applications of UML State Machines

There are various applications of UML state machines to systems and software applications. Harel and Gery (1997) describe the executable object modeling of a rail-car system using UML state machines. The authors in Trowitzsch and Zimmermann (2006) present an approach for the evaluation of performance and dependability for the European Train Control System (ETCS). The dynamic behavior of the system is represented using UML state machines that are transformed into stochastic Petri nets to allow quantitative

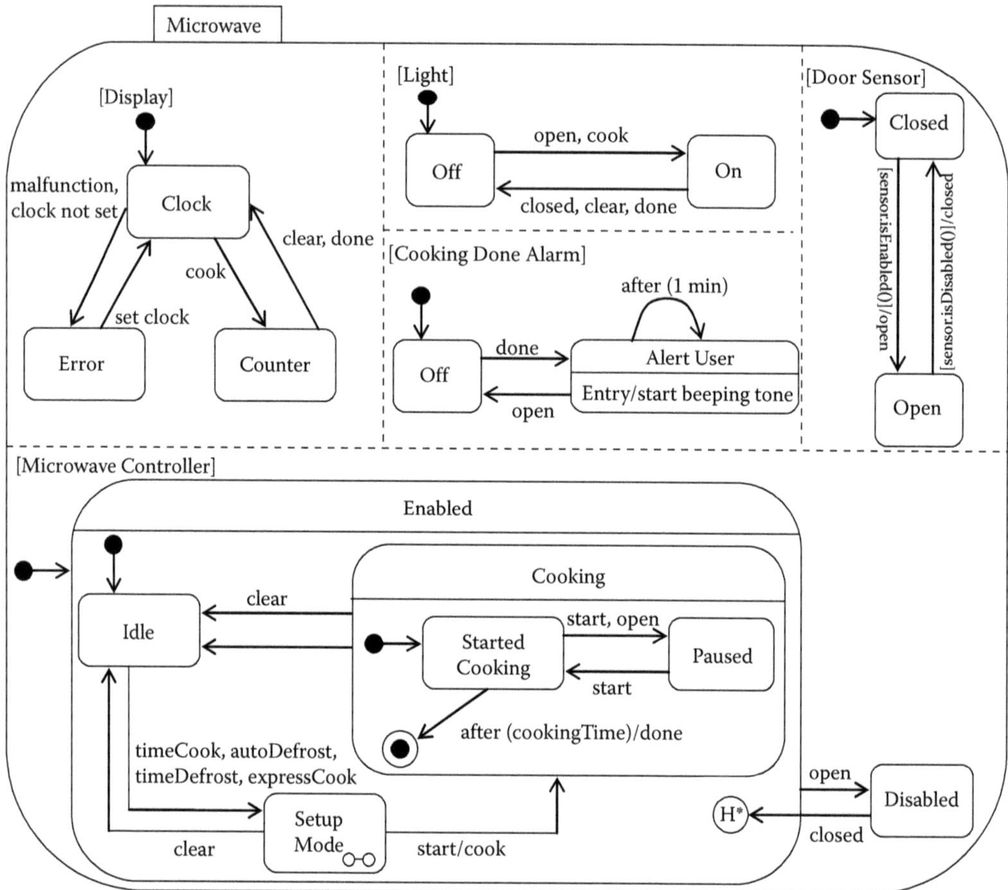

FIGURE 18.16 The complete microwave state machine representation.

analysis. El Ariss and Xu in (El Ariss and Xu 2011) propose to use statecharts to depict the dynamic, low-level behavior of security attacks. On the other hand, El Ariss et al. make use of statecharts to integrate fault trees (El Ariss et al. 2011b) and attack trees (El Ariss et al. 2011a) into the behavioral specification of a software system for safety and threat analysis.

There has been some work done to automate the generation of design models expressed using UML state machines directly from system requirements. In Whittle and Schumann (2000), Whittle and Schumann propose a technique that automatically generates UML state machines (statecharts) from UML sequence diagrams. They apply their approach on an ATM system. On the other hand, Yue et al. (2011) present an approach that automatically generates UML state machines from use case diagrams. Their proposed tool allows the transformation of the system requirements into dynamic behavioral models. These models can then be used to validate the correctness of design and to verify the implementation. The effectiveness of the approach is then demonstrated through its application on three case studies: an elevator system, an ATM, and Saturn—a video conferencing system developed by Cisco Norway.

Moreover, there has been various research done on source code generation directly from state machines to automate the design process. For example, IBM Rational allows C, C++, Java, or Ada code generation from UML state machines (IBM). Samek (2008) gives a detailed description of how to generate C/ C++ code from state machine through numerous examples of event-driven systems such as a calculator, and the Dining Philosopher problem. Further, Benowitz et al. (2006) present a process to generate C/C++

code from UML state machines for flight software in the space interferometer mission. Finally, Derezinska and Pilitowski (2009) automatically generates and executes C# code from UML class diagrams and state machines. The authors demonstrate their work on a web page, an object-oriented modeling language meta model, an airport control system, and other different models.

Amstel et al. (2008) present an approach that transforms system models of embedded systems written in Process Algebra into UML state machines to allow automatic software generation. In Knapp et al. (2002), an approach is proposed to transform UML collaborations and UML state machines into UPPAAL timed automata for model verification. The authors demonstrate their work using a case study on a generalized railroad crossing system. On the other hand, Choppy et al. (2011) transform UML state machines into Colored Petri nets to perform model checking in order to ensure automatic verification of safety properties. The work is demonstrated on a gas station case study.

There also has been some work that introduced extensions to UML state machines. Hirsch, Henkler, and Giese use Mechatronic UML, a refined version of UML 2.0, state machines to depict the behavior of multi-ports in mechatronic systems (Hirsch et al. 2008). The work by Mcintosh et al. (2008) extends UML state machines with 3D visualization to enable sequential evaluation. The authors demonstrated their work on a traffic light case study. Drusinsky (2006) introduces assertion statecharts, an extension to UML state machines for the verification of reactive systems. While Drusinsky et al. (2008) make use of UML state machines to represent reusable executable assertions for mission and safety critical behavior.

18.10 Summary

In this chapter, we described and discussed UML behavioral state machines and how they can be used in system modeling. The different types of states, transitions, events, and pseudostates supported by UML state machines were explained in the context of the light controller, microwave, or the phone system. We also described how the system modeling process commences. We concluded this chapter with a discussion of how orthogonal components collaborate and communicate through broadcasting of events.

References

Amstel, M. F., M. G. Brand, Z. Protic, and T. Verhoeff. Transforming process algebra models into UML state machines: Bridging a semantic gap? *Proceedings of the 1st International Conference on Theory and Practice of Model Transformations (ICMT'08)*. Zurich, Switzerland, 2008, pp. 61–75.

Benowitz, E., K. Clark, and G. Watney. Auto-coding UML statecharts for flight software. *Proceedings of the 2nd IEEE International Conference on Space Mission Challenges for Information Technology (SMC-IT)*. Pasadena, USA, 2006, pp. 413–417.

Choppy, C., K. Klai, and H. Zidani. Formal verification of UML state diagrams: A Petri net based approach. *SIGSOFT Softw. Eng. Notes* 36, 2011, 1–8.

Derezinska, A. and R. Pilitowski. Realization of UML class and state machine models in the C# code generation and execution framework. *Informatica (Slovenia)*, 33(4), 2009, 431–440.

Drusinsky, D., J. B. Michael, T. W. Otani, and M. Shing. Validating UML statechart-based assertions libraries for improved reliability and assurance. *Proceedings of the 2008 Second International Conference on Secure System Integration and Reliability Improvement (SSIRI'08)*. Yokohama, Japan, 2008, pp. 47–51.

Drusinsky, D. *Modeling and Verification using UML Statecharts: A Working Guide to Reactive System Design, Runtime Monitoring and Execution-based Model Checking*. Elsevier, Burlington, USA, 2006.

El Ariss, O. and D. Xu. Modeling security attacks with statecharts. *Proc. of the 2nd International ACM Sigsoft Symposium on Architecting Critical Systems (ISARCS 2011)*. Boulder, USA., June 2011, pp. 20–24.

El Ariss, O., D. Xu, and J. Wu. Building a secure system model: Integrating attack trees with statecharts. *Proc. of the 5th International Conference on Secure Software Integration and Reliability Improvement (SSIRI 2011)*. Korea, June 2011a, pp. 27–29.

El Ariss, O., D. Xu, and W. E. Wong. Integrating safety analysis with functional modeling. *IEEE Transactions on Systems, Man, and Cybernetics–Part A: Systems and Humans* 41(4), 2011b, 610–624.

Harel, D. Statecharts: A visual formalism for complex systems. *Science of Computer Programming* 8, 1987, 231–274.

Harel, D. and E. Gery. Executable object modeling with statecharts. *IEEE Computer* 30(7), 1997, 31–42.

Hirsch, M., S. Henkler, and H. Giese. Modeling collaborations with dynamic structural adaptation in mechatronic UML. *Proceedings of the 2008 International Workshop on Software Engineering for Adaptive and Self-Managing Systems (SEAMS'08)*. Leipzig, Germany, 2008, pp. 33–40.

Hopcroft, J. E., R. Motwani, and J. D. Ullman. *Introduction to Automata Theory, Languages, and Computation*. 3rd edition. Addison-Wesley, Boston, USA, 2006.

IBM. *Rational Software*. n.d. http://www-01.ibm.com/software/rational/.

Knapp, A., S. Merz, and C. Rauh. Model checking—Timed UML state machines and collaborations. *Proceedings of the 7th International Symposium on Formal Techniques in Real-Time and Fault-Tolerant Systems*. Oldenburg, Germany, 2002, pp. 395–416.

Mcintosh, P., M. Hamilton, and R. Schyndel. X3DUML: 3D UML state machine diagrams. *Proceedings of the 11th International Conference on Model Driven Engineering Languages and Systems (MoDELS'08)*, Toulouse, France, 2008, pp. 264–279.

Object Management Group. OMG Unified Modeling Language (OMG UML), Superstructure. Version 2.3, May 2010. http://www.omg.org/spec/UML/2.3/Superstructure/PDF/.

Samek, M. *Practical UML Statecharts in C/C++: Event-Driven Programming for Embedded Systems*. 2nd Edition. Newton: Newnes, 2008.

Trowitzsch, J. and A. Zimmermann. Using UML state machines and petri nets for the quantitative investigation of ETCS. *Proceedings of the 1st International Conference on Performance Evaluation Methodolgies and Tools (valuetools'06)*, Pisa, Italy, 2006.

Whittle, J. and J. Schumann. Generating statechart designs from scenarios. *Proceedings of the 22nd International Conference on Software engineering*. Limerick, Ireland, 2000.

Yue, T., S. Ali, and L. Briand. Automated transition from use cases to UML state machines to support state-based testing. *Proceedings of Seventh European Conference on Modelling Foundations and Applications (ECMFA 2011)*. Birmingham, UK, 2011.

Index

A

B

C

G

H

I

Q

R

S

T

U

V

W

X

Z